Lecture Notes in Networks and Systems

Volume 224

The series "Lecture Notes in Networks and Systems" publishes the latest developments in Networks and Systems—quickly, informally and with high quality. Original research reported in proceedings and post-proceedings represents the core of LNNS.

Volumes published in LNNS embrace all aspects and subfields of, as well as new challenges in, Networks and Systems.

The series contains proceedings and edited volumes in systems and networks, spanning the areas of Cyber-Physical Systems, Autonomous Systems, Sensor Networks, Control Systems, Energy Systems, Automotive Systems, Biological Systems, Vehicular Networking and Connected Vehicles, Aerospace Systems, Automation, Manufacturing, Smart Grids, Nonlinear Systems, Power Systems, Robotics, Social Systems, Economic Systems and other. Of particular value to both the contributors and the readership are the short publication timeframe and the world-wide distribution and exposure which enable both a wide and rapid dissemination of research output.

The series covers the theory, applications, and perspectives on the state of the art and future developments relevant to systems and networks, decision making, control, complex processes and related areas, as embedded in the fields of interdisciplinary and applied sciences, engineering, computer science, physics, economics, social, and life sciences, as well as the paradigms and methodologies behind them.

Indexed by SCOPUS, INSPEC, WTI Frankfurt eG, zbMATH, SCImago.

All books published in the series are submitted for consideration in Web of Science.

More information about this series at http://www.springer.com/series/15179

Dalia A. Magdi · Yehia K. Helmy ·
Mohamed Mamdouh · Amit Joshi
Editors

Digital Transformation Technology

Proceedings of ITAF 2020

Springer

Editors
Dalia A. Magdi
Canadian International College
Sadat Academy for Management Sciences
New Cairo, Egypt

Yehia K. Helmy
Department of Information Systems
Helwan University
Cairo, Egypt

Mohamed Mamdouh
Department of Computer Sciences
Ahram Canadian University
Cairo, Egypt

Amit Joshi
Global Knowledge Research Foundation
Ahmedabad, India

ISSN 2367-3370 ISSN 2367-3389 (electronic)
Lecture Notes in Networks and Systems
ISBN 978-981-16-2274-8 ISBN 978-981-16-2275-5 (eBook)
https://doi.org/10.1007/978-981-16-2275-5

This Springer imprint is published by the registered company Springer Nature Singapore Pte Ltd.
The registered company address is: 152 Beach Road, #21-01/04 Gateway East, Singapore 189721, Singapore

Preface

This volume contains the papers presented at ITAF 2020, Digital Transformation Technology: Informatics Insights, held online due to COVID-19 pandemic at Cairo, Egypt, 26 and 27 January 2021, collaborated by Global Knowledge Research Foundation. The associated partners were Springer and InterYIT IFIP. The 2nd ITAF Congress featured two days of focused networking and information sharing at the IoT cutting edge. This first edition brought together researchers, leading innovators, business executives, and industry professionals to examine the latest advances and applications for commercial and industrial end users across sectors within the emerging Internet of things ecosphere. It targeted the state-of-the-art as well as emerging topics related to the Internet of things such as big data research, emerging services and analytics, Internet of things (IoT) fundamentals, electronic computation and analysis, big data for multi-discipline services, security, privacy and trust, IoT technologies, and open and cloud technologies.

The main objective of the conference was to provide opportunities for the researchers, academicians, industry persons, students, and expertise from all over the world to interact and exchange ideas and experience in the field of Internet of things. It also focused on innovative issues at the international level by bringing together the experts from different countries.

It introduced emerging technological options, platforms, and case studies of IoT implementation in the areas by researchers, leaders, engineers, executives, and developers who will present the IoT industry which are dramatically shifting business strategies and changing the way we live, work, and play.

The ITAF Conference incited keynotes, case studies, and breakout sessions, focusing on smart solutions leading Egypt in IoT technologies into 2030 and beyond.

The conference started with the welcome speech of Assoc. Prof. Dalia A. Magdi, Conference Chair, ITAF 2020; Prof. Inas Ezz, Acting President of Sadat Academy for Management Sciences; Dr. Amit Joshi, Organizing Secretary, ITAF 2020, Director of Global Knowledge Research Foundation; and Mr. Aninda Bose, Sr. Publishing Editor, Springer Nature.

On behalf of ITAF 2020 board, we thank all respectable keynote speakers, Senator Hayam Farouk, Member of the Egyptian Senate; Mike Hinchey, Ph.D. Professor and

Director—LERO, University of Limerick, Ireland, and President, IFIP—International Federation for Information Processing; Sir Aninda Bose, Sr. Publishing Editor, Springer Nature, Germany; Prof. Mohamed Zahran, Computer Science Department, Courant Institute of Mathematical Sciences, New York University; Nilanjan Dey, Ph.D. Associate Professor—JIS University, Kolkata, India; Mihir Chauhan, Director—Global Knowledge Research Foundation, India; Dr. Khaled Elbehiery, Professor of Computer Networking and Cloud Engineering, Park University, USA, and Senior Director, Charter Communications, Inc., USA; Dr Ahmed Samir, Data Science Senior Manager, Vodafone; Prof. Magdy Abo el Ela, Computer and Information Systems Department, Sadat Academy for Management Sciences; Prof. Nevine Makram, Chair of Computer and Information Systems Department, Sadat Academy for Management Sciences; and Prof. Amal Elsayed Aboutabl, Professor, Computer Science Department, and Vice Dean for Community Service and Environmental Development, Helwan University.

A lot of researches were submitted in various advanced technology areas, and 37 researches were reviewed and accepted by the committee members to be presented and published. There were four technical sessions in total, and talks on academic and industrial sector were focused on both the days as follows:

- Cloud Security and DT Applications
- Big Data and AI in Digital Transformation
- Digital Transformation in Business
- AI and IoT in Health Care

On behalf of the conference chairs and editors, we owe a special thanks to Prof. Inas Ezz, Acting President, Sadat Academy for Management Sciences, for her support and participation, and we also thank Senator Hayam Farouk, Member, Egyptian Senate, for her speech; many thanks to all the keynote speakers, researchers and attendees of this conference, wishing to see you soon in ITAF 2021.

Dalia Magdi

New Cairo, Egypt

Associate Professor Dalia A. Magdi
Conference Chair and Editor

Contents

Editors and Contributors

About the Editors

Associate Professor Dalia A. Magdi is Vice Dean of School of Computer Sciences, Canadian International College. She acted as Dean of Faculty of Management and Information Systems, and she was Head of Information System Department, Faculty of Management and Information Systems, Vice-Director of CRI (Centre de Recherche Informatique), X Coordinator of Univrsite de Paris-Sud, French University in Egypt and X Coordinator of University of New Brunswick Program at Sadat Academy for Management Sciences. She is Chair and Editor of ITAF2019. She is Member of the Editorial Board of many international journals and reviewer of many international journals. She published many books internationally such as A Proposed Enhanced Model for Adaptive Multi-agent Negotiation Applied On E-commerce, Dalia A. Magdi, LAP LAMBERT Academic Publishing, a trademark of: OmniScriptum GmbH & Co. KG Bahnhofstraße 28, D-66111 Saarbrücken, ISBN (978-3-659-96009-3). She was invited as a keynote speaker and participated in many national and international conferences.

Dr. Yehia K. Helmy is Professor of Information Systems, Faculty of Business, Helwan University. He is Supervisor of the Department of Information Systems, Faculty of Commerce, Helwan University, Academic Supervisor for the Graduate Program in Business Information Systems, Former Dean of the Faculty of Computers and Information, Helwan University, Former coordinator of the Business Information Systems Program at the Faculty of Commerce, Helwan University. He is also Member of the permanent scientific committee for the promotion of professors and assistant professors, and management information systems for computers and information. Also, he is Member of the Commercial Studies Sector Committee at the Supreme Council of Universities and Member of the Board of Directors of the Canadian International College.

Dr. Mohamed Mamdouh is Assistant Professor of Software Engineering at Faculty of Computer Science and Information Technology, Ahram Canadian University, Egypt, and earned a Ph.D. in Information Systems at Mansoura University, Egypt. He has more than 14 years of teaching experience in different educational institutes. His research interests are artificial intelligence, machine learning, data science, bioinformatics and Internet of things (IoT). He has published extensively as author and co-author of many papers in highly regarded, peer-reviewed journals and international conferences. He is Researcher in Centre for Research and Interdisciplinary CRI at French University in Egypt, an environment for experimental research in the above-mentioned areas, and is involved in several open-source software projects. He is an active member in International Rough Set Society IRSS. He was co-chair of Internet of things application and future international conference 2019.

Amit Joshi is currently Director of Global Knowledge Research Foundation and also Entrepreneur Researcher who has completed his masters and research in the areas of cloud computing and cryptography in medical imaging. He has an experience of around 10 years in academic and industry in prestigious organizations. He is an active member of ACM, IEEE, CSI, AMIE, IACSIT, Singapore, IDES, ACEEE, NPA and many other professional societies. Currently, He is International Chair of InterYIT at International Federation of Information Processing (IFIP, Austria). He has presented and published more than 50 papers in national and international journals/conferences of IEEE and ACM. He has also edited more than 40 books which are published by Springer, ACM and other reputed publishers. He has also organized more than 50 national and international conferences and programs in association with ACM, Springer and IEEE to name a few across different countries including India, UK, Europe, USA, Canada, Thailand, Egypt and many more.

Contributors

Hammam Abdelaal Department of Information Technology, Faculty of Computers and Information, luxor University, luxor, Egypt

Almoataz Y. Abdelaziz Faculty of Engineering and Technology, Future University, New Cairo, Egypt

Muhammed Abd-elnaby Faculty of Computer and Information Sciences, Ain Shams University, Cairo, Egypt

A. Abdo Department of Information Systems, Faculty of Computers and Artificial Intelligence, Helwan University, Helwan, Egypt

Amany Abdo Information Systems Department, Faculty of Computers and Artificial Intelligence, Helwan University, Helwan, Egypt

Mohamed A. Abourezka Faculty of Computing and Information Technology, Arab Academy for Science, Technology & Maritime Transport, Cairo, Egypt

Ahmed R. Adly Nuclear Research Center, Atomic Energy Authority, New Cairo, Egypt

Lamia Aladel Department of Computers and Information Systems, Sadat Academy, Cairo, Egypt

Nermin Ibrahim Alarabi Sadat Academy for Management Sciences, Cairo, Egypt

Marco Alfonse Computer Science Department, Faculty of Computer and Information Sciences, Ain Shams University, Cairo, Egypt

Mostafa Al-Gabalawy Power Engineering and Automatic Control Department, Pyramids Higher Institute for Engineering and Technology, Giza, Egypt

Abdelmgeid A. Ali Department of Computer Science, Faculty of Computers and Information, Minia University, AL Minia, Egypt

Nabil Amein Ali Suez Institute of Management Information Systems, Suez, Egypt

Mahmoud Allam The Knowledge Hub Universities, Cairo, Egypt

Mona H. Alnaggar Kafrelsheikh University, Kafr El-Shaikh, Egypt

Omar Alomari Department of Computer Science, Jordan University of Science and Technology, Irbid, Jordan

Samar Aly Computer Science Department, Faculty of Computer and Information Science, Ain Shams University, Cairo, Egypt

Omar AlZoubi Department of Computer Science, Jordan University of Science and Technology, Irbid, Jordan

Olha Anisimova Vasyl' Stus Donetsk National University, Vinnytsia, Ukraine

Mostafa Aref Computer Science Department, Faculty of Computer and Information Sciences, Ain Shams University, Cairo, Egypt

Mohamed Ashraf Faculty of Computing and Information Technology, Arab Academy for Science, Technology & Maritime Transport, Cairo, Egypt

Vitalina Babenko V. N. Karazin, Kharkiv National University, Kharkiv, Ukraine

Iryna Balabanova Kherson State Agricultural University, Kherson, Ukraine

Razan Bayoumi Computer Science Department, Faculty of Computer and Information Sciences, Ain Shams University, Cairo, Egypt

Mohammed M. Eisa Ahram Canadian University, Cairo, Egypt

Mohammed M. Eissa Ahram Canadian University, Cairo, Egypt

Hussam Elbehiery October 6 University (O6U), Giza, Egypt

Khaled Elbehiery DeVry University, Denver, CO, USA

El-Sayed A. El-Dahshan Egyptian E-Learning University, Cairo, Egypt

Adly S. Tag Eldien Faculty of Engineering at Shoubra, Benha University, Cairo, Egypt

Ahmad M. Gamal El-Din Computer and Information Systems Department, Sadat Academy for Management Sciences, Cairo, Egypt

Yasmin Elgazzar Canadian International College, Cairo, Egypt

Sherif Elhennawy Faculty of Computers and Artificial Intelligence, Cairo University, Giza, Egypt

Hassan ElMadany Computer Science Department, Faculty of Computer and Information Sciences, Ain Shams University, Cairo, Egypt

Emad Elsamahy Arab Academy for Science, Technology, and Maritime Transport, Cairo, Egypt

Islam ElShaarawy Department of Electrical Engineering, Faculty of Engineering at Shoubra, Benha University, Cairo, Egypt

Rehab El-Shahawy Canadian International College, Cairo, Egypt

Mohammed A. Farahat Department of Information Systems, Faculty of Computers and Artificial Intelligence, Helwan University, Helwan, Egypt

Sara Ahmed Farghal Department of Software Engineering, Faculty of Computer Science, Nile University, Giza, Egypt

Alexandr Fedorchuk Kherson State University, Kherson, Ukraine

Mona A. Fouda Faculty of Engineering at Shoubra, Benha University, Cairo, Egypt

Ahmed Gaber Department of Computers and Systems Engineering, Faculty of Engineering, Minia University, Minia, Egypt

Mohamed Gamal Faculty of Computers and Artificial Intelligence, Cairo University, Giza, Egypt

Ahmad M. Gamal El-Din Department of Computers and Information Systems, Sadat Academy, Cairo, Egypt

Jeanne Georges French University in Egypt, Cairo, Egypt

Kareem Kamal A. Ghany Faculty of Computers and Artificial Intelligence, Beni-Suef University, Beni, Suef, Egypt;
College of Computing and Informatics, Saudi Electronic University, Riyadh, Kingdom of Saudi Arabia

Noha A. Haikal Faculty of Computers and Information Science, Mansoura University, Mansoura, Egypt

Alaa Hamdy Department of Communications, Electronics and Computer, Faculty of Engineering, Helwan University, Cairo, Egypt

M. Hassaballah Department of Computer Science, Faculty of Computers and Information, South Valley University, Qena, Egypt

Abeer Hekal Faculty of Computers and Information, Mansoura University, Cairo, Egypt

Iman M. A. Helal Faculty of Computers and Artificial Intelligence, Cairo University, Giza, Egypt

Ismail Hmeidi Department of Computer Science, Jordan University of Science and Technology, Irbid, Jordan

Nesreen S. Hosny Power Engineering and Automatic Control Department, Pyramids Higher Institute for Engineering and Technology, Giza, Egypt

Aisha Mohamed Hussein Information Systems Department, Faculty of Computers and Artificial Intelligence, Helwan University, Helwan, Egypt

Ahmed F. Ibrahim Department of Computer Science, Faculty of Computer Science, Nahda University, Banisuef, Egypt

Ibrahim A. Ibrahim Department of Computer Science, Faculty of Computers and Information, Minia University, AL Minia, Egypt

Sanaa Moharram Kamal Tropical Medicine and Gastroenterology Department, Faculty of Medicine, Ain Shams University, Cairo, Egypt

Samar K. Kassim Department of Biochemistry and Molecular Biology, Faculty of Medicine, Ain Shams University, Cairo, Egypt

Dragana Krstic Faculty of Electronic Engineering, University of Niš, Niš, Serbia

Nevien Makram Labib Sadat Academy for Management Sciences, Cairo, Egypt

Dalia Magdi Canadian International College, Cairo, Egypt;
Sadat Academy for Management Sciences, Cairo, Egypt

Fahima A. Maghraby Faculty of Computing and Information Technology, Arab Academy for Science, Technology & Maritime Transport, Cairo, Egypt

Ahmed Mahmoud Arab Academy for Science, Technology, and Maritime Transport, Cairo, Egypt

Hala A. K. Mansour Faculty of Engineering at Shoubra, Benha University, Cairo, Egypt

Wail Mardini Department of Computer Science, Jordan University of Science and Technology, Irbid, Jordan

Sherif A. Mazen Faculty of Computers and Artificial Intelligence, Cairo University, Giza, Egypt

Eman Ali Metwally Faculty of Computers and Information Science, Mansoura University, Mansoura, Egypt

Sherief H. Murad Electrical Engineering Department, Faculty of Engineering, Minia University, Minia, Egypt

Oleksandr Nakisko Kharkiv Petro Vasylenko National Technical University of Agriculture, Kharkiv, Ukraine

Maryna Nehrey National University of Life and Environmental Sciences of Ukraine, Kyiv, Ukraine

Silvia Parusheva University of Economics—Varna, Varna, Bulgaria

Daniela Pencheva University of Economics—Varna, Varna, Bulgaria

Iryna Perevozova Ivano-Frankivsk National Technical University of Oil and Gas, Ivano-Frankivsk, Ukraine

Oleksandr Prystemskyi Kherson State Agricultural University, Kherson, Ukraine

Kamel Hussein Rahouma Electrical Engineering Department, Faculty of Engineering, Minia University, Minia, Egypt

Zeina Rayan Department of Computer Science, Faculty of Computer and Information Sciences, Ain Shams University, Cairo, Egypt

Mohamed Roushdy Faculty of Computers and Information Technology, Future University in Egypt, New Cairo, Egypt

Heba M. Sabri Department of Computers and Information Systems, Sadat Academy, Cairo, Egypt

Abdel-Badeeh M. Salem Computer Science Department, Faculty of Computer and Information Science, Ain Shams University, Cairo, Egypt

Ahmed Salem Arab Academy for Science, Technology, and Maritime Transport, Cairo, Egypt

M. B. Senousy Computer and Information Systems Department, Sadat Academy for Management Sciences, Cairo, Egypt

Youssef Senousy National Organization for Social Insurance, Cairo, Egypt

Ahmed Sharaf-Eldin Faculty of Information Technology and Computer Science, Sinai University, Cairo, Egypt;
Faculty of Information Technology and Computer Science, Helwan University, Helwan, Egypt

Hassan Hussein Soliman Faculty of Computers and Information Science, Mansoura University, Mansoura, Egypt

Sarah A. Soliman Department of Computer Science, Higher Technological Institute, Cairo, Egypt

Michael Gr. Voskoglou Graduate Technological Educational Institute, Patras, Greece

Muneer Bani Yassein Department of Computer Science, Jordan University of Science and Technology, Irbid, Jordan

Hassan Youness Department of Computers and Systems Engineering, Faculty of Engineering, Minia University, Minia, Egypt

Dalia Youniss Arab Academy of Science Technology and Maritime Transport (AASTMT), Alexandria, Egypt

Magdy Zakaria Faculty of Computers and Information, Mansoura University, Cairo, Egypt

Hossam M. Zawbaa Faculty of Computers and Artificial Intelligence, Beni-Suef University, Beni, Suef, Egypt

Larysa Zomchak Ivan Franko National University of Lviv, Lviv, Ukraine

Cloud-Securiy and DT Applications

Gravitational Energy; Aerospace's Future Technology

Hussam Elbehiery and Khaled Elbehiery

Abstract Gravity is one of the four fundamental interactions (Gravitational, Electromagnetic, Strong, and Weak) that exist in nature and is essential for understanding the behavior of the universe, and all astrophysical bodies within it. Those particles hang together or break apart by swapping other types of particles, giving rise to forces of attraction and repulsion. The ongoing search for the graviton which is the proposed fundamental particle carrying gravitational force—is a crucial step in physicists' long journey toward a theory of everything. For the time being, space exploration is the environment that could be the hope despite it is an expensive arena because of the fuel costs and the technological challenge of operating it. Space exploration and in particular gravity is considered promising field, even entrepreneurs predict it will enable thousands of humans not only to travel into space like never before but also live and work in the space. This research paper introduces the foundations of gravity since the early observations of Kepler, Newtonian theory, and Einstein's theory of gravity. It also covers Quantum gravity that was established as a branch of modern theoretical physics that tries to unify its guiding principles with regard to quantum mechanics and general relativity. The association of the graviton to the geometry of the universe that controls the geometry of the space-time is also outlined. The paper will end with an explanation of space research future particularly in vehicle designs with respect to space-time theories along with applied Gravity research.

Keywords Gravitons · Quantum gravity · Zero-gravity · Microgravity · Artificial gravity · Space-time · Space exploration · Warp drive real · Teleportation

1 Introduction

Gravity was the first fundamental force that humanity recognized, yet it remains the least understood. Nearly, the century-long search for a theory of quantum gravity

H. Elbehiery
October 6 University (O6U), Giza 12572, Egypt

K. Elbehiery (✉)
DeVry University, Denver, CO 80126, USA

D. A. Magdi et al. (eds.), *Digital Transformation Technology*, Lecture Notes in Networks and Systems 224, https://doi.org/10.1007/978-981-16-2275-5_1

that describes how the force works for the universe's smallest pieces is driven by the simple expectation that one gravitational rulebook should govern all galaxies, quarks, and everything in between. Albert Einstein's theory of general relativity replaced Isaac Newton's notion of simple attraction between objects with a description of matter or energy bending space and time around it, and nearby objects following those curved paths, acting as if they were attracted to one another, which is demonstrated in Einstein's equations through gravity is the shape of space itself [1].

A small amount of gravity can be found everywhere in space which holds the moon into orbit around Earth, it causes Earth to orbit the sun, and in turn, it keeps the sun in place in the Milky Way galaxy. Gravity, however, does become weaker with distance, when the spacecraft goes far enough from Earth, the astronaut inside would feel very little gravity, but this is not why things float on a spacecraft in orbit. The International Space Station orbits Earth at an altitude approximately 250 miles, at this altitude, Earth's gravity is about 90 percent of what it is on the surface, in other words, if a person who weighed 100 lb on Earth's surface could climb a ladder all the way to the space station, that person would weigh 90 lb at the top of the ladder [2].

Sections 1 and 2 of the research paper discuss the theory of gravity and the concepts of the Gravitons. Sections 3 and 4 cover the gravity control, quantum gravity, and the effects of microgravity on astronauts in space. Section 5 explains the principles of black holes. Artificial Gravity will be explained in Sects. 6 and 7 goes over the space-time concepts. Lastly, Sect. 10 shade light on what is the Warp Drive Real which is a new technological trend, and the research paper ends up with Sect. 12 that discuss the space future (Vehicle Designs).

2 Gravitons

In the late 1600s, Isaac Newton devised the first serious theory of gravity. He described gravity as a field that could reach out across great distances and dictate the path of massive objects like the Earth. In 1915, Albert Einstein's theory of general relativity gave theorists their first look "under the hood" of gravity. What we call now gravity, Einstein argued, it is actually the distortion of space and time. Einstein's theory of gravity is very good at explaining the behavior of large objects. But just few years later, physicists opened up the world of the ultra-small, revealing that the other fundamental forces are due to the exchange of specialized force-carrying particles: photons convey electromagnetism, the strong nuclear force is transmitted by gluons and the weak nuclear force is imparted by the movement of bosons, there is a name for that hypothetical particle if it does exist; it is called the graviton, even though we have never observed it before. Since the range of the force due to gravity is infinite and it weakens as one over the square of the distance between two objects (i.e. $1/r^2$), the graviton must have zero mass, it is a fact because if the photon had a mass, it would change the "2" in the exponent and that "2" has been established with incredible precision, and so, like massless photons, gravitons should travel at the speed of light.

In general relativity, the distribution of mass and energy in the universe is described by a four-by-four matrix that mathematicians call a tensor of rank two. This is important because if the tensor is the source of gravitation, you can show that the graviton must be a particle with a quantum mechanical spin of two. Quantum mechanics tells us that every particle is also a vibrating wave, and it has been proposed that gravitons could vibrate in these extra dimensions, wrapping around the small dimension like bracelets encircling a slender wrist. However, the cyclical nature of the extra dimension imposes limits on how a graviton can vibrate. Only an integer number of wavelengths can fit evenly in the extra dimension. And this brings us to a couple of interesting consequences. In theories with extra dimensions, more than one type of graviton can exist. One way to see that is to imagine taking a sine wave and wrapping it around a cylinder. In order for it to fit perfectly, you must use one wavelength or two or three or any integer number of wavelengths. Each of these instances is a distinct graviton; the ones with more vibrations can actually have mass. Particles of this kind are called Kaluza-Klein gravitons after physicists Theodor Kaluza and Oskar Klein, who first proposed the idea of additional small spatial dimensions. On tiny scales, Kaluza-Klein gravitons can have mass, but on larger scales, they reduce to the familiar massless gravitons of classical theory [3].

The graviton is expected to be massless because the gravitational force is very long range and appears to propagate at the speed of light. The graviton must be a spin-2 boson because the source of gravitation is the stress–energy tensor, a second-order tensor (compared with electromagnetism's spin-1 photon, the source of which is the four-current, a first-order tensor). While gravitons are presumed to be massless, they would still carry energy, as does any other quantum particle. Photon energy and gluon energy are also carried by massless particles. It is unclear which variables might determine graviton energy; the amount of energy carried by a single graviton. Alternatively, if gravitons are massive at all, the analysis of gravitational waves yielded a new upper bound on the mass of gravitons. The graviton's Compton wavelength is at least 1.6×10^{16} m, or about 1.6 light-years, corresponding to a graviton mass of no more than 7.7×10^{-23} eV/c^2. This relation between wavelength and mass-energy is calculated with the Planck–Einstein relation, the same formula that relates electromagnetic wavelength to photon energy [4].

However, if gravitons are the quanta of gravitational waves, then the relation between wavelength and corresponding particle energy is fundamentally different for gravitons than for photons, since the compton wavelength of the graviton is not equal to the gravitational-wave wavelength. Instead, the lower-bound graviton compton wavelength is about 9×10^9 times greater than the gravitational wavelength for the GW170104 event, which was ~ 1700 km. The report did not elaborate on the source of this ratio. It is possible that gravitons are not the quanta of gravitational waves, or that the two phenomena are related in a different way [13].

3 Gravity

Gravity is a force like all other forces that we are aware of (electromagnetic force, weak decay force, strong nuclear force). In quantum theory, each particle acts both as a particle and a wave. This is called duality. So, if there is a graviton, we expect it to behave both as particle and as a wave as well. The electromagnetic force, for example, is transmitted by photons, and light is nothing but a large number of photons. Photons/light show wave and particle properties [5].

Scientists currently try to verify whether gravity produces classical (not quantum) waves, something that nobody has observed, but Einstein's theory of general relativity allows for such gravitational waves. In the same way that you can study the wave phenomena of light without knowing about the existence of light particles (photons), one expects to be able to detect some sort of gravity waves without producing evidence for a graviton. It takes huge massive objects, such as stars spinning around each other, to produce gravitational waves that might be strong enough to be detected on earth.

3.1 "Gravity Particles" Unite All of Physics

Theoretical physicists claim that it may be possible to draw energy from a vacuum using gravity, if researchers succeed in showing that this can happen, it could prove the long-postulated existence of the graviton, the particle of gravity, and perhaps bring scientists one step closer to developing a "theory of everything" that can explain how the universe works from its smallest to largest scales. The new research specifically found that it might be possible to show that gravitons do exist by using superconducting plates to measure a phenomenon with the esoteric name of "the gravitational Casimir effect."

Showing that gravitons exist would help scientists who have long sought to develop a "theory of everything" that can describe the workings of the cosmos in its entirety. Currently, they use the theory of quantum mechanics to explain the universe at its tiniest level, and the theory of general relativity to explain the universe at its largest level. Whereas quantum mechanics can explain the behavior of all the known particles, general relativity describes the nature of space-time and gravity.

Quantum mechanics suggests that particles—including the elusive graviton—can behave both like a particle and a wave. But quantum mechanics also reveals that the world becomes a fuzzy, surreal place at its very smallest levels. For instance, atoms and other fundamental building blocks of the universe actually exist in states of flux known as "superpositions," meaning they can seemingly be located two or more places at once, or spin in opposite directions at the same time.

Since quantum mechanics suggests that any given particle may not be where one thinks, but rather could essentially be anywhere, one of the many weird consequences of this theory is that what might seem like vacuum (completely empty space) may

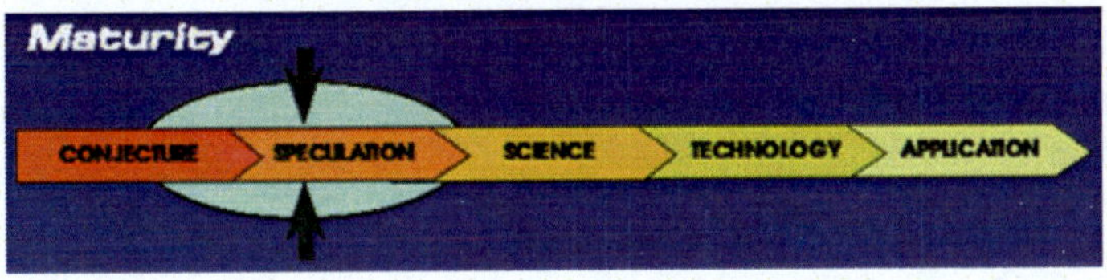

Fig. 1 Maturity sequence of gravity control in space research projects

actually contain "virtual particles" that regularly pop in and out of existence. These ghostly entities are more than just theory—they can generate measurable forces. It may be possible to draw energy from a vacuum using gravity, a theoretical physicist says. If researchers succeed in showing that this can happen, it could prove the long-postulated existence of the graviton, the particle of gravity, and perhaps bring scientists one step closer to developing a "theory of everything" that can explain how the universe works from its smallest to largest scales [6].

3.2 *Gravity Control*

Controlling gravity or inertial forces, you would have a propulsion breakthrough (thrusting without rockets), a means to create synthetic gravity environments for space crews, a means to create zero-gravity environment on Earth (see Fig. 1).

We do know that gravity and electromagnetism are linked phenomena, the scientists become quite adept at controlling electromagnetic phenomena, so one can presume that such a connection might eventually lead to using the control of electromagnetism to control gravity. General Relativity, another one of Einstein's doings, is one way to describe such connections. Another way is through new theories from quantum mechanics that link gravity and inertia to something called "vacuum fluctuations." [7].

3.3 *Quantum Gravity*

> "If there is no theory [of quantum gravity], then the universe is just chaos. It's just random," said Netta Engelhardt, a theoretical physicist at the Massachusetts Institute of Technology. "I can't even say that it would be chaotic or random because those are actually legitimate physical processes."

Electric and Magnetic forces come from objects exchanging particles known as virtual photons. For example, the force sticking a magnet to the fridge can be described as a smooth, classical magnetic field, but the field's fine details depend on the quantum particles that create it. Of the Universe's four fundamental forces only gravity lacks the "Quantum" description. So, no one knows for sure (although there are plenty of ideas) where gravitational fields come from or how individual particles act inside them (see Fig. 2).

Even though gravity keeps us stuck to the ground and generally acts as a force, general relativity suggests it's something more—the shape of space itself. Other quantum theories treat space as a flat backdrop for measuring how far and fast particles fly. Ignoring the curvature of space for particles works because gravity is so much weaker than the other forces that space looks flat when zoomed in on something as small as an electron. The effects of gravity and the curvature of space are relatively obvious at more zoomed-out levels, like planets and stars. But when physicists try to calculate the curvature of space around an electron, slight as it may be, the math becomes impossible (see Fig. 3).

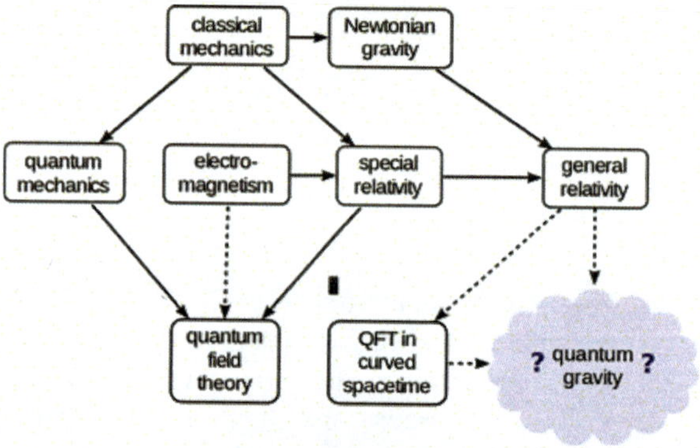

Fig. 2 Quantum gravity in the hierarchy of physics theories

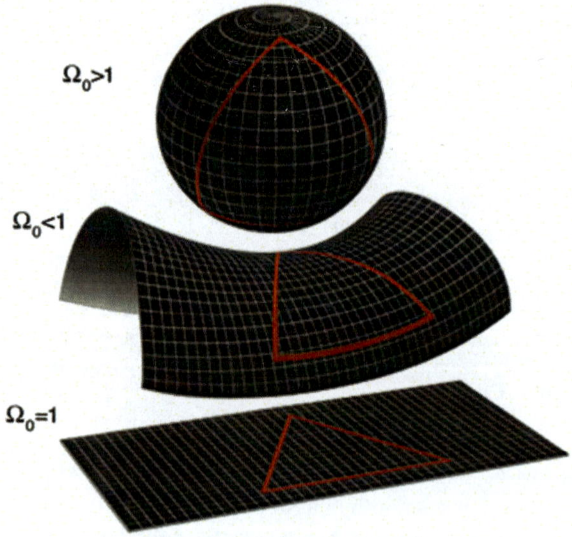

Fig. 3 Overall curvature of space (closed, open, and flat universe)

Using an approximation of general relativity, physicists have developed a notion of what gravitons might look like, but no one expects to see one anytime soon. One thought experiment suggests it would take 100 years of experimentation by a particle collider as heavy as Jupiter to detect one. So, in the meantime, theorists are rethinking the nature of the universe's most fundamental elements. One theory, known as loop quantum gravity, aims to resolve the conflict between particles and space-time by breaking up space and time into little bits—an ultimate resolution beyond which no zooming can take place.

Theorized by Einstein and confirmed in 2016, gravitational waves have finally been observed from the merger of two neutron stars—ultradense stellar zombies left over from the explosive deaths of giant stars. Two neutron stars rotate around each other; the closer they get, the faster they spin. Eventually, they collide. The energy from their spiraling and merging releases energy in the form of gravitational waves, or ripples in space-time. The merger most likely resulted in a black hole, although it's also possible it created an abnormally massive neutron star. Regardless, the final object is less massive than the two combined neutron stars. The equivalent mass of 25 Jupiters was converted into gravitational waves. The collision also ejected 50 Jupiters' worth of heavy elements such as gold and silver [1] (see Fig. 4).

All the fundamental forces of the universe are known to follow the laws of quantum mechanics, save one: gravity. Finding a way to fit gravity into quantum mechanics would bring scientists a giant leap closer to a "theory of everything" that could entirely explain the workings of the cosmos from first principles. A crucial first step in this quest to know whether gravity is quantum is to detect the long-postulated elementary particle of gravity, the graviton. In search of the graviton, physicists are now turning to experiments involving microscopic superconductors, free-falling crystals, and the afterglow of the big bang. Quantum mechanics suggests everything is made of quanta, or packets of energy, that can behave like both a particle and

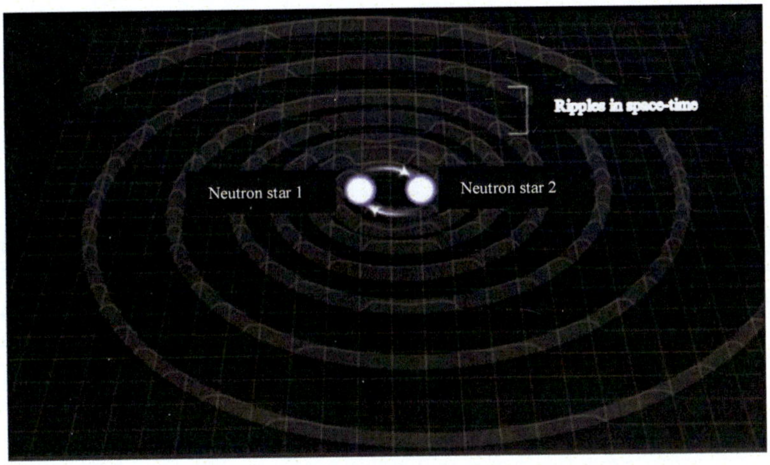

Fig. 4 Gravitational waves linked to neutron star crash

a wave—for instance, quanta of light are called photons. Detecting gravitons, the hypothetical quanta of gravity, would prove gravity is quantum. The problem is that gravity is extraordinarily weak. To directly observe the minuscule effects a graviton would have on matter, physicist Freeman Dyson famously noted, a graviton detector would have to be so massive that it collapses on itself to form a black hole.

"One of the issues with theories of quantum gravity is that their predictions are usually nearly impossible to experimentally test," says quantum physicist Richard Norte of Delft University of Technology in the Netherlands. "This is the main reason why there exist so many competing theories and why we have not been successful in understanding how it actually works." The researchers would seek out entanglement by shining lasers into each diamond's heart after the drop. If particles in the crystals' centers spin one way, they would fluoresce, but they would not if they spin the other way. If the spins in both crystals are in sync more often than chance would predict, this would suggest entanglement. "Experimentalists all over the world are curious to take the challenge up," says quantum gravity researcher Anupam Mazumdar of the University of Groningen in the Netherlands, co-author of one of the entanglement studies.

Another strategy to find evidence for quantum gravity is to look at the cosmic microwave background radiation, the faint afterglow of the big bang, says cosmologist Alan Guth of M.I.T. Quanta such as gravitons fluctuate like waves, and the shortest wavelengths would have the most intense fluctuations. When the cosmos expanded staggeringly in size within a sliver of a second after the big bang, according to Guth's widely supported cosmological model known as inflation, these short wavelengths would have stretched to longer scales across the universe. This evidence of quantum gravity could be visible as swirls in the polarization, or alignment, of photons from the cosmic microwave background radiation [8].

3.4 Quantum Computing in Cosmological Evolution

In 2001, Seth Lloyd presented how to model the evolution of the universe as a quantum computer. This modeling allows one to reconcile the existence of a huge burst of gravitons with the fact that cosmological CMB (Cosmic Microwave Background) is limited (see Fig. 5).

First of all, we need to consider if there is an inherent fluctuation in early universe cosmology which is linked to a vacuum state nucleating out of 'nothing'. The answer we have is yes and no. The vacuum fluctuation leads to production of a dark energy density which we can state is initially due to contributions from an axion wall, which is dissolved during the inflationary era. What we will be doing is to reconcile how that wall was dissolved in early universe cosmology with quantum gravity models, brane world models, and Weinberg's prediction (published as of 1972) of a threshold of 10 to the 32 power Kelvin for which quantum effects become dominant in quantum gravity models. All of this leads up to conditions in which we can expect relic graviton production which could account for the presence of strong gravitational fields in the

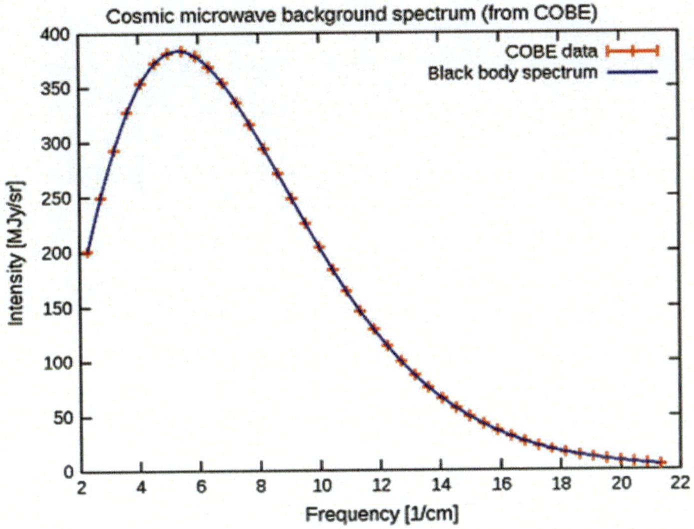

Fig. 5 Graph of cosmic microwave background spectrum

onset of Guth style inflation, would be in line with Penrose's predictions via the Jeans inequality as too low temperature, low entropy conditions for pre inflationary cosmology [9].

3.5 *The Future of Zero-Gravity Living*

The body drifts up from the floor, and there is no force on me at all from any direction. That is over the Gulf of Mexico in G-Force One, a vintage Boeing 727 that belongs to the Zero Gravity Corporation. The plane, which provides scientists and thrill-seekers with the chance to experience weightlessness without going to space, has just seven rows of seats, way at the back. Instead, there is 66 feet of wide-open space, the better to make the most of the kind of acrobatic flying that shakes passengers loose from gravity (see Fig. 6).

G-Force One gives scientists their best chance to work in zero gravity without having to go to the Space Station, and they pay tens of thousands of dollars, often using grants from NASA, for the privilege of performing experiments 27 s at a time. Researchers from Carthage College, in Wisconsin, are testing a new method to use sound waves to gauge the fuel in a spacecraft's tank, which is notoriously hard to measure in zero gravity. A group from the Applied Physics Laboratory at Johns Hopkins University is testing a technology to allow small probes that land on asteroids to reposition themselves in ultra-low gravity without pogoing back into space (see Fig. 7).

Fig. 6 A mid-air tourist flight

Fig. 7 Interpretation of a future space station captures

Planetary Resources is a few years from launching its first prospecting satellite, which will scout for water on nearby asteroids. And Lewicki acknowledges that a series of technological innovations, from robot asteroid miners to refillable rocket fuel tanks, need to be developed before a self-sufficient space economy takes hold. But he insists it will happen, and asteroid mining will play a critical role: "The leap we're making is that this is all going to scale one day to millions of people living and working in space. And the only way to do that is to use 'on-site' resources." [10] (see Fig. 8).

4 Photons; Massive Gravity

This is unusual new design compared to the world's most sensitive gravitational wave detectors. Gravitational waves, or ripples in space-time, slip through Earth all the time, carrying secrets about the universe. But until a few years ago, these waves could not detect, but the most basic ability which is available in the detecting of the stretching and squeezing of the cosmos. However, a proposed new gravitational

Fig. 8 NASA G-Force one (loading scientific cargo)

wave hunter, which would measure how particles of light and gravity interact, could change that. In the process, it could answer big questions about dark energy and the universe's expansion.

The three detectors on Earth today, all together called Laser Interferometer Gravitational-Wave Observatory (LIGO) and Virgo, operate according to the same principle: As a gravitational wave moves through the Earth, it faintly stretches and squeezes space-time. By measuring how long a laser light takes to travel over long distances, the detectors notice when the size of that space-time changes. But the changes are minute, requiring extraordinarily sensitive equipment and statistical methods to detect.

Three researchers proposed a radical new method: hunting gravitational waves by looking for effects of direct interactions between gravitons—theoretical particles that carry gravitational force—and photons, the particles that make up light. By studying those photons after they've interacted with gravitons, you should be able to reconstruct the properties of a gravitational wave, according to Subhashish Banerjee, a co-author of the new paper and physicist at the Indian Institute of Technology in Jodhpur, India. Such a detector would be much cheaper and easier to build than existing detectors, Banerjee said. "Measuring photons is something which people know very well," Banerjee told Live Science. "It's extremely well-studied, and definitely it is less challenging than a LIGO kind of setup."

No one knows exactly how gravitons and photons would interact, largely because gravitons are still entirely theoretical. No one's ever isolated one. But the researchers behind this new paper made a series of theoretical predictions: When a stream of gravitons hits a stream of photons, those photons should scatter. And that scattering would produce a faint, predictable pattern—a pattern physicists could amplify and study using techniques developed by quantum physicists who study light. Linking the physics of the tiny quantum world with the large-scale physics of gravity and relativity has been a goal of scientists since Albert Einstein's time. But even though the newly suggested approach to studying gravitational waves would use quantum methods, it wouldn't fully bridge that tiny-to-large-scale gap on its own, Banerjee said.

"It would be a step in that direction, however," he added. Probing the direct interactions of gravitons might solve some other deep mysteries about the universe, though, he said. The authors showed that the way the light scatters would depend on the specific physical properties of gravitons. According to Einstein's theory of general relativity, gravitons are massless and travel at the speed of light. But according to a collection of theories, together known as "massive gravity," gravitons have mass and move slower than the speed of light. These ideas, some researchers think, could resolve problems such as dark energy and the expansion of the universe. Detecting gravitational waves using photon scattering, Banerjee said, could have the side effect of telling physicists whether massive gravity is correct. No one knows how sensitive a photon-graviton detector of this kind would end up being, Banerjee said. That would depend a lot on the final design properties of the detector, and right now, none are under construction. However, he said, he and his two co-authors hope that experimentalists will start putting one together soon [11].

5 Microgravity

Microgravity is the condition in which people or objects appear to be weightless. The effects of microgravity can be seen when astronauts and objects float in space. Microgravity can be experienced in other ways, as well. "Micro-" means "very small," so microgravity refers to the condition where gravity seems to be very small. In microgravity, astronauts can float in their spacecraft—or outside, on a spacewalk. Heavy objects move around easily. For example, astronauts can move equipment weighing hundreds of pounds with their fingertips. Microgravity is sometimes called "zero gravity," but this is misleading (see Fig. 9).

Microgravity affects the human body in several ways. For example, muscles and bones can become weaker without gravity making them work as hard. Astronauts who live on the space station spend months in microgravity. Astronauts who travel to Mars also would spend months in microgravity traveling to and from the Red Planet. NASA must learn about the effects of microgravity to keep astronauts safe and healthy. In addition, many things seem to act differently in microgravity. Fire burns differently. Without the pull of gravity, flames are more round. Crystals grow better.

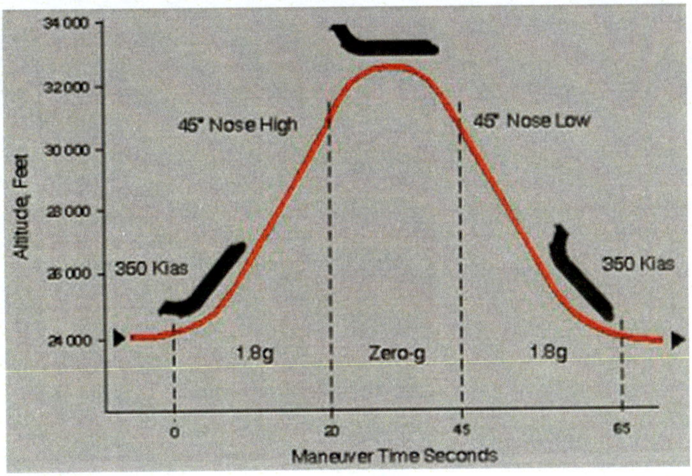

Fig. 9 On reduced-gravity flights, airplanes fly in large arcs called parabolas

Without gravity, their shapes are more perfect. NASA performs science experiments in microgravity. These experiments help NASA learn things that would be hard or perhaps impossible to learn on Earth [2].

5.1 Biological Science in Microgravity

Astronaut Serena Auñón-Chancellor discusses her experience in microgravity and doing biological experiments in space. Microgravity, or very weak gravity, on the International Space Station (ISS) is what lets astronauts glide and somersault around effortlessly as they orbit Earth. It is also a useful environment for gaining insights into human health, both in terms of the impacts of long-duration spaceflight and new perspectives on diseases that afflict people on our planet (see Fig. 10).

Space-based biomedical research was one of the key topics discussed in September 2019 at the ISS R&D Conference in Atlanta. Researchers highlighted some of the current work on the Space Station, as well as further studies NASA and the ISS National Laboratory hope to do while seeking to commercialize low-Earth orbit. They also aim to use the ISS as a stepping-stone to landing back on the Moon and eventually Mars. Changes in the immune system have been noticed. We see what they call latent viral reactivation (when dormant viruses begin reproducing), and that is measured in the saliva, feces, urine to blood. But it's interesting how quickly things do revert almost back to normal once getting down to Earth. The biggest health challenges—certainly for exploration-class missions, longer and longer missions—number one is radiation. The ISS is well protected on.

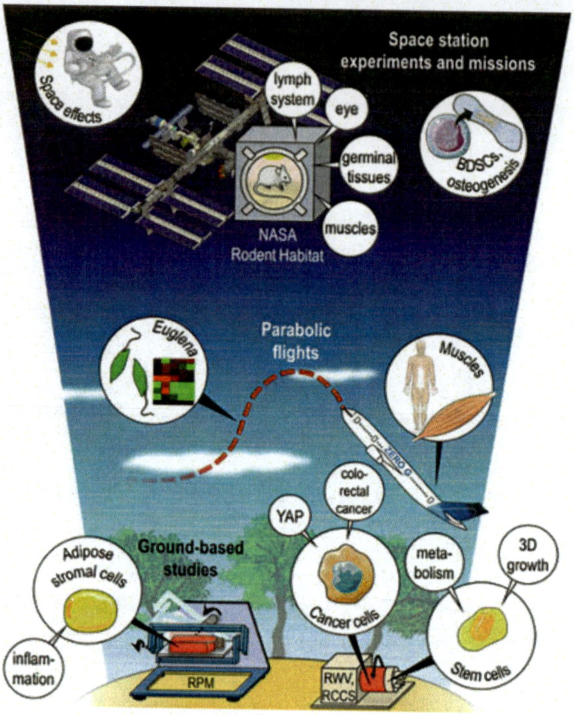

Fig. 10 Gravitational biology and space medicine

Cell growth differs up in microgravity. Scientists are able to culture cells such as endothelial cells (which line the inside of blood and lymphatic vessels) for a little bit longer. They grow in a better, more three-dimensional fashion than growing them on a flat plate on Earth, which allows scientists to study different things. The other thing that changes is that it is sort of like a rapid aging process that occurs in orbit. So, we look at all the molecular markers and the way cells also change in orbit. And processes that take years on the ground, such as osteoporosis, happen much more quickly up there. So, scientists see it as a test bed. And finally, the third thing that was protein crystal experiments. Whether it was a protein involved in Parkinson's disease or a drug that a pharmaceutical company was studying to improve, these protein crystals are structures that grow better (on the ISS) [12].

6 Extra Dimensions, Gravitons, and Tiny Black Holes

In our everyday lives, we experience three spatial dimensions, and a fourth dimension of time. Einstein's general theory of relativity tells us that space can expand, contract, and bend. Now if one dimension were to contract to a size smaller than an atom, it would be hidden from our view. But if we could look on a small enough scale, that hidden dimension might become visible again. Imagine a person walking on a tightrope. He can only move backward and forward; but not left and right, nor up and down, so he only sees one dimension. How could we test for extra dimensions? One option would be to find evidence of particles that can exist only if extra dimensions are real.

Theories that suggest extra dimensions predict that in the same way as atoms have a low-energy ground state and excited high-energy states, there would be heavier versions of standard particles in other dimensions. These heavier versions of particles—called Kaluza-Klein states—would have exactly the same properties as standard particles (and so be visible to our detectors) but with a greater mass. Such heavy particles can only be revealed at the high energies reached by the Large Hadron Collider (LHC).

Some theorists suggest that a particle called the "graviton" is associated with gravity in the same way as the photon is associated with the electromagnetic force. If gravitons exist, it should be possible to create them at the LHC, but they would rapidly disappear into extra dimensions. Collisions in particle accelerators always create balanced events—just like fireworks—with particles flying out in all directions. A graviton might escape our detectors, leaving an empty zone that we notice as an imbalance in momentum and energy in the event. Another way of revealing extra dimensions would be through the production of "microscopic black holes".

What exactly we would detect would depend on the number of extra dimensions, the mass of the black hole, the size of the dimensions, and the energy at which the black hole occurs. If micro black holes do appear in the collisions created by the LHC, they would disintegrate rapidly, in around 10–27 s. They would decay into Standard Model or supersymmetric particles, creating events containing an exceptional number of tracks in our detectors, which we would easily spot. Finding more on any of these subjects would open the door to yet unknown possibilities [14].

7 Artificial Gravity

The only physically possible way to create a force as strong as earth's gravity that acts on all objects in a ship is through acceleration. Acceleration always creates inertial forces. Inertial forces such as the centrifugal force or Coriolis force are very real in the accelerating reference frame. If the acceleration is held constant and at the right value, the inertial force will behave identically to earth's gravity and will, in fact, be equivalent to earth's gravity. This fact is actually a basic tenet of General Relativity (see Fig. 11).

There are two kinds of accelerations, rotational and linear. A ship could achieve artificial gravity by rotating about its axis. To be practical, the radius of rotation would have to be quite large. Additionally, a ship could create artificial gravity by constantly accelerating forwards. Shows that portray artificial gravity without rotation or constant forward acceleration are simply non-physical [15].

Fig. 11 In outer space, all the masses in the universe gravitate just as normal

Space station residents currently rely upon different exercises to keep themselves fit for the eventual return to Earth. But a spinning centrifuge device could create artificial gravity, which simulates the gravitational tug a planet, like Earth, has. The giant spinning device will give astronauts a healthy break from the weightlessness of space [16] (see Fig. 12).

If antimatter has negative gravitational mass, then by setting up a ceiling of anti-matter and a floor of normal matter, we could create an artificial gravity field that always pulled you down. By building a gravitationally conducting shell as the hull of our spacecraft, everyone inside would be protected from the forces of ultra-rapid acceleration which would otherwise prove lethal. And most spectacularly, humans in space would no longer suffer the negative physiological effects, from balance disorders to the atrophy of your heart muscle, that currently plagues today's astronauts. But until we discover a particle (or set of particles) with negative gravitational mass, artificial gravity will only be brought about through acceleration, no matter how clever we are (see Fig. 13).

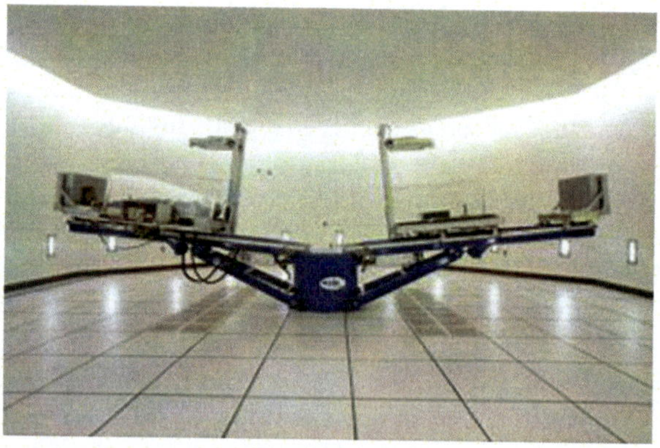

Fig. 12 Short radius centrifuge at UTMB in Galveston by NASA

Fig. 13 Artificial gravity is tantalizing, predicated on the existence of negative gravitational mass

8 Hypergravity

There is a circular ride there that spins dizzyingly fast. Standing inside it, your back is pressed against the wall. It spins faster and faster until, suddenly, the floor falls away. But you do not fall with it. You remain in place, pinned to the wall by forces "as great as 3-g—or three times the normal force of gravity," says Malcolm Cohen, chief of the Human Information Processing Research Branch at NASA Ames.

NASA is interested because it is not just microgravity that astronauts experience in space. They are exposed to hypergravity, too: up to 3.2-g at launch, and about 1.4-g on reentry. "Under these conditions," Cohen points out, "fluid weighs more." The heart has to change the way it operates, pumping faster, and working harder to push the blood all the way to the brain. This could cause astronauts to become dizzy or even, in extreme cases, to pass out. By spinning people in his centrifuge, Cohen hopes to learn whether the heart's response can be conditioned. Perhaps if astronauts were exposed to controlled doses of hypergravity before launch or reentry, then they might be able to tolerate high g-forces better than they otherwise would have (see Fig. 14).

Fig. 14 Hypergravity experiment for astronauts

Fig. 15 Artificial gravity equipment by NASA

An easier ride to space is not the only potential benefit. Here on Earth, hypergravity could be used to train athletes, providing an environment in which exercises could be conducted with more benefit in shorter time. People who suffer from muscle atrophy might be exposed to it, to stress their muscles more effectively. Centrifuges could be key to long-term space travel, too. That's because microgravity causes the body to deteriorate in a multitude of ways: cardiovascular deconditioning, loss of muscle mass, loss of bone density, and a host of other problems. Artificial gravity could prevent all that—and centrifuges are one plausible way to generate artificial gravity (see Fig. 15).

Cohen ticks off ways to make centrifugal gravity feasible: Perhaps engineers could develop a centrifuge with a radius of several kilometers, large enough to generate high artificial gravity without rotating fast enough to trigger the tumbling illusion. Rather than using small onboard centrifuges, space travelers might slowly rotate their entire spaceships instead.

Cohen found that his centrifuge riders spent a lot of time lying down, in part because it was more comfortable, and in part, because spinning made them drowsy—an effect called "the sopite syndrome." Cohen noted that he was surprised at how strong it was. Going forward, he'd like to examine what happens when they perform a range of predetermined activities, such as standing, in which the g-force places more stress on the heart [17].

9 Planets Motion; Graviton Mass Limitation

The motions of the planets have been used to make the best estimate yet of the upper limit of the mass of the graviton—a hypothetical particle that is a quantum of the gravitational field. That is the claim of Leo Bernus at the Paris Observatory and colleagues, who used over a century's worth of data in their calculations (see Fig. 16).

Fig. 16 Long-distance force: planets has put a new upper limit on the mass of the graviton

In theories that try to provide a quantum description of gravity, the graviton mediates the gravitational force between massive objects. It can be thought of as a gravitational version of the photon, which mediates the electromagnetic force between charged objects. A correct theory of quantum gravity has yet to be developed, but it is possible to test some aspects of nascent theories including their predictions of whether the graviton has a mass.

If gravitational fields have an infinite range—as Einstein's general theory of relativity dictates—gravitons must be massless and travel at the speed of light. However, some theories of quantum gravity suggest that the graviton could have an extremely small mass. If this were true, it would limit the range of the gravitational force and impose a subluminal speed limit on the graviton [18].

10 Space-Time Concepts

The fabric of space-time is a conceptual model combining the three dimensions of space with the fourth dimension of time. According to the best of current physical theories, space-time explains the unusual relativistic effects that arise from traveling near the speed of light as well as the motion of massive objects in the universe. Nowadays, when people talk about space-time, they often describe it as resembling a sheet of rubber. This, too, comes from Einstein, who realized as he developed his theory of general relativity that the force of gravity was due to curves in the fabric of space-time (see Fig. 17).

Massive objects—like the Earth, sun, or you—create distortions in space-time that cause it to bend. These curves, in turn, constrict the ways in which everything in the universe moves because objects have to follow paths along this warped curvature. Motion due to gravity is actually motion along the twists and turns of space-time [19] (see Fig. 18).

Fig. 17 Fabric of space-time (weirdly-shaped wormholes might work better than spherical ones)

Fig. 18 Space-time curvature caused by earth

Wherever matter exists, it bends the geometry of space-time. This results in a curved shape of space-time which can be understood as gravity. In general relativity, space-time is thought of as smooth and continuous. However, in the theory of quantum mechanics, space-time is not always continuous [20]. Matter determines how space-time curves. General Relativity explains how a spaceship traveling any speed slows down as it passes through a gravitational field, such as passing the planet Earth [21].

11 Warp Drive Real

While NASA is not pursuing interstellar flight, scientists continue to advance ion propulsion for missions to deep space and beyond using solar electric power. This form of propulsion is the fastest and most efficient to date. There are many theories that have become reality over the years of scientific research. But for the near future, warp drive remains a dream [22].

A maverick research team at NASA's Johnson Space Center has reached a milestone that many experts thought was impossible. The team formally published their experimental evidence for an electromagnetic propulsion system that could power a spacecraft through the void—without using any kind of propellant.

According to the team, the electromagnetic drive, or EmDrive, converts electricity into thrust simply by bouncing around microwaves in a closed cavity. In theory, such a lightweight engine could one day send a spacecraft to Mars in just 70 days. (Find out why Elon Musk thinks a million people could live on Mars by the 2060s.) Now, though, the latest study has passed a level of scrutiny by independent scientists that suggest the EmDrive really does work. Is this the beginning of a revolution in space travel—or just another false start for the "impossible" spaceship engine?

Weirdly, the EmDrive does not expel anything at all, and that does not make sense in light of Newton's third law or another tenet of classical mechanics, the conservation of momentum. If the EmDrive moves forward without expelling anything out the back, then there's no opposing force to explain the thrust. It's a bit like arguing that a person inside a car could propel it forward by repeatedly hitting the steering wheel, or that the crew of a spaceship could fly the craft to their destination simply by pushing on the walls [23].

12 Space Future (Vehicle Designs)

For many people, space tourism and even colonization are attractive ideas. But in order for these to start we need vehicles that will take us to orbit and bring us back. Current space vehicles clearly cannot. Only the Space Shuttle survives past one use, and that's only if you ignore the various parts that fall off (intentionally!) on the way up. You could be forgiven for thinking that space is therefore an impossibly expensive place to get to. But this need not be the case. Launch to orbit requires accelerating to Mach 26, and so it uses a lot of propellant—about 10 tons per passenger. But there is no technical reason why reusable launch vehicles could not come to be operated routinely, just like aircraft. The only reason why this has not been done yet is that launch vehicle development has been left to government space agencies. And they have had neither the priority nor the will to achieve it—they don't use even 2% of their budgets (of $25 billion per year) to study the design of launch vehicles suitable for passenger service!

So, it may well turn out to be private enterprise that is the solution—plenty of ideas for reusable launch vehicles exist, and with incentives like the X-Prize, there's going to be fierce competition to see who can be first. Space Vehicles presents some of the ideas that could change the meaning of "Space" from being a remote place where government staff carries out "missions" to being a weekend destination, just a few minutes' flight away.

The following is a list of projects underway today and some significant previous projects. Some are aimed initially at sub-orbital flights—a much easier target than getting to orbit, as demonstrated by SpaceShipOne. Others are designs for orbital vehicles. Ultimately, the only ones of importance are the piloted, passenger-carrying vehicles [24]:

- SPACESHIPONE & WHITE KNIGHT
- KANKOH-MARU
- DRAGON & FALCON 9
- SPACEBUS & SPACECAB
- ASTRIUM SPACE JET
- KELLY ECLIPSE
- STAR-RAKER

13 Teleportation

Teleportation in its simplistic meaning is just like when we scan an object, it converts it to a digitized form then email it or send it as a normal data transmission to a destination. In turn, if the capability to scan a human being or just their mind exists and send the digitized human copy some other place, we have achieved human teleportation. The journey of teleportation starts from the human brain which is very complex organ and in particular the human consciousness which is considered mathematically complicated, obviously the journey will face challenges.

If cloning of embryonic stem cell has changed not only the course of human life but ultimately the course of humanity, scientists have actually made some significant strides when it comes to teleportation to make it physically possible, teleportation will allow humanity to make major leaps forward. At the present time we can only teleport photons (particles of light) and atoms like Cesium and Rubidium, however, in the coming years, scientists expect to teleport molecules such as water and carbon dioxide, perhaps afterwards DNA and organic molecules. Now to teleport a human, atoms will fall apart then to rebuilt again with the same memories and personality of the original human, basically the original dies first then a new one somewhere else a carbon copy of the original comes to life, with this theology it raises an important question about the human soul, could it be copied, could it be altered, could it be multiplied, could it be hacked, is it just information. Physicists say it is physically possible to teleport an entire human being across the room or maybe to Mars but the concern about soul will always remain until actually human teleportation happens.

All in all, there is undeniable progress with quantum physics and quantum computing, this progress is measured in results, the results are obvious and promising, teleportation is feasible if we look how far the computers have come just in the past fifty years, teleportation is possible, Human desperation has driven many great men to accomplish the impossible [25].

14 Conclusion

Despite the Gravity abundance of the outer space, the lack of solid ground in space means that objects without thrust are in a continual state of free fall, and free fall feels just like zero gravity, so physicists are searching for the particle responsible for the gravitational force which was the derive of Einstein's theory of general relativity. General relativity also gives us some insight into the nature of gravitons. In this research paper, the theories of quantum gravity, facts about the graviton which is the hypothetical quantum of gravity, and the elementary particle that mediates the force of gravity were summarized. The working in zero gravity without having to go to the space station has been achieved by G-Force One which gave scientists that best chance, NASA's scientists studying microgravity to learn what happens to people and equipment in space and also trying to determine how different kinds of activities done in hypergravity affect cardiovascular conditioning. In the history of modern physics, the familiar and vague notion of force through the work of Galileo, Huygens, Newton, and others, became a physical concept with a constructive spatio-temporal definition; the one that did not really violate the common notion but that rendered it a powerful tool of physical investigation, and thereby made the discovery of physical forces a clear and attainable goal. Science fiction writers have given us many images of interstellar travel, and traveling at the speed of light has not been achieved yet for the time being. Scientific researches and experiments are working progress, scientists continue to advance ion propulsion for missions to deep space and beyond using solar electric power, this form of propulsion is the fastest and most efficient to date. Warp Drive Real is one of these applicable projects which was described in the introduced paper. In order for people to be able to travel economically to space for the purpose of tourism, exploration, mining, and more, we are in a need for reusable launch vehicles. The research paper ends with listing the projects that are underway today with regard to the future of space, and in particular the Vehicle Designs.

Acknowledgements The authors would like to thank ***October 6 University (O6U), Egypt*** and ***Devry University, USA*** for their support to introduce this research paper in a suitable view and as a useful material for researchers. The authors also would like to thank their colleagues who provided insight and expertise that greatly assisted the research.

References

1. Wood C (2019) What is quantum gravity? Space.com contributor, ©Future US Inc. New York, US, 27 Aug 2019 [Online]. Available https://www.space.com/quantum-gravity.html
2. May S (2017) What is microgravity? National Aeronautics and Space Administration, US, 7 Aug 2017 [Online]. Available https://www.nasa.gov/audience/forstudents/5-8/features/nasa-knows/what-is-microgravity-58.html
3. Lincoln D (2014) What are gravitons? The Nature of Reality, WGBH Educational Foundation, 15 May 2014 [Online]. Available https://www.pbs.org/wgbh/nova/article/what-are-gravitons/

4. Wikipedia, the free encyclopedia, "Graviton." [Online]. Available: https://en.wikipedia.org/wiki/Graviton
5. Davis GA (1999) Inquiring minds: light, wave or particle? Fermi National Accelerator Laboratory, Office of Science/U.S. Department of Energy, US, June, 1999 [Online]. Available https://www.fnal.gov/pub/inquiring/questions/light_dual.html
6. Choi C (2015) Physicists are searching for 'gravity particles' that could help unite all of physics. Copyright © 2020 Insider Inc., US, Mar 2015 [Online]. Available https://www.businessinsider.com/looking-for-gravity-particles-2015-3
7. Zona K (2008) Status of gravity control. Glenn Research Center, National Aeronautics and Space Administration, US, May, 2008 [Online]. Available https://www.nasa.gov/centers/glenn/technology/warp/gravstat.html
8. Choi CQ (2018) Is gravity quantum? © 2020 Scientific American, A Division of Springer Nature America, Inc., US, Aug 2018 [Online]. Available https://www.scientificamerican.com/article/is-gravity-quantum/
9. Beckwith AW (2007) Using quantum computing models for graviton communication/information processing in cosmological evolution. arXiv, July 2007 [Online]. Available https://www.researchgate.net/publication/1893282_Using_quantum_computing_models_for_graviton_communicationinformation_processing_in_cosmological_evolution
10. Fishman C (2017) The future of zero-gravity living is here. Smithsonian Magazine, © 2020 Smithsonian Magazine, CA, US, June 2017 [Online]. Available https://www.smithsonianmag.com/science-nature/future-zero-gravity-living-is-here-180963243/
11. Letzter R (2020) Photons could reveal 'massive gravity,' new theory suggests. Space.com contributor, ©Future US Inc. New York, US, Feb 2020 [Online]. Available https://www.space.com/graviton-photon-interaction-hunting-machine.html
12. Thompson A (2019) Medicine in space: what microgravity can tell us about human health. © 2020 Scientific American, a Division of Springer Nature America, Inc., US, Aug 2019 [Online]. Available https://www.scientificamerican.com/article/medicine-in-space-what-microgravity-can-tell-us-about-human-health/
13. AWS (2020) AWS graviton processor; enabling the best price performance in amazon EC2. © 2020, Amazon Web Services, Inc., US [Online]. Available https://aws.amazon.com/ec2/graviton/
14. CERN Accelerating Science (2020) Extra dimensions, gravitons, and tiny black holes. Copyright © 2020 CERN, Switzerland [Online]. Available https://home.cern/science/physics/extra-dimensions-gravitons-and-tiny-black-holes
15. Baird CS (2013) How do space ships make artificial gravity? Science Questions with Surprising Answers, Feb 2013 [Online]. Available https://wtamu.edu/~cbaird/sq/2013/02/14/how-do-space-ships-make-artificial-gravity/
16. Hsu J (2010) New artificial gravity tests in space could help astronauts. © 2020 Scientific American, a Division of Springer Nature America, Inc., US, May 2010 [Online]. Available https://www.space.com/8384-artificial-gravity-tests-space-astronauts.html
17. Ryba J (2007) The pull of hypergravity. National Aeronautics and Space Administration, US, Nov 2007 [Online]. Available https://www.nasa.gov/missions/science/hyper.html
18. Jarman S (2019) Motions of the planets put new limit on graviton mass. IOP Publishing, Physics World portfolio, UK, Nov 2019 [Online]. Available https://physicsworld.com/a/motions-of-the-planets-put-new-limit-on-graviton-mass/
19. Mann A (2019) What is space-time? A simple explanation of space-time, live science contributor. Future US Inc., Dec 2019 [Online]. Available https://www.livescience.com/space-time.html
20. Kids Encyclopedia Facts (2020) Space-time facts for kids. Powered by MediaWiki, © 2020 Kiddle.co, Aug 2020 [Online]. Available https://kids.kiddle.co/Space-time
21. Physics-Astronomy Team (2016) Here are 10 facts about space and time that everyone should know about. Physics-astronomy.com, Mar 2016 [Online]. Available http://physics-astronomy.com/2016/03/here-are-10-facts-about-space-time-that.html

22. NASA Content Administrator (2017) Is warp drive real? Glenn Research Center, National Aeronautics and Space Administration, US, Aug 2017 [Online]. Available https://www.nasa.gov/centers/glenn/technology/warp/warp.html
23. Drake N, Greshko M (2016) NASA team claims 'impossible' space engine works—get the facts. Copyright © 1996–2015 National Geographic Society, Copyright © 2015–2020 National Geographic Partners, LLC. All rights reserved, Nov 2016 [Online]. Available https://www.nationalgeographic.com/news/2016/11/nasa-impossible-emdrive-physics-peer-review-space-science/
24. Space Future (2019) Space vehicles. US, 2019 [Online]. Available http://www.spacefuture.com/vehicles/designs.shtml
25. University of Tokyo (2020) Large scale quantum computing with quantum teleportation. © 2000–2020 Furusawa & Yoshikawa Lab, 2020 [Online]. Available http://www.alice.t.u-tokyo.ac.jp/index-en.php

Entropy Detection-Based Spectrum Sensing in Frequency Domain Under Noise Uncertainty for Cognitive Radios Networks

Mona A. Fouda, Adly S. Tag Eldien, and Hala A. K. Mansour

Abstract Noise uncertainty is one of the important spectrum sensing threats. Many techniques are provided for spectrum sensing in cognitive radio (CR), but these techniques have a bad performance at low SNR due to the noise uncertainty effect. To counteract noise uncertainty, an entropy-based spectrum sensing scheme is introduced in this letter, where the performance of the entropy detection technique is estimated at frequency domain with a probability space partitioned into fixed dimensions to minimize the effect of noise uncertainty on the detection performance gain. This performance is compared by the energy detection performance with/without noise uncertainty effect where the adverse effect of noise variance uncertainty is much less with the entropy detector than with that of the energy detector. The simulation results proved that entropy detection has a good performance at low SNR up to -25 dB under the assumption of noise uncertainty in the frequency domain. Also, the performance of the entropy detector in the time domain with different sampling size is analyzed by the simulation which proved that the value of entropy is constant and invariant against SNR, so the entropy in the time domain cannot distinguish between the signal and noise, and this is not applicable. Also, the entropy detection in the frequency domain did not need to increase the sampling size to improve its performance such as in energy detection.

Keywords Cognitive radios · Spectrum sensing · Energy detection · Entropy detection · Noise uncertainty.

M. A. Fouda (✉) · A. S. T. Eldien · H. A. K. Mansour
Faculty of Engineering at Shoubra, Benha University, Cairo, Egypt
e-mail: Mona.A.Fouda@ieee.org

A. S. T. Eldien
e-mail: adlytag@feng.bu.edu.eg

H. A. K. Mansour
e-mail: hala.mansour@feng.bu.edu.eg

D. A. Magdi et al. (eds.), *Digital Transformation Technology*, Lecture Notes in Networks and Systems 224, https://doi.org/10.1007/978-981-16-2275-5_2

1 Introduction

Cognitive radio aims to improve the spectrum utilization to overcome the rareness problem of the spectrum, and CR technology is emerged according to rigid licensing policies [1]. So, CR maximizes the throughput of secondary users (SU) coexisting with primary users without any harmful interference [2]. This assumption requires the SU to collect cognition about the spectrum environment, and this is known as spectrum sensing. So, the spectrum sensing is a crucial issue in cognitive radio [3]. There are various methods for spectrum sensing that include energy detection (ED) [4, 5], matched filtering detection [6], cyclo-stationary feature detection [7] and waveform-based sensing [8], and each one has its benefits and drawbacks [9]. Most existing spectrum sensing methods suffer from the noise uncertainty, and it can hardly acquire promising performances in the adverse situation [10]. These techniques try to improve its performance by increasing the sampling size as in the conventional energy detection the performance is improved, but at a certain SNR wall, the performance is unreliable, and the increase in sampling size will be not applicable [11]. So, many studies focused on this problem and provide a new technique that has robustness against the effect of noise. This technique is known as entropy detection technique. The performance of entropy detection is reliable in case of the frequency domain where can distinguish between the noise and the signal, but in case of time domain, the feature of entropy remains constant for different values of SNR, so it is not applicable in the detection process. In information theory, entropy is a measure of the uncertainty associated with a discrete random variable. The term by itself usually refers to the Shannon entropy, which the information contained in a message [12]. Recently, the Shannon entropy has been used for spectrum sensing [13]. Approaches to entropy estimation generally include histogram method [14], model-based methods [15] and kernel-based methods [16]. Many types of researches seek to improve the performance of entropy detection so find many types of entropy detection techniques, and each one tries to improve the detection performance. In [17, 18], the Shannon and differential entropy is provided in the frequency domain, but the performance did not achieve the robustness against the noise as in [19] which the authors provided Renyi and Tsallis are modified entropy techniques, the performance of detection is improved. In [20], the authors discuss the modification of Shannon entropy to improve the detection performance by providing a new technique called Parzan entropy technique. To address this limitation, in this paper, the performance of Shannon entropy detection for $QPSK$ signal under AWGN is investigating in the case of frequency domain by applying Fourier transform (FT) to the sensed signal, and the probability space is partitioned into fixed dimensions as the spectrum magnitude is regarded as a random variable. The Shannon entropy is calculated as an information measure of the received signal to detect the presence/absence of the primary signal. Simulation results verify the robustness of entropy detection in the frequency domain against noise uncertainty, and the performance in the time domain is unreliable. Also, the relation between SNR and probability of detection P_d is simulated with/without noise uncertainty, and this performance is compared by the conventional ED. The receiver operating characteristics ROC are drawn for entropy detection. So, the simulation results confirm that the entropy detection is robust against the noise uncertainty compared by the ED. The simulation results are based on the Monte Carlo with 100,000 iterations. The simulation tool is MATLAB version R2014a. The rest of this paper is organized as follows: In Sect. 2, system model is formulated. Sect. 3 describes simulation results. Sect. 3.2 gives comparison between ED and entropy. Finally, the conclusion is given in Sect. 4.

2 System Model

2.1 Binary Hypothesis Testing Problem

To avoid interfering with the primary users when the frequency bands are already occupied, detection should be made before the CR accesses the bands [21], so the most critical technique is the spectrum sensing which decides the success or failure of the following steps. The target of spectrum sensing in CR is to determine whether a licensed band is currently occupied by its primary user or not. This can be formulated into a binary hypotheses testing problem, let $n(t)$ denote white Gaussian noise of mean $\mu_n = 0$ and variance σ_n^2, and $s(t)$ denotes also Gaussian signal of mean $\mu_s = 0$ and variance σ_s^2. We can state the binary hypothesis in the following [22]:

$$y(t) = n(t) \quad H_0, \quad y(t) = s(t) + n(t) \quad H_1 \tag{1}$$

where H_0 denotes the absence of primary signal, H_1 denotes the presence of primary signal, $y(t)$ is the received signal, $n(t)$ denotes the additive σ^2, and $s(t)$ is the transmitted signal PU. The two hypotheses indicate the presence of PU when the entropy value (E) is greater than the threshold (λ) and vice versa. If $y(t)$ is sampled, then the nth sample $y(n)$ is given as

$$y(n) = n(n) \quad \leq n \leq M \quad H_0, \quad y(n) = s(n) + n(n) \quad 1 \leq n \leq M \quad H_1 \tag{2}$$

where M is the total number of samples. In spectrum sensing, the performance metrics of detection are two probabilities:

- **probability of detection** P_d: is the probability of the algorithm correctly detecting the presence of primary signal under hypothesis H_1. So that a higher P_d is preferred for a good performance.
- **probability of false alarm** P_f: is the probability of algorithm falsely declaring the presence of a primary signal. So that the P_f leads to poor spectrum usage and a higher P_f is not preferred for a good performance.

2.2 Noise Uncertainty Model

One of the limitations of spectrum sensing is the sensitivity to noise uncertainty. The noise uncertainty exists in every practical system, and it is caused by many factors like thermal noise, quantization noise and interference between PUs or between SUs, temperature changes and filtering effect. All of these can be caused by the noise power estimation error which is referred to as noise uncertainty. Due to noise uncertainty, the performance of detection deteriorates rapidly at low SNR. The threshold selection depends on the accurate noise power estimation, and it is not always possible, so the error estimation must be taken in simulation considerations so that assume there is uncertainty $\pm x$ dB in noise model, and in linear scale terms defined $\rho = 10^{x/10}$ with SNR and large M, the probability of false alarm and detection is given by [22]

$$P_\mathrm{f} = 1 - Q\left(\frac{\lambda - \sigma_n^2\rho}{\sqrt{\frac{2}{M}}\rho\sigma_n^2}\right) \tag{3}$$

$$P_\mathrm{d} = 1 - Q(\frac{\lambda - \sigma_n^2(1/\rho + \mathrm{SNR})}{\sqrt{\frac{2}{M}}\sigma_n^2(1/\rho + \mathrm{SNR})}) \tag{4}$$

where

$$Q(x) = \frac{1}{\sqrt{2\pi}}\int_x^\infty \exp(\frac{-u^2}{2})\mathrm{d}u$$

is the Gaussian Q-function.

2.3 Entropy Detection in Time Domain

The differential entropy, denoted by $b(X)$, for continous random variable X is defined as

$$b(X) \cong \int_{-\infty}^{\infty} f_x(x)\log(f_x(x))\mathrm{d}x$$

where $f_x(.)$ is the probability density function of X, we estimate the differential entropy in the observations and use it as a test statistic to carry out spectrum sensing, so let σ_x^2 denote variance of the received signal, and the analytical expression of differential entropy is

$$\ln\left(\sqrt{\sigma_x^2 2\pi\gamma}\right),$$

so the Shannon entropy of the received signal $H(X)$ in the time domain can be approximated by differential entropy $b(X)$ [18] as

$$H(X) \simeq b(X) - \log \Delta = \frac{1}{2} \ln 2\pi\gamma\sigma_x^2 - \ln \Delta \tag{5}$$

where Δ is bins width, γ is Euler's constant number, and for each random value X_m, the cumulative distribution function of X_m is

$$\rho = \text{Prob}(X < X_m)(e.g., 0.99 < \rho < 1),$$

then we have

$$Q(X_m - \mu_x)/\sigma_x = 1 - \rho,$$

so X_m can obtained as

$$X_m = \mu_x + \sigma_x Q^{-1}(1 - \rho)$$

Substituting by

$$\Delta = X_m/L = (\mu_x + \sigma_x Q^{-1}(1 - \rho))/L$$

into Eq. (5)

$$H(X) = \ln \frac{\sigma_x L \sqrt{2\pi\gamma}}{\mu_x + \sigma_x Q^{-1}(1 - \rho)} \tag{6}$$

It is seen that the expression of estimated entropy in time domain is a function of the mean μ_x and variance σ_x^2 of the received signal. For simplicity, we assume the measured μ_x = zero, so this shows that, the estimated Shannon entropy in time domain is a constant in both hypotheses under the given bin number L and confidence coefficient ρ [18]:

$$H(X) \simeq \frac{\ln(L\sqrt{2\pi e})}{Q^{-1}(1 - \rho))} \tag{7}$$

where Q^{-1} is the inverse of Q-function.

In information theory, entropy is a measure of uncertainty associated with a random variable; according to Shannon entropy, in time domain, we find that the estimated entropy is constant at the SNR regardless the increasing of number of samples values, and it leads to estimated entropy that cannot distinguish between noise and signal, so the performance of the detector is not applicable in this case. By simulation for $QPSK$ signal for different number of sampling size [N = 16, 32 and 64], we provided the performance of the entropy in the time domain in Fig. 1 compared to [18].

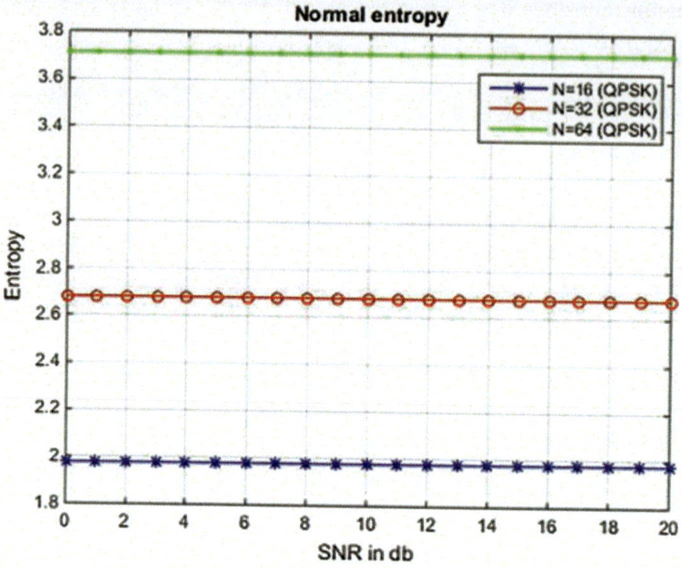

Fig. 1 The relation between Entropy and SNR in time domain

2.4 *Entropy Detection in Frequency Domain*

By applying discrete Fourier transform (DFT) to Eq. (2), we have the following hypotheses:

$$H_0: \mathbf{X}(K) = \mathbf{W}(K),\ K = 0, 1, 2, \ldots, K - 1$$

$$H_1: \mathbf{X}(K) = \mathbf{S}(K) + \mathbf{W}(K),\ K = 0, 1, 2, \ldots, K - 1$$

where K is the length of DFT equal to sample size N, and $\mathbf{X}(K)$, $\mathbf{S}(K)$ and $\mathbf{W}(K)$ denote the complex spectrum of the received signal, primary signal and noise, respectively. Information entropy is a measure of uncertainty associated with a random variable. The random variable Y for which we want to estimate the PDF represents the spectrum magnitude of the measured signal. Hence, a possible detection strategy consists of testing information entropy based on the frequency model as

$$H(Y): H_0 \text{ aganist } H(Y): H_1$$

Let y_i denote a possible state value, and p_i denote corresponding probability in state y_i, let L represent he total number of countable states, referred to as dimension of probability space. We then have [17, 18]:

$$\sum_{i=1}^{L} p_i = 1, (i = 1, 2, ...L)\text{with}0 \le p_i \le 1$$

Information entropy is a measure of the uncertainty associated with a random variable. It quantifies information contained in a message and can be written as

$$H_L(Y) = -\sum_{i=1}^{L} p_i \log_b p_i \tag{8}$$

where b is the base of the logarithm, and p_i is the probability in state y_i. The discrete entropy is approximated by a constant for a given bin number L which implies that the false alarm ratio is almost fixed for a given threshold. In this sense, the proposed detector is robust to noise uncertainty, where

$$H_L(Y) = \ln \frac{L}{\sqrt{2}} + \frac{\gamma}{2} + 1 \tag{9}$$

where L is a dimension of the probability space, and γ is Euler's constant number.

For a given bin number L, let K_i denote the total number of occurrences in the ith bin width

$$\sum_{i=1}^{L} K_i = N,$$

the probability in each state p_i is the frequency of occurrence in ith bin, that is, $p_i = K_i/N$.

The bin width Δ can expressed by

$$\Delta = \frac{Y_{\max} - Y_{\min}}{L} \tag{10}$$

where $Y_{\max}$ and $Y_{\min}$ denote the maximum and minimum value of random variable Y, respectively. Once bin number L is fixed, and bin width Δ varies with the range of spectrum magnitude.

Substituting p_i into Eq. (8), the entropy is estimated for both hypotheses, and the test statistic is obtained as

$$T(Y) = H_L(X) = -\sum_{i=1}^{L} \frac{k_i}{N} \log \frac{k_i}{N} \tag{11}$$

where k_i is the number of occurrences.

Assuming that the estimated noise entropy follows a Gaussian distribution with theoretical value H_L due to noise term in practical not be perfectly canceled, so the operating value of the threshold based on the probability of false alarm P_f as the following equation:

Fig. 2 Architecture of entropy detection technique in frequency domain

$$\lambda = H_L + Q^{-1}(1 - P_f)\sigma_e \tag{12}$$

where Q^{-1} is the inverse of Q-function, and σ_e^2 is variance signal.

The following Fig. 2 shows the architecture of entropy detection technique in the frequency domain where this model does not need any prior information of the primary user [24]:

- **FFT Block**: discrete signal is performed by an N point fast Fourier transform (FFT) to obtain the frequency domain of the received signal,
- **Histogram Unit**: calculates both binary hypotheses,
- **Entropy Estimator**: used to obtain the estimated entropy value,
- **Threshold Unit**: used to obtain the value of λ depends on P_{f},
- **Decision Unit**: used to compare the value of λ and the value of the entropy estimated.

In Fig. 3, the log-likelihood of estimated entropy is obtained against SNR for a $QPSK$ signal with different sampling sizes. From this figure, the value of log-likelihood entropy is decreased by increasing SNR, so the entropy detection in the frequency domain can distinguish between the noise and signal, so in this case, the detection performance will be applicable; however, the entropy in the time domain cannot distinguish between noise and signal, so it is not applicable as detection technique [18],

3 Simulation Results

3.1 Entropy Detection Without Noise Uncertainty Effect

This section will show all simulation results for Section II. MATLAB program version R2014a is used as a simulation tool. In the simulation, the Mont Carlo method is used with 1000 iterations. In the first case of the simulation result, the assumption of Shannon entropy without any noise uncertainty effect is taken. The simulation results achieved the desired $P_{\text{d}} \geq 0.9\, and\, P_{\text{f}} \leq 0.1$ at low SNR up to -16 dB, and the performance metrics are simulated to obtain ROC curve and the relation between SNR and P_{d}. Fig. 4 shows the relation between SNR and P_{d} at P_{f}= 0.45 and number of samples $N = 5000$. We noticed that the entropy detection can detect signal at low SNR by reliable performance.

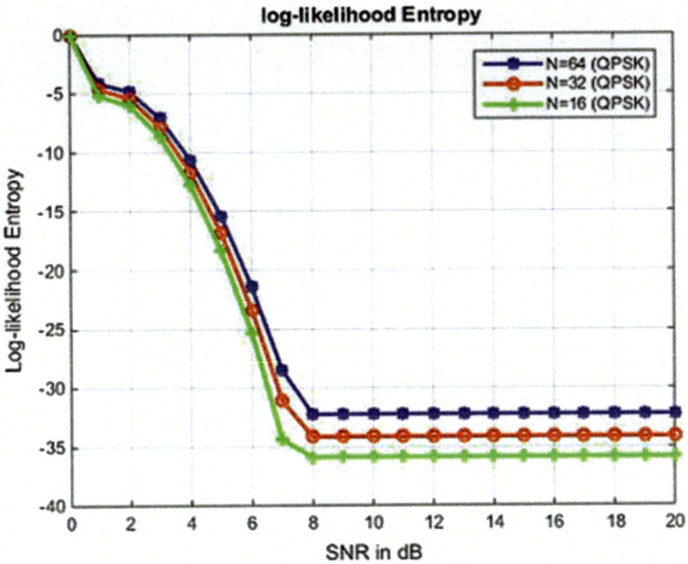

Fig. 3 Relation between entropy and SNR in frequency domain

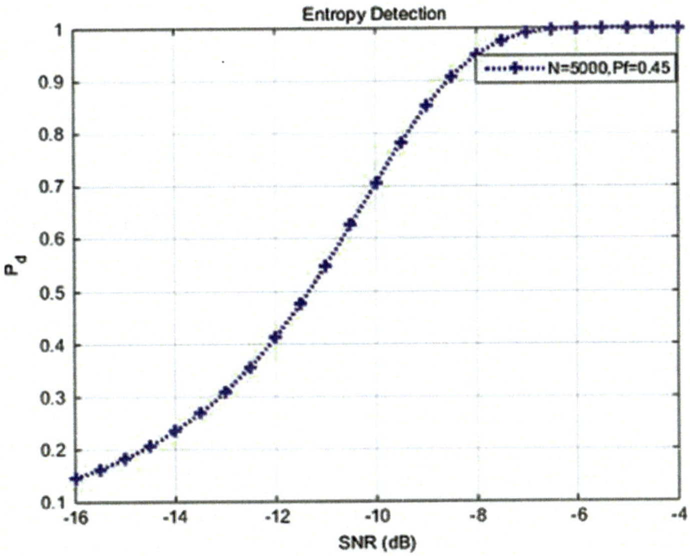

Fig. 4 Relation between SNR and P_d for entropy detection

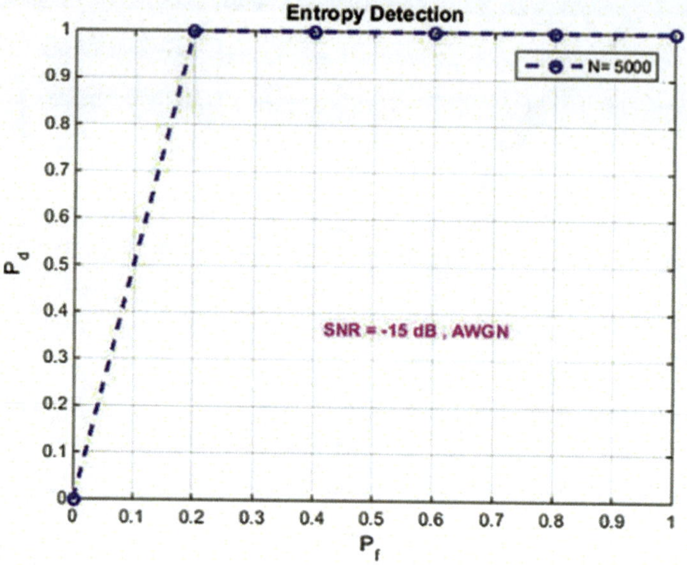

Fig. 5 Relation between P_f and P_d under $SNR = -15$ dB

Fig. 5 shows the trade-off between the P_f and P_d at $SNR = -15$ dB and $N = 5000$. This relation represents the ROC curve which measures the performance gain.

3.2 Comparison Between Energy Detection and Entropy Detection With/without Noise Uncertainty Effect

In the second case of the simulation result, the assumption of the noise uncertainty effect is taken into consideration. Our results will study this case in detail where the number of iteration is 100,000 and the simulation occurred at different values of N, SNR and P_f. The simulation results achieved the desired $P_d \geq 0.9 \, and \, P_f \leq 0.1$ at low SNR up to -25 dB and the value of noise which the $\pm x = -5$ dB for Entropy detection and ED. The effect of noise uncertainty is studied on the relation between SNR and P_d.

Fig. 6 shows the relation between entropy detection and ED without any noise at N =5000, and we observe that entropy detector performance is clearly better and can be measured by numbers, for example, at SNR equal to -5 dB, the probability of detection is 0 and 0.6 for ED and entropy detector, respectively, so the entropy detection performance is more reliable than ED, but in Fig. 7, we considered the assumption of noise uncertainty by $\pm x = -5$ dB and $N = 1000$, we noticed that the ED performance with noise is destroyed, but the entropy detection performance is still robust against this effect.

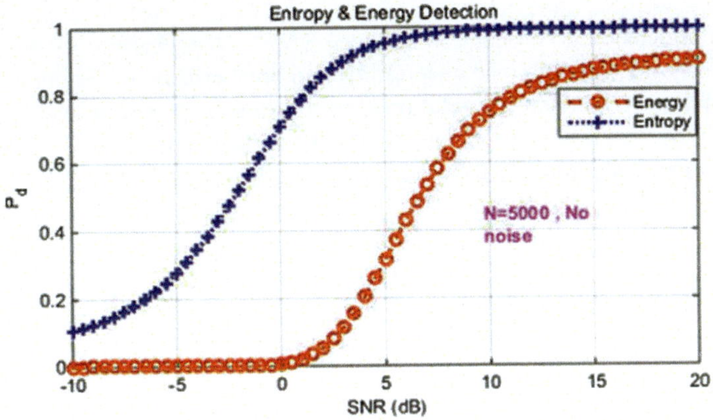

Fig. 6 Relation between entropy detection and ED without noise uncertainty effect at N =5000

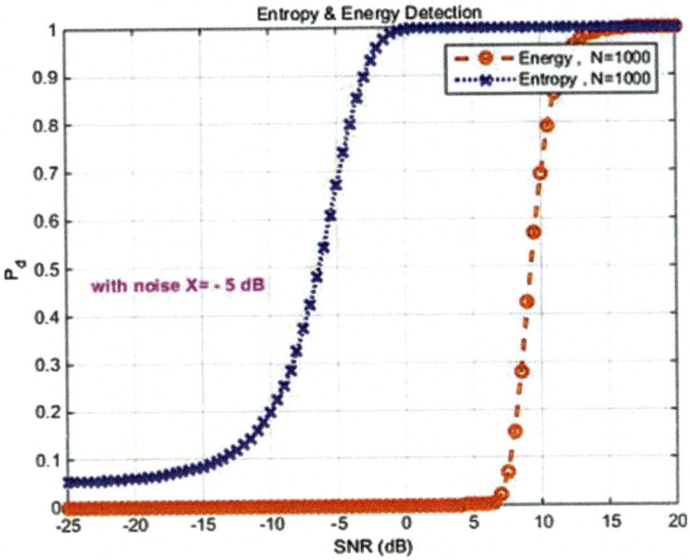

Fig. 7 Relation between entropy detection and ED with noise uncertainty effect at N =1000

4 Conclusion

CR aims to maximize the throughput of secondary users (SU) coexisting with primary users without any harmful interference. Spectrum sensing techniques have more challenges and limitations such as sensitivity to noise uncertainty. In this paper, the performance of Shannon entropy detection for $QPSK$ signal under AWGN is investigating in the case of frequency domain, and simulation proved that the performance in the time domain is unreliable. The probability space is partitioned into fixed dimensions. The relation between SNR and probability of detection P_d is simulated with/without noise uncertainty, and this performance is compared with the conventional ED. Also, the relation between P_f and P_d is simulated under Low SNR (-15 dB). From the simulation results, the entropy detection is robust against the noise uncertainty compared by the ED.

References

1. Spectrum policy task force report. In: Federal communication commission, Washington, DC, USA, Tech. Rep. 02-155, Nov 2002
2. Zheng S, Chen S, Qi P, Zhou H, Yang X (2020) Spectrum sensing based on deep learning classification for cognitive radios. China Commun 17(2):138–148
3. Ali A, Hamouda W (2017) Advances on spectrum sensing for cognitive radio networks: theory and applications. IEEE Commun Surv Tutor 19(2):1277–1304. https://doi.org/10.1109/COMST.2016.2631080
4. Lopez-Benitez M, Casadevall F (2012) Improved energy detection spectrum sensing for cognitive radio. IET Commun 6(8):785–796
5. Abdullah Hikmat N, Abed Hadeel S (2019) Implementation of selected spectrum sensing systems for cognitive radio networks using FPGA platform. J Telecommun Inf Technol. https://doi.org/10.26636/jtit.2018.125018
6. Zhang Z, Yang Q, Wang L, Zhou X (2010) A novel hybrid Matched Filter structure for IEEE 802.22 standard. In: Proceedings of IEEE Asia-Pacific conference on circuits system (APCCAS), Kuala Lumpur, Malaysia, Dec 2010, pp 652–655
7. Sabat SL, Srinu S, Raveendranadh A, Udgata SK (2012) Spectrum sensing based on entropy estimation using cyclostationary features for cognitive radio. In: Proceedings of 4th international conference on communication system networking (COMSNETS), Bangalore, India, Jan 2012, pp 1–6
8. Tang H (2005) Some physical layer issues of wide-band cognitive radio systems. In: Proceedings of IEEE international symposium on new frontier dynamic spectrum access networking (DySPAN), Baltimore, MD, USA, Nov 2005, pp 151–159
9. Yücek T, Arslan H (2009) A survey of spectrum sensing algorithms for cognitive radio applications. IEEE Comun Surv Tutor 11(1)
10. Tandra R, Sahai A (2008) SNR walls for signal detection. IEEE J Sel Top Signal Process 2(1):4–17
11. Fouda MA, Eldien AST, Mansour HAK (2017) FPGA based energy detection spectrum sensing for cognitive radios under noise uncertainty. In: 2017 12th international conference on computer engineering and systems (ICCES), Cairo, 2017, pp. 584–591.https://doi.org/10.1109/ICCES.2017.8275374
12. Cover TM, Thomas JA (1991) Elements of information theory. Wiley, pp 228–229
13. Nagaraj SV (2009) Entropy based spectrum sensing in cognitive radio. Signal Process 89(2):174–180
14. Bercher J-F, Vignat C (2000) Estimating the entropy of a signal with applications. IEEE Trans Signal Process 48(6):1687–1694
15. Kay S (1998) Model-based probability density function estimation. IEEE Signal Process Lett 5(12):318–320
16. Jo S (2005) A robust approach to empirical PDF estimate. Neurocomputing 67:288–296
17. Zhang YL, Zhang QY, Melodia T (2010) A frequency-domain entropy-based detector for robust spectrum sensing in cognitive radio networks. IEEE Commun Lett 14(6):533–535. https://doi.org/10.1109/LCOMM.2010.06.091954

18. Zhang Y, Zhang Q, Wu S (2010) Entropy-based robust spectrum sensing in cognitive radio. IET Commun 4(4):428–436, 5 Mar 2010. https://doi.org/10.1049/iet-com.2009.0389
19. Vaidehi G, Swetha N, Sastry PN (2015) Entropy based spectrum sensing in cognitive radio networks. IJARCCE 4(11)
20. Rajasree Rao Y, Sabat SL, Swetha N, Sastry PN (2016) Parzen window entropy based spectrum sensing in cognitive radio, Feb 2016
21. Struzak R (2018) Cognitive radio, spectrum, and evolutionary heuristics. IEEE Commun Mag 56(6):166–171. https://doi.org/10.1109/MCOM.2018.1700388
22. Atapattu S, Tellambura C, Jiang H (2014) Energy detection for spectrum sensing in cognitive radio
23. Sabat SL, Srinu S, Raveendranadh A, Udgata SK (2012) Spectrum sensing based on entropy estimation using cyclostationary features for Cognitive radio. In: 2012 fourth international conference on communication systems and networks (COMSNETS 2012), Bangalore, 2012, pp 1–6. https://doi.org/10.1109/COMSNETS.2012.6151311
24. Xia H, Zhang G, Ding Y (2009) Spectral entropy based primary user detection in cognitive radio. In: 2009 5th international conference on wireless communications, networking and mobile computing, Beijing, 2009, pp 1–4

Detecting Semantic Social Engineering Attack in the Context of Information Security

Eman Ali Metwally, Noha A. Haikal, and Hassan Hussein Soliman

Abstract In IEEE 802.11 standard, the management frames are sent unencrypted, so the network name (SSID), MAC address (BSSID), and/or IP address can be easily spoofed. Impersonating existing AP with faked one to steal sensitive information from the connected devices is known as an evil twin attack. The current approaches for detecting Evil-twin AP depends on techniques as clock skew, route option, IP packet header, and data frame statistics. **The relevant literature approaches are either outdated, limited in their detection methods, architecture, and/or scope of detection. This research proposed an admin and user tool that can detect the evil twin attack**. In this paper, we detect the de-authentication and disassociation packets or both (mixed frames), as it is an essential part of evil twin attack. By using a low-cost microcontroller capability to detect and classify frames and then trigger different lighting alert for each type of frames. The main contribution of this paper does not lie only in its ability to detect different types of attack but also in detecting them in real-time and determining the attacker's MAC address. It is prototyped under real attack as it is implemented over two different scenarios, in both admin and user side then compared with other detection method. Experimental results show accuracy rate of 95.30% for the admin side in (DE authentication attack—disassociation attack—mixed attack—normal packets). While it proves accuracy rate of 88.18% for the user side.

Keywords Social engineering · Wi-Fi attacks · Evil twin

E. A. Metwally (✉) · N. A. Haikal · H. H. Soliman
Faculty of Computers and Information Science, Mansoura University, Mansoura, Egypt
e-mail: eman.ali.metwally@gmail.com

N. A. Haikal
e-mail: dr_nahikal@mans.edu.eg

H. H. Soliman
e-mail: hhsoliman@yahoo.com

D. A. Magdi et al. (eds.), *Digital Transformation Technology*, Lecture Notes in Networks and Systems 224, https://doi.org/10.1007/978-981-16-2275-5_3

1 Introduction

Social Engineering in information security systems is the art of availing the weakest point. It is a general term for a wide range of exploiting a computer that depends on a set of attack methods in order to deceive the user. This technique is used for bypassing the IDS (intrusion detection system), access control systems, and firewalls. The main danger here is in its legitimate appearance; because the victim could not recognize that he is victimized and that leads to many security breaches [1, 2].

A particular kind of social engineering that overcomes the defensive layers by presenting a misleading object characteristic, as appearing to be a legitimate software, to launch an attack over the client is called a semantic social engineering attack. For instance, phishing mails, misleading URLs, drive-by download, spoofed websites, WI-FI evil twin, and scareware. For distinguishing between attacks that can exceed system's security from those that exist in a nontechnical field or area, for instance, a scamming letter that expresses a prize-winning or physically mimic an authority figure, the term semantic has been proposed. A semantic attack means deceiving the user-computer interface to mislead the client and violate the security of the system or information. As an example, the evil twin attack is a fraudulent Wi-Fi access point that appears to be legitimate, and set up to eavesdrop on wireless communications and it is also called as rouge access point [3].

These days, public places as hotels or coffee shops allow their clients to use the Wi-Fi service, also do stores and workspaces. The client is considered the "weakest link" in the spectrum of information security systems. However, evil twin rogue access Point attack can clone the SSID and MAC Address to appear legitimate [4].

Rogue Access Point (RAP) is a serious wi-fi security violation that can be built up without a frank allowance of the network's admin in which the intruder can steal passwords and gain payment credentials. The attacker can spoof the Domain Name System (DNS) by eavesdropping over the medium of the communication. The wireless networks are identified not only by the Service Set Identifier (SSID) but also with the (MAC) address. Upon that, the attacker sniffs the Legitimate Access Point's (LAP) SSID to attract the user to connect by launching a Rogue Access Point (RAP) [5].

There are two forms of existence in which the Evil Twin RAP exists. These two forms are coexistence and replacement. In the first form, the genuine access point and the malicious one coexists at the same location. It could be with higher signal strength to derive the users to connect to it, as the IEEE 802.11 standard states that WLAN clients must connect to APs that have the strongest signal, the attacker will benefit from it. The second form of the rouge access point existence is the replacement, as it replaces the legitimate AP by shutting it down. In this case, the rouge access point needs to have an internet connection for his own, while in the first case it could depend on the legitimate AP's internet [6].

Actually, the Evil Twin RAPs are mainly used to listen to user's traffic, then the attacker launches attacks as interception, replaying, traffic manipulation, and many more attacks [7]. Setting up an Evil Twin AP may need more steps to avoid (IDSs)

Intrusion Detection Systems, the attacker masquerades the MAC address of the access point and the SSID, configures a Domain Name Server (DNS) server to connect to the internet and a Dynamic Host Configuration Protocol (DHCP) server to assign IP addresses. In modern age, rouge APs can be deployed using smartphones and that makes it harder to be identified [8].

Rouge access point based on De-authentication/disassociation attack can be launched in different ways as:

1. The attacker can simulate being the client device by spoofing its MAC address and send DE authentication frames to the access point as if he is the original client, so the access point considers that a connected client wants to leave the network and disconnected.
2. The attacker can simulate being the access point and DE authenticates all the network's clients by sending a broadcast of (FF: FF: FF: FF: FF: FF) as the destination mac address. This attack floods the network and disconnects all legitimate clients and has a tremendous DOS impact.
3. The intruder can spoof the MAC address of the access point, then inject the DE authentication/disassociation frames over the network to disconnect the user [9].

For the reason that DE authentication/disassociation frames can be sent in place of the legitimate AP to kick all users from the network, it indicates an Evil-twin attack, as the attacker spoofs the genuine AP's MAC-address. The attacker who is masked as the real AP, tries to interrupt the message exchange of the four-way handshaking between the client and the AP by sending fake messages for the four-way handshaking [10]. Attackers aim to exploit the vulnerabilities of the used protocol by falsifying their source address and spreading traffic to put the service down and then launch their rouge access point [11].

Figure 1 shows the DE authentication attack over a network of two users connected to a W-LAN and an access point that is connected to the internet in public place to provide the service to all the authenticated clients. The attacker operates on Kali Linux version 2019.4 [12] specially in Airplay-ng toolkit [13] as used to disconnect clients to launch the attack by spreading DE authentication/disassociation frames. Under attack, the user's nodes are kicked out of the network and be unable to reconnect, then set up the evil twin rouge access point with the same name which is malicious.

Detecting RAP attacks' solutions are categorized into two groups. The initial group of solutions is the administrator-based one. It always depends on monitoring the radio frequency signal, need to be applied on switches, servers, routers, and special devices. Moreover, it needs a pre-prepared AP's authorized list that depends on fingerprinting to decide if the connected AP is a real or a rogue one. It also does traffic monitoring at aggregation point as the gateway, to decide the way of connectivity used. It is used by the network admin. But it has its limitations as it does not support real-time detection, needs protocol modification, or depends on a server.

The other group of solutions is a client-based solution. It is applied on users' devices to detect the attack. Using the connection of TCP, Clock Skew, route option, IP packet header, and data frame statistics are its main ways. The pros of this solution

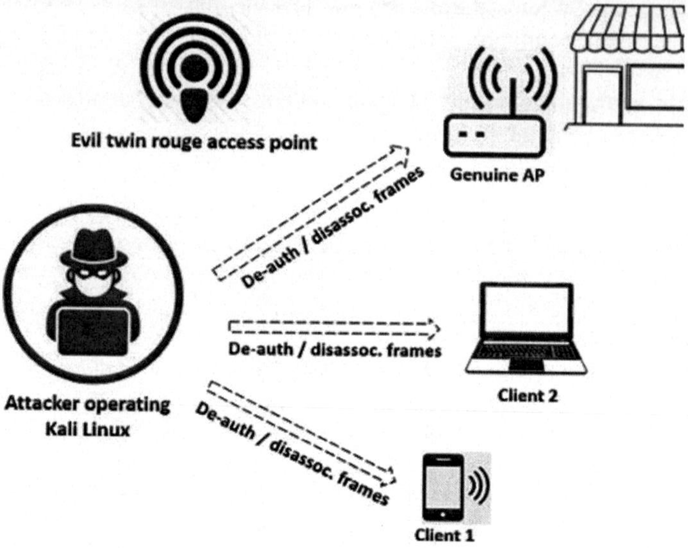

Fig. 1 Practical way for DE authentication and disassociation. Attack in public place with installation of evil Twin rouge access point

are that the client can perform the detection by himself to make sure that he is secured, and they can protect their information timelier. But it has its con as it may need a pre-defined data about the network [14].

This paper considers 802.11 Standard which is a group of rules on implementation and usage of the W-LAN. There are three types of packet frames in 802.11 networks: control frames, data frames, and management frames. Each of these frames has its subtypes. The type of frames that is used for admission control, managing the network, and responsible for initiating and maintaining connection with users over the WLAN is called management frames. The control frames used for access control as allowing or preventing access to stations, while the data frames are used in transmitting data. De-authentication and disassociation frames are classified under the management frames as in Table 1 [15]. For the reason that the DE authentication and disassociation frames are being not secured, the attacker spoofs these types of frames [16]. All the schemes of encryption in 802.11. As WEP, Wi-Fi Protected Access (WPA), WPA2 encrypts only the data frames, While The management and control frames used for data exchange are always sent in a clear text, it still considered as a weakness point [17].

The IEEE 802.11 standard mentioned that DE authentication frames are a non-rejected notification that received by the wireless client. Hence, the attacker can pretend to be the legitimate AP, and send DE authentication frames on behalf of the AP to the users in order to terminate the connection [18]. As each Wi-Fi network sends and receives data over a certain frequency, or channel, regarding the IEEE

Table 1 802.11 frames subtypes [15]

Management	Control	Data
Association request	PS-Poll (power save poll)	Data
Association response	RTS (request to send)	Data + CF-ACK
Reassociation request	CTS (clear to send)	Dara + CF-Poll
Reassociation response	ACK (acknowledgement)	Data + CF-ACK + CF-Poll
Probe request	CF end (content free end)	Poll
Probe response	CF end + CF-ACK	Null function (no data)
Beacon		CF-ACK (no data)
ATIM		CF-Poll (no data)
Disassociation		
Dequthentication		

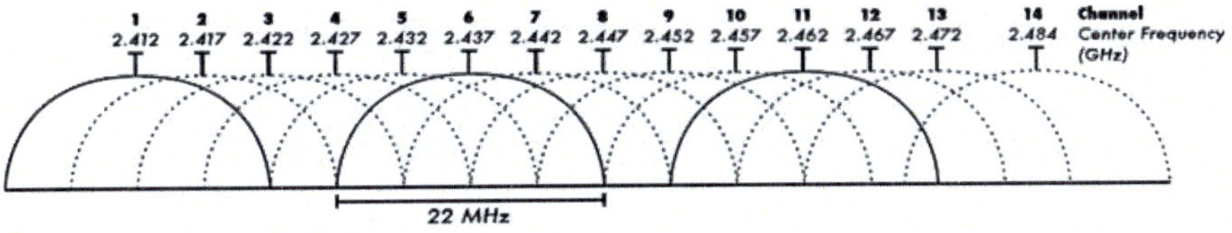

Fig. 2 802.11 channels frequencies in 2.4 GHZ band [20]

802.11 standard [6, 15, 19], we focus on 2.4 GHZ range and channels, and Fig. 2 shows the number of channels in 2.4 GHZ [20].

Table 2 shows the number of channels operates in different countries because it varies from 11 to 14 regarding to the IEEE Standards of 802.11 (2.4 GHz) [19].

The key contribution of this paper is as follows: Diverse detection logic to detect different types of attack signature. An empirical implementation of the proposed scheme for detecting the evil twin attack (DE authentication—disassociation—both DE authentication and disassociation frames) are prototyped under real attack; we recommend a better detection method for the evil twin attack by detecting the de-authentication and disassociation frames in 802.11. Using a low-cost (microcontroller capability) to detect and classify frames type by lighting a different alert for (DE authentication frames, disassociation frames, both DE authentication and disassociation frames (mixed frames or mixed attack).

The relevant approach in a research conducted by Afzal et al. in [21] is close to our work, it depends on Snapshotting Frames, logging, and analyzing the wireless frames. It stores the SSID and signal strength in database and compared it with the new values of the access points in addition to using the time difference technique.

Table 2 The physical frequency and channel plan in different countries according to the IEEE Standards [19]

CHNL_ID	Frequency (MHz)	Requlatory domins							
		X″10″ FCC	X″20″ IC	X″30″ ETSI	X″31″ Spain	X″32 France	X″40″ Japan	X″41″ Japan	X″51″ China
1	2412	X	X	X	–	–	–	X	X
2	2417	X	X	X	–	–	–	X	X
3	2422	X	X	X	–	–	–	X	X
4	2427	X	X	X	–	–	–	X	X
5	2432	X	X	X	–	–	–	X	X
6	2437	X	X	X	–	–	–	X	X
7	2442	X	X	X	–	–	–	X	X
8	2447	X	X	X	–	–	–	X	X
9	2452	X	X	X	–	–	–	X	X
10	2457	X	X	X	X	–	X	X	X
11	2462	X	X	X	X	–	X	X	X
12	2467	–	–	X	–	X	–	X	X
13	2472	–	–	X	–	X	–	X	X
14	2484	–	–	–	–	–	–	–	–

As for the challenges in this approach, it does not support the real-time detection as it depends on Snapshotting Frames to generate the alert, and as a criticism this could be useful in case of tracing back an attack issue, but our proposed method can work in real-time detection, determine the attacker's MAC address and it is used by admin or user sides which is a comparative advantage. It can work for the admin side by specifying the mac address of the APs and/or only the specified channel but it returns more accurate value with less false positive. It also can be used as client-side with higher false positive by scanning all channels without specifying the APs MAC address. Moreover, the proposed algorithm is not in need to be established in each device in the network. It fits to discover two famous Layer two attack in 802.11 networks, which are executed in real scenario; Also, it needs no Wi-Fi network pre-configuration, training data nor fingerprinting as required by the solution proposed by Zeeshan Afzal. Our detection method is not expensive, can be deployed in home networks to protect the IoT devices, security cameras, and doorbell cameras from being attacked or leak sensitive information. By the end, an alarm is produced to distinguish de-authentication, disassociation, the mixed frames (DE authentication and disassociation frames) from the normal frames or packets.

The other parts of the research are organized as the following: the related work in Sect. 2. Section 3 describes the proposed method and the prototype. In Sect. 4 we present the results of the experiment. Then Sect. 5 provides the conclusion and the future work.

2 Related Work

Most of the detection techniques focus on the detection of the Evil-twin which indirectly detect RAP-based DE authentication/disassociation attacks. Some techniques. There is no single technique that detects all RAP types. The ideal method is the one that can detect all types of RAP, is passive, does not require protocol modification nor special hardware. All existing techniques have one or more of these features, but none of them has all four [10].

2.1 Admin Side

The paper [22] proposed by Burns et al. is based on traceroute. It presented a novel detection scheme for evil twin using a bidirectional traceroute and a remote server for detection to compare values of both client-to-server and server-to-client traceroute, mainly the hops in the LAN on both paths. The difference between the two recorded value of traceroute indicates the presence of evil twin attack. On the other hand, this solution depends on a server which is costly and a one point of failure. In addition, that the approach has nothing to do if the attacker used his own internet connection as 4G.

Bhatia et al. in [23] proposed an algorithm for detecting a wireless evil twin attack, to precise the location of the evil twin transmitter by using four square antennas. Using the relation between RSS (received signal strength) values to determine the evil twin transmitter's location. This approach can recognize the evil twin assault and reveal real access point. The outcome identified that fifty-three percent of assaults are identified if transmitters are accessible in indistinguishable zones and about 77% the transmitters are correctly localized but detects only 53% of the attacks when the transmitters are at the same zone. It is complex and costly because of the four-square antenna used.

Sachin Sonawane et al. introduced a methodology of detecting fake access points by preparing a white list of legitimate access points' MAC and IP addresses. Then send a broadcasting packets over a central service to reveal the evil twin attack over the WLAN by capturing all replies from all access point devices and compare them with the white list. But its drawback is that it works over wireless network only if all devices have been at the same network IP range. It does not work on different network IDs. It works if attacker's access point is associated with the business network to use its resources in addition to depending on the whitelisting and does not provide real-time detection [24, 25].

In Gonzales et al. [26] presented a Context-leashing method that depends on recording the nearby access points when first associating with an access point and an Authentication Protocol of Secure Shell is presented in as a revealing method for evil twin.

It creates a secure channel and generates two keys one is public and the other is private and all for encrypting the connection. This Context leashing gives a path to the client to be protected from the malicious evil twin, but this methodology has a few constraints that do not guarantee integrity, privacy or authentication. However, the main limitation of this detection approach is that it does not provide authentication, confidentiality, and integrity that can prevent eavesdropping attacks. In fact, regardless of whether evil twin access points are detected or ignored, any arbitrary wireless device can inject wireless frames and eavesdrop over the network.

Fabian lanze et al. presented a science-based model by using the root-mean-square error (RMSE) of an ordinary least square regression (LSR) fitted into the (x, y)-points of a trace. For utilizing the airbase-ng software which is a component in several tools that could be used for this malicious purpose. The method tries to sense the presence of the malicious access point and identifying evil twins operated by software. In any case, this methodology is constrained to a particular tool (airbase-ng) that considered an issue too [27].

Diogo Monica et al. presented a way to detect a multi-hop evil twin with the usage of a device used by the user. This device is a real-time detection method and not subjected to latency or the network's bandwidth.it does not need a pre-prepared list. In this method, users are able to switch channels in less than 500 ms. But it is limited as a normal access point could be identified as a malicious one.it detects evil twin in 30 s which is a big issue [28].

2.2 Client-Side Solutions

Kumar et al. prevent users from accessing the evil twin by modifying the communication protocol between AP and terminal devices. This solution adds a new identifier called 'COUNT', which records the number of successful connections between each client and each AP, to the information list in AP and client. Before the client establishes connection with target AP, the values of 'COUNT' respectively stored in AP and client are compared through the request-response frames. Evil twins can be detected if these two values are different. However, such detection method based on modification of 802.11 protocol is not practical because it needs to change the existing driver and firmware in large scales. It still has its drawback as it needs protocol modification to be done by the user and it is not easy [29].

Y. Song et al. designed new, active, statistical and anomaly detection algorithms to detect evil twins by differentiating the wireless hops (one or two hops). that uses the (IAT) Inter-packet Arrival Time between two sequential data packets originated from the same device and sent to the hosts as an indicator for detecting the evil twin. In addition to considering a variance in throughput due to wireless network saturation. This methodology depends on two algorithms: one is named Trained Mean Matching (TMM), requiring training knowledge of one-hop and two-hop wireless channels. The second algorithm is the Hop Differentiating Technique (HDT), for making the final detection. HDT and TMM combine the Inter-packet Arrival Time statistics of the wireless network and the Sequential Probability Ratio Test to detect the rogue AP. It depends on a (remote) server for detecting hops from and to the access point as more hops mean more noise. It detects evil twin with a high detection rate and a low false positive rate of 2.4%. But It has its limitation that it requires training knowledge of Server IAT (Inter-packet Arrival Time) in wireless channels, depends on a server and still it is a drawback as once the remote server is not available, the method cannot operate in a good way. In addition to it is not user-oriented because of the need for the authorized list [30, 31].

The approach proposed by Zhuang et al. scans all available APs to find the presence of multiple APs having the same SSID at the same W-LAN environment, in case of not finding, it informs the user that there are no evil twin attacks. Otherwise, it records the MAC addresses (BSSID) of the two Aps that are with the same name and monitors the sent traffic rate/time. A filter statement is used to find out control frames. Then, data frames flow rate sent by the target APs to each user is recorded in an array consecutively to determine the presence of a forwarding behavior in the network. Pearson correlation coefficients are calculated by using the values between each user and target AP recorded in the array. If the correlation coefficient value exceeds the threshold, it indicates MAC addresses of the evil twin and then the rouge access point can be located with respect to the signal strength. This experiment returns accuracy rate of 96% in distinguishing evil twins from legitimate APs. This approach is implemented in ET-spotter which is a Python tool depending on the forwarding behavior to detect evil twins. Its drawback is that there must be at least one user connected with the evil twin. And finally, in case the user connects only to the AP without surfing the internet, the approach has nothing to do with it [32].

In Nakhila et al. [33] presented a real-time evil twin detector that operates on the client side by making two (VWCs) Virtual Wireless Clients. The first virtual wireless client randomly monitors the multiple channels of wireless searching for a data packets that are specific and sent by an online server. Meanwhile, the second virtual wireless client alert the wireless client in case of using different gateways by switching from an access point to another during a secure connection. This method of detection returns rate close to 100% but still have its limitations as this method depends on virtual wireless clients that is not simple for a user to operate on. In addition to its need for an online server.

In Modi et al. [34] depends on using the Round-Trip Time (RTT), the number of hops between client and server-based techniques, and de-authentication detection between client and legitimate access point. The attacker kicks out all users from the real access point by initiating a de-authentication attack. The method depends on two access points and in case the SSID of both access points are similar and BSSID of them are different, the algorithm checks for DE authentication attack on legitimate access point, then ban it from reconnecting and gives alert. And if the round-trip time and the number of hops for access points that were previously connected to AP1 is greater than the currently connected access point AP2 then AP1 is an Evil Twin. The system bans it from reconnecting and give alert. The experiment used the Airmon ng on kali Linux. But for its limitation it needs more variables to consider various possibilities of the evil twin attack.

To sum up, the main drawbacks of the existing solutions for detecting the evil twin attacks are as follows: (1) costly deployment; (2) needs protocol alternation or editing; (3) depending on a one point of failure as a server; (4) needs authorized list or a white list.

In this paper, we proposed both a client-based and admin-based solution for detecting the evil twin attack. And Unlike the previous solutions, our proposed method has the following advantages: (a) it does not need any authorized list or a white list of access points. (b) It provides a real-time detection for both user and admin. (c) it does not need any training data of the targeted wireless network. (d) it does not depend on a remote server, but it depends on a handy chipset which is very cheap, and its cost is not exorbitant. (e) No need for protocol modification. (f) determine the attacker's MAC address. (g) it does not need to connect to any access point for detection.

3 Proposed Method

This paper's section explains the used indicators for the presented module by detecting DE authentication/disassociation frames, both DE authentication and disassociation frames (the mixed frames) as they are main characteristics of evil twin. In the proposed method, we have previously mentioned many techniques given by the researchers as some helped to protect from the evil twin attack but had their own limits. Here we are proposing an approach to detect DE authentication/disassociation frames, both DE authentication and disassociation frames (the mixed frames) and normal ones. In a very handy, cheap, and effective way by using an electronic stability programable 8285H controller.

3.1 Detecting DE Authentication/Disassociation Frames

As a preparation phase, we used Wireshark tool for capturing the packets on wireless network. As shown in Table 3 [35], we capture the DE authentication frames that are represented by (0x0C) and the disassociation frames that are represented by (0x0A). For the reason that IEEE 802.11 specifies the various reason codes to determine the reason for the station's disconnection [36], It was noted that most of the attack tools uses a Fixed Reason Code, for example, Code 7 in all the DE authentication frames [37]. This observation can be considered as an attack indicator. One can argue that the same reason can be used during legitimate authentication, but all legitimate DE authentication. Frames in one analyzed sequenced packet using the same reason code are not normal. Therefore, it is a good indicator of a DE authentication attack. Then we start the algorithm as in Fig. 3.

As In Fig. 3, there are 3 steps in the proposed method: monitoring, detection, and alerting. In the (monitoring phase), we perform channel hopping after putting the interface into monitor mode and we can define the number of channels to scan

Table 3 Filtering frames in 802.11 [35]

Frame type/subtype	Filter syntax
Disassociate	wlan.fc.type_subtype == 0x0A
Authentication	wlan.fc.type_subtype == 0x0B
Deauthentication	wlan.fc.type_subtype == 0x0C
Action frame	wlan.fc.type_subtype == 0x0D

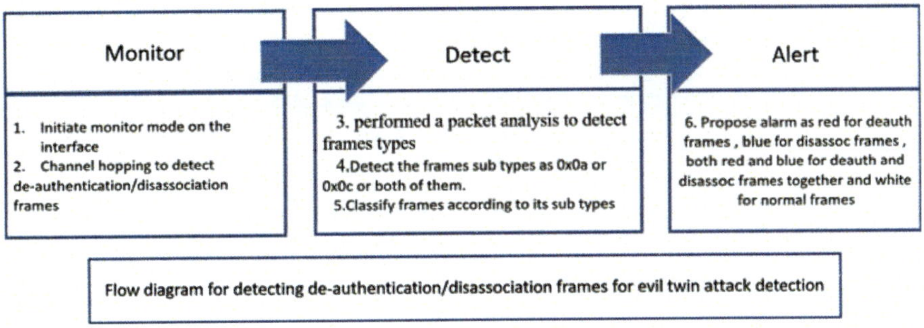

Fig. 3 Flow diagram for detecting de-authentication and disassociation

in our experiment as 11 channels because it is common except in (Japan is 14) as mentioned previously in Sect. 1 Table 2.

In the (detection phase), we performed a packet analysis for not only legitimate but also malicious AP. We filter packets based on its types and subtypes as in Table 3. The method used the chipset to detects the malicious packets as each kind of packets turn on a corresponding LED to alert the user in the phase of alert which is the third and final phase.

3.2 Detection Algorithm and Pseudo Code for the Proposed Method

This section clarifies the proposed algorithm module to find out the DE authentication/disassociation attack frames and the occurrence of both DE authentication and disassociation frames (mixed attack) as in Fig. 3.

As shown in Fig. 4, we start the detection by setting the interface into monitoring mode and define the number of channels that are defined in our experiment to be 11 shown previously in Table 2 [19], Then we will start a continuous channel hopping that goes to each channel and scan it from channel 1 to channel 11. We will define the threshold as 5 packets per second.

Based on the network needs, a dynamic or a static threshold can be set by the admin. In the proposed experiment, we have set the value of the threshold to 5 as it achieves an accuracy of 95.3% for the reason that different threshold values return various rate of detection and accuracy as shown in Table 4.

```
1. Start
2. Put interface in monitoring mode
3. Set number of channels that are going to be scanned 1-11
4. Start channel hopping
5. Define threshold as 10 concurrent malicious packets / second
6. Define Counter C_deauth
7. Define Counter C_disass
8. Define attacker MAC address M_att
9. Define receiver MAC address M_rec
      a. IF C_deauth >= threshold, turn on the corresponding alarm 1
      b. IF C_disass >= threshold, turn on the corresponding alarm 2
      c. IF C_deauth + C_deauth >= threshold, turn on the corresponding
         alarm 1 and alarm 2

         Print "Attacker MAC address = "+M_att +
         "Receiver MAC address= "+M_rec

10. else stop alarm
```

Fig. 4 Pseudo code of proposed method

Table 4 Threshold value

Threshold	Number of launched attacks	Detection rate (%)	Accuracy (%)
2	50	96.4	93.2
3	50	95.15	93.7
4	50	92.54	94.6
5	50	91	95.3
6	50	89.15	96.2
7	50	83.54	97.8
8	50	80.41	98.4
9	50	76.3	99.1
10	50	74.15	99.9

It shows the reflection of using different number of disassociation/de-authentication frames sent on the rate of detection and accuracy. We can conclude from Table 4 that incase of increasing the threshold, the returned accuracy value increases but the detection rate decreases. That is for the reason that if the attacker is capable of launching a real attack with a number of packets which is less than the threshold, he will be able to bypass detection. The accuracy raises with the increase in the value of threshold, as higher threshold means a higher amount of de-authentication/disassociation frames sent and that is a clear indicator of evil twin attack. On the other side, the lower threshold value can detect the evil twin attack faster, but with a small value of false positives. This can be debriefed from the accuracy results of the small values of threshold.

De-authentication attacks that are launched using a single de-authentication frame are never detected, as setting a threshold of 1 is impractical and would generate a large number of false positives. As a result, those de-authentication attack that are caused due to a single de-authentication frame are never detected and consequently the detection rate remains below 100% [9].

In this approach, we used a sniffer function that counts the number of DE authentication frames incrementally that is expressed by (0x0C) as shown in Table 3 and generate alarms. Depending on an if condition, if the number of DE authentication frames equal or exceed the threshold. It generates alarm 1 So that the light stays on continuously when an attack is underway, and it turns off as soon as the attack stops and normal traffic resumes. Considering the pseudo code shown in Fig. 4 in line 9b, the same if condition and a counter are used to count the number of disassociation frames that is expressed by (0x0A) as shown in Table 2, then it is compared to the threshold and generate alarm 2 in case of the number is exceeding or being equal to the threshold value.

As for the mixed attack frames of both (DE authentication and disassociation), if the number of both DE authentication + disassociation frames (0x0C & 0x0A) is equal or exceeds the threshold value, generate alarm 1 and alarm 2 together. And in all the cases of detection, the algorithm will print the MAC address of both the

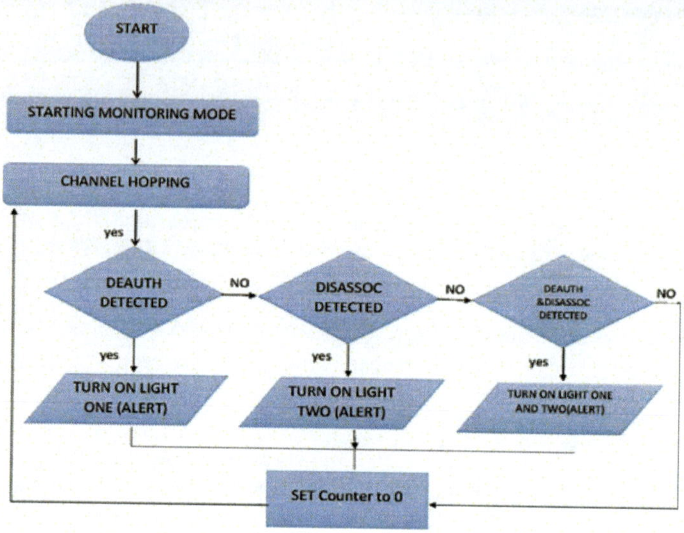

Fig. 5 Flowchart of the proposed method

attacker and the receiver. In case of not finding the previous types of frames, the algorithm will classify frames as normal ones and does not start any alarm. In order to loop again, it set the counter to 0 and start from line 4 (channel hopping) to return the process again. Figure 5 expresses the flow chart of the proposed logic.

3.3 Flow Chart of the Proposed Method

See Fig. 5.

4 Results

This section is specified for the proposed method evaluation. It contains the lab description, proposed detector design, detector efficiency, evaluation measures, and finally the analysis of results over the confusion matrix.

4.1 Lab Description

We have implemented the experiment on a wireless network named Zyxel to evaluate the proposed method of detection. The Wireshark software has been used for

Fig. 6 The used controller

monitoring and analyzing the packets sent and received over each channel. We have used the chipset and a 3-color led to generate the alert that reflects the type of attack underway. The attack was launched using a Core I5 laptop operating on Kali Linux OS. The wireless interface card used by the attacker is ALFA model AWUS036H. The programable 8285H controller detector is shown in Fig. 6.

4.2 Design of the Proposed Detector

Our proposed detection method overcomes the vulnerabilities in admin side solutions discussed in Sect. 2 as the efficiency of the proposed detector does not depend on servers, protocol modification, white list of trusted access points, or round-trip time parameters. Furthermore, the proposed detection is a real-time that also can be used by users as a client-side way of detection that does not operate on training data or fingerprint of the Wi-Fi network.

The attack was launched by cloning the MAC address of the legitimate access point and start sending DE authentication/disassociation frames over the network to disconnect clients and force them to connect to the fake access point. On that point, the proposed method of detection consists of wireless clients, a sniffing function, if condition, and detector. First, by doing channel hopping to listen to the Wi-Fi DE authentication/disassociation frames over all the Wi-Fi channels for the connected access point. This step does not need any interaction or communication between the user and the AP.

Second, the frame counter starts counting the DE authentication/disassociation frames and compare it with a threshold, depending on that it generates alarm. We have operated over 5 packet/s. The chipset can receive and detect the DE authentication/disassociation frames over the air and generate the alarm. It prints the attacker's and the user's MAC address. Then the counter goes back to zero in order to start the process of channel hopping again and start searching for the DE authentication/disassociation frames to loop again.

4.3 *Proposed Detector Efficiency*

This section is dedicated for evaluating the proposed method. The section is divided into two main parts. The first part evaluates the performance of the proposed detector (from the admin and the user side) (scenario A, scenario B) against the evil twin attack. The second part compares our results to the method in proposed by Afzal et al. [21].

Evaluation Measures

In order to evaluate the quality of the proposed method, we have used: Accuracy, Precision, Recall and *F*-Measure. These measures are applied and calculated over a confusion matrix classification based on Eqs. 1–4. In which TP, FN, and FP represent numbers of true positives, false negatives, and false positives, respectively.

As for the accuracy value, it refers to how accurate the proposed method can classify frames types in a correct way, and this is expressed by Eq. 1 that is applied to return the accuracy value. The accuracy value expresses a comparison between frames that are correctly classified with the whole frames.

$$\text{Accuracy} = \frac{\text{TP} + \text{TN}}{\text{TP} + \text{FN} + \text{FP} + \text{TN}} \tag{1}$$

The value of precision refers to the number of DE authentication/disassociation category frames that are classified correctly divided by the total frames classified as DE authentication/disassociation. Precision is calculated by Eq. 2.

$$\text{Precision} = \frac{\text{TP}}{\text{TP} + \text{FP}} \tag{2}$$

Nevertheless, recall shows how many percent of the DE authentication/disassociation category frames are correctly classified by the classification. Equation 3 is used for resulting the value of recall.

$$\text{Recall} = \frac{\text{TP}}{\text{TP} + \text{FN}} \text{Recall} = \frac{\text{TP}}{\text{TP} + \text{FN}} \tag{3}$$

*F*1 Score. The *F*1 Score is the 2 * ((precision * recall)/(precision + recall)). It is also called

$$F\text{ - score} = \frac{2 * \text{TP}}{\text{TP} + \text{FN} + \text{FN}} F\text{ - score} = \frac{2 * \text{TP}}{\text{TP} + \text{FN} + \text{FN}} \tag{4}$$

Table 5 User side detection with respect of 5 frame per second (threshold)

	DE authentication packets	Disassociation packets	DE authentication and disassociation packets	Normal packets
DE authentication packets	79	2	7	12
Disassociation packets	0	80	6	14
DE authentication and disassociation packets	6	5	77	12
Normal packets	8	7	6	79

4.4 Scenario A: Evaluation of the Proposed Detection Method for User Side

For evaluation, a comprehensive analysis is conducted. Results showed the false positive, true positive, false negative, and true negative, presented and analyzed via a confusion matrix.

These classifications are performed based on the number of each frame type by classifying each type of DE authentication/disassociation frames whether to be an evil twin attack frame or not (DE authentication, disassociation, mixed attack), other packets are considered as normal packets. As shown in Table 5 it expressed the evaluation of the user side detection with respect of 5 frame per second.

Analysis under different attacks are performed using the kali Linux tools: The performed evaluation is for testing the detector's accuracy. Assuming that under the evaluation process, no-other attack frames exist over the medium except the injected ones for testing.

The used set of data for evaluation is outlined in Table 5. To make sure that the predicted frames of different attacks are the actual ones that were sent by the attacking tool. This analysis is shown in Table 5: the calculations done for the user side, Table 6 shows results with more than 75% precision, 0.80% recall.

4.5 Scenario B: Evaluation of the Proposed Detection Method-Admin Side

For evaluation, a comprehensive analysis is conducted but from the admin side. Several possible results of the false positive, true positive, false negative, and true negative are presented and analyzed via a confusion matrix. As shown in Table 7: it expressed the evaluation of the admin side detection with respect of the following factors: considering 5 frame per second as a threshold, ignoring channel hopping,

Table 6 User side real-time detection analysis with respect of 5 frame (threshold)

	TP	TN	FP	FN	Accuracy (%)	Precision (%)	Recall (%)	*F*1 score (%)
DE authentication packets	79	236	14	21	90	84.95	79	81.87
Disassociation packets	80	235	14	20	90.26	85.11	80	82.47
DE authentication + disassociation packets	77	238	19	23	88.24	80.21	77	78.57
Normal packets	79	236	38	21	84.22	67.52	79	72.81
Average	79	236	21	21	88.18	79.45	78.75	78.93

Table 7 Admin side detection with respect of 5 frame (threshold)

	DE authentication packets	Disassociation packets	DE authentication + disassociation packets	Normal packets
DE authentication packets	89	1	4	6
Disassociation packets	3	92	3	2
DE authentication + disassociation packets	3	2	91	5
Normal packets	2	4	1	93

specifying the channel, and the AP BSSID. Attacks are performed using the kali Linux tools. Assuming that under the evaluation process, no-other attack frames exist over the medium except the injected ones for testing.

As in Table 8: Similar to the calculations done for the user side, Table 8 shows results for the admin side detection with about 91.08% precision, 91.02% recall.

4.6 Testing

We tested our solution against Airgeddon [38] that uses (MDK 4, airplay-ng) [13] and other kali Linux tools in order to run some hostile packets against a network we have permission to, and calculate the response. The alert reflects the kind of the program. Figures 7 and 8 show the captured frames by wire shark.

Table 8 Admin side real-time detection analysis with respect of 5 frame (threshold)

	TP	TN	FP	FN	Accuracy (%)	Precision (%)	Recall (%)	$F1$ score (%)
DE authentication packets	89	276	8	11	95.05	91.75	89.00	90.36
Disassociation packets	92	273	7	8	96.05	92.93	92.00	92.46
DE authentication + disassociation Packets	91	274	8	10	95.30	91.92	90.10	91.00
Normal packets	93	272	13	7	94.81	87.74	93.00	90.29
Average	91	274	9	9	95.30	91.08	91.02	91.03

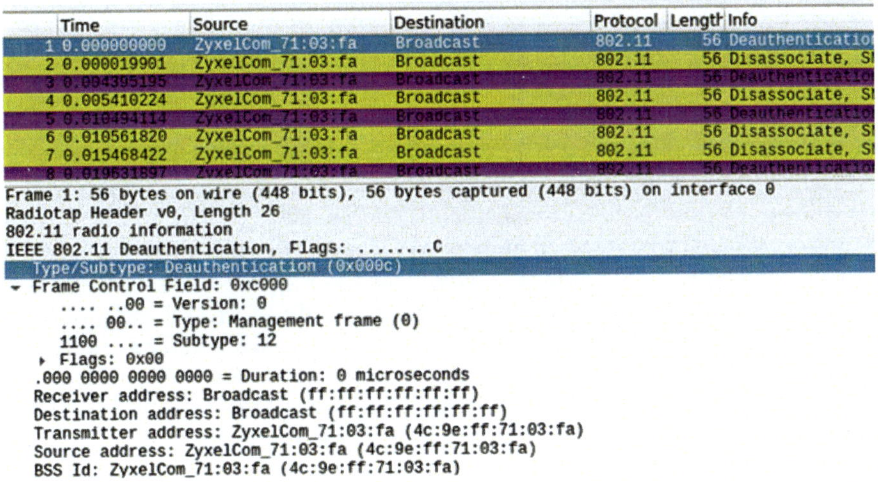

Fig. 7 DE authentication frames by Wireshark

4.7 *Proposed Method Comparison with Another Method*

Here is a comparison between our detection method and the proposed method by Afzal et al. in [21]. This approach depends on Snapshotting Frames, logging to analyze the wireless frames. It stores the SSID and signal strength in database and compare it with the new values of the access points in addition to using the time difference. In his future work, he recommended finding a way to decrease the false positive rate. But it does not provide real-time detection as it stores SSID of all access points in a data base. For that reason, our proposed method can work in real-time

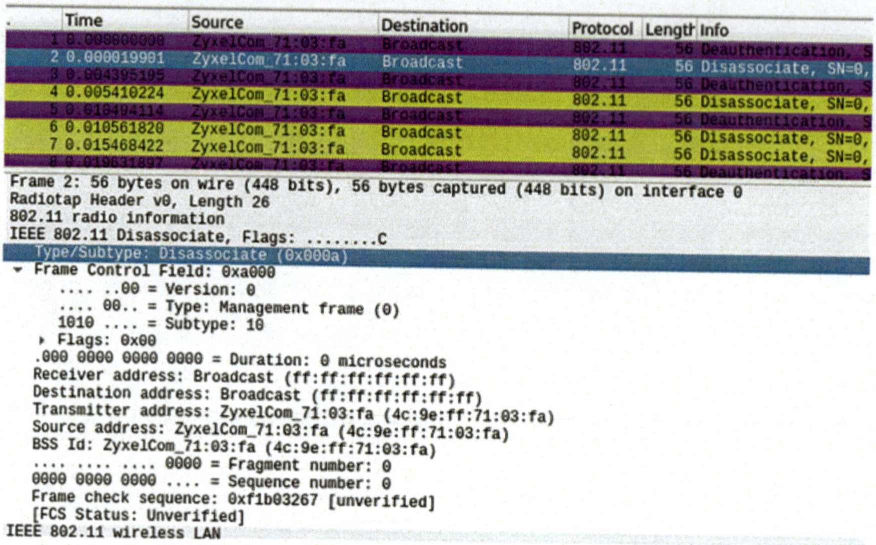

Fig. 8 Disassociation frames by Wireshark

detection, determine the attacker's MAC address and it is used by admin or user sides.it can work for the admin side defender by specifying the mac address of the APs and/or only the specified channel but it returns more accurate with less false positive. It also can be used as client side with higher false positive by scanning all channels without specifying the APs MAC address. It provides Passive detection for DE authentication frames and it does not detect disassociation frames or the mixed frames (disassociation + DE authentication frames). It depends on Collecting, logging, and analyzing wireless traffic and it uses a host file system to store the captured traffic after performing a bookkeeping of all the APs in the neighborhood. It Checks Access point's transmission power level difference. It uses time stamp as an indicator. This method is an admin side way of detection that provides accuracy of 89–93.3% with false-positive rate of 14.6% in detecting DE authentication and 20% in detecting evil twin. It does not detect attacker's MAC address.

Our proposed method provides real-time detection, it is not passive but it is active, it can be used as admin or user side solution, it is very cheap, it does not store SSID in DB nor perform bookkeeping of all the APs in the neighborhood. It does not Check Access point's transmission power level difference nor using host file system to store the captured traffic.it detects Detecting mixed type of attack disassociation packets, DE authentication packets, mixed frames (disassociation packets, DE authentication) by depending on Sniffing and analyzing the wireless frames. It proofs accuracy rate of 95.30% for the admin side in (DE authentication attack—disassociation attack—mixed attack—normal packets). While it proofs accuracy rate of 88.18% for the user side. The false-positive rate of admin side: about 9 and 21% in DE authentication for user side and it also detects the attacker's MAC address. The results proofs that the detector's accuracy is quite high and provide most of the expected features. It also shows that the proposed system can be used both in admin and user side. While in admin side it runs on 5 packets per second it needs no channel hopping, less time, specified channel, and specified BSSID and it fingerprints the attacker's MAC address. It is also used by any user (user side) as it is cheap, handy, needs no previous configuration, it runs too on 5 packets rate per second, it uses channel hopping so it will take longer time.

5 Conclusion

The wireless network is considered to be a main part in our world, due to its usage in all life aspects. An attacker can set up a rouge access point, that has the same name of the network but with stronger signal. Then, the attacker can re-direct clients to fake login portal and steal password. There are two ways of solution to detect this kind of attacks, whether to be an administrator-based solution or a client-side one. For the reason that the DE authentication and disassociation frames are unencrypted, an attacker can easily spoof these frames and launch the attack.

In addition to his ability in spoofing reauthentication's frame reason code to mislead the indicator. The proposed detector works in real time. It works for both client and admin side.it needs no pre-configuration for the network, nor training data and authorized list as required by other solutions. In addition to being not expensive, it is used in home networks to protect the IOT devices, security Cameras, doorbell cameras. If an attacker tries sending fewer frames in order not to be detected, this trial will also be detected because of the data frames indicator as the attacker does not have any control on it. But If the attacker used a program to change his MAC address or spoofed it from a legitimate client, the proposed method could not know the attacker's MAC but still the proposed method can detect attack existence.

Future work: As an improvement for channel hopping step the model can be applied with more chips number maximum 14, each for one channel, but the cost will be higher. For further improvement a feature of making a counterattack after detecting the attacker's MAC address can be considered.

References

1. Hatfield JM (2018) Social engineering in cybersecurity: the evolution of a concept. Comput Secur 73:102–113
2. Jakobsson M (2016) Understanding social engineering based scams
3. Heartfield R, Loukas G (2018) Detecting semantic social engineering attacks with the weakest link: Implementation and empirical evaluation of a human-as-a-security-sensor framework. Comput Secur
4. Agarwal M, Biswas S, Nandi S (2018) An Efficient scheme to detect evil twin rogue access point attack in 802.11 Wi-Fi networks. Int J Wirel Inf Netw 25(2):130–145

5. Noman SA, Qasaimeh M, Al-qassas R, Noman JA, Lumpur K (2017) Mitigating evil twin attacks in wireless 802.11 Networks 14(1):60–69
6. Gonçalves R, Correia ME, Brandão P (2018) A flexible framework for rogue access point detection. ICETE 2018 Proc 15th Int Jt Conf E-bus Telecommun 2
7. Song Y, Yang C, Gu G (2010) Who is peeping at your passwords at Starbucks? To catch an evil twin access point 10:323–332. ieeexplore.ieee.org
8. Jorge R, Gonçalves E (2017) A flexible framework for Rogue access point detection
9. Agarwal M, Biswas S, Nandi S (2013) Detection of de-authentication denial of service
10. Alotaibi B, Elleithy K (2016) Rogue access point detection: taxonomy, challenges, and future directions. Wirel Pers Commun 90(3):1261–1290
11. Loukas G, Öke G (2010) Protection against denial of service attacks: a survey. Comput J 53(7):1020–1037
12. Kali Linux 2019.4 ReleaselKali Linux [Online]. Available https://www.kali.org/news/kali-linux-2019-4-release/. Accessed 29 Aug 2020
13. de Bouvette TD (2018) aireplay-ng [Aircrack-ng]. Aireplay-ng. [Online] Available https://www.aircrack-ng.org/doku.php?id=aireplay-ng
14. Lu Q, Qu H, Ouyang Y, Zhang J (2019) SLFAT: client-side evil twin detection approach based on arrival time of special length frames. Secur Commun Networks
15. IEEE Standard for Information technology-telecommunications and information exchange between systems-local and metropolitan area networks-specific requirements. Part 11: Wireless LAN Medium Access Control (MAC) and Physical Layer (PHY) Specifications, p 351, 352, Dec 2012
16. Mar J, Yeh YC, Hsiao IF (2010) An ANFIS-IDS against deauthentication DOS attacks for a WLAN . ISITA/ISSSTA 2010–2010 Int Symp Inf Theor Appl 2:548–553
17. Agarwal M, Biswas S, Nandi S (2016) Detection of de-authentication DoS attacks in Wi-Fi networks: a machine learning approach. In: Proceedings of 2015 IEEE international conference on system man, cybernetics SMC 2015, pp 246–251
18. Alotaibi B (2016) Empirical techniques to detect rogue wireless devices
19. IEEE Standard for information technology—telecommunications and information exchange between systems—local and metropolitan area networks—specific requirements—ANSI/IEEE Std 802.11. In: Part 11: wireless LAN medium access control (MAC) and physical layer (PHY) specifications, p.447 table 18-9, vol 1999, June 2007
20. Butler J, Pietrosemoli E, Zennaro M, Fonda C, Okay S, Aichele C (2013) Wireless networking in the developing world, p 9
21. Afzal Z, Rossebo J, Talha B, Chowdhury M (2016) A wireless intrusion detection system for 802.11 networks. In: Proceedings of 2016 IEEE international conference on wireless communication signal processing networking, WiSPNET 2016, pp 828–834
22. Burns A, Wu L, Du X, Zhu L (2018) A novel traceroute-based detection scheme for Wi-Fi Evil twin attacks. In: 2017 IEEE global communication conference on GLOBECOM 2017—proceedings, vol 2018, pp 1–6
23. Bhatia P Detecting and localizing transmitters in a wireless evil-twin attack background of HPB
24. Sonawane SR, Vanjale S (2013) Wireless LAN intrusion prevention system (WLIPS) for evil twin access points. IJCST Int J Comput Sci Technol 4–8491(2):2–5
25. Sonawane SR, Chavan SP, Ghodeswar AA (2013) Study of different rogue access point detection and prevention techniques in WLAN. Int J Adv Res Comput Sci Softw Eng 3(10):1232–1237
26. Gonzales H, Bauer K, Lindqvist J, McCoy D, Sicker D (2010) Practical defenses for evil twin attacks in 802.11. In: GLOBECOM—IEEE global telecommunications conference
27. Lanze F, Panchenko A, Ponce-Alcaide I, Engel T (2014) Undesired relatives: protection mechanisms against the evil twin attack in IEEE 802.11. In: Proceedings of 10th ACM symposium QoS security wireless mobile networks
28. Mónica D, Ribeiro C (2011) WiFiHop—Mitigating the evil twin attack through multi-hop detection. In: Lecture notes in computer science (including subseries lecture notes in artificial intelligence and lecture notes in bioinformatics), vol 6879 LNCS, pp 21–39

29. Kumar A, Mobility A (2016) Security analysis and implementation of a simple method for prevention and detection against evil twin attack in IEEE 802.11 Wireless LAN. In: International conference on computer technology information communication technology, pp 367–372
30. Yang C, Song Y, Guofei Gu (2012) Active user-side evil twin access point detection using statistical techniques. IEEE Trans Inf Forensics Secur 7(5):1638–1651
31. Song Y, Yang C, Gu G (2010) Who is peeping at your passwords at starbucks? To catch an evil twin access point. In: Proceedings of the international conference on dependable systems and networks
32. Lu Q, Qu H, Zhuang Y, Lin XJ, Zhu Y, Liu Y (2017) A passive client-based approach to detect evil twin attacks. In: Proceedings of 16th IEEE international conference on trust, security private computing communication 11th IEEE international conference on big data science engineering 14th IEEE international conference on embedded software systems, pp 233–239
33. Nakhila O, Amjad MF, Dondyk E, Zou C (2018) Gateway independent user-side wi-fi evil twin attack detection using virtual wireless clients. Comput Secur 74:41–54
34. Modi V, Tech M, Security C, Parekh C (2017) Detection and analysis of evil twin attack in wireless network. Int J Adv Res Comput Sci 8(5):774–777
35. 802.11 Data Types and Wireshark filters [Online]. Available http://www.netprojnetworks.com/802-11-data-types-and-wireshark-filters/. Accessed 13 Sep 2020
36. Wright J (2005) Weaknesses in wireless LAN session containment
37. IEEE 802.11 Working Group, Wireless LAN MAC and physical layer specifications. IEEE Std., 802 (2010)
38. Airgeddon, a multi-use bash script for Linux systems to audit wireless networks—page 3 [Online]. Available https://forums.kali.org/showthread.php?35556-airgeddon-a-multi-use-bash-script-for-Linux-systems-to-audit-wireless-networks/page3. Accessed 14 Sep 2020

IS Risks Governance for Cloud Computing Service

Mohamed Gamal, Iman M. A. Helal, Sherif A. Mazen, and Sherif Elhennawy

Abstract Cloud computing services have become important to IS/IT recently. Cloud service providers (CSPs) support different types of cloud computing services (Infrastructure as a Service, Platform as a Service, and Software as a Service). There are many challenges and concerns in using cloud computing services that may cause many risks. These risks need addressing while using cloud computing. CSPs provide procedures in order to control risk areas while providing their services. There are many studies that assess risks in cloud computing services. They provide evaluation tools and/or frameworks to manage the risk in cloud computing services. Moreover, IS/IT governance is one of the most powerful ways to achieve IS/IT business alignment. The main goals of IS/IT governance are to ensure that the investments in IS/IT generate business values and to mitigate the risks associated with IS/IT. Even though there are various risks in cloud computing services, there is no comprehensive framework to support risk assessment according to user's needs. In this paper, we aim to provide a conceptual framework for governance in cloud computing services. This framework will focus on cloud client (CC) requirements to achieve the governance.

Keywords Cloud computing · Governance · Risk assessment · SLA · Risk control

1 Introduction

Nowadays, cloud computing services are used in construction of IT infrastructure in all fields business, academic, and governments. These services became a valid solution for data storage and processing. Many organizations now depend on these services. Cloud computing provides many benefits such as sharing resources among

M. Gamal (✉) · I. M. A. Helal · S. A. Mazen · S. Elhennawy
Faculty of Computers and Artificial Intelligence, Cairo University, Giza, Egypt
e-mail: mohamed.gamal@acu.edu.eg

I. M. A. Helal
e-mail: i.helal@fci-cu.edu.eg

S. A. Mazen
e-mail: s.mazen@fci-cu.edu.eg

D. A. Magdi et al. (eds.), *Digital Transformation Technology*, Lecture Notes in Networks and Systems 224, https://doi.org/10.1007/978-981-16-2275-5_4

many users, efficiency, optimizing IT costs, business process efficiency, data security, scalability, mobility, providing business continuity, employee optimization, and flexibility of work practices [1].

One of the major concerns of every business, regardless of size and industry, is the security of its data. The company's revenue can be diminished by data breaches and other cybercrimes, and data accessibility, speed processing, regulations compliance, and/or reputation must also be concerned. While moving to the cloud, the cloud customer (CC) loses some or all controls according to the service model. CSPs provide services controls to assure that the services are performed properly to save their reputation as service providers. Moreover, CC must understand the provided controls, and the needed controls too.

Moving to the cloud is surrounded by many risks. Therefore, it is important to perform risk assessment tasks to ensure that all assets are controlled and protected. IS/IT governance has also to confirm that these IS/IT assets are implemented and used according to policies and procedures, ensure that these assets are properly controlled, maintained, and supporting the organization's strategy and business goals.

Many frameworks and models were developed for risk assessment and governance in the cloud computing service. In this paper, the authors intend to provide a conceptual framework to help the auditor in selecting the CSP by showing the risk areas related to the selected service model, and the general threats that are in need to control. That will help the auditor to focus on the needed controls and SLA contractual requirements to achieve the governance.

The rest of this paper is presented as follows. Section 2 is covering the basic concepts of cloud computing and their services, risk management, governance, and service level agreement. Section 3 will highlight the concerns of using cloud services, risk assessment and governance, and frameworks for risk assessment, and governance in cloud computing. Section 4 will present the suggested conceptual framework for achieving governance while using cloud computing services. Finally, Sect. 5 concludes the paper and suggests future work.

2 Background

Cloud computing is a model for enabling ubiquitous, convenient, on-demand network access to a shared pool of configurable computing resources (e.g., networks, servers, storage, applications, and services) that can be rapidly provisioned and released with minimal management effort or service provider interaction. Cloud model is composed of three service model and four deployment models [2].

Cloud delivery models are as follows: (1) infrastructure as a service (IaaS), (2) platform as a service (PaaS), and (3) software as a service (SaaS) [3]. The four models to deploy these services are as follows: (1) *private,* in which the organization owns its cloud data center, (2) *public,* the CSP gives specific service to many organizations, (3) *hybrid,* it consists of the private and public models, and (4) *community,* in which many organizations have same business core tasks, and it must be supervised [4].

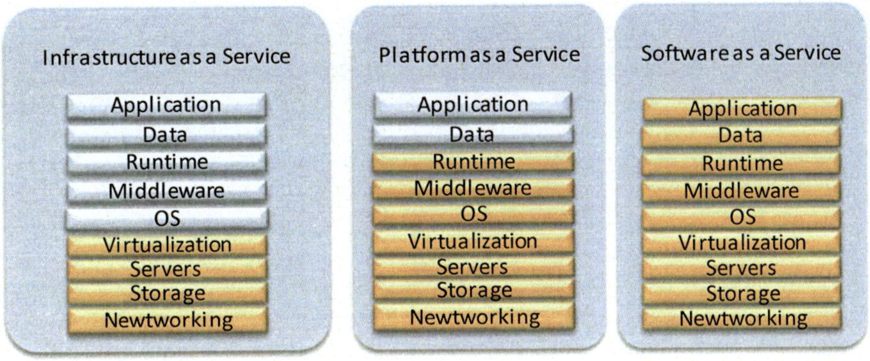

Fig. 1 Controls between CSP and customer service

Risk management can be defined as an organized and comprehensive method tailored toward "analyzing," "identifying," and "responding" to risk factors in order to achieve the project goals [5]. Governance ensures that stakeholder needs, conditions, and options which are evaluated to determine balanced, agreed-on enterprise objectives to be achieved. Also, it sets directions through prioritization and decision making. Finally, it monitors the performance and investigates compliance against agreed-on direction and objectives [6]. While the organization decides to use one of these services, some of the controls on all assets must be applied. Controls on the resources are divided between the CSP and CC Fig. 1.

Service level agreement (SLA) is the essential basis for the legal contract between supplier and customer. It guarantees the quality of the service and is used by both CC and CSP for clarifying responsibilities between them. SLA [7] describes the minimum performance criteria and quality that provider promises to deliver. It suggests corrective measures and penalties imposed on the provider. In [8], the authors made an architecture that categorizes the SLA articles to performance (response time), customer level satisfaction (CLS), pricing, and security.

3 Related Work

3.1 Concerns in Cloud Computing

Many threats face the CC without respecting the model used. Control loss is one of the major threats as some or all controls move from the CC to the CSP. Multitenancy is a big issue as many CCs share the same resources that may lead to denial attacks and reputation fate scaling if one of the other CCs did bad behavior [9]. Also, availability that CC expects is to have no down time [10] but that may not happen as expected. SLA [11] will be a weak point for CC if it does not cover all requirements. Performance in cloud computing is unpredictable as the I/O operations time will differ because of the VM schedules and type of memory [11]. An issue that may appear is quick scaling. It can trigger an increase in costs. Also, scaling down reduces the cost. CC may have no idea about these cost variations. Moreover, data locality is very important too, as in some organizations and countries, putting data out of the country will violate the compliance or regulations [12].

Even in IaaS model, most of controls are related to CC but there are many concerns. All threats associated with the Internet like IP spoofing, ARP spoofing, and DNS poisoning are considered a concern in IaaS [12, 13]. Defending the physical layer against unauthorized access is a big concern too [9].

PaaS may cause an attack if it is not secured [9]. Most of CSPs use software development lifecycle (SDLC) instead of the secured software development lifecycle (SSDLC) [10] and that may lead to security issues. Vendor lock-in is another threat as the CSP uses specific tools and software for database, frameworks, and storage structure. Therefore, the CC must use the same or compatible tools. CSPs' used API and browser vulnerabilities may affect the whole business process [11]. The users of CC will need licenses for the software, and they may pay by the user so it will be big cost [11]. There are many threats related to PaaS like the bugs in interfaces between CSP and CC. Also, interfaces that use suspicious tools that access the data files may lead to vulnerabilities.

Authentication and authorization are one of the major controls to protect the data in cloud computing. They manage who can access the data and what data can be accessed [12]. The level of data encryption and the way of destroying data files after ending contract ensure CSP, and the next CC will not be able to access these files. Also, the way of destroying data files after finishing contract specifies how it will be deleted [14]. CC must ensure that all controls related to physical and logical security of data are applied [12]. CC and CSP sides must be integrated and protected from breaches while transmitting among the two sides [12]. A control that ensures that there is a backup of data in more than a site and data transmission from site to site is secured which is very important [14]. Malware is malicious code or software that can be remotely injected to the cloud by using the various methods that can affect the data [9]. Finally, each service type has many risk areas. There are general risk areas which will face the CC while deciding moving to the cloud. These concerns are as follows:

- **Infrastructure as a service (IaaS)**

  1. Internet protocol
  2. Infrastructure failure
  3. Physical security
  4. Scalable storage
  5. Virtual machine escape
  6. Insecure VM migration
  7. Shared technology vulnerabilities
  8. TCP/session hijacking
  9. Improper virtual machine management

- **Platform as a service (PaaS)**

  1. Security threats in addition to the web-based services
  2. Lack of secure software development process
  3. Tools and services access to user files
  4. Bugs in large distributed systems
  5. Software licensing
  6. Using suspicious software
  7. Account or service hijacking
  8. Costumer data manipulation
  9. Eavesdropping
  10. Trusted transaction
  11. Insecure APIs
  12. Injection and XSS attack
  13. The feat of unauthorized access
  14. Attacks against virtualization
  15. API and browser Vulnerabilities

- **Software as a service (SaaS)**

  1. Access control
  2. File storage way and its encryption level
  3. Availability
  4. Data security
  5. Data confidentiality
  6. Data integrity
  7. Port scanning
  8. Malware injection
  9. Data transfer bottlenecks
  10. Data breaches
  11. Data destroying way after finishing service
  12. Data leakage
  13. Roll back attack
  14. Data corruption
  15. File storage way and its encryption level

- **General**

  1. Authentication and authorization
  2. Multitenancy
  3. Integrity
  4. Weak service level agreement
  5. Availability of services
  6. Performance unpredictability
  7. Scaling quickly
  8. Reputation fate sharing
  9. Difficulty in detecting problems

10. Loss of control
11. Changes of business model
12. Abuse use of cloud computational resources
13. Denial of service
14. Data locality
15. Cloud security attacks

3.2 *Risk Assessment and Governance in Cloud Computing*

C. Tunc et al. provided a taxonomy covering security issues in cloud computing (security, privacy, and compliance) in IaaS. It has a configuration for continuously reporting any irregular behavior to the CC to check the controls [15]. E. Cayirci et al. provided a model for selecting CSP by evaluating all risk scenarios and assessing these risks to facilitate selecting CSP [16].

An approach was proposed by R. Kalaiprasath et al. which classifies the threats facing cloud and specifies the controls of protection and enforcement policies that must be triggered for each threat [17]. Tsaregorodtsev et al. developed an approach that assesses the risk in cloud through defining the harms that may happen and their impact in order to check financial loss in case of risk appearance [18].

M. Medhioub et al. proposed a framework that is based on gathering cloud environment background details, estimating risks, and deciding which protection mechanisms are appropriate. The process of monitoring and updating the risk situations provides a better understanding of the evolving risk situation and enhances the respose of the changing risk areas [19]. Another developed framework focuses on assessing risks of the needed service model for the CC and checking if it can reduce these risks or mitigate them in order to accept this model [20]. Another research aimed to do an assessment of vulnerability when a code is deployed into the cloud environment based on vulnerability database and using similarity [21]. Another framework which was proposed in 2019 defined five main attributes (strategy, technology, organization, people, and environment) and defined the risks related to them and their controls [22].

E. Kristiani et al. proposed a guide for applying the risk management framework to federal information systems for governance in cloud computing. They defined the key security and privacy issues. Then, they defined the preliminary activities which are as follows: (1) *planning,* which defines requirements, performs risk assessment, and checks CS's competency, (2) *initiating and coincident activities*, which are concerned with evaluating the ability of CSP to deliver the service in compliance regulations and law and achieve security and privacy, and (3) *concluding activities* [23].

M. Al-ruithe et al. provided a framework that determines the responsibilities between CSP and CC and then evaluates the business process to achieve the governance while moving to the cloud. Also, they focused on negotiation for control and legal terms after that and developing SLA [24]. K. Brandis et al. developed a framework to show the effect of using standards in cloud computing in mitigating risks according to five aspects which are legal, organizational, technical, cultural, and financial in order to use the cloud without risk in these aspects [25].

S. M. Faizi et al. divided cloud computing security domains into storage, software, network security, trust management, Internet and services, compliance, legal, and virtualization [26]. M. Al-Ruithe et al. provided a strategy for determining the critical success factors (CSF) in implementing cloud services that include (1) *organizational* (defining the responsibilities, business, and IT alignment), (2) *technological* (automation of data integration life-cycle, having data metrics), (3) *strategic points of control*, (4) *training and awareness of data stakeholders*, and (5) *monitoring the compliance* [27]. We can conclude that every framework or model focuses on a specific area and having strength points and drawbacks that will be highlighted in Table 1.

3.3 Current Challenges in Cloud Computing

After investigating the concerns related to the cloud computing services, we can categorize challenges in cloud computing services into seven main points:

1. *Security*. It is the most critical concern facing cloud computing customers.
2. *Compliance*. Using cloud computing will lead to lose some or all of controls and may violate regulations for the CC.
3. *Performance*. After disaster, while moving to backup will lead to low performance. Also data transfer bottlenecks may lead to more response time.
4. *Cost model*. Cloud computing services usage increases cost of data communication and cost of license that will be used for each CC's user.
5. *Charge model*. Additional cost for redesign and redevelopment while moving to the cloud especially if the customer chose to use SaaS model.
6. *Cloud interoperability*. It ensures smooth data flow across different clouds and applications within that or data flow between local applications.
7. *Down-time*. It is a downtime for any disaster or losing connection between CSP and CC that may affect business reputation or cause financial loss.

Risk assessment models and frameworks determine all risk areas, financial, infrastructure, or security issues. The governance frameworks try to work on the whole controls. All of that may need to pay more for unneeded controls, as well as SLA is very important while moving to the cloud as it ensures that all needed controls are implemented, and the process is compliant with the regulations and rules.

4 Proposed Framework

This research proposes a conceptual framework based on defining the cloud service model and its risk areas. It helps in selecting a trusted CSP who can give appropriate controls for these risks and developing a suitable SLA. Negotiating with CSP and

Table 1 Existing frameworks evaluation

The study	Methodology	Benefits	Drawbacks
Risk assessment model for selecting cloud service providers	Qualitative model	Make automatically risk assessment and select CSP	Cloud service customer compares the risk profile of any two CSPs and select the suitable CSP. And that needs expert to compare
Cloud security and compliance	Qualitative framework	Define the cloud security and compliance policies needed for CC and CSP that provide them	Focuses on the security issues only
Information security risk estimation for cloud infrastructure	Quantitative approach	Define the opportunity of having a harm and financial loss of it	It assesses the losses financially only, and the financial loss is not the only concern for large organizations and differs from customer to another
Risk management tool for cloud computing environments	Qualitative tool	Experts define all risks and related controls and do iterative monitoring for evaluating controls	It operates on small data. Applying on bigger data sample will be hard as the controls may have conflicts
Cloud computing security risks: identification and assessment	Framework	According to service model and check risk areas and control and accept model or refuse it if there is unacceptable risk	It focuses on the deployment model as a unit and do not focus on every risk and take decision according to this model not the risk and available controls
Continuous security assessment of cloud-based applications using distributed hashing algorithm in SDLC	Hashing algorithm	Predict changes in code and check if it causes a risk and possible results	Needs to solve first vulnerabilities to use these solutions in the future
Context establishment framework for securing cloud computing through IT governance	Framework study	Define all risks areas and controls to be applied	It will be hard for CC to check all controls or pay for it
Analysis and classification of barriers and critical success factors for implementing a cloud data governance strategy	Conceptual framework	Define the cloud data governance CSF and evaluate them then implement the governance if acceptable	No risk assessment and provide headlines not in details
Governance, risk, and compliance in cloud scenarios	Framework	Check applying standards will mitigate risk areas in cloud computing	Need increase sample size and make it random to generalize the result

Fig. 2 Steps to apply the proposed framework

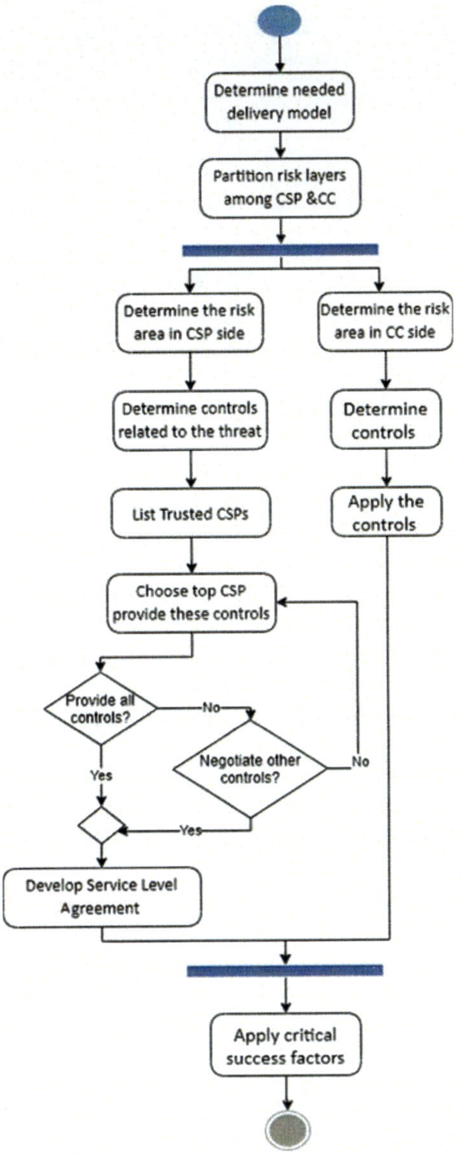

developing the SLA will need experts to perform these tasks in an appropriate way. The auditor must be knowledgeable about risk controls and the complete process of governance. The SLA must cover all needed controls to ensure that no regulations are violated Fig. 2.

Step (1) Define the cloud delivery model: There is a concept in cloud service called pay-as-you-go. The CSPs provide CPU, memory, storage, operating system, security, networking capacity and access controls, and any additional software needed to run their environment. CC chooses the required delivery model (IaaS, PaaS, or SaaS) from the cloud according to its needs and budget.

Step (2) Partition risk layers among CSP and CC: As shown in (Fig. 1), moving to the cloud will lead to dividing layers to be controlled between CSP and CC. Auditor determines the risk layer distribution between the CSP and CC. Auditor must know which layers are mandatory to develop controls for on their side and the layers that CSP will need to have controls on to check. So, in this step, for illustrating, the distribution of these risks to be controlled.

For **CSP side steps are**:

Step (3) Determine the risk areas in CSP's side: Every layer which will be controlled by the CSP may be surrounded by concerns and risks. Thus, in this step, CC's auditor will define all risks that relate to these layers. Risk areas are categorized into eight domains (storage, software, network security, trust management, Internet services, compliance, legal, and virtualization) [26]. So determining risks must cover these domains in the CSP's side.

Step (4) Determine controls related to the threats: In this phase, the client's auditor defines all controls needed to be applied in order to avoid the risk areas in the CSP's side.

Step (5) List trusted CSPs: List the CSPs that have a good reputation and provide the needed service. Knowing trusted CSPs can be performed by contacting the current clients of the CSP and evaluate their customer level of satisfaction (CLS). The client can also evaluate the CSP by checking

Step (6) Choose most of CSPs that provide these controls: Identify CSP who provides most of the needed controls to CC on the cloud layers according to its documentation. Also it will be checked according to governance policies, principles, process, roles and responsibilities, communication, and change management plan.

Step (7) If there are missing controls CC will negotiate with the CSP for applying these missed controls or check if they can mitigate the risk consequences of missing the control. If negotiation succeeded, CC would accept the model and this CSP. If not, CC will choose the next most appropriate CSP. As well as, repeat the same step till finding the appropriate CSP who provides all required controls.

Step (8) Develop service level agreement: Develop SLA, as the quality of the provided services is defined using SLA. It ensures for the CC a good level of service and gives them confidence that if something goes wrong the CSP will respond quickly. The auditor who negotiates with CSP must be involved while developing an SLA to make sure that SLA includes contractual requirements that are required while getting services and also while service terminate.

And for **CC side** steps are:

Step (9) Determine the risk areas in CSP's side: CC may have layers that include risks in case of using IaaS or PaaS. Also, in case of using SaaS, CC may have risk areas to control. Examples of covering missing controls in CSP's side are if CSP does not provide data encryption or data redundancy in many sites.

Step (10) Determine needed controls: Auditor determines controls that must be applied in CC side to avoid the risk areas in CC's side according to the risks determined to cover all security domains [26] in CC's side. In previous step examples, the CC's auditor must recommend encrypting data before sending to CSP's side. For redundancy issue, CC may deal with more than one CSP.

Step (11) Apply the controls: Auditor will apply all the needed controls and make sure that they cover all risk areas in the side of CC.

Step (12) Apply some of CSFs: Apply some CSFs [27] that define the responsibilities of both CC and CSP, training, and improve the awareness of data stakeholders, and monitoring the compliance.

5 Conclusion

Many organizations already moved to use cloud computing services with its different delivery model and deployment models because of its benefits. Using cloud computing services may lead to much vulnerability. The organization loses some of its applied controls when moving to the cloud environment. Cloud service provider (CSP) will be the responsible for applying these controls.

Every service model has specific threats. In addition, using cloud computing services has general threats. Cloud client (CC) must focus on both of them. CCs priorities differ according to business type and needs. Determining risk areas while using cloud services must be performed based on: (1) CC rules for assuring the business process requirements and (2) abiding to regulations.

Applying governance while using the cloud computing services will protect the CC from many risks. Governance also ensures applying the business strategy and achieving compliance with regulations when applying the right controls. SLA is very critical for reaching governance. Its weakness will lead to violating the regulations, violate the controls, and/or lead to extra fees.

The proposed conceptual framework aims to help CC's auditor to identify possible threats while using cloud services. Thus, it helps in applying all controls on both CC and CSP sides. It aids in achieving governance. It helps the auditor in applying all the needed controls and SLA contractual requirements. Framework's controls are based on the service model provided. They are also based on CC needs in their business process that complies with laws and regulations.

The proposed future work is to implement an automated framework. The framework's database will contain controls for both of general threats and each service model threats. Also, it will contain SLA contractual requirements of general concerns and for each service model. The automated framework will provide basic controls and SLA contractual requirements for CC's auditor. That should help the auditor in selecting appropriate CSP.

References

1. Müller SD, Holm SR, Søndergaard J (2015) Benefits of cloud computing: literature review in a maturity model perspective. Commun Assoc Inf Syst 37(1):851–878. https://doi.org/10.17705/1cais.03742
2. Mell P, Grance T (2011) The NIST-National Institute of Standards and Technology definition of cloud computing. NIST Spec Publ 7:800–145
3. Ashraf I (2014) An overview of service models of cloud computing. Int J Multidiscip Curr Res 2:779–783
4. Diaby T, Rad BB (2017) Cloud computing: a review of the concepts and deployment models. Int J Inf Technol Comput Sci 9(6):50–58. https://doi.org/10.5815/ijitcs.2017.06.07
5. Guide P, Edition F A guide to the project management body of knowledge
6. Lainhart JW (2012) COBIT 5: a business framework for the governance and management of enterprise IT COBIT 5
7. Lourenco J, Santos-Pereira C, Rijo R, Cruz-Correia R (2014) Service level agreement of information and communication Technologies in Portuguese hospitals. Proc Technol 16:1397–1402. https://doi.org/10.1016/j.protcy.2014.10.158
8. Sultana A, Raghuveer K (2017) Security risks in cloud delivery models
9. Sultana A (2017) Security risks in cloud delivery models, pp 2637–2640
10. Singh S, Jeong YS, Park JH (2016) A survey on cloud computing security: issues, threats, and solutions. J Netw Comput Appl 75:200–222. https://doi.org/10.1016/j.jnca.2016.09.002
11. Fox A et al (2009) Above the clouds: a berkeley view of cloud computing. Department of Electrical Engineering and Computer Sciences, University of California, Berkeley, Rep. UCB/EECS 28.13 (2009)
12. Sharma PK, Kaushik PS, Agarwal P, Jain P, Agarwal S, Dixit K (2017) Issues and challenges of data security in a cloud computing environment. In: 2017 IEEE 8th annual ubiquitous computing electronics mobile communication conference UEMCON 2017, vol 2018 Jan, pp 560–566. https://doi.org/10.1109/UEMCON.2017.8249113
13. Ramachandra G, Iftikhar M, Khan FA (2017) A comprehensive survey on security in cloud computing. Proc Comput Sci 110(2012):465–472. https://doi.org/10.1016/j.procs.2017.06.124
14. Sen AK, Tiwari PK (2018) Security issues and solutions in cloud computing. Security issues and solutions in cloud computing, Mar 2018. https://doi.org/10.9790/0661-1902046772
15. Tunc et al C (2017) Cloud security automation framework. In: Proceedings of 2017 IEEE 2nd international workshop foundations application Self* Systems FAS*W 2017. IDC, pp 307–312. https://doi.org/10.1109/FAS-W.2017.164
16. Cayirci E, Garaga A, Santana de Oliveira A, Roudier Y (2016) A risk assessment model for selecting cloud service providers. J Cloud Comput 5(1):1–12. https://doi.org/10.1186/s13677-016-0064-x
17. Kalaiprasath R, Elankavi R, Udayakumar R (2017) Cloud security and compliance–a semantic approach in end to end security. Int J Smart Sens Intell Syst 2017:482–494. https://doi.org/10.21307/ijssis-2017-265

18. Tsaregorodtsev AV et al (2018) (2018) Information security risk estimation for cloud infrastructure. Int J Inf Technol Secur 10(4)
19. Medhioub M, Hamdi M, Kim TH (2017, Mar) Adaptive risk management framework for cloud computing. In: 2017 IEEE 31st international conference on advanced information networking and applications (AINA). IEEE, pp 1154–1161
20. Patel K, Alabisi A (2019) Cloud computing security risks: identification and assessment. ERA J Ideas 17(2):11–19
21. Vijayakumar K, Arun C (2019) Continuous security assessment of cloud based applications using distributed hashing algorithm in SDLC. Cluster Comput 22(s5):10789–10800. https://doi.org/10.1007/s10586-017-1176-x
22. Chiou S-F, Pan H-T, Cahyadi EF, Hwang M-S (2019) A context establishment framework for cloud computing information security risk management based on the STOPE view. Int J Netw Secur 21(1):100. https://doi.org/10.6633/IJNS.201901
23. Kristiani E, Yang CT, Wang YT, Huang CY (2019) Implementation of an edge computing architecture using openstack and kubernetes. Lect Notes Electr Eng 514:675–685. https://doi.org/10.1007/978-981-13-1056-0_66
24. Al-ruithe M, Benkhelifa E, Hameed K (2016) A conceptual framework for designing data governance for cloud computing. Proc Comput Sci 94:160–167. https://doi.org/10.1016/j.procs.2016.08.025
25. Brandis K, Dzombeta S, Colomo-Palacios R, Stantchev V (2019) Governance, risk, and compliance in cloud scenarios. Appl Sci 9(2):1–21. https://doi.org/10.3390/app9020320
26. Faizi SM, Rahman SSM (2019) Securing cloud computing through IT governance. SSRN Electron J 7(1). https://doi.org/10.2139/ssrn.3360869
27. Al-Ruithe M, Benkhelifa E (2017) Analysis and classification of barriers and critical success factors for implementing a cloud data governance strategy. Proc Comput Sci 113:223–232. https://doi.org/10.1016/j.procs.2017.08.352

Blockchain Technology in Cloud Computing: Challenges and Open Issues

Muneer Bani Yassein, Ismail Hmeidi, Omar Alomari, Wail Mardini, Omar AlZoubi, and Dragana Krstic

Abstract Blockchain technology (BC) has recently emerged as a potential solution of security and privacy issues in cloud computing systems. Particularly, the decentralization concept is driven by Blockchain technology. This paper aims to define the cloud computing and Blockchain technologies as recent technological trends. It also surveys the latest trends of Blockchain inclusion within cloud computing. A systematic review method is conducted and evaluated in a dozen studies in time period between 2016 and 2019. The methodologies and results of these studies were explored, and it was found that these studies focused on two main aspects. First, does the data provenance, integrity, and privacy tackle in real data sets such as medical sets and business datasets? Second aspect is the data security in which various studies suggested and proposed for handling this issue. It has been noticed that there is a great variety of designs that are not contradictory, as well as variation of the interests of studies and adopted data sets. In this paper, the authors will analyze the Blockchain technology within cloud computing environments to demonstrate advantages and disadvantages in distributed data and protection.

Keywords Cloud computing · Blockchain technology · Security · Data integrity · Data provenance

1 Introduction and Literature Review

Blockchain technology (BC) has emerged recently as a developed generation of the financial technologies that adopts the electronic cashes and coins without the needs of third-party inclusions. The BitCoins is the foremost common example of such applications [1]. Accordingly, Blockchain gains the research interests due to its efficiency of providing higher security compared to the current security technologies

M. B. Yassein · I. Hmeidi · O. Alomari · W. Mardini · O. AlZoubi
Department of Computer Science, Jordan University of Science and Technology, Irbid, Jordan

D. Krstic (✉)
Faculty of Electronic Engineering, University of Niš, Aleksandra Medvedeva 14, Niš, Serbia
e-mail: Dragana.Krstic@elfak.ni.ac.rs

D. A. Magdi et al. (eds.), *Digital Transformation Technology*, Lecture Notes in Networks and Systems 224, https://doi.org/10.1007/978-981-16-2275-5_5

utilized in the huge data storage, as well as other potentials of BC in data transparency [2]. Therefore, many of studies exploited the idea of Blockchain to provide solution for cloud computing technology's risks.

Pradip et al. [3] proposed a new distributed architecture model of Blockchain cloud based on three emerging technologies: Blockchain, fog computing, and software-defined networking (SDN), designed to increase security, high scalability, low latency, real-time data delivery, resiliency, and high availability. BCPay, as a BC based fair payment framework introduced for outsourcing services in cloud computing, is presented in [4]. That analysis has proved that BCPay is very effective in the field of the number of engaged transactions and computation expense as well is characterized by robust fairness and soundness.

Xia et al. suggested in [2] a Blockchain-based technique to interpose access between a pool of sensitive participated data and users. They compared it with BitCoin BC network and built a lightweight and scalable BC to prove the effectiveness of their design that allows data involvement in a safe method and keeps data confidentiality. Blockchain-based cloud data provenance framework including proof of stake (PoS) engine, called BlockCloud, is analyzed in [5]. In spite of the qualification of PoS powerful Blockchain, it came over various design challenges when merging to distribute general ledger technology within the cloud computing.

The group of authors discussed in [6] an electronic bill payment (BPay). This is a frame of outsourcing service fair payment based on BC within cloud computing to achieve safe and equitable payment of outsourcing services without dependence on any third-party, whether trusted or not. In [7], the state-of-the-art research papers that associated with Blockchain technology are discussed and reviewed. A group of papers is selected comprehensively from various domains, from the online database, and classified.

Aniello et al. [8] have evaluated a prototype of two-layer blockchain architecture (2LBC). A two-layered architecture for a Blockchain-based database is capable to support both, high data integrity guarantees and high performance in a completely decentralized environment. They suggested a solution dealing with issues of availability. To analyze the feasibility of Blockchain in cloud, the authors have comparatively analyzed the cloud and BC in [9]. They introduced a four-layered sample structure of cloud that depends on Blockchain technology. Also, they suggested BC based cloud utilizing the advantages of BC to preserve transparency, security, and trust in the cloud, what decreases fraud and hazard.

A model of sharing data between cloud service providers using the Blockchain is presented in [10]. It utilizes the smart contracts and an entrance to monitoring technique to trace the behavior of the data efficiently in addition to prevent access to violated rules and authorization on data. The applicability and protection implications of BC in investigation of BC cloud are examine in [11]. The case of block withholding attacks widespread in mining pools based on proof of work (PoW), to realize the attacker's way toward grabing the pool members' remunerations. The results of simulation explain that attacker's arrive to additional computational authorities could damage the sincere mining process in Blockchain cloud. In [1], the authors debated the BC technology and core technologies and surveyed the trend of studies to date in order to look further areas to be discussed. Also, the ways of providing protections by introducing a technique of secure Blockchain utilization and removal protocol are looked.

In this study, research trends pertaining emphasis on cloud auditing, with conserved user privacy and raising the availability, are analyzed. Using Blockchain technology, the record with immutable timestamp and produce for each data records for validation a BC receipt was made. Also, we try to scrutinize briefly the definition of cloud computing and BC technologies and survey the recent trends of the research in the inclusion of BC in the cloud computing environments.

This paper is organized through three main sections. After Introduction with analysis of related papers, in Sect. 2, cloud computing and Blockchain are described; research methodology is involved in Sect. 3, and analysis of recent studies is given in Sect. 4. Conclusion presents last part of the paper.

2 Background

2.1 *What is the Cloud Computing?*

Cloud computing is a technology which comprises tremendous number of heterogeneous entities, devices, and software from a dynamic wide vendor, all combined in one huge distributed environment [12]. Cloud computing provides uncountable benefits and advantages to all genres of industries such as the IT environments [1], military [12], business, medical sectors [2, 10], and so forth. Cloud computing is considered a wide platform that shares resources and offers everything as services [13]. Consequently, various industries and companies working in same fields collaborate and support interoperability providing unified goal-oriented cloud environments, such as the EU—SUNFISH project, which provides a cloud federation platform as service enabling ease data management, and secure service providing [14].

The main advantages of the Cloud computing is the ease and remote data sharing, on-demand services from a tremendous shared resource pool [11]. As well as, the security issues are considered a critical dimension of using cloud computing, in addition to the privacy issues [1]. Such risks add a new dimension to the struggle of cloud computing service providers to gain collaboration of the industries and business owners [2]. Assured secure sharing, privacy, transparency, trustworthiness and integrity of data are defined as the cloud computing critical issues. Furthermore, the security is not a demand for data in cloud sharing, it also required for data out-cloud such as provenance data, which is a metadata of the sharing data [12].

2.2 What is Blockchain?

Blockchain is defined as a technology of distributed data structure, "public ledger," which can hold any information that is stimulated and shared securely between the memberships in the network without any authority.

Blockchain technology is anticipated to be the next big metamorphose. The network of peer-to-peer contributor protects an immutable public record of data. BC is composed of blocks chained together as a ledger. Blockchain is a decentralized and distributed digital ledger that is used to record transactions across many computers, so that the record cannot be altered retroactively without the alteration of all subsequent blocks and the collusion of the network. Due to the success of BITCOINs, first application of BC in 2008, BC technology can now be used in different areas such as financial markets, Internet of things, supply chains, voting, health care, and smart cities [15].

Blockchain is not a single one technique, but has combinations of other technique such as mathematics, algorithms, cryptography, peer-to-peer networking [13]. BC technology has many attributes; (1) decentralized, which means data can be recorded, stored, and updated, and distributed at every block, (2) transparent, which means the data can be seen by any node on the BC, (3) open source, which means the people can use BC to build any application they want, (4) autonomy (based of consensus), (5) immutable, which means records cannot be altered unless someone can take control of more than 51% of the nodes at the same time.

Figure 1 shows the structure of the Blockchain in details [Fig. 1; 1]. Thus, this technology founded as a solution of dealing with sensitive information problem without any deployed verification, in which the data of the transaction is recorded on the ledger [16].

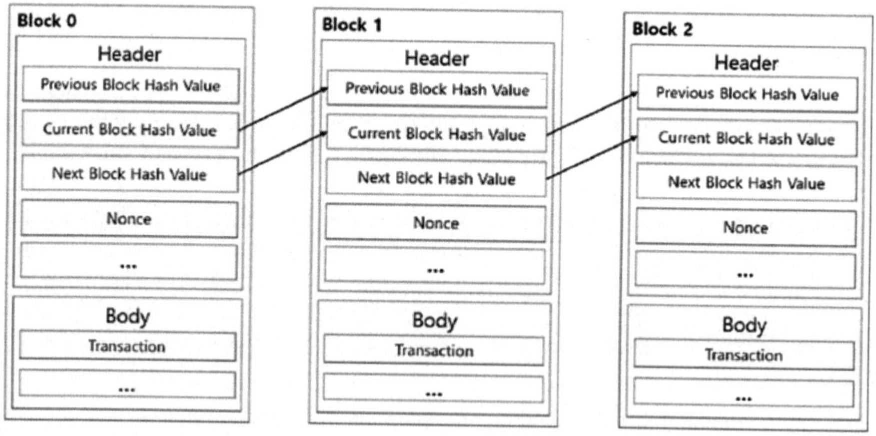

Fig. 1 Blockchain structure

The Blockchain based on the decentralization includes distributed nodes based on a set of rules instead of centralized node. These rules consider a platform of decentralization that enables evaluation of data exchange without any authority [2] to be secure platform that distributed database containing a list of records linked together through chains on blocks [10]. This system, characterized by decentralization, maintains the security of databases from cyber-attacks in cloud, and depends on assured data provenance and supports cloud auditing. So, it attracted many researchers to develop many applications such as smart contracts and identity management [12].

The Blockchain structure is based on the peer-to-peer (p2p) infrastructures "broker-free," with cryptographic primitives and consensus protocol [13]. Each peer has its version of the ledger, where each peer deploys a consensus protocol for a validation purposes that all transaction or exchange data are accumulated in one block to create a hash chain [1]. Each block in chain is defined as individual component that contains information relating to a particular transaction [10].

The Blockchain system is described as data storage, which is distributed through networks among peers with main ledger, containing a series of blocks. Each block has a bundle of time stamped with two hashes: cryptographic hash as identifier and previous block hash as a reference of the previous block, as shown in Fig. 1. The structure of each block is divided into two main parts [16]. Header, consisting of metadata (previous hash, mining competitions, Merkle tree root), and body, consisting of data.

The process of Blockchain explicates through the following: suppose that node sends a new data over the network. The destination node checks the incoming data according to the data. If it validates, then the new data will be stored in the block [13].

3 Research Method

3.1 Systematic Review Process

This study adopts the systematic review method in order to achieve the review purposes. In the field of the cloud computing, this method is extremely utilized among researchers, in terms of in-depth exploration of this field. Systematic review is a method of revealing the research interest to determine relevant subject associated with current research purposes and objectives. It is also known as evidence-based research practice utilized in various field categories of research such as, but not limited to: education, software engineering, psychological research, applied science, and so forth. Systematic review is an established, organized process which follows a specific defined protocol, as shown in Fig. 2.

The systematic review process includes four main phases, which are initiated with objectives and research questions formulating, followed by the pertinent studies collection from the specific databases providing studies and researches as electronic documentations sources. The third phase is evaluation of the accessed studies and

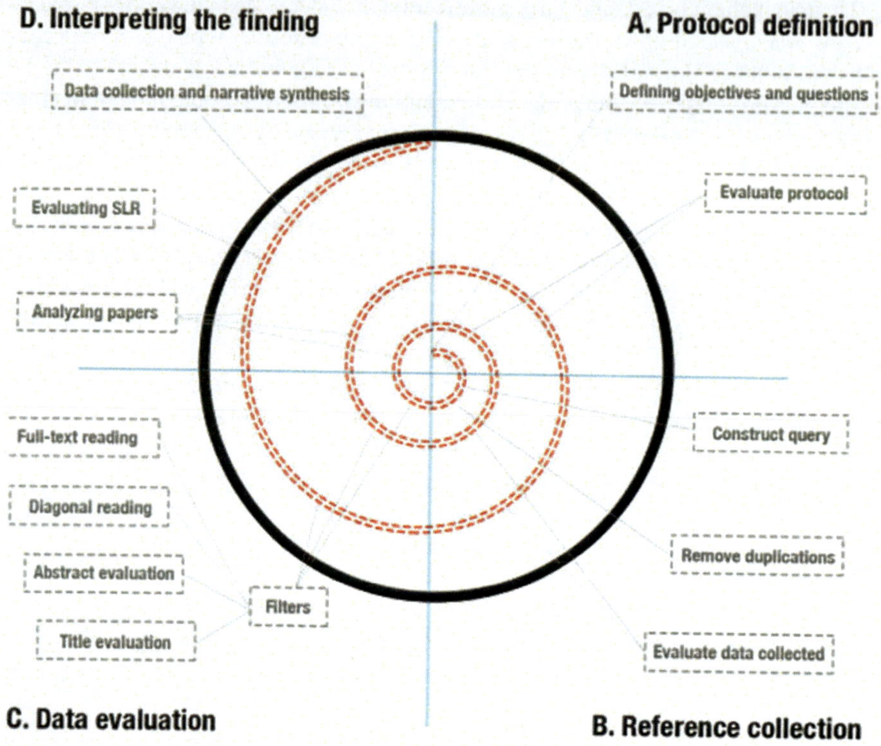

Fig. 2 Systematic review process

research based on the specific inclusion and exclusion criteria that will be scrutinized in the following sections. The final phase is interpretation of the results of the included studies in depth and synthesizing the information.

The process definition identifies the systematic review objectives and controls questions that determine the study orientation. The research study question is described as control due its meaning of limited going on the studies and researches of the research subject. References and studies collection comprise all studies exploring in the subject area, which mainly anchored in the effective meaning of electronic resource exploiting the electronic databases selected through the study.

In the current systematic review, numerous electronic databases are used for relevant studies extraction process. Approximately eight databases were searched for relevant studies, and research includes: IEEE database, ACM database, Research Gate, ArXiv, ISI library, Google Scholar, Metapress, and Citeseerx. Various and numerous studies were extracted through such tremendous databases during research process. However, not all of studies considered in this research included information that answered current research questions. Because of that the inclusion and exclusion criteria were defined in the incoming part of the paper. The main purpose of the current study is conducting a review of the inciting and previously published researches about integrating Blockchain technology within cloud computing environments. Further, study is extending to reveal the challenges, requirements, and authors perspectives in the investigated subject.

3.2 Selecting Criteria

As research question requests, the vital point in the study is the Blockchain inclusion within cloud environment and its related dimensions. Findings of appropriate studies mostly depend on the selecting criteria, as listed below.

(1) **Inclusion Criteria**: Inclusion criteria mainly stand on the research purposes, objectives, titles, abstracts, and keywords that oriented selected papers and studies, in addition to the following extra parameters:

   (a) The language of studies and researches must be English language.
   (b) Only studies of the Blockchain with cloud computing environments are conducted.
   (c) Studies with the Blockchain structure, challenges, opportunities, requirements, implications, and cloud computing applications are requested.
   (d) Studies which suggest a Blockchain-based cloud structure are required.
   (e) Works evaluating Blockchain security performance within cloud environments were looked for.
   (f) Papers published from 2016. to 2019. are analyzed.

(2) **Exclusion Criteria**: Basic parameters that determine the exclusion criteria are following:

   (a) Research papers published in languages other than English, such as Arabic or Dutch.
   (b) Any non-academic articles such as those published on unofficial Web sites.
   (c) Papers printed before 2016.
   (d) Studies published within white, yellow, or gray papers.
   (e) Papers or studies dealing with cloud computing without Blockchain adoptions, or vice versa.
   (f) Duplicated studies within multiple platforms.

(3) **Search process**: The research process stands on the three pillars: the eight databases aforementioned for extraction purposes, research question, and main research keywords. The research nature is described as versatile search method.

The list of keywords was used within research process to guarantee appropriated results of searching counters with study objectives in conjunction of numerous available potential options within research electronic bases functions, in order to enable roundness of research process.

The main keywords used in the search phase are:

- Cloud computing.
- Cloud computing challenges.
- Cloud computing via Blockchain.
- Blockchain within cloud computing.
- Blockchain-based cloud.
- Blockchain security in cloud computing.
- Blockchain structure in cloud or fog environment.

Through searching process, authors have found some obstacles in access to the scientific papers, in which access was allowed only to abstracts. Such papers were excluded from the study.

(4) **Studies retrievals**: The number of obtained studies reaches 50 papers found within research process. Sixteen papers remained after adopting inclusion and exclusion criteria, 14 papers were discarded due to duplication, 6 were turned off due to access barriers, while the remaining works were excluded due to unofficial publishing and time constraints. However, these 16 studies were carefully analyzed, summarized, and interpreted in the next part.

(5) **Information synthesizing**: The synthesizing process mainly established on the collected information from studies, which are about Blockchain inclusion within cloud computing, which will be examined from different aspects and dimensions such as structure, challenges, implications, and so forth.

The information extracted by research questions handling provide a screen of the accessible and current research in adopting Blockchain in cloud computing for various purposes, such as data integrity, or security enhancements, perspective of inclusion, and so on.

4 Analysis of Recent Studies

The surveys on the research and efforts of scientists were conducted to meet the current goal of the study.

4.1 Medical Data Sharing

The concept of Blockchain was associated with data privacy of various data, as many studies confirmed. For example, Xia et al. [10] were proposed cloud computing system adopting Blockchain to keep privacy data for patients' medical records called "MeDShare" system. This system is characterized by stability and centralization to audit and control medical data that shared in cloud repositories and ensure the source of security data. To achieve this aim, the researchers built a system for sharing

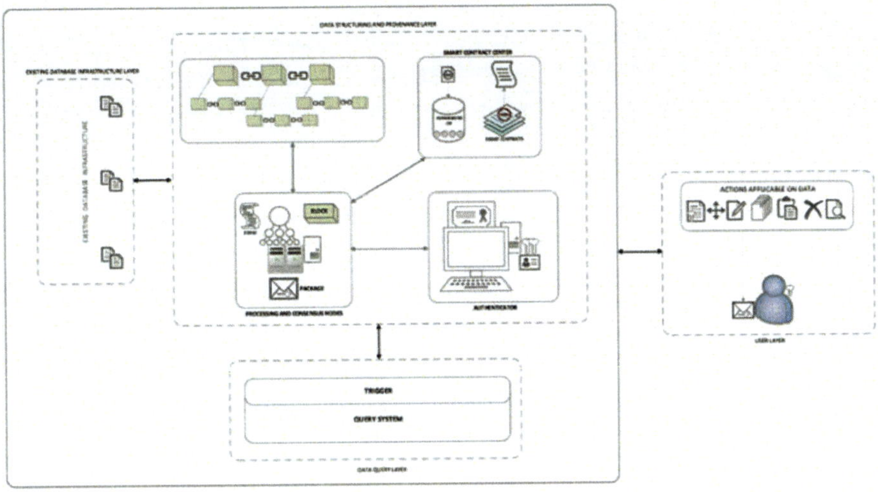

Fig. 3 System design with four layers

information through four layers (user layer, data query layer, data structuring and provenance layer, and existing database infrastructure layer) as shown in Fig. 3 [Fig. 1; 10].

The system includes next: creation the user request for data using private key that was created previously; then, the request is moved to data structure source to translate the request; after, the verification stage of the legitimacy of the signature, and finally, the data are sent to the system to create a smart authentication contract and the encryption tag to distribute applicants.

The proposed system has been evaluated by the level of data privacy and the data access time by JMeter.

The evaluation found that the system can detect many unauthorized access attacks, but if data requests are increased, the efficiency of detecting all security vulnerabilities may be reduced, and the time for retrieving data will increase. The time of retrieval data is affected by number of requests. If the number of requests for data retrieval increases, the data retrieval time increases, especially as processing and hiding the identity of the data contribute to increasing the time of access to the data, as it is presented in Fig. 4 [Fig. 6; 10].

Also, proposed system was evaluated by comparing with many of the systems in studies through Blockchain-based, identity management, decentralized access, centralized access, distant access, tamper proof data audit, data access revocation, everywhere there was a similarity between the proposed system and others in the usage and accumulation of data. Therefore, it was concluded that the proposed system MeDShare is better than several other systems, so it can be used to access medical data securely, and data can be shared with research and medical institutions without any risk.

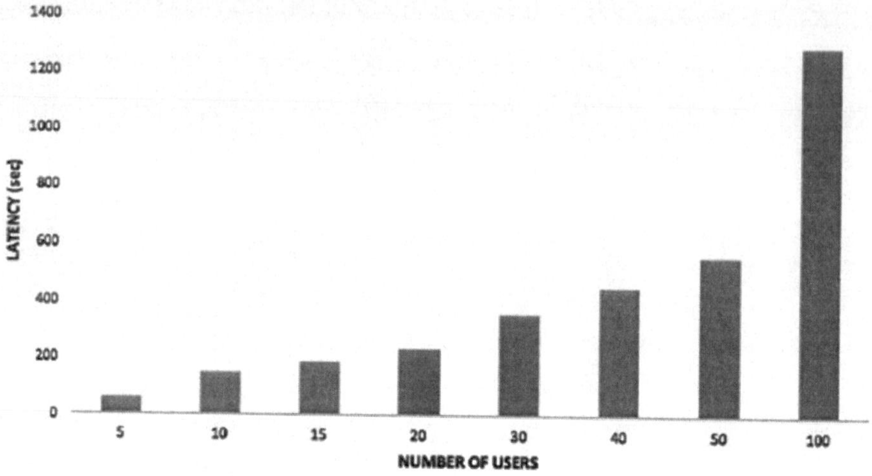

Fig. 4 Latency per number of requests

4.2 Blockchain-Based Data Sharing (BBDS) Ecosystem

Likewise, the authors of [2] proposed cloud computing system based on Blockchain to use and share medical data in a safety way called "Blockchain-based data sharing (BBDS) ecosystem." The model of that system was created to sharing data by Blockchain, so there are three entities: user, system management, and storage. This scenario is shown in Fig. 5 [Fig. 1; 2].

The system creates a set of keys to confirm membership when requesting and publishing data by the verifier. When creating a request, the user is asked for the access key, and then creates entry mark. After confirming the user, the verifier creates a special membership key and maintains it in the database.

Therefore, the system depends on three steps: Request file, Grant request, and Access file. This is presented in Fig. 6 [Fig. 5; 2].

To complement the system, two types of protocols were adopted:

1-User-Issuer Protocol,

2-User-Verifier Protocol.

Finally, the proposed system was evaluated by comparing it with the bitcoin system by calculation: Tx(B_ts) used to calculate data size per second, Tx(B_60(tm)) for data size per minute, Tx(B_36_102(th)) for data size per hour, and Tx(B_864_102(td)) for data size per day.

After comparison, it was found that the proposed system was more scalable than bitcoin, which relied on low time to processing request, thus increasing the volume (in gigabytes) of the Bitcoin series. BBDS system adopting Blockchain architecture supports large amounts of data every second without worrying about scalability.

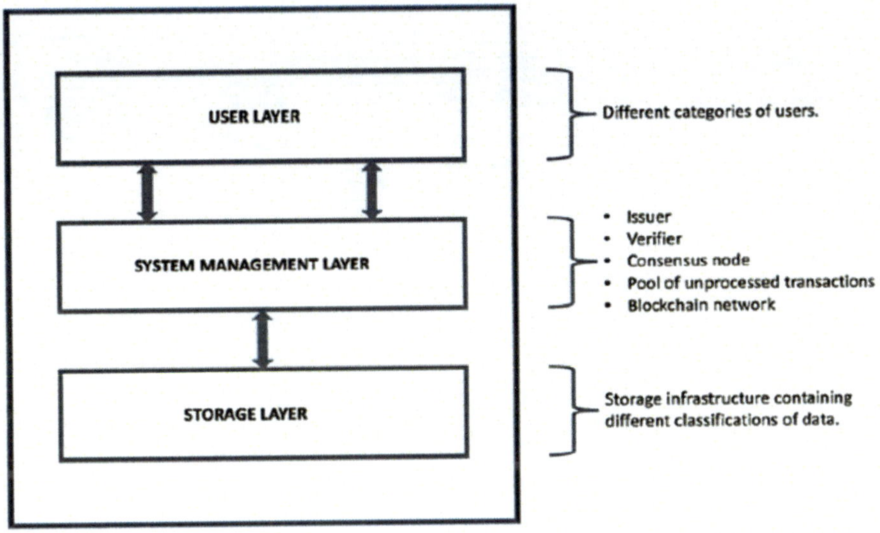

Fig. 5 BBDS ecosystem model

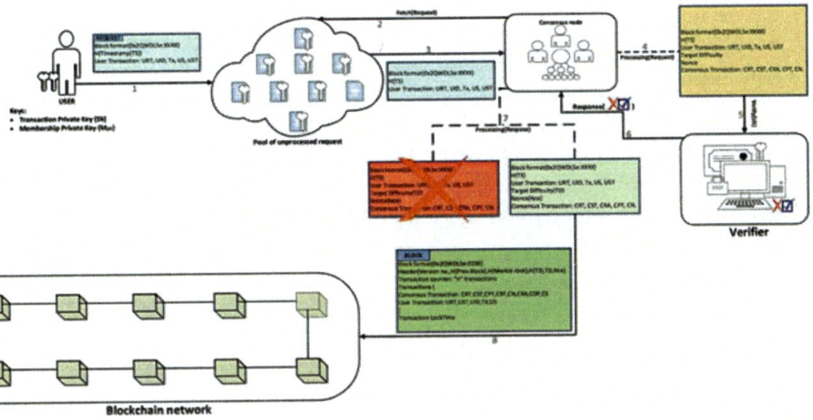

Fig. 6 BBDS system setup

4.3 ProvChain

In [12], "ProvChain" system based on Blockchain technology for providing data protection from tampering, and enhance the privacy of source data in the cloud of military environment, was proposed and visible in Fig. 7 [Fig. 1; 12]. ProvChain is consisting of Cloud User, Cloud Service Provider (CSP), Provenance Database, Provenance Auditor (PA), and Blockchain Network. The system operation was in

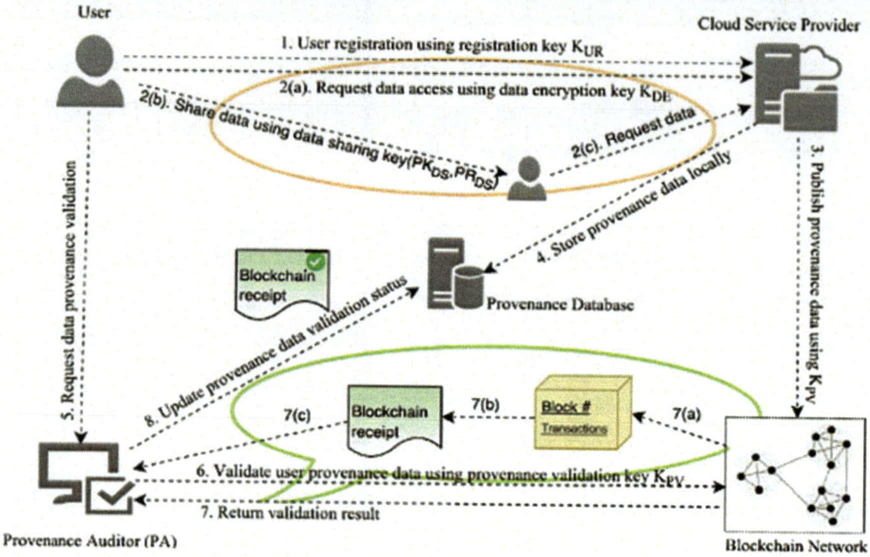

Fig. 7 Components of provChain

three steps on data provenance: data collection, data storage, and data validation, as drawn in Fig. 8 [Fig. 2; 12].

ProvChain system was evaluated by JMeter to show the average response time with different file size when using a system shown in Fig. 9 [Fig. 6; 12].

After evaluating, provChain was observed as system providing real-time auditing of data access in the cloud and record them in a directory to monitor them and keep secure. In addition, the system adopted decentralization to ensure data integrity, so

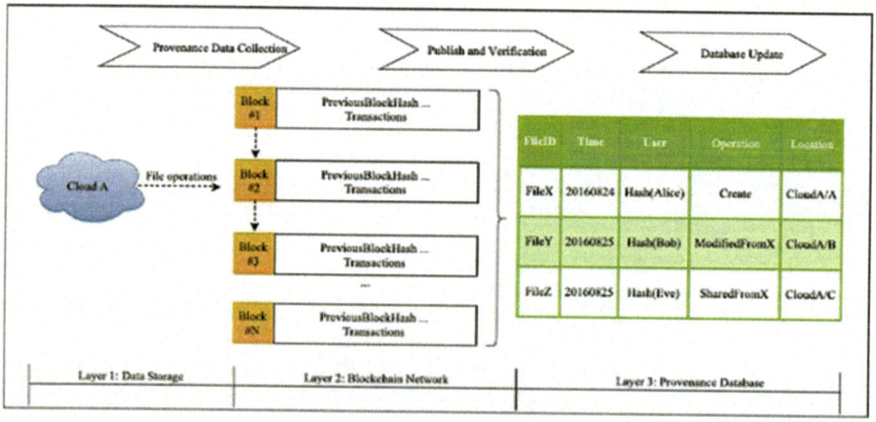

Fig. 8 Operation of provChain

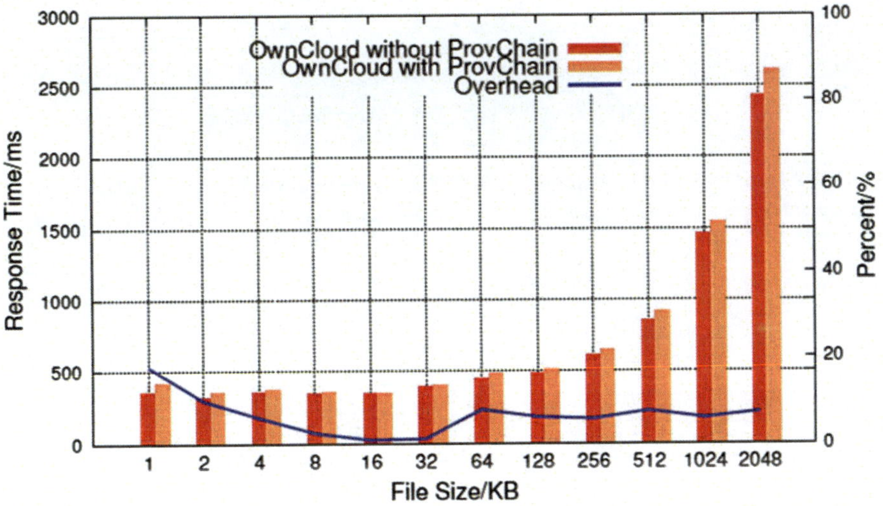

Fig. 9 Response time with file size with/without provChain

each record has a copy with each node in the Blockchain network. Thus, users can subscribe to service of data provenance without any threat to their privacy.

4.4 *SUNFISH Project*

Gaetani, et al. confirmed in their study [14] that Blockchain has properties of data integrity and security vulnerabilities, but there are limitations that lead to low productivity, high latency, and poor stability that hinder systems based on Blockchain. Thus, this study discussed SUNFISH European project that relies on Blockchain to be a secure cloud containing public sector databases, especially as it is exposed to threats such as: change part of the database directly, update data by a member without informing others, and multiple members change part of database maliciously.

So, proposed Blockchain-based database increases efficiency of system, depending on data integrity, performance, and stability. This Blockchain-based database is presented in Fig. 10 [Fig. 1; 14].

An explanation of the system identifies the actual safety and integrity data needs of cloud computing environments. Finally, study emphasizes the importance to rely on the Blockchain as a reliable infrastructure to store data in cloud computing environments.

Further, in [1], the authors deliberated the Blockchain concept and last trends in the research field and expanded the discussion of Blockchain adaption to the cloud computing. The studies revise previous studies that analyzed the Blockchain and their trends in research.

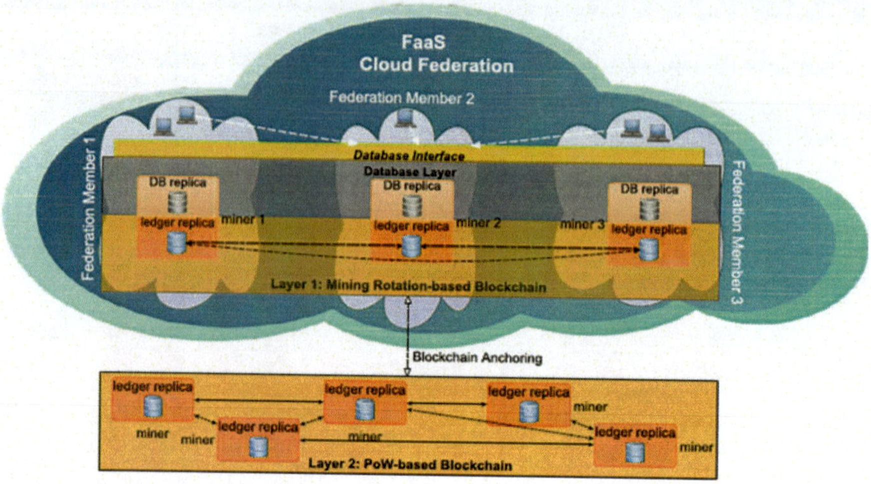

Fig. 10 Blockchain-based database proposal

The results reveals that the security challenges are still growing challenges in the Blockchain technology. As well, the anonymity of user information is a critical issue in the inclusion of Blockchain.

The authors suggested to use Blockchain technology to delete user data in cloud computing environments, which are installed an electronic wallet securely removed when the message was sent. Such wallet removing technology is used for user anonymity and privacy conservation. Authors compared proposed methods in the previous studies based on the confidently, anonymity, integrity, privacy, and residual information. The analysis found that other studies did not discuss the residual information protection such as authors of this study did.

4.5 *Block with Holding Attack (BWH)*

In same context, [11] discussed the security implication with Blockchain cloud system. Despite the advantages brought by BC to cloud, such as data provenance, auditing, and so forth, there are several vulnerabilities in Blockchain cloud such: block with holding attack (BWH), double sending attack, selfish mining attack, Eclipse attack, block discarding attack, and difficulty raising attack. The authors modeled BWH attack—"Sabotage" through the mining process, as one of the most significant attack that causes loose for all system members. The main idea of this attack is to enter the mining pool as one of the authorized members, and wait good opportunity to get any member pool in order to demotivate it.

The aim of the model is to analyze the attacker's behaviors and its strategy with pool category variance. The study negotiated the BWH in different pools. First is

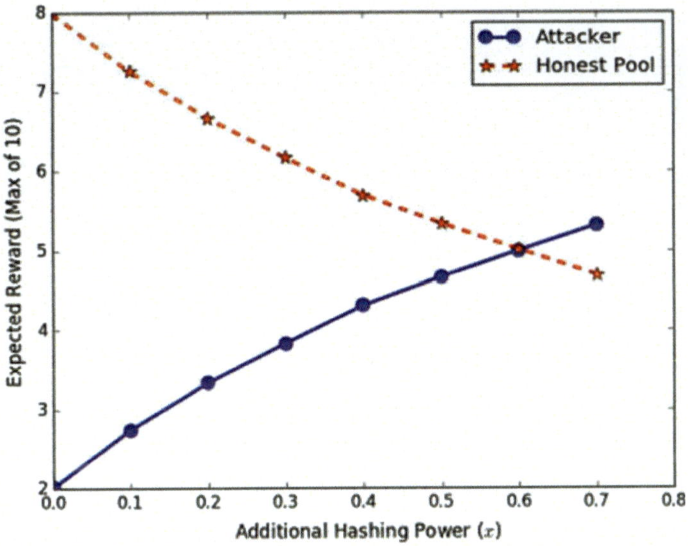

Fig. 11 Impact of hashing power on the reward in proportional pool

the proportional reward pool, in which attacker needs to submit an efficient number of shares in pool. Accordingly, the analysis shown that increasing of the hashing competence will affect the reward, in which increasing the attacker's hashing power negatively influence the honest member rewards. Further exponential increasing of hashing power scenario definitely will impact the honest member's existence in pool, which counter the status in the second pool. This model is shown in Fig. 11 [Fig. 2; 11].

The second is the BWH with in pay-per-Last N-Shares (PPLNS) pool, in which increasing the number of shares was not sufficient as attacker strategy. Thus, attacker has to adopt different strategies such as increasing the speed of mining process compared to others honest members and get reset in the last N windows of mining. The hashing power of the attacker in average of the honest miners hashing power due to large number of hashes induced by the honest members in overall overcomes the proportional power.

The attacker plans is redundant to take away the reward of honest members which is represented in Fig. 12 [Fig. 13; 16]. Results approved that there is a linear relation within number of users and the process; thus, the developed system can be described as stable system.

A tabular overview of the studies with their most important characteristics is given in Table 1.

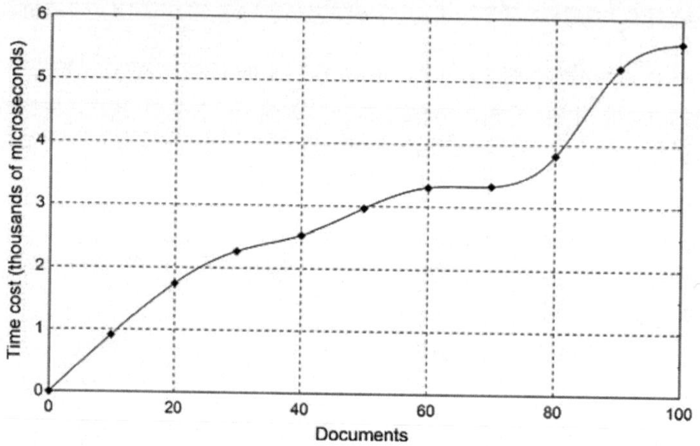

Fig. 12 Hashing power of attacker in PPLNS pool

5 Conclusion

The cloud computing platform is a new advanced environment using in various fields that rely on large databases. However, this cloud was encountered under cyber-attacks and security breaches, what requires to find a protection system. So, the Blockchain technology within cloud computing environments to demonstrate advantages and disadvantages in distributed data and protection was talking about in this paper. The discussion of the latest studies of BC confirms that the trend in most of them was within two main aspects. The first one is the provenance and integrity and privacy in the data sets. The second aspect is the data security. It was suggested to address this problem, as well as indicating the time of process for the system to retrieve data. At the end, it was concluded that the Blockchain system is important in securing the provenance data and privacy of distributed and sharing data from penetration and manipulation. It is useful to say that proposed system can apply in many fields of medicine, business, management, etc.

This paper is a review of Blockchain architecture and its applications in cloud computing. It also illustrates the challenges and issues that are facing BC. Finally, the aspects of BC security and the advantages of integrating BC with cloud computing are discussed. Also, brief survey of recent research trends of the inclusion of Blockchain in cloud computing environments is given.

In future work, we will discuss in more details the challenges and issues of the applications of BC technology and cloud computing, and we will present some solutions to these issues. Also, we will apply the applications mentioned in this paper in a real world.

Table 1 Blockchain in cloud computing through studies

References	Aims	Design-structure	Evaluation
[10]	Proposed system for the share medical data between cloud service providers by Blockchain, called "MeDShare"	Include four layers • User layer • Data query layer • Data structuring provenance layer • Existing database infrastructure layer	-MeDShare can detect vulnerabilities to keep security of data -The efficiency of detecting vulnerabilities may be reduced if increased number of request data -Data access time effected by number of request data
[2]	Proposed cloud computing system based on Blockchain to use and share medical data in a safety way called "BBDS ecosystem"	Three entities • User • System management • Storage	BBDS system supports large amounts of data every second without worrying about scalability
[12]	Proposed "ProvChain" system to provide data protection from tampering, and enhance the privacy of source data in the cloud	ProvChain consisted of Cloud User, Cloud Service Provider (CSP), Provenance Database, Provenance Auditor (PA), Blockchain network	Provides real-time auditing of data access in the cloud and record it adopted decentralized to ensures data integrity
[14]	Identify secure database requirements in cloud computing environments, based on the real needs of the SUNFISH European project	The SUNFISH European project	Emphasizes the importance to rely on the Blockchain as a reliable infrastructure to store data in cloud computing environments • Proposed Blockchain-based data management

References

1. Park JH, Park JH (2017) Blockchain security in cloud computing: use cases, challenges, and solutions. Symmetry 9(8):164, 13 p. https://doi.org/10.3390/sym9080164
2. Xia Q, Sifah E, Smahi A, Amofa S, Zhang X (2017) BBDS: blockchain-based data sharing for electronic medical records in cloud environments. Information 8(2):44, 16 p. https://doi.org/10.3390/info8020044
3. Sharma PK, Chen M-Y, Park JH (2018) A software defined fog node based distributed Blockchain cloud architecture for IoT. IEEE Access 6:115–124. https://doi.org/10.1109/access.2017.2757955
4. Zhang Y, Deng R, Liu X, Zheng D (2018) Blockchain based efficient and robust fair payment for outsourcing services in cloud computing. Inf Sci 462:262–277. https://doi.org/10.1016/j.ins.2018.06.018
5. Tosh D, Shetty S, Liang X, Kamhoua C, Njilla L (2017) Consensus protocols for blockchain-based data provenance: challenges and opportunities. In: 2017 IEEE 8th annual ubiquitous

computing, electronics and mobile communication conference, UEMCON.https://doi.org/10.1109/uemcon.2017.8249088
6. Zhang Y, Deng R, Liu X, Zheng D (2018) Outsourcing service fair payment based on blockchain and its applications in cloud computing. IEEE Trans Serv Comput:1–1. https://doi.org/10.1109/tsc.2018.2864191
7. Tama B, Kweka B, Park Y, Rhee K (2017) A critical review of blockchain and its current applications. In: 2017 International conference on electrical engineering and computer science, ICECOS, pp 109–113. https://doi.org/10.1109/icecos.2017.8167115
8. Aniello L, Baldoni R, Gaetani E, Lombardi F, Margheri A, Sassone V (2017) A prototype evaluation of a Tamper-resistant high performance blockchain-based transaction log for a distributed database. In: 13th European dependable computing conference, EDCC, pp 151–154. https://doi.org/10.1109/edcc.2017.31
9. Singh L, Lee S (2018) Comparative requirements analysis for the feasibility of blockchain for secure cloud. In: Communications in computer and information science, pp 57–72. https://doi.org/10.1007/978-981-10-7796-8_5
10. Xia Q, Sifah E, Asamoah K, Gao J, Du X, Guizani M (2017) MeDShare: trust-less medical data sharing among cloud service providers via blockchain. IEEE Access 5:14757–14767. https://doi.org/10.1109/access.2017.2730843
11. Tosh D, Shetty S, Liang X, Kamhoua C, Kwiat K, Njilla L (2017) Security implications of blockchain cloud with analysis of block withholding attack. In: 17th IEEE/ACM international symposium on cluster, cloud and grid computing, CCGRID.https://doi.org/10.1109/ccgrid.2017.111
12. Liang X, Shetty S, Tosh D, Kamhoua C, Kwiat K, Njilla L (2017) ProvChain: a blockchain-based data provenance architecture in cloud environment with enhanced privacy and availability. In: 17th IEEE/ACM international symposium on cluster, cloud and grid computing, CCGRID, Spain, pp 468–477. https://doi.org/10.1109/ccgrid.2017.8
13. Li Z, Barenji AV, Huang GQ (2018) Toward a Blockchain cloud manufacturing system as a peer to peer distributed network platform. Robot Comput Integr Manuf 54:133–144. https://doi.org/10.1016/j.rcim.2018.05.011
14. Gaetani E, Aniello L, Baldoni R, Lombardi F, Margheri A, Sassone V (2017) Blockchain-based database to ensure data integrity in cloud computing environments. In: Italian conference on cybersecurity, Italy
15. Lin IC, Liao TC (2017) A survey of blockchain security issues and challenges. Int J Netw Secur 19(5):653–659. https://doi.org/10.6633/IJNS.201709.19(5).01
16. Zhu L, Wu Y, Gai K, Choo KKR (2019) Controllable and trustworthy blockchain-based cloud data management. Future Gener Comput Syst 91:525–537. https://doi.org/10.1016/j.future.2018.09.019

Micro-energy Systems (MESs); Modeling and Optimal Operating Based on NSGA-II

Mostafa Al-Gabalawy, Ahmed R. Adly, Almoataz Y. Abdelaziz, and Nesreen S. Hosny

Abstract In this paper, an optimization tool for Micro-energy systems (MESs) is designed and implemented in a modular and scalable architecture. A flexible modeling tool suitable for optimizing a microgrid setup under diverse conditions is developed. In addition to modeling of the small-scale generation or storage unit, where the modeling tool reflects the trade between connected entities as well as behavior such as load shifting and curtailment. It is then applied to a case study set in a village in North Rhine-Westphalia which is due to be destroyed by lignite mining. The case study furthermore indicates the potential of a decentralized energy system, especially as the investment costs of the applied technologies further decrease, and 'energy quality' factors, such as CO_2 emissions are becoming increasingly more relevant. Finally, the objective of the model will be to meet the demand of all connected entities at a minimum cost. Moreover, with these functionalities at a small solving time, the modeling tool should meet the criteria of being scalable, flexible in regard to the form and quality of input data and as user-friendly as possible. At the last, a multi-objective optimization has been executed based on the non-dominated sorting genetic algorithm-II (NSGA-II).

Keywords Micro-energy systems · Hybrid energy systems · PV · Storage batter · Fuel cell · Modeling · Optimal operating · Multi-objective optimization

1 Introduction

To achieve the highly ambitious goal of the Paris Agreement to keep global temperature increase 'well below two degrees Celsius' [1] in this century, an unparalleled

M. Al-Gabalawy · N. S. Hosny (✉)
Power Engineering and Automatic Control Department, Pyramids Higher Institute for Engineering and Technology, Giza, Egypt

A. R. Adly
Nuclear Research Center, Atomic Energy Authority, New Cairo, Egypt

A. Y. Abdelaziz
Faculty of Engineering and Technology, Future University, New Cairo, Egypt

D. A. Magdi et al. (eds.), *Digital Transformation Technology*, Lecture Notes in Networks and Systems 224, https://doi.org/10.1007/978-981-16-2275-5_6

transformation of the energy sector is required. Germany sees itself as a leader in this effort, as one of the first countries to encourage a large-scale shift of electricity generation towards renewable sources. However, the country is presently set to miss its 2020 goals for CO_2 mitigation [2]. While Germany is phasing out relatively CO_2-neutral nuclear power [3], over 30% of electricity production still comes from coal and lignite plants, which are expected to remain running until 2035. Experts claim a coal exit by 2030 is necessary for the Paris goals to remain attainable [4], and a severe reduction necessary to achieve the national goals for CO_2-mitigation already set [5]. Despite this, the continually falling costs for small-scale energy generation equipment, such as photovoltaic (PV) panels and fuel cells, and the rising cost of electricity have spawned a new movement favoring decentralized energy generation. This trend is further propelled by ever more affordable energy storage devices, which enable solar energy to be used throughout the night. The concept of a Microgrid or micro-energy system proposes to take the idea of decentralization a step further by creating local autonomous grids that only interact with the main grid if necessary or advantageous. A Microgrid is defined as a small geographically bounded zone with clear electrical boundaries, which manages local loads and possibly contains generation units and storage. Furthermore, it can be connected to the grid, from whose perspective it is seen as a single entity.

Lázár et al. employ both a Mixed Integer Linear Programming (MILP) model and a Genetic Algorithm (GA) and come to the conclusion that while both deliver accurate and robust results the MILP model is faster [6]. Nemati et al. mostly concur. While in this case, the results of the GA were better in two scenarios, it was outperformed by the MILP model in the remaining three. However, a key problem of MILP seems to be its deterministic nature, which anticipates perfect knowledge of all the parameters involved. Especially for optimization problems concerned with short timeframes, such as electricity dispatch this is a significant problem, since actual parameters may be different [7]. Several methods addressing this problem; one such method is rolling time horizons. This approach optimizes the dispatch of a Microgrid for a fixed time horizon based on steadily updated forecasts of the uncertain parameters. The optimization is repeated periodically to reflect the updated forecasts and increase dispatch accuracy in the nearer future [8, 9]. But rolling horizon optimization is unfit to optimize investment since it is not possible to adapt investment decisions ex-post to changed conditions. While the rolling time horizon method helps to limit uncertainty by reacting to changes of input parameters, Robust Optimization and Stochastic Optimization try to proactively account for a variety of possible scenarios. Robust Optimization achieves this by optimizing for a number of scenarios deemed equally likely, as in [10] using ensemble weather forecasts or in [11, 12] using upper and lower boundaries for uncertain parameters. Stochastic Optimization on the other hand uses detailed probability distributions to weigh the probability of each scenario occurring as explained in [13]. To arrive at these distributions, secondary tools are usually needed. Shams et al. use a simple Gaussian randomization to make their demand data reflect uncertainty, as well as more specific distributions for irradiation and wind speeds [14]. A number of other methodologies appear in literature, such as employing a Monte Carlo simulation [15] or deriving multiple scenarios and corresponding realization probabilities from historical data [16]. The scope of the microgrid models found in literature varies greatly. A constant is the definition of a Microgrid as bounded, operating in a small geographical zone and with clear electrical boundaries [17–19]; managing local loads [20, 21]; possibly containing various generation units and storage [9, 22] and possibly being able to exchange power with the main grid [7, 23], from whose perspective it is seen as a single entity [24]. While most of the selected literature considers cases where a grid connection exists and can be used at all times, there are several publications, which consider either completely islanded Microgrid [8, 21, 25, 26] or Microgrid that can sustain themselves in islanded mode for extended periods for example in the case of natural disasters [17]. The clearest distinctions between Microgrid models, after the object and applied methodology is properly defined, is based on the aspect or aspects that the model is supposed to optimize. The literature can be split in two groups, optimizing either dispatch only or both investment (e.g., the planning phase) and dispatch. The former group is definitely the larger, with only 12 out of 61 publications considering investment. It is notable, that only one of these, Moshi et al., uses any of the methods to model uncertainty discussed above. There is also significant diversity in literature when it comes to the technologies considered in modeling [13]. Modeling dispatchable (i.e. non-renewable and biofuels) as well as non-dispatchable (i.e. renewable) generation is rather common, but the specifics differ: While most models consider PV or wind as well as CHP generators, the included generation technologies are as diverse as geothermal generators [6] and gas turbines [7, 27]. In addition to modeling electricity, some publications also consider heat generation and transmission [28, 29]. This is valuable, because the economic performance of a nonrenewable generation unit (usually CHP) present in most models depends heavily on whether and how the heat is used, as in [30]. The types of storage used in literature also vary. Although battery storage [27, 31, 32] or an abstract storage device [33, 34] are the most common choices, there are a number of papers modeling heat storage [12, 20, 28, 29], and some with more exotic technology choices such as flywheels [34] or an electrolyzer [35]. In addition to these generation and storage technologies, some publications also model the network topology [14, 18, 36] and some even consider electrical phenomena such as active and reactive power losses [22, 37] and voltage deviation [22]. Demandside management, which is implemented in some models, is usually represented by dividing loads into different categories. Chen et al. distinguish between' critical loads', meaning loads that absolutely have to be satisfied, 'shiftable loads', meaning loads that have to be satisfied, although there is a time window, rather than an exact point in which they can be serviced, and 'adjustable' loads, which can be dropped if needed [36]. Although the terms may vary, and many publications do not introduce the 'shiftable loads' category altogether, these conceptual distinctions are made by a number of other authors, such as in [9, 12, 38]. The model results depend almost as much on the given input parameters used in case studies as on the modeling approach itself. There are, however, only a limited number of publications with detailed documentation of the parameters used. The parameters can be broadly categorized as either economic parameters, such as capital and operation and maintenance cost for the technologies used, generation-related parameters such as irradiation and wind speeds, and load data representing consumer behavior. In publications documenting cost parameter there is a clear trend towards a split of technology cost into investment and maintenance cost. However, whereas Lauinger et al. define maintenance cost as a fixed cost per unit of time, Wouters et al. define it as a function of kWh produced, a measurement which Lauinger et al. call 'fuel costs' [28, 29]. Atia and Yamada choose a definition closer aligned with the former, but they define yearly maintenance cost simply as a flat percentage of the units' investment cost [31]. Koltsaklis et al. seem to neglect maintenance costs altogether, focusing only on investment costs [24]. While some models document economic parameters such as discount rates [31], many don't and it is unclear if future flows of value are discounted in these models. Furthermore, many of the publications declare cost parameters as assumptions rather than trying to derive them from sources, although a few such as in [28, 29]. The selected literature offers multiple ways of obtaining input parameters of meteorological data. Henao-Muñoz et al. use historical meteorological data [39], while Palma-Behnke et al. build a prediction model for forecasting irradiation and wind speeds [8]. Similarly, there are some authors who use historical data for load parameters such as in [39] and others who generate synthetic data based on appliance use and probabilistic models [12, 20, 34] or neural networks using empirical data [8]. When it comes to modeling tools, Distributed Energy Resources–Customer Adoption Model (DER-CAM) [15, 18, 28] and HOMER [15, 28, 40] are the most often referenced. DER-CAM or Distributed Energy Resources Customer Adoption Model is an optimization tool developed at Berkeley Labs that uses Mixed Integer Linear Programming to optimize portfolio, placement, sizing, and dispatch of Microgrid Energy Systems [41]. HOMER Grid is a commercial optimization tool for behind-the-meter systems. Its main advantage is its large database of components and tariff rates in the United States and Canada [42]. The most often mentioned solver is the CPLEX commercial solver [7, 22, 43, 44]. Most of the models are written in GAMS [9, 10, 44], although many are written in MATLAB [45, 46].

The aim of this paper is to develop a flexible modeling tool suitable for optimizing a Microgrid setup under diverse conditions. In addition to modeling any kind of small-scale generation or storage unit, the modeling tool should be able to reflect trade between connected entities as well as behavior such as load shifting and curtailment. The objective of the model will be to meet the demand of all connected entities at a minimum cost. In addition to these functionalities at a small solving time, the modeling tool should meet the criteria of being scalable, flexible in regard to the form and quality of input data and as user-friendly as possible. To demonstrate the modeling tool performance, a case study will be conducted for a small community in the German state of North Rhine-Westphalia, where also the majority of the German lignite capacity is located. The robustness of the results will be confirmed through a sensitivity analysis. It is expected that the case study would show whether such a system is currently commercially viable in North Rhine-Westphalia and, to a lesser extent, if decentralized energy management in the form of Microgrid is competitive compared to the prevailing centralized solution.

A literature review has been represented in the above section to reflect the current state of Microgrid research. Then the model constructed will be described in mathematical terms, followed by a brief overview of the technical implementation. Finally, a case study including a sensitivity analysis is conducted, before conclusions are drawn from its result.

2 Mathematical Formulation

The range of scenarios and complexity the model can handle is, among other things, dependent on its implementation of discrete time. More and smaller time-steps and a representative sample of diverse expectable conditions regarding the parameters lead to higher model accuracy. Of course, the number of time-steps is limited by processing power and therefore by model complexity. To enable both diverse and therefore potentially discontinuous parameter sets and a reasonable length for each set, the length of 1 h is chosen for the basic time-step $t; t \in [1, T]; T \in \mathbb{N};$. T denotes the number of hours in each discontinuous set of supply and demand parameters. Furthermore, the ability of including multiple discontinuous sets of supply and demand parameters of the same length in one scenario is gained by introducing a second iterator $s; s \in [1, S]; S \in \mathbb{N};$. S denotes the number of discontinuous sets of supply and demand parameters. In addition, a vector $SW \in \mathbb{N}^S$ can be defined, which will be used by the model for weighing the costs incurred in each season S differently. The vector will be normalized to S to not interfere with the scaling of the results. This normalized weight vector is called SWN.

To properly conduct trading and for the future possibility of adding voltage and topological constraints to the model, the households are modeled as independent entities, each with its own demand, generation, and storage. Therefore, a third iterator $u; u \in [1, U]; U \in \mathbb{N};$ is introduced. U denotes the number of households modeled. An abstract household H is defined as a tuple of storage and generation capacities as well as demand curves. For discrete storage and generation devices several sets of separate sets of positive integers define how much of each option available in the scenario was installed. Additionally, each household has a price shift-P and curt-P at which they are willing to shift or curtail a unit of their load respectively.

$$H := (GEN, ST, dGEN, dST, DEM, \text{ shift-P, curt-P}) \tag{1}$$

A concrete household H_u implements the values of; $GEN \in \mathbb{R}^K; GEN_k \geq 0 \forall k; k \in [0, K]$, $ST \in \mathbb{R}^L; ST_l \geq 0 \forall l; l \in [0, L]$, $dGEN \in \mathbb{N}^M$, and $dST \in \mathbb{N}^N$. With, $k, l, K, L, M, N \in \mathbb{N}$, and shift-P and curt-P are greater than 0. K and L denote the amount of generation and storage devices in household H_u.This, in practice, is equal to the amount of linear scaling investment options for each category, since a zero capacity device is still modeled as a device. M and N denote the amount of discrete investment options in their respective category. GEN_k and ST_l are the capacities installed of the kth and lth linear investment option respectively. GEN_m

and dST_n are the amount installed of the mth and nth discrete investment option available. DEM is further defined in the subsection demand. A list of instances of the structure H and its length U are both model parameters.

The objective of the model is to find optimal investment, dispatch, and trade. Therefore, the following sets of internal parameters are introduced $\forall u, s, t$:

$$\text{to}TR_{u,s,t},\ \text{from}TR_{u,s,t} \geq 0 \tag{2}$$

The trade supply variables $\text{to}TR_{u,s,t}$ and $\text{from}TR_{u,s,t}$ reflect the amount of traded power to the trading pool and from the trading pool respectively. They are defined for each household at each time-step $\forall u, k, m, s, t$.

$$genS_{u,k,s,t}, dgenS_{u,m,s,t} \geq 0 \tag{3}$$

The generation supply variables $genS_{u,k,s,t}$ and $dgenS_{u,m,s,t}$ describe the amount of energy produced by each linear or discrete generation option respectively. They are defined for each generation option in each household and at each time-step. It is useful to keep in mind that not every household needs to implement all the generation options. In case of zero instances of the discrete generation option $dGEN_2$ installed in household H_5, for example, $\text{dgen}S_{5,2,s,t} = 0$ for all s and t.

$$\text{from}ST_{u,l,s,t}, \text{to}ST_{u,l,s,t}, \text{from}DST_{u,n,s,t}, \text{to}DST_{u,n,s,t} \geq 0 \tag{4}$$

The storage supply variables $\text{from}ST_{u,l,s,t}$ and $\text{to}ST_{u,l,s,t}$ indicate the amount of power fed into or withdrawn from a linear storage investment option. They are defined for each linear investment option and each household at each time-step. The variables $\text{from}DST_{u,n,s,t}$ and $\text{to}DST_{u,n,s,t}$ have the equivalent function for all discrete storage investment options.

$$\text{to}SC_{u,s,t}, \text{from}SC_{u,s,t}, \text{ and } ncS_{u,s,t} \geq 0 \tag{5}$$

The shifted consumption supply variable $\text{to}SC_{u,s,t}$ represents the amount of power consumption shifted by each household at each time-step. $\text{from}SC_{u,s,t}$ indicates the amount of power consumption that has previously been shifted and is now consumed. The non-consumption supply variable $ncS_{u,s,t}$ designates the amount of power consumption curtailed for each household at each time-step. $\text{From}GR_{u,s,t}$ and $\text{to}GR_{u,s,t} \geq 0$. Finally, the grid supply variables $\text{from}SC_{u,s,t}$ and $\text{to}GR_{u,s,t}$ denote the amount of power supplied by and to the grid respectively by each household at each time-step. Only power flowing out of and into the Microgrid is included in this definition, power traded internally, while using the Microgrid is reflected in the trade variables; $H_u \rightarrow GEN_k \geq 0$ and $H_u \rightarrow ST_l \geq 0$, while, $H_u \rightarrow dGEN_m$ and $H_u \rightarrow dST_n \in \mathbb{N}$. Where, $H_u \rightarrow GEN_k$, $H_u \rightarrow ST_l \geq 0$, $H_u \rightarrow dGEN_m$, $H_u \rightarrow dST_n$ are variables describing the realized capacities of each linear and discrete investment option in each household.

For a model that is explicitly built to optimize decentralized generation in Micro-grid a robust and flexible mathematical representation of generation options is required. Therefore, an abstract linear generation device GEN and an abstract discrete generation device $dGEN$, which can represent a range of technology options, are defined as

$$GEN := \left(C_{\text{Cap}}, CopFix, C_{\text{OPVar}}, T_{\text{Life}}, EFF_{\text{el}}, PR, \max Fl\right) \tag{6}$$

$$dGEN := \left(CAP, C_{\text{Cap}}, C_{\text{OpFix}}, C_{\text{OpVar}}, T_{\text{Life}}, EFF_{\text{el}}, PR, \max Fl\right) \tag{7}$$

$CAP \geq 0$, which is only required for discrete investments, defines the generation capacity of the discrete investment option. $C_{\text{Cap}} \geq 0$ defines the total investment cost for discrete generation devices. For linear generation devices, the investment cost is expressed in terms of one unit of capacity. $C_{\text{OpFix}} \geq 0$ expresses the fixed maintenance cost per year. In the case of linear generation, they are expressed in proportion to capacity, while for discrete generation they express the maintenance cost of one discrete device. $C_{\text{OpVar}} \geq 0$ expresses the variable maintenance cost per unit of input consumed. $T_{\text{Life}} \in \mathbb{N}$ defines the life expectancy of a generation device, i.e. the time before it is replaced. $EFF_{\text{el}} \in [0, 1]$ defines the ratio of supplied energy, for example, sunlight or gas, to produce electricity. $PR \in [0, 1]$ is another definable penalty to electricity production, which can, for example, express the degradation of solar cells.;max $Fl \in \mathbb{R}^{S \times T}$ and max $Fl \in \mathbb{R}^{S \times T} \geq 0 \forall s, t$ denotes the maximal possible flow of input energy, e.g., sunlight in case of solar PV or gas in case of a fuel cell. This enables the representation of intermittent availability for example for PV installations. It could nevertheless be used to model any condition that would restrict a conventional generation device from running at capacity at certain times, for example, air quality regulation. All instances of the GEN structure are collected in one list. The same is true for all instances of the $dGEN$ structure. These lists and their sizes K and M are parameters. The model uses the iterators k and m to represent individual instances of these structures. In addition, to model generation appropriately, several sets of constraints need to be defined as in Eqs. 8 and 9.

$$\begin{aligned} genS_{u,k,s,t} \leq \min\left(GEN_k \rightarrow EFF_{el} * GEN_k \rightarrow \max Fl_{s,t}, 1\right) \\ * H_u \rightarrow GEN_k * GEN_k \rightarrow PR \end{aligned} \tag{8}$$

$$\begin{aligned} dgenS_{u,m,s,t} \leq \min\left(GEN_m \rightarrow EFF_{el} * GEN_m \rightarrow \max Fl_{s,t}, 1\right) \\ * H_u \rightarrow dGEN_m * dGEN_m \rightarrow CAP * dGEN_k \rightarrow PR \end{aligned} \tag{9}$$

The capacity constraints, Eqs. 8 and 9 restrict the use of any linear generation device GEN_k or any discrete generation device $dGEN_m$ to its capacity rating and the maximum flow of input energy supplied to it. The constraints differ due to the capacity of linear generation devices being stored in the household structure directly, while for discrete generation devices the household structure only contains an integer representing the amount implemented. This value is then multiplied with the capacity of one unit of the concerned discrete generation device to compute the total installed capacity. In both cases, the result is multiplied with the performance ratio PR of the concerned device, which acts as a penalty to capacity. The minimized term calculates the ratio of maximum output achievable with the available maximum input to the capacity of the device. If the input suffices for running at capacity, the term returns a maximum of 1, not restricting supply further. Otherwise, it restricts capacity to the maximum possible under current resource input.

The introduction of the concept of energy storage is necessary if the use of a large proportion of intermittent power production while maintaining intermittent power demand is to be seriously explored. Therefore, an abstract linear storage device ST and an abstract discrete storage device dST, which can represent a range of technology options, are defined:

$$ST := \left(C_{\text{Cap}}, T_{\text{Life}}, EFF_{+-}, \max P_{+}, \max P_{-}\right) \tag{10}$$

$$dST := \left(CAP, C_{Cap}, T_{Life}, EFF_{+-}, max\, P_{+}, max\, P_{-}\right) \tag{11}$$

$CAP \geq 0$, which is only required for discrete investments, defines the usable capacity of the discrete investment option. $C_{\text{Cap}} \geq 0$ defines the total investment cost for discrete storage devices. For linear storage investment options, the investment cost is expressed in terms of one unit of capacity. $T_{\text{Life}} \in \mathbb{N}$ defines the life expectancy of a storage device, i.e. the time before it is replaced. To be able to better estimate this value it is assumed, that storage is, on average, cycled once per day. $EFF_{+-} \in [0, 1]$ defines the round trip efficiency, i.e., what proportion of one unit of power is left after charging and discharging inefficiencies. It is used to compute the charge and discharge efficiencies. The charge efficiency EFF_{+} is defined as $EFF_{+} = \sqrt{EFF_{+-}}$ and the discharge efficiency EFF_{-} as $EFF_{-} = 1/EFF_{+}$. While, $\max P_{+} \geq 0$ and $\max P_{-} \in [0, 1]$ denote the maximum charge rate and discharge rate respectively, relative to devices capacity. All instances of the ST structure are collected in one list. The same applies for all instances of dST. The two lists and their sizes L and N are model parameters. The model uses the iterators l and n to represent individual instances of these structures. To model storage appropriately several sets of constraints need to be specified as $\forall u, l, s, t$;

$$\text{to}ST_{u,l,s,t} \leq ST_{l} \rightarrow \max P_{+} * H_{u} \rightarrow ST_{l} \tag{12}$$

$$\text{from}ST_{u,l,s,t} \leq ST_{l} \rightarrow \max P_{-} * H_{u} \rightarrow ST_{l} \tag{13}$$

The flow constraints for linear storage, Eqs. 12 and 13 constrain the maximum in- and outflow for each linear storage device to its maximum in- and outflow rate relative to capacity $\forall u, n, s, t$;

$$\text{to}DST_{u,n,s,t} \leq dST_{n} \rightarrow \max P_{+} * H_{u} \rightarrow dST_{n} \tag{14}$$

$$\text{from } DST_{u,n,s,t} \leq dST_n \rightarrow \max P_- * H_u \rightarrow dST_n \tag{15}$$

The flow constraints for discrete storage, Eqs. 14 and 15 implement the same functionality as the linear flow constraints for discrete storage devices $\forall u, l, s$.

The model assumes the storage to be full at the start of each simulated discontinuous period s and requires it to be full at the end of each period. This means inflows minus outflows weighted by the respective efficiencies need to sum up to zero for the entirety of each period s as in Eqs. 16 and 17.

$$\sum_{t=1}^{T} \left(ST_l \rightarrow EFF_+ * \text{to}ST_{u,l,s,t'} - ST_l \rightarrow EFF_- * \text{from}ST_{u,l,s,t'} \right) = 0 \tag{16}$$

$$\sum_{t=1}^{T} \left(dST_n \rightarrow EFF_+ * \text{to}DST_{u,n,s,t'} - dST_n \rightarrow EFF_- * \text{from}DST_{u,n,s,t'} \right) = 0 \tag{17}$$

The storage capacity constraints, Eqs. 18 and 19, prohibit storage usage beyond the storage device's capacity. The remaining storage capacity at time (s, t) is in theory calculated as the capacity minus current storage level, which is the sum of all inflows and outflows weighted by in- and outflow efficiency.

$$\begin{aligned} &\text{to}ST_{u,l,s,t} * ST_l \rightarrow EFF_+ \\ &\leq \sum_{t'=1}^{t-1} \left(-ST_l \rightarrow EFF_+ * \text{to}ST_{u,l,s,t'} + ST_l \rightarrow EFF_- * \text{from}ST_{u,l,s,t'} \right) \end{aligned} \tag{18}$$

$$\begin{aligned} &\text{to}DST_{u,n,s,t} * dST_n \rightarrow EFF_+ \\ &\leq \sum_{t'=1}^{t-1} \left(dST_n \rightarrow EFF_+ * \text{to}DST_{u,n,s,t'} - dST_n \rightarrow EFF_- * \text{from}DST_{u,n,s,t'} \right) \end{aligned} \tag{19}$$

Since the storage at each time-step $(s, 0)$ is required to be full, the capacity is subtracted once, which eliminates it and leaves only the sum. The constraint is implemented equivalently for discrete storage. Finally, the storage level constraints, Eqs. 20 and 21 prohibit the storage level to fall below zero. This means there can be no more power withdrawn, weighted by outflow efficiency than is currently stored in the device. Because it is assumed that the storage is full at the start of each period s, the current charge amounts to one full charge (storage capacity) plus the sum of all transactions during the current period up to the current time-step t.

$$\text{from}ST_{u,l,s,t} * ST_l \rightarrow EFF_- \leq H_u \rightarrow ST_l$$

$$+\sum_{t'=1}^{t-1}\left(ST_l \rightarrow EFF_+ * \text{to}ST_{u,l,s,t'} - ST_l \rightarrow EFF_- * \text{from}ST_{u,l,s,t'}\right) \tag{20}$$

$$\text{from}DST_{u,n,s,t} * dST_n \rightarrow EFF_- \le H_u \rightarrow dST_n * dST_n \rightarrow CAP$$
$$+\sum_{t'=1}^{t-1}\left(dST_n \rightarrow EFF_+ * \text{to}DST_{u,n,s,t'} - dST_n \rightarrow EFF_- * \text{from}DST_{u,n,s,t'}\right) \tag{21}$$

Trade is implemented as an exchange platform, which sums up deposits and withdrawals of power at each time-step and always needs to be balanced (22). This approach has the advantage of greatly reducing the number of constraints and therefore complexity, although it is not very well suited to handle grid topology constraints, which might be added in the future. The trade constraint is defined as:

$$\sum_{u=1}^{U}\left(\text{to}TR_{u,s,t} - \text{from}TR_{u,s,t}\right) = 0 \tag{22}$$

Demand Abstract demand data DEM is defined as a set of the size S of demand profile sets of the size T:

$$\begin{aligned} DEM &:= \{DEM_s | s \in [1, S]; s \in \mathbb{N}; \forall DEM_s, \\ DEM_s &:= \left\{DEM_{s,t} \mid t \in [0, T]; t \in \mathbb{N}\right\} \end{aligned} \tag{23}$$

As frequently applied in literature, $DEM_{s,t}$ is not a demand value, but rather a tuple of three demand values, critical demand, shift-able demand, and curtail-able demand

$$DEM_{s,t} := \left(\text{crit}DEM_{s,t}, \text{shiftDEM}_{s,t}, \text{curt}DEM_{s,t}\right) \tag{24}$$

And $\forall u, s, t$:

$$H_u \rightarrow \text{crit}DEM_{s,t}, H_u \rightarrow \text{shift}DEM_{s,t}, H_u \rightarrow \text{curt}DEMs, t \ge 0 \tag{25}$$

Critical demand will be met by the model as it is seen as a constraint. Shift-able demand can be shifted by one hour at a fixed rate defined for each household. Curtail-able demand can be dropped at a fixed rate defined for each household. The restrictions on shift-able and curtail-able demand are expressed by the following sets of constraints $\forall u, s, t$. The shifted demand limit constraint, Eq. 26 restricts the amount of shift-able demand to the shift-able part of household demand for the current time-step. To this, the demand shifted in previous time-steps, Eq. 28, which can be shifted further, is added. The curtail-able demand limit constraint, Eq. 27 restricts the amount of curtailable demand to the curtailable part of household demand for the current time-step $\forall u, s; t \in [2, T]$.

$$\text{to}SC_{u,s,t} \leq H_u \rightarrow \text{shift}DEM_{s,t} + \text{from}SC_{u,s,t} \tag{26}$$

$$ncS_{u,s,t} \leq H_u \rightarrow \text{curt}DEM_{s,t} \tag{27}$$

$$\text{from}SC_{u,s,t} = \text{to}SC_{u,s,t-1} \tag{28}$$

Demand shifted in the previous time-step has to be consumed in the current time-step or shifted further. This is ensured by Eq. 28. It does not apply for the first time-step, but the model still does not set from$SC_{u,s,0}$ to greater than 0, because it would raise cost $\forall u, s$. Equation 29 ensures that there is no consumption shifted at the last time-step of each season as it would not have to be consumed at all.

$$\text{to}SC_{u,s,T} = 0 \tag{29}$$

The modeled Microgrid is connected to the grid via a common point of coupling. The grid is defined as:

$$\begin{aligned}
&\text{GRID} := (\max S, \max D, \text{grid}C, \text{feed}C) \\
&\text{With } \max S, \max D, \text{grid}C, \text{feed}C \in \mathbb{R}^{S\times T}, \\
&\max S_{s,t}, \max_{s,t}, \text{grid}C_{s,t}, \text{and feed}C_{s,t} \geq 0
\end{aligned} \tag{30}$$

where, max S describes the maximum amount of power supplied by the grid at each time-step (s, t), max D describes the maximum amount power that can be fed into the grid at time-step (s, t). While, grid$C_{s,t}$ expresses the price at which this power can be bought, and feed$C_{s,t}$ expresses the price at which the grid buys this power. And to model the expected behavior of the grid several sets of constraints need to be introduced:

$$\sum_{u=1}^{U} \left(\text{from}GR_{u,s,t} - \text{to}GR_{u,s,t}\right) \leq \max Ss, t \tag{31}$$

The first grid constraint, Eq. 31 restricts the sum of each households 'trade balance' with the grid to the maximum power the grid can supply at each time-step. The second grid constraint, Eq. 32 does the reverse, in that it restricts the sum of each households 'trade balance' with the grid to the maximum power the Microgrid can feed into the grid at each time-step.

$$\sum_{u=1}^{U} \left(\text{to}GR_{u,s,t} - \text{from}GR_{u,s,t}\right) \leq \max Ds, t \tag{32}$$

The power balance constraints describe the need for a balance between supply and demand at each node at each time-step:

$$H_u \rightarrow \text{crit}DEM_{s,t} + H_u \rightarrow \text{shift}DEM_{s,t} + H_u \rightarrow \text{curt}DEM_{s,t}$$
$$= \text{from}GR_{u,s,t} - \text{to}GR_{u,s,t} + \sum_{k=1}^{K} genS_{u,k,s,t}$$
$$+ \sum_{m=1}^{M} dgenS_{u,m,s,t} + \text{from}TR_{u,s,t} - \text{toT}R_{u,s,t} + \text{to}SC_{u,s,t}$$
$$- \text{from}SC_{u,s,t} + ncS_{u,s,t} + \sum_{l=1}^{L} \text{from}ST_{u,l,s,t} - \text{to}ST_{u,l,s,t}$$
$$+ \sum_{n=1}^{N} \text{from}DST_{u,n,s,t} - \text{to}DST_{u,n,s,t} \quad (33)$$

The power balance constraint Eq. 33 requires electricity supply and demand to be balanced for each household at each time-step. On the demand side, the three types of demand of the concerned household and the current time-step are summed up. The supply side is made up of several terms: $\text{from}GR_{u,s,t} - \text{to}GR_{u,s,t}$ expresses the momentary trade balance with the grid, while $\text{from}TR_{u,s,t} - \text{to}TR_{u,s,t}$ denotes the momentary trade balance with the microgrid's trading pool. $\text{from}SC_{u,s,t} - \text{to}SC_{u,s,t}$ is the balance of consumption shifted from the current time-step into the future and previously shifted demand realized now. $ncS_{u,s,t}$ reflects loads dropped in the current time-step. Finally, the power generated by all linear and discrete generation devices is summed up, as well as the balance of storage use for all linear and discrete storage devices.

2.1 Cost Minimization Formula

The cost minimization formula calculates the cost of all investment and dispatch decisions made:

$$\min C_{\text{total}} = C_{\text{Investment}} + C_{\text{Operation}} + C_{\text{Dispatch}} \quad (34)$$

The total cost can be broken down into investment, operation, and dispatch costs.

$$C_{\text{Investment}} = \sum_{u=1}^{U} \left[\sum_{k=1}^{K} \frac{GEN_k \rightarrow C_{\text{Cap}} * H_u \rightarrow GEN_k * S * T}{8760 * GEN_k \rightarrow T_{\text{Life}}} \right.$$

$$+\sum_{l=1}^{L}\frac{ST_l \rightarrow C_{Cap} * H_u \rightarrow ST_l * S * T}{8760 * ST_l \rightarrow T_{\text{Life}}}$$

$$+\sum_{m=1}^{M}\frac{dGEN_m \rightarrow C_{Cap} * H_u \rightarrow dGEN_m * S * T}{8760 * dGEN_m \rightarrow T_{\text{Life}}}$$

$$\left.+\sum_{n=1}^{N}\frac{dST_n \rightarrow C_{Cap} * H_u \rightarrow dST_n * S * T}{8760 * dST_n \rightarrow T_{\text{Life}}}\right] \quad (35)$$

The investment costs consist of all investment into generation and storage devices throughout the Microgrid. In the context of the model, this means that for linear investment options the installed capacity times the costs of capital per unit of capacity, and for discrete investment options the costs of capital for a single instance of a device multiplied with the number of instances installed. All values are multiplied by the total amount of time-steps $S * T$ and then divided by the entire lifetime of the device in hours. This is inaccurate to some degree as it resembles a leasing model with negligible interest rates rather than an investment, but is vastly easier to compute and very close to the truth for most of Europe and the US.

$$C_{\text{Operation}} = \sum_{u=1}^{U}\left[\sum_{k=1}^{K}\frac{GEN_k \rightarrow C_{\text{OpFix}} * H_u \rightarrow GEN_k * S * T}{8760}\right.$$

$$\left.+\sum_{m=1}^{M}\frac{dGEN_m \rightarrow C_{\text{opFix}} * H_u \rightarrow GEN_m * S * T}{8760}\right] \quad (36)$$

The operation costs are the sum of all fixed operation costs of all generation devices deployed. They are calculated for each investment option in each household by multiplying its fixed operation costs with the installed capacity (or pieces in the case of discrete investment options) and the total number of time-steps. This value is then divided by the length of a year in hours to adjust the yearly values to the timespan considered in the conducted optimization. The sum of all of these investment options are the total maintenance costs for generation in the Microgrid. Since there is no fixed maintenance cost defined for storage this value is equal to all the maintenance costs incurred.

$$C_{\text{Dispatch}} = \sum_{s=1}^{S} SWN_s * \sum_{t=1}^{T}\sum_{u=1}^{U}[\text{grid}C_{s,t} * \text{from}GR_{u,s,t} - \text{feed}C_{s,t} * \text{to}GR_{u,s,t}$$

$$+ H_u \rightarrow \text{curt}P * ncS_{u,s,t} + H_u \rightarrow \text{shift}P * scS_{u,s,t}$$

$$+\sum_{k=1}^{K}\left(GEN_k \rightarrow C_{\text{OpVar}} * genS_{u,k,s,t}\right)$$

$$+\sum_{m=1}^{M}\left(dGEN_m \rightarrow C_{\text{OpVar}} * dgenS_{u,m,s,t}\right)\Bigg] \qquad (37)$$

The dispatch cost consists of the fuel costs (or variable operation costs C_{OPVar}) incurred by the use of all deployed generation devices as well as the value of all shifted and curtailed loads. In addition, all transactions with the main grid are valued at the power and feed-in rates for their respective time-step and summed up. The result in each season is then weighted according to the value of the normalized weight vector for that season SWN_s.

3 Proposed Algorithm

In the NSGA-II technique, the offspring population is produced firstly by using the parent population then the old and off-spring populations are combined together to form the total population. Then, the non-dominated criterion is utilized to sort the total population. The new population is composed of diverse non-dominated fronts. Firstly, the best non-dominated fronts are occupied then, the filling remains with solutions of the second non-dominated front, then the third, and so on, as shown in Fig. 1. All fronts which could not be accommodated are simply canceled. When the last allowed front is being considered, there may exist more solutions in the last front than the remaining slots in the new population. Instead of arbitrarily discarding some members from the last front, a niching strategy is used to choose the members from the last front, which reside in the least crowded region in the front. The algorithm guarantees that crowding (niching) will select a diverse set of solutions. When the

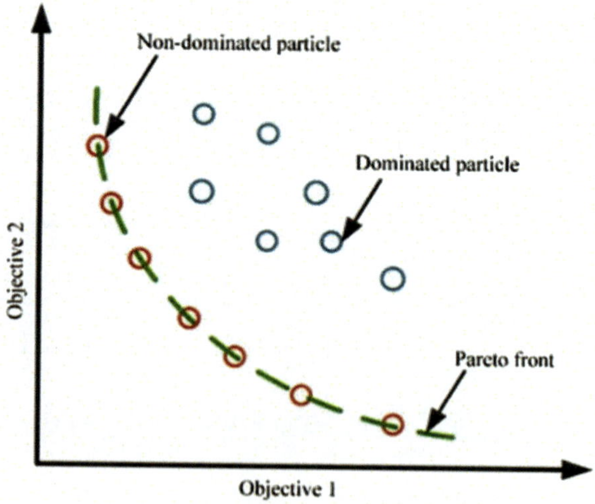

Fig. 1 Pareto front plot

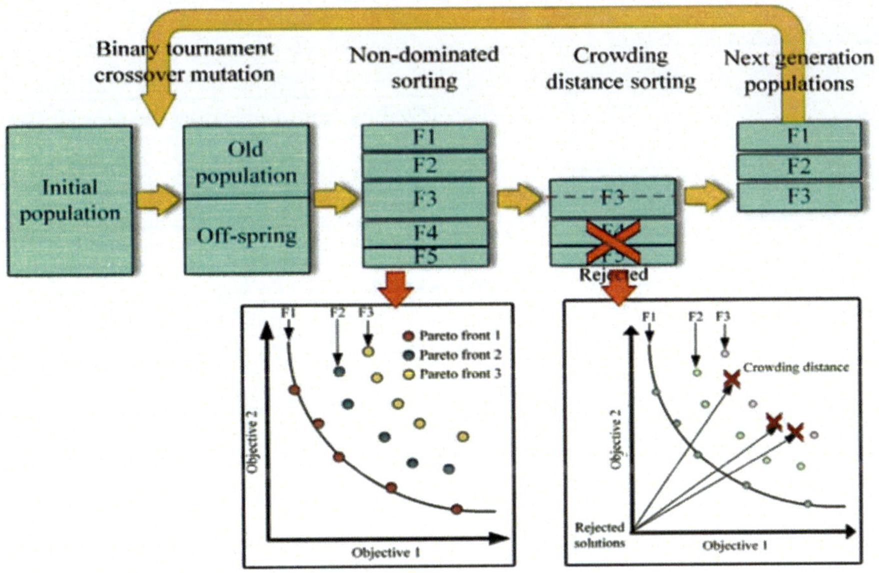

Fig. 2 Schematic procedure of NSGA-II

whole population converges to the Pareto-optimal front, the continuation of this algorithm will promise a better spread among the non-dominated solutions.

The binary tournament selection randomly selected two solutions from the current population then selects the best one with respect to the non-domination rank. The schematic procedure of NSGA-II is illustrated in Fig. 2 and the flowchart of NSGA-II is presented in Fig. 3.

Crowded distance is the mean distance between two solutions on either side of a particular solution along with each of the objectives. The crowded distance calculation is illustrated in Fig. 4 and the following steps are utilized to determine the crowded distance of every solution in the F_r set [41].

Step 1: Solutions are arranged in each objective domain.

Step 2: Crowded distances of the first solution and the last solution in the rank are selected as to infinity.

Step 3: For each of the other solutions, the crowded distance will be calculated in Eq. 38:

$$d_i = \sum_{m=1}^{M} \frac{f_m^{i+1} - f_m^{i-1}}{f_m^{\max} - f_m^{\min}}, i \in [2, j-1] \tag{38}$$

where, M, i, and j represent the objectives number, the number of the solution, and the total number of solutions in the set F_r, respectively. f_m^{i+1} and f_m^{i-1} represent the

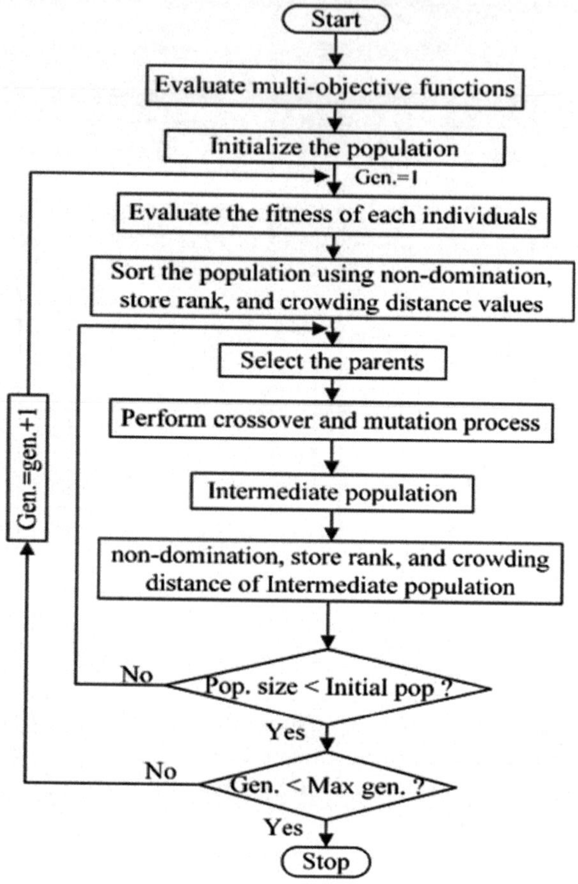

Fig. 3 Flowchart of NSGA-II [41]

mth objective functions of solution number $(i + 1)$ and $(I - 1)$ in the set F_r, respectively. $f_m^{\min}$ and $f_m^{\max}$ denote the minimum and maximum values of the mth objective function in the set F_r, respectively. After that, elitism is implemented by utilizing a crowded distance comparator to choose solutions from the last front to be passed over to the next generation. The crowded comparison operator monitors the selection process during the different stages of the algorithm in order to obtain a uniformly spread-out and produce a Pareto-optimal front based on two aspects; Non-domination rank (i_{rank}) and crowded distance (i_{distance}). A comparison is performed between two solutions with different non-domination ranks and if a solution has a better rank (i.e. $i_{\text{rank}} < j_{\text{rank}}$), it will be chosen. If both solutions have the same rank, then the solution with a higher crowded distance is chosen. i.e., ($i_{\text{rank}} = j_{\text{rank}}$) and ($i_{\text{distance}} > j_{\text{distance}}$).

The performance of a modeling tool depends as much on its implementation, as on mathematical; soundness. This section outlines the choices made implementing the model described in the previous section. The implementation of the modeling tool pursues multiple objectives; Performance, where it should have the ability to solve

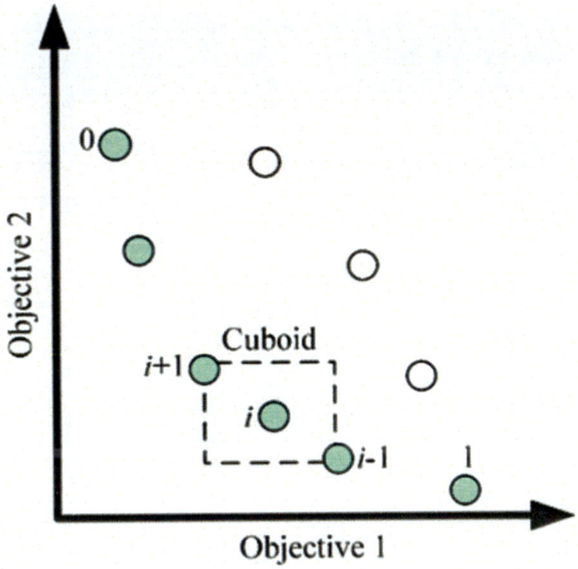

Fig. 4 Calculation of crowded distance

the problem reasonably fast. Scalability that determines the ability for the system to be used by many users concurrently and without encountering performance issues. Flexibility that shows the ability to receive a wide range of differently formatted input data and still functions adequately. Sparse or otherwise inadequate input data is preprocessed and can be supplemented by default values.

To address these objectives, the modeling tool was designed to work as a multi-layer architecture as in Fig. 5, with each layer performing a distinct part of the work flow. It is important to note, that these layers can run on separate machines to provide scalability but they do not have to do so. In the setup used for conducting the case study all layers were run on a single machine.

4 Application to Case Study

The case study will examine the business case of the establishment of an autonomous Microgrid in the selected area. Therefore, investment decisions and dispatch are modeled for three discontinuous periods of one week each to reflect the different conditions encountered throughout the year. The designated buildings will be assumed to be the participants of the considered Microgrid. The different classes of buildings are modeled with different parameters to accurately represent their distinct sizes and uses. The following sections document the assumed input parameters the model requires for the calculation of a base scenario and the way in which they were derived. In order to function properly, the model requires meteorological data for constraints on intermittent power generation, as well as data on investment options,

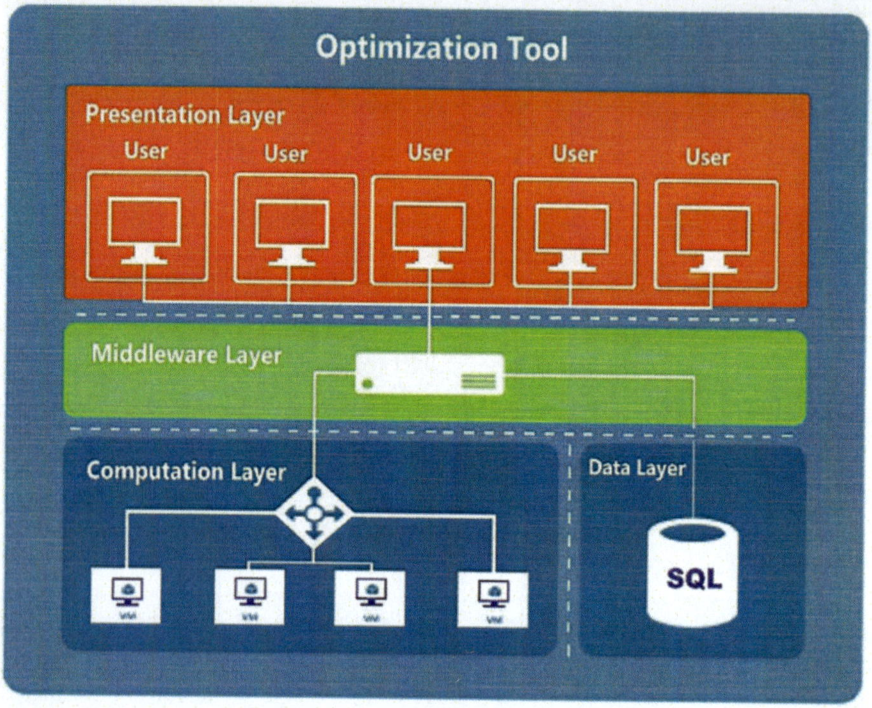

Fig. 5 The proposed architecture for scaling the optimization tool

household demand, and grid supply. The final section will introduce a number of further scenarios in which some of the assumptions are changed. An hourly profile of the capacity factor for a PV installation in Morschenich is sourced from [47]. From the profile, which includes an entire year of data, three samples are extracted: One 144-h sample representing the transitional season, i.e. spring and fall, and two 72 h samples representing summer and winter. The difference in length is not relevant for use in the model, as the data preprocessing only uses these profiles as input for generating the profiles used in the optimization itself. The resulting vector of values is used as the parameter $GEN_1 \rightarrow \max Fl$, which is interpreted as the maximum input flow for the solar PV installation. In addition the parameter EFF_{el} is set to 1.0, which means the model effectively treats the vector as a capacity factor.

To test the different implementations of discrete and linear investment options the case study incorporates two linear-scaling generation devices as possible investments, as well as two discretely scaling storage devices. The generation devices consist of a photovoltaic system as a renewable option and a combined heat and power fuel cell as a conventional but decentralized and highly efficient option. It is also capable of using biofuels, which can be expressed through higher fuel prices. Parameters describing the generation technologies are listed in Table 1 [48, 49].

Table 1 Parameters describing the generation technologies

Parameter	Name	Unit	Value
Photovoltaic system (GEN_1)			
Installation cost	C_{Cap}	€ KW^{-1}	1200
Maintenance cost	$C_{\text{op, Fix}}$	€$KW^{-1}yr^{-1}$	50
Fuel cost	$C_{\text{op, Var}}$	€KW^{-1}	0.0
Lifetime	T_{Life}	yr	25
Electric efficiency	EFF_{el}	%	100
Performance ratio	PR	%	85
CHP micro fuel cell (GEN_2)			
Installation cost	C_{Cap}	€ KW^{-1}	3500
Maintenance cost	$C_{\text{op, Fix}}$	€$KW^{-1}yr^{-1}$	175
Fuel cost	$C_{\text{op, Var}}$	€KW^{-1}	0.0608
Lifetime	T_{Life}	yr	15
Electric efficiency	EFF_{el}	%	60
Performance ratio	PR	%	100
Maximum input flow	$\max Fl$	KW	999

The storage investment options given are two differently sized lithium-Ion batteries: A 4 KWh option manufactured by Victron Energy and a 13.5 KWh option manufactured by Tesla. While the two-generation devices are hypothetical, these storage devices can actually be bought, which leads to lower error due to false assumptions and linearization of non-linear relationships (such as storage size and price). This is the main advantage discrete modeling of investment options has over linear modeling. Parameters describing the storage technologies are listed in Table 2 [49, 50]. The state of charge (SOC) of the lithium-ion (Li-ion) battery has been studied and modeled in many literature [50, 51].

As the case study distinguishes between three types of buildings, three demand profiles where compiled. Each profile is made up of three hourly load vectors spanning one or several days. Demand vectors for the two types of residential units are sourced from [52]. The choice was made because these profiles are specifically created for optimization of decentralized generation. They are samples of measurements taken on several houses and therefore feature a much higher variability than standard profiles, which are usually averages of many measurements. In the context of this case study, it is critical to accurately reflect this high variability, which a Microgrid can do a lot to manage. For the commercial units, profiles were sourced from the association of the German power industry (BDEW) [53, 54]. These are standard profiles specifically from commercial consumers. The samples were of varying sizes of one to several days, so as to reflect the approximate ratios of occurrences of the different model days provided for the chosen meteorological region. They were further preprocessed by the model itself. The noise filter of the model was provided with a variance value of 0.25 for the smaller residential households and a variance

Table 2 Parameters describing the storage technologies

Parameter	Name	Unit	Value
Li-ion Battery-1 (dST_1)			
Usable capacity	CAP	kWh	13.5
Installation cost	C_{Cap}	€	10,000
Lifetime	T_{Life}	yr	10
Round-trip efficiency	EFF_{+-}	%	90
Max. charging power	$\max P_+$	kW	4.6
Max. discharging Power	$\max P_-$	kW	4.6
Initial SOC	–	%	100
Terminal SOC	–	%	100
Li-ion Battery-2 (dST_2)			
Usable capacity	CAP	kWh	4
Installation cost	C_{Cap}	€	5000
Lifetime	T_{Life}	yr	10
Round-trip efficiency	EFF_{+-}	%	92
Max. charging power	$\max P_+$	kW	4.8
Max. discharging Power	$\max P_-$	kW	7.2
Initial SOC	–	%	100
Terminal SOC	–	%	100

of 0.15 for the larger residential and commercial buildings. This additional noise is crucial for distinguishing the demands of different instances of one household type from one another, as would occur in reality. Furthermore, the noise filter makes the load profile more realistic when scaled beyond its original length, since it makes each repetition of the profile differ notably from the others as in Fig. 6. For the residential buildings, it is assumed that at any time 10% of loads can be curtailed at a flat remuneration rate of 0.35 € per kWh. A further 20% of total loads are assumed to be shiftable at a flat rate of 0.15 € per kWh per hour shifted. The remaining 70% of loads are deemed essential. The commercial units can also shift up to 20% of their load for 0.15€ per kWh per hour, but no curtailment is allowed. The profiles are scaled for the total yearly demand of the smaller residential buildings to be around 5000 kWh. The multi-family residential and commercial buildings' profiles are scaled to a yearly demand of approximately 10,000 and 16,000 kWh respectively.

For the base scenario, the grid supply is assumed to be capped at a constant value of 500 KW, which guarantees that all supply can be met solely by the grid. It is further assumed that the price of this energy is set at 0.3 € at any time [13]. The feed-in tariff is assumed to be constant at 0.08 € and the feed-in is capped at a constant 20 kW.

To keep the model from trading randomly inside the Microgrid, a trade fee of 0.001 € per kWh is imposed on all trade within the Microgrid. To accurately reflect the prevalence of the weather conditions and demand profiles represented by the three seasonal periods modeled, the summer, winter and transitional periods have

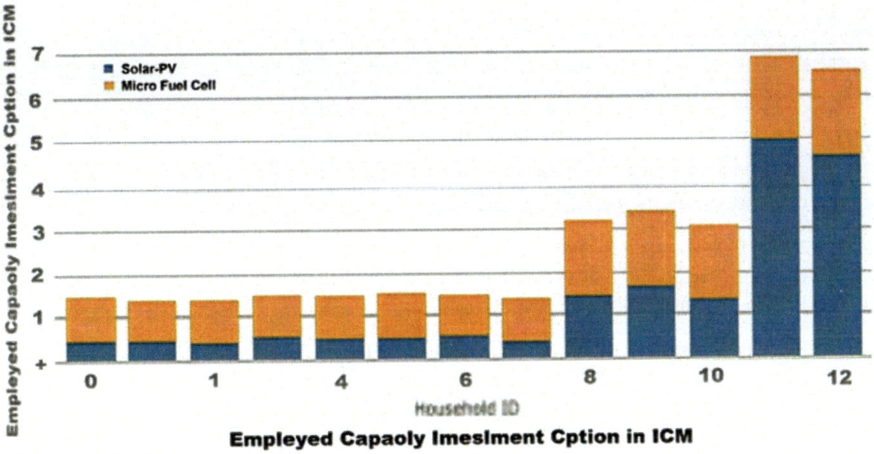

Fig. 6 Generation investment in the base setup of the case study

been weighted with the factors 169, 81, and 115, respectively. These factors are normalized during preprocessing and all cost incurred during each period is multiplied with the normalized corresponding weight. The weights have been chosen according to the prevalence of different weather conditions in the meteorological region of Morschenich [52].

5 Results and Discussions

Surprisingly, even in the base scenario, the Microgrid seems to be commercially viable, as an autarky level of 94% is reached. Investment is heavily biased towards Fuel Cells, which are installed at a smaller capacity than PV installations but operate at a capacity factor of up to 100% versus around 12 for PV. Therefore, most power is generated by flexible gas fuel cells (dark blue) and only supplemented by a small amount of PV generation (light blue) as in Figs. 7 and 8. The ratio of fuel cell versus PV installation varies depending on the total amount and temporal distribution of demand: While the small residential residences install mostly fuel cells, the larger residential units invest in an almost equal capacity of both and the commercial units install a much more PV capacity than fuel cells. This is probably to keep the self-consumption ratio high, while at the same time minimizing trade, which is cheap but not free. The commercial units have by far the highest demand during midday, which can be served by solar energy without demand shifting or storage.

With this setup, the model stays at a self-consumption rate of 100%, as feed-in remuneration seems to be smaller than the marginal cost of generating power with the fuel cells. Furthermore, no storage is needed to achieve these levels of autarky and self-consumption. As can be observed in Figs. 6 and 7 trading (orange and pink)

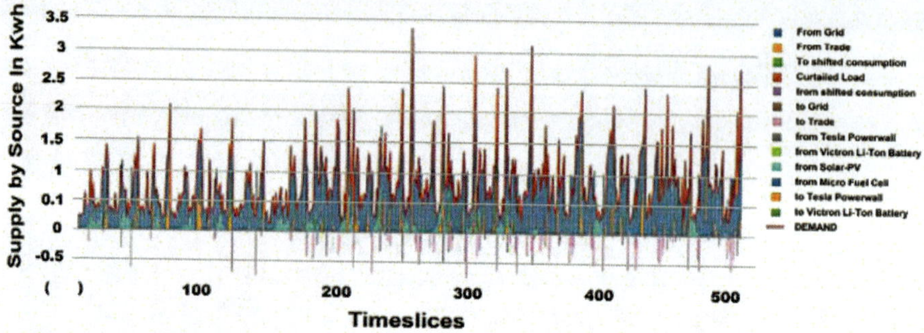

Fig. 7 The dispatch of a residential building in the base setup of the case study

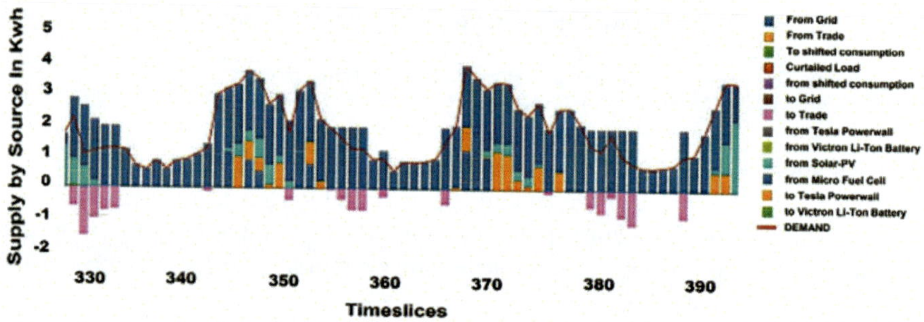

Fig. 8 The dispatch of a commercial building in detail for two winter days

is heavily utilized for balancing, as the load profiles of participants differ enough to enable this. Also, load shifting (green and purple) is sporadically used as in Figs. 6 and 7, but loads are rarely shifted for more than one hour.

The use of grid energy (also dark blue, but below orange and light blue) occurs during the winter week (the last third of the dispatch chart), when generation and internal dispatch are insufficient to meet peak demand, most likely due to a lack of irradiation as in Fig. 8. This indicates that the levelized cost of solar energy is significantly below that of the electricity generated by fuel cells, as it would otherwise be cheaper to satisfy demand entirely with them instead of adding PV and as a consequence having to rely on some grid imports during winter.

Overall, the result is surprisingly cheap, with the average cost per kWh being at about 0.164 e and the total cost at approx. 1259.624 e. The model solves in approx. 4–6 min on a macOS 10.14 systems with the Gurobi Commercial Solver and a 3.3Ghz Quad-Core Intel Core i5 processor with 24 GB of 1867 MHz DDR3 RAM. To test the validity and robustness of the solution obtained above, sensitivity analysis is conducted. To test the importance of the use of fuel cells for the result and its robustness, first the price of gas is raised by 50%. This could reflect the CO_2 neutral use of more expensive biogas or the insecurity of the price of fossil fuels

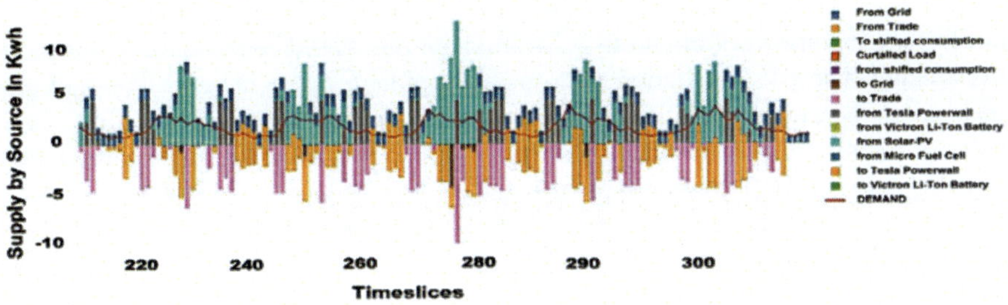

Fig. 9 Effects of a gas price increase of 50%

in general, as the investment horizon considered in the case study is up to 25 years (Fig. 9).

The increased gas price results in a drop of installed fuel cell capacity by around a quarter. PV deployment consequently increases by over 300%. One of the commercial buildings is equipped with both 17.7 kW_p of PV and a Tesla Power wall, which is utilized heavily to store solar energy traded to the Microgrid during the evening as in Fig. 8. Autarky remains high at 92%, while self-consumption drops slightly to 96%. The grid is still not utilized much since the average price per KWh at 0.199 € is still well below grid power prices. Still the objective value rises approx. 20% to 1524.547 €. Another factor which seems critical to the base result is the reliance on trading. To test the robustness of trading, the price per traded KWh of electricity is raised from a merely symbolic 0.001 € to 0.03 €. This could be considered the cost of maintaining the infrastructure of the grid, or the energy lost in cross-Microgrid transactions. This has only a negligible effect, as the objective value is only raised by about 10 €. Raising the fee to 0.06 € per KWh has a similarly small effect, as it increases the objective value by another 18 €. A further increase to 0.09 € fails to have a much greater impact at an objective value of approx. 1293 €.

Another possibility is a change in the price of PV panels and batteries in the coming years. To test the effect these changes could have on the optimal result, the base case is altered by reducing the investment price of PV installations by 25% and the investment price of storage by 35%. This has a similar effect to increased gas prices, in that fuel cell deployment falls, whereas PV installations increase. Additionally, as before, one Tesla Power wall is installed by a commercial entity. However, the effect on the objective value is not significant, as it is only decreased by 46 € or 0.005 € per KWh. Adding the 50% increased gas prices to the alterations from base case results in a total installation of 64 PV, but only a single Power wall, which seems to suffice, as self-consumption is still at 92%. The average electricity price rises to 0.19 € decreasing battery ˙ prices by 50% relative to the base case result in the installation of a second Power wall in one of the multi-family residential buildings. The solution, however, is hardly superior at 0.188 € per KWh. Finally, a major weakness of the base cases result is the linear scaling of the price of generation technology, which leads to the installation of small capacities throughout the Microgrid. In reality, this would be very expensive, as the prices do not in fact scale linearly. To test the robustness to non-linear generation options, both generation options would be converted to discrete investments. The PV installation will only be available in increments of 5 and the Fuel Cell in increments of 2.5. The computation is run with a 15 min time limit, as it does not terminate in a reasonable amount of time otherwise. The gap to optimality after this period is at 0.26% or approx. 4 €.

While the objective value at approx. 1257 € does not greatly differ from the base case, the added constraints led to a very different behavior. Households are now clearly separated into producers and consumers, some generating power with a PV installation, some with a fuel cell as in Figs. 10 and 11. This leads to a strong increase in trading activity between the households as the Microgrid tries to distribute the now more centralized generation.

This is conducted quite successfully, as autarky remains high at 94%. Electricity from the grid is only imported during the winter period as Fig. 12. The model still does not construct any storage.

Although the sensitivity analysis showed the model result to be quite robust, there are weaknesses to point out: The cost computation is simplified and does not take into account borrowing rates, and does not discount future earnings. The result rather resembles a situation in which all equipment is leased and in which interest rates on this leasing contract are negligible. While this is very close to the truth in most of

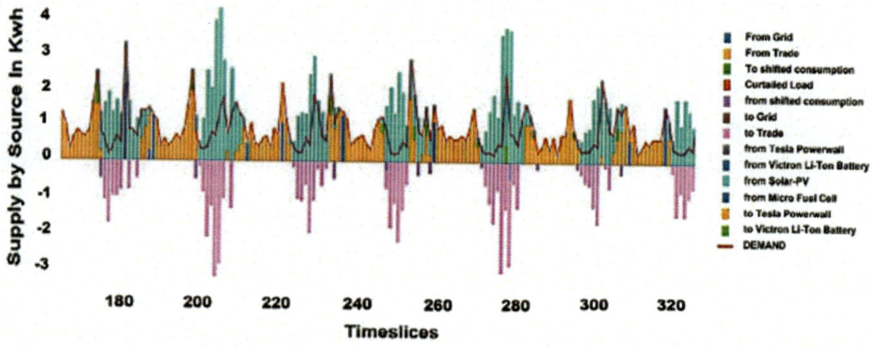

Fig. 10 A household with a PV installation under an all-discrete constraint

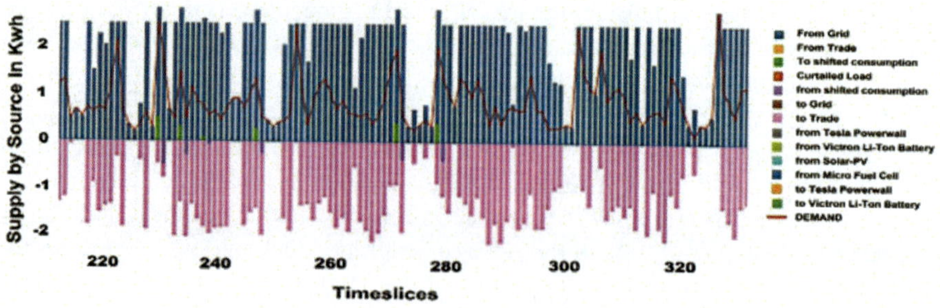

Fig. 11 A household with a fuel cell under an all-discrete constraint

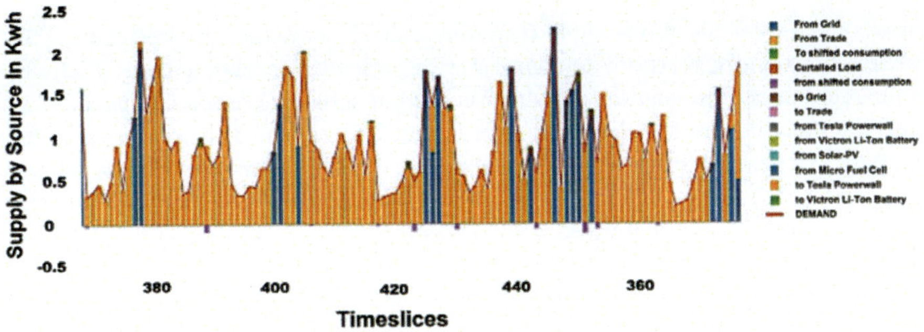

Fig. 12 A household without generation under an all-discrete constraint during winter

Europe and the US, it is still a minor distortion and a major one anywhere else. Another weakness is the modeling approach itself. By its very nature LP assumes perfect knowledge of the input parameters involved. This includes all parameters describing demand and supply, which in reality cannot be known at the point of an investment decision. Furthermore, dispatch decisions in an LP are also made with perfect knowledge of the future, so that every time storage is filled or demand shifted, the perfect amount is certain, which of course is not the case in a real-world scenario. This affects only dispatch decisions with an impact on future dispatch decisions, which do not feature heavily in the base case, since it does not install any storage. As a consequence, the fact that the results arrived at are the product of an LP should only imply a minor advantage in investment optimality, since this can be predicted with quite high accuracy by means of statistics. It also implies only a minor advantage in dispatch optimality, since demand shifting is the only dispatch method utilized that is impacted by perfect foresight. Moreover, demand shifting is mostly utilized for only one hour into the future, for which quite accurate forecasting tools are available, which further reduces the information disadvantage of a real-life scenario. Any minor inaccuracies aside the model results clearly conclude that a Microgrid, as described in the base case, if regulatory feasible, would be highly profitable with a margin of over 0.13 € per KWh. The sensitivity analysis further shows that even under significant deterioration of some of the assumptions made the margin still remains higher than 0.10 € per KWh. There is, however, some uncertainty involved, as being raised by RWE would significantly worsen the profitability of the system. To the question of the bigger picture of decentralization, the answer is less clear: While clearly profitable in the context of German electricity retail prices, the generation costs of 0.164 € per KWh are still significantly higher than the current generation costs included in the retail price of 0.067 € per KWh. This would conclude that decentralized generation is still far off, when it comes to large-scale competitiveness, it however fails to consider some important factors: First, there is an additional cost of 0.0719 € per KWh incurred by transmission, when buying electricity. In a decentralized scenario of 94% autarky, as exhibited in the base results, these costs would most likely be vastly smaller, if at all significant. This brings us to a total of 0.1389 € per KWh which is significantly closer to the 0.164 € result of the base case. Secondly, there are another 0.0917 € of levies (excluding taxes) included in the retail price, the majority of which is used to subsidize renewable generation. Still, the present German electricity mix is much dirtier, than the one used in the Microgrid of the base case, which only consists of solar and highly efficient natural gas generation. Taking this 'power quality' consideration into account the total price of equivalent electricity delivered by a centralized system is (at least in Germany) closer to 0.2305 €, which is significantly higher than even the results reached in the sensitivity analysis. Furthermore, it does not even take into account the heat output of the fuel cells, which can be used without losses for heating water or air. In addition, it is highly likely that a major uptake of decentralized generation methods would reduce investment costs, especially of fuel cells, further bringing down the estimated price of electricity. Of course, a decentralized system would incur other costs, as a rudimentary grid would still be needed and industry would possibly suffer from the loss of access to subsidized electricity [54]. But overall, the results arrived at in this case study suggest a bright future for decentralized generation systems, as they close—or even have already closed—the cost gap to current centralized systems.

6 Conclusion

In this paper a mathematical modeling tool for Microgrid energy systems has been designed, taking the current literature on the subject into account. The modeling tool has then been used for a case study on a village in North Rhine-Westphalia, in which several buildings of the village form a hypothetical Microgrid. The investment choices available in the case study consisted of a PV installation and a fuel cell as generation technologies and two different discrete battery investments as energy storage. In addition, the participants were able to trade energy within the Microgrid as well as shift or curtail some of their demands. There also was the possibility to draw power from the main grid at standard retail rates. Special attention was given to the realism of the demand profiles, as standard profiles used elsewhere are averaged and would have distorted the model's results significantly. The results showed the clear profitability of the base case at an average electricity cost of about 0.164 € per kWh. The Microgrid had an ratio of 94%, while 100% self-consumption was reached. This was due to significant investment in fuel cells, which produced the majority of electricity in the base case. There was no need for storage installation. The sensitivity analysis showed the astounding robustness of the base case. An increase in fuel prices of 50% lead to the biggest additional cost, with the price per kWh rising to 0.199 €. Even a full discretization of all investment options merely resulted in a vast increase of trading activity within the Microgrid. It has been shown clearly, that the base case, if regulatory feasible, would be highly profitable. Based on these results Microgrid system architecture could even be considered on par with, or superior to the current grid performance excluding taxes and levies: It hardly needs any transmission infrastructure outside Microgrid boundaries and the power generated comes exclusively from renewable or highly efficient conventional sources that could also use biofuels or hydrogen in the future. Considering these results, the concept of a decentralized multi-Microgrid energy system should be explored further, e.g. by means of a larger scale pilot project. Furthermore, a partially or fully decentralized option should be added to the plans to decarbonize the German electricity system by 2050, as only heavily centralized solutions are currently discussed by planners. Fuel cell technology as a small scale, flexible way of generating power and heat should receive more attention. It has the potential to be a key technology enabling a faster transition to renewable generation without relying on lignite and coal plants. The next steps in the development of the modeling tool include improvement of the user interface, including the possibility for users to add their own investment options and demand profiles. Microgrid energy systems are becoming serious competitors for conventional centralized power systems.

References

1. The Paris Agreement (2018) https://unfccc.int/process-and-meetings/the-paris-agreement/the-parisagreement
2. Oei PY, Brauers H, Kemfert C, Kittel M, Goke L, von Hirschhausen C, Walk P (2018) Kohleausstieg in NRW im deutschen und europaischen Kontext Energiewirtschaft, Klimaziele und wirtschaftliche Entwicklung
3. Nuclear Power in Germany (2019) http://www.world-nuclear.org/information-library/country-profiles/countriesg-n/germany.aspx
4. Yanguas Parra P, Roming N, Sferra F, Fuentes Hutfilter U, Zimmer A, Aboumahboub T, Schaeffer M, Hare B (2018) Science based coal phase-out pathway for germany in line with the paris agreement 1.5 °C Warming Limit. Tech. rep. Climate Analytics
5. Goke L, Kittel M, Kemfert C, Lorenz C, Oei P-Y, von Hirschhausen C (2018) Erfolgreicher Klimaschutz durch zügigen Kohleausstieg in Deutschland und Nordrhein- Westfalen. DIW Wochenbericht 33
6. Lazar E, Petreuş D, Etz R, Patarau T (2017) Minimization of operational cost for an Islanded microgrid using a real coded genetic algorithm and a mixed integer linear programming method. In: Proceedings of 2017 international conference on optimization of electrical and electronic equipment (OPTIM), Intl Aegean Conference on Electrical Machines and Power Electronics (ACEMP), pp 693–698
7. Nemati M, Braun M, Tenbohlen S (2018) Optimization of unit commitment and economic dispatch in microgrids based on genetic algorithm and mixed integer linear programming. Appl Energy 210:944–963
8. Palma-Behnke R, Benavides C, Lanas F, Severino B, Reyes L, Llanos J, Saez D (2013) A microgrid energy management system based on the rolling horizon strategy. IEEE Transactions on Smart Grid 4(2):996–1006
9. Silvente J, Kopanos GM, Pistikopoulos EN, Espuña A (2015) A rolling horizon optimization framework for the simultaneous energy supply and demand planning in microgrids. Appl Energ 155:485–501
10. Craparo E, Karatas M, Singham DI (2017) A robust optimization approach to hybrid microgrid operation using ensemble weather forecasts. Appl Energy 201:135–147
11. Hussain A, Bui V, Kim H (2018) Robust optimal operation of AC/DC hybrid microgrids under market price uncertainties. IEEE Access 6:2654–2667

12. Zhang Y, Zhang T, Wang R, Liu Y, Guo B (2015) Optimal operation of a smart residential microgrid based on model predictive control by considering uncertainties and storage impacts. Sol Energy 122:1052–1065
13. Moshi GG, Bovo C, Berizzi A, Taccari L (2016) Optimization of integrated design and operation of microgrids under uncertainty. In: Proceedings of 2016 power systems computation conference (PSCC), pp 1–7
14. Shams MH, Shahabi M, Khodayar ME (2018) Stochastic day-ahead scheduling of multiple energy carrier microgrids with demand response. Energy 155:326–338
15. Zheng Y, Jenkins BM, Kornbluth K, Træholt C (2018) Optimization under uncertainty of a biomass-integrated renewable energy microgrid with energy storage. Renewable Energy 123:204–217
16. Nugraha PY, Hadisupadmo S, Widyotriatmo A, Kurniadi D (2015) Optimization of capacity and operational scheduling for grid-tied microgrid using pumped-storage hydroelectricity and photovoltaic. In: Proceedings of 2015 10th Asian Control Conference (ASCC), pp 1–6
17. Tavakoli M, Shokridehaki F, Akorede MF, Marzband M, Vechiu I, Pouresmaeil E (2018) CVaR-based energy management scheme for optimal resilience and operational cost in commercial building microgrids. Int J Electr Power Energy Syst 100:1–9
18. Mashayekh S, Stadler M, Cardoso G, Heleno M, Madathil SC, Nagarajan H, Bent R, Mueller-Stoffels M, Lu X, Wang J (2018) Security-constrained design of isolated multi-energy microgrids. IEEE Trans Power Syst 33(3):2452–2462
19. Oei P-Y, Rieve C, von Hirschhausen C, Kemfert C (2019) Weichenstellung Kohlekonsens: Kohlevorrate ermoglichen den Erhalt des Hambacher Waldes und aller noch bedrohten Dorfer
20. Zhang D, Shah N, Papageorgiou LG (2013) Efficient energy consumption and operation management in a smart building with microgrid. Energy Convers Manage 74:209–222
21. Zheng Y, Jenkins BM, Kornbluth K, Kendall A, Traeholt C (2018) Optimization of a biomass-integrated renewable energy microgrid with demand side management under uncertainty. Appl Energy 230:836–844
22. Liu G, Starke M, Zhang X, Tomsovic K (2016) A MILP-based distribution optimal power flow model for microgrid operation. In: Proceedings of IEEE power and energy society general meeting (PESGM), pp 1–5
23. Farsangi AS, Hadayeghparast S, Mehdinejad M, Shayanfar H (2018) A novel stochastic energy management of a microgrid with various types of distributed energy resources in presence of demand response programs. Energy 160:257–274
24. Koltsaklis NE, Giannakakis M, Georgiadis MC (2018) Optimal energy planning and scheduling of microgrids. Energ Syst Eng Chem Eng Res Des 131:318–332
25. Sechilariu M, Wang B, Locment F (2014) Power management and optimization for isolated DC microgrid. In: Proceedings of international symposium on power electronics, electrical drives, automation and motion, pp 1284–1289
26. Santos LTD, Sechilariu M, Locment F (2015) Prediction-based optimization for Islanded microgrid resources scheduling and management. In: Proceedings of IEEE 24th international symposium on industrial electronics (ISIE), pp 760–765
27. Umeozor EC, Trifkovic M (2016) Energy management of a microgrid via parametric programming. In: 11th IFAC symposium on dynamics and control of process systems including biosystems DYCOPS-CAB, IFAC, vol 49, no 7, pp 272–277
28. Lauinger D, Caliandro P, Herle JV, Kuhn D (2016) A linear programming approach to the optimization of residential energy systems. J Energy Storage 7:24–37
29. Wouters C, Fraga ES, James AM (2015) An energy integrated, multi-microgrid, MILP (mixed-integer linear programming) approach for residential distributed energy system planning–a South Australian case-study. Energy 85:30–44
30. Costa A, Fichera A (2014) A mixed-integer linear programming (MILP) model for the evaluation of CHP system in the context of hospital structures. Appl Therm Eng 47(2):921–929
31. Atia R, Yamada N (2016) Sizing and analysis of renewable energy and battery systems in residential microgrids. IEEE Trans Smart Grid 7(3):1204–1213

32. Wu X, Wang X, Bie Z (2012) Optimal generation scheduling of a microgrid. In: Proceedings of 2012 3rd IEEE PES innovative smart grid technologies Europe (ISGT Europe), pp 1–7
33. Parisio A, Glielmo L (2013) Stochastic model predictive control for economic/environmental operation management of microgrids. In: Proceedings of European Control Conference (ECC)
34. Rigo-Mariani R, Sareni B, Roboam X (2013) A fast optimization strategy for power dispatching in a microgrid with storage. In: Proceedings of IECON, 39th annual conference of the IEEE industrial electronics society, pp 7902–7907
35. Gulin M, Matusko J, Vasak M (2015) Stochastic model predictive control for optimal economic operation of a residential DC microgrid. In: Proceedings of IEEE international conference on industrial technology (ICIT), pp 505–510
36. Chen J, Zhang W, Li J, Zhang W, Liu Y, Zhao B, Zhang Y (2018) Optimal sizing for grid-tied microgrids with consideration of joint optimization of planning and operation. IEEE Trans Sustain Energy 9(1):237–248
37. Mashayekh S, Stadler M, Cardoso G, Heleno M (2017) A mixed integer linear programming approach for optimal DER portfolio, sizing, and placement in multi-energy microgrids. Appl Energy 187:154–168
38. Zhaoxia X, Jiakai N, Guerrero JM, Hongwei F (2017) Multiple time-scale optimization scheduling for islanded microgrids including PV, wind turbine, diesel generator and batteries. In: Proceedings of IECON—43rd Annual Conference of the IEEE Industrial Electronics Society, pp 2594–2599
39. Henao-Muñoz AC, Saavedra-Montes AJ, Ramos-Paja CA (2017) Energy management system for an isolated microgrid with photovoltaic generation. In: Proceedings of 14th international conference on synthesis, modeling, analysis and simulation methods and applications to circuit design (SMACD), pp 1–4
40. Amrollahi MH, Bathaee SMT (2017) Techno-economic optimization of hybrid photovoltaic/wind generation together with energy storage system in a stand-alone micro-grid subjected to demand response. Appl Energy 202:66–77
41. Distributed Energy Resources—Customer Adoption Model (DER-CAM). https://buildingmicrogrid.lbl.gov/projects/der-cam
42. HOMER Grid Behind-the-Meter Optimization Software for Demand Charge Reduction and Energy Arbitrage (2018) https://www.homerenergy.com/products/grid/index.html
43. Sechilariu M, Wang BC, Locment F (2014) Supervision control for optimal energy cost management in DC microgrid: design and simulation. Int J Electr Power Energy Syst 58:140–149
44. Umeozor EC, Trifkovic M (2016) Operational scheduling of microgrids via parametric programming. Appl Energy 2016(180):672–681
45. Duong MQ,Nguyen HH, Le T, Mussetta M (2019) New planning for the 500 kV vietnamese grid with high penetration of renewable energy sources. IEEE Milan PowerTech, Milan, Italy, pp 1–6.https://doi.org/10.1109/PTC.2019.8810879
46. Nguyen TT, Nguyen TT, Duong MQ, Doan AT (2019) Optimal operation of transmission power networks by using improved stochastic fractal search algorithm. In: Neural Computing and Applications, pp 1–36
47. Pfenninger S, Staffell I (2016) Long-term patterns of european PV output using 30 years of validated hourly reanalysis and satellite data. Energy 114:1251–1265
48. Omar AI, Ali ZM, Al-Gabalawy M, Abdel Aleem SHE, Al-Dhaifallah M (2020) Multi-objective environmental economic dispatch of an electricity system considering integrated natural gas units and variable renewable energy sources. Mathematics
49. Mekhamer S, Abdelaziz A, Algabalawy M (2018) Design of hybrid power generation systems connected to utility grid and natural gas distribution network: a new contribution. Eng Rev 38(2):204–214
50. Lithium-Ionen HE (High Energy) Batteries und Lynx Ion BMS (2018)
51. Al-Gabalawy M, Hosny NS, Hussien SA (2020) Lithium-ion battery modeling including degradation based on single-particle approximations. Batteries 6(37)

52. Referenzlastprofile von Ein- und Mehrfamilien- häusern für den Einsatz von KWK-Anlagen (2008)
53. Fünfgeld C, Tiedemann R (2000) Anwendung der Reprasentativen VDEW-Lastprofile
54. Küchler S, Wronski R (2014) Industriestrompreise in Deutschland und den USA

Hybrid Cryptography for Cloud Security: Methodologies and Designs

Sherief H. Murad and Kamel Hussein Rahouma

Abstract Cloud computing is a trending information technology applied for storing and accessing information over the Internet. Probably, sensitive information is stored on remote servers that are not managed nor controlled by the customers. Therefore, potential attacks may be launched against stored information from either inside the cloud service provider or outsider attackers. Cryptography is the fundamental mechanism that provides enough level of security to the cloud. Hybrid cryptography endeavors to enhance security and performance by integrating more than one cryptographic algorithm. In our study, we conducted a survey on applied hybrid cryptographic models for data security in the cloud between 2013 and 2020. We have presented the design, the implementation methodology, limitations found, and the suggested applications for each proposal. We finalized this paper with a comparison summary table. We hope to make a scientific contribution to secure the cloud.

Keywords Cloud computing · Information security · Symmetric ciphers · Asymmetric ciphers · Hybrid cryptography

1 Introduction

Cloud computing is considered a new model for hosting and providing IT services through the Internet. It allows access to a shared collection of computational resources with minimal interactions with service providers [1]. Cloud services fall into three main categories: software as a service (SaaS) like Gmail, platform as a service (PaaS) like GoogleApp Engine, and infrastructure as a service (IaaS) like Microsoft Azure [2].

Reducing the implementations and maintenance costs, flexibility and scalable infrastructures, and high availability of well-performance applications have motivated governments and individuals to move their data on the cloud servers. These benefits are limited due to several issues concerning security [3, 4].

S. H. Murad (✉) · K. H. Rahouma
Electrical Engineering Department, Faculty of Engineering, Minia University, Minia, Egypt
e-mail: Sh1547@fayoum.edu.eg

D. A. Magdi et al. (eds.), *Digital Transformation Technology*, Lecture Notes in Networks and Systems 224, https://doi.org/10.1007/978-981-16-2275-5_7

Cryptography or secret writing could be utilized to provide confidentiality, integrity, and availability to the data stored or accessed through the cloud. It transforms plain data into unreadable or encrypted form to the unintended users. Both encryption and decryption take place with an extra input called the key [5]. Mainly, cryptography is divided into symmetric-key cryptography and public-key cryptography [6]. Hybrid cryptography implies that two or more cryptographic mechanisms are blended to enhance security. In this paper, we propose a survey study of cryptographic algorithms applied to secure data on the cloud.

2 Literature Survey

Symmetric or secret-key cryptosystems utilize a single key in both encryption and decryption phases. Communicating parties must share this key via a secure channel before any encryption or decryption. An example of symmetric cryptosystems is the data encryption standard (DES), triple-DES (3DES), blowfish, and advanced encryption standard (AES) [5]. On the contrary, asymmetric or public-key cryptosystems utilize two keys; a publically shared key used in encryption, and the other is kept private for decryption. The public key is sent over the network and not necessarily over a secured channel, but the private key must be kept safe from disclosure. An example of asymmetric cryptosystems is Rivest–Shamir–Adleman (RSA), ElGamal, and digital signature algorithm (DSA) [6].

Symmetric and asymmetric cryptosystems have their advantages and disadvantages. Secret key cryptosystems are fast but suffer the secret key exchanging. Public key cryptosystems are secure and solve the key exchange problem but they are slow. Therefore, in practices, hybrid cryptographic schemes, that is a combination of both, are used to exploit the efficiency of symmetric-key algorithms and the simplicity of asymmetric algorithms [4]. The main objective of this approach is to produce a more secure, better performance, and robust algorithm than applying them individually.

A survey on different security issues related to cloud computing was presented by Nigoti et al. [21]. They focused on solving these issues using hybrid cryptographic algorithms and concluded that DES was easier to implement on the cloud than AES. They used RSA and Diffie-Hellman to generate keys to be utilized with the symmetric ciphers.

Another survey was conducted by Sajjan et al. [22] to analyze multilevel encryption used in cloud data security. After various ciphers have been investigated, they implemented a two-layer encryption algorithm composed of DES and RSA ciphers. Their study concluded that applying multilayer encryption provides more security than single-level models.

A novel proposal by Sinchana and Savithramma [4] has examined different hybrid cryptographic models and emphasized the design, implementation, and features of these models. They have enhanced both security and efficiency via integrating symmetric and asymmetric algorithms.

3 Methodology

Hybrid cryptography is mainly categorized into two schemes: The first scheme uses a symmetric algorithm to encrypt data and an asymmetric algorithm to encrypt the secret key. The other scheme performs two layers of symmetric or asymmetric encryption where data is double encrypted by applying either two consecutive symmetric or asymmetric ciphers. Other researchers have applied encryption using symmetric algorithms followed by asymmetric algorithms, both for data encryption, then applied asymmetric algorithms to exchange the secret keys. Many studies used to achieve data security on the cloud as described below:

Sengupta and Jeffrey [7] have proposed a hybrid model to secure cloud infrastructure based on combining Caesar and Vigenere algorithms [5]. In Fig. 1, the plaintext is double encrypted using Caesar cipher. The resulting ciphertext is encrypted using Vigenere cipher with a keyword then encrypted again but with the keyword reversed. The decryption phase is the same as encryption but applied in reversed order.

Vishwanath and Aniket [8] have proposed a hybrid model to enhance data security in the cloud, based on RSA and AES algorithms. As shown in Fig. 2, the data is encrypted using AES then encrypted again using RSA. The system timing is used in the key generation phase.

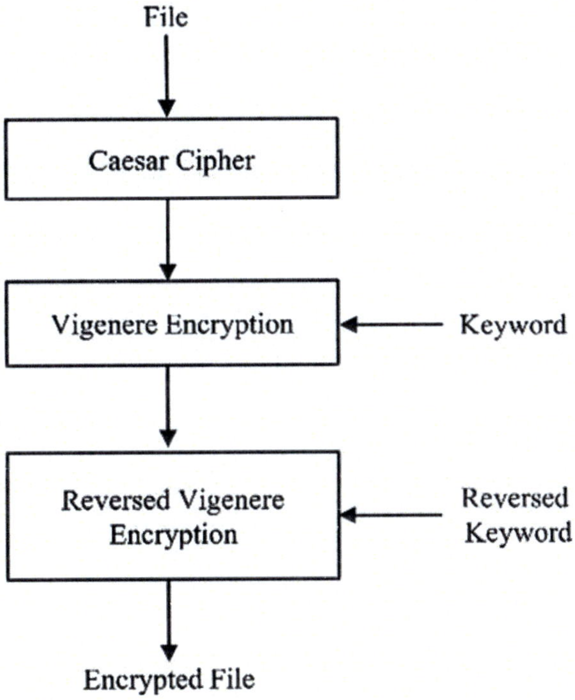

Fig. 1 Hybrid cryptographic model proposed by Sengupta and Jeffrey

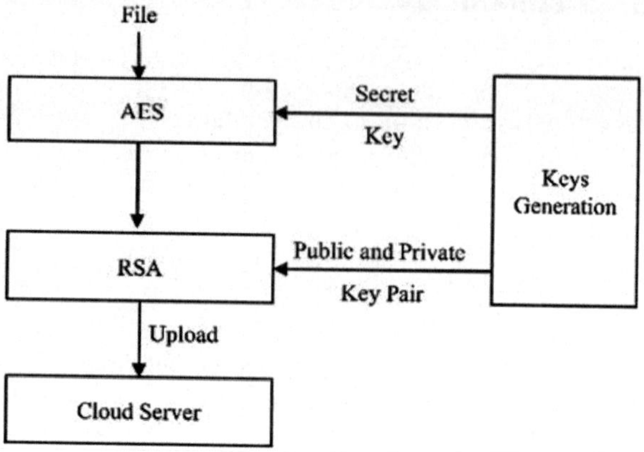

Fig. 2 Hybrid cryptographic model proposed by Vishwanath and Aniket

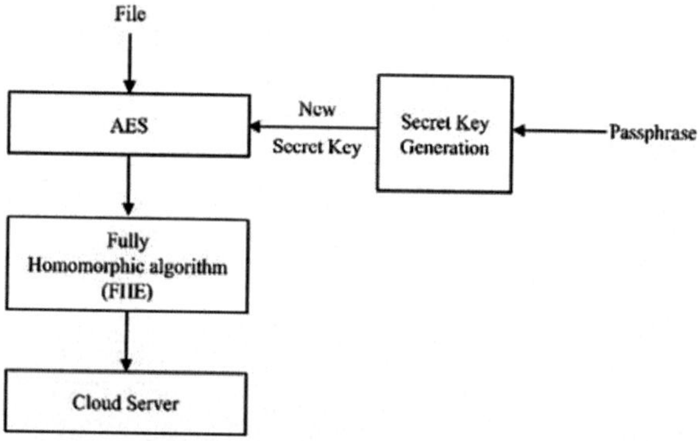

Fig. 3 Hybrid cryptographic model proposed by Atewologun et al.

Atewologun et al. [9] have proposed a hybrid model to provide secure transmission of data on the cloud by using AES and fully homomorphic encryption (FHE) algorithms. As shown described in Fig. 3, the file is firstly encrypted using AES, and the resulting ciphertext is encrypted using FHE cipher. The secret key is derived from a passphrase provided by the sender.

Punam and Aruna [10] proposed a hybrid model to secure cloud storage using AES, blowfish, RC6, and byte rotation algorithm (BRA) ciphers. In Fig. 4, data is split into eight parts, each part is encrypted using a different cipher. Encrypted blocks and secret keys are embedded into an image then uploaded to the cloud using the least significant bit steganography [5, 6, 16].

Adviti and Jyoti [11] have enhanced cloud security using Blowfish and MD5 hybrid model. In Fig. 5, data is encrypted using blowfish cipher, and integrity is provided with message digest 5 (MD5). Finally, encrypted parts and MD are uploaded to the cloud.

Rohini and Sharma [12] have proposed a hybrid model to secure data over the cloud, based on RSA and hashed message authentication code (HMAC) [5] ciphers. As shown in Fig. 6, the data is encrypted using RSA cipher, and the HMAC is

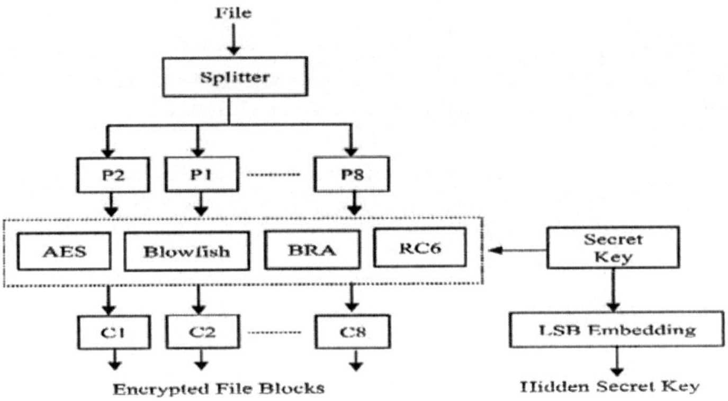

Fig. 4 Hybrid cryptographic model proposed by Punam and Aruna

Fig. 5 Hybrid cryptographic model proposed by Adviti and Jyoti

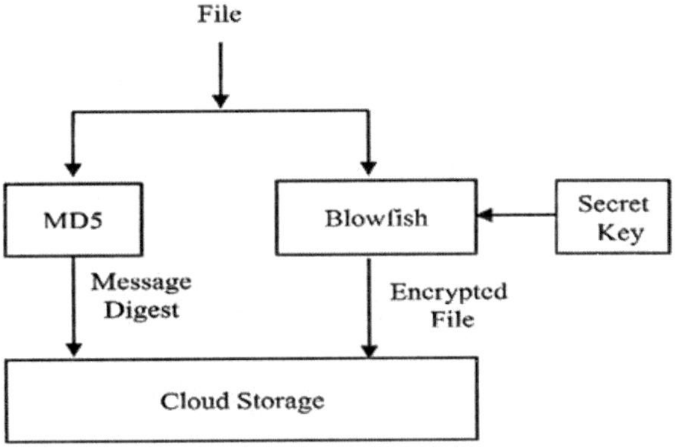

Fig. 6 Hybrid cryptographic model proposed by Rohini and Sharma

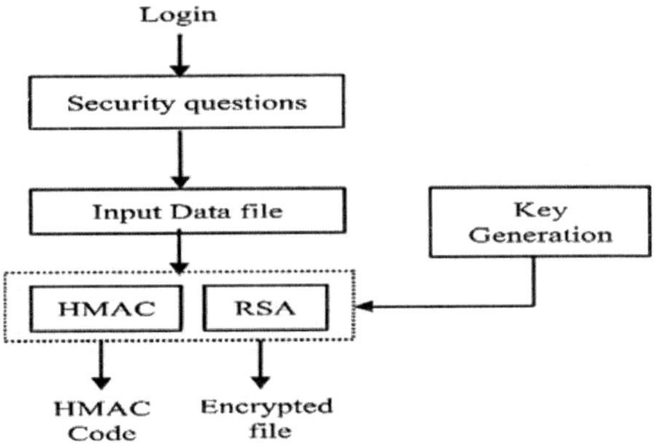

generated to provide integrity. Both message digest and encrypted data are uploaded to the cloud.

Salma et al. [13] have proposed a hybrid model to enhance cloud data security based on blowfish and AES ciphers. As shown in Fig. 7, after a successful login, the uploaded file is twice encrypted using dynamic AES (DAES) then blowfish. Secret keys used are encrypted using blowfish. All encrypted data and keys are uploaded to the cloud.

Sherief et al. [14] have proposed a hybrid model to enhance data security based on blowfish, visual cryptography [15], and steganography. As shown in Fig. 8, data is first encrypted using blowfish followed by visual cryptography, producing two shares. These shares are then embedded into two images using LSB before transmission.

Fig. 7 Hybrid cryptographic model proposed by Salma et al.

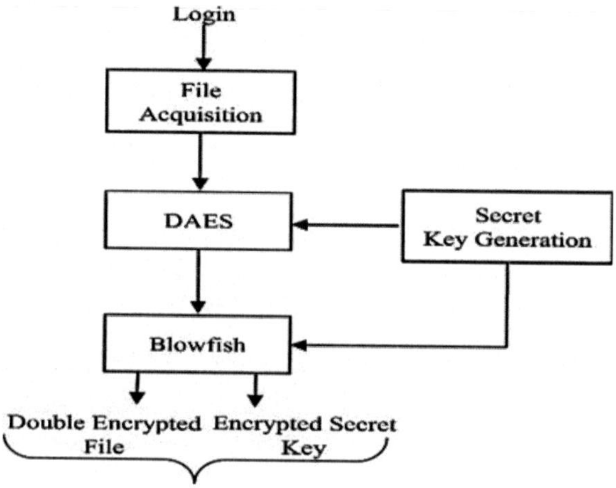

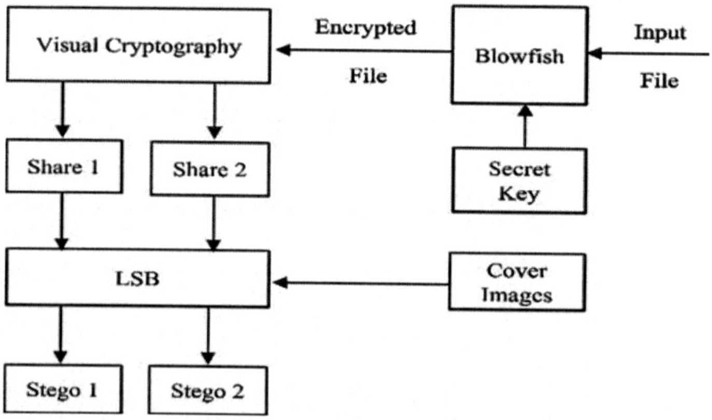

Fig. 8 Hybrid cryptographic model proposed by Sherief et al.

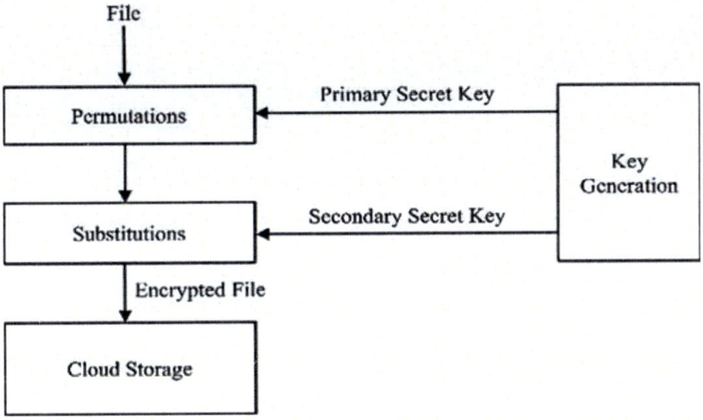

Fig. 9 Hybrid cryptographic model proposed by Shweta and Ashish

Shweta and Ashish [16] have proposed a hybrid model to secure data on the cloud based on transposition and substitution algorithms. As shown in Fig. 9, the input file is permuted using a primary secret key. Before uploading to the cloud, another encryption is made using substitution cipher with a secondary secret key. The auditor is responsible for generating the secret keys and sharing them between the cloud users.

A proposal to secure robots on the cloud by Huili et al. [17] based on the hybrid of RSA and MD5 ciphers. As shown in Fig. 10, data is first encrypted with RSA cipher. Integrity is achieved via MD5. Finally, the encrypted data and message digest are uploaded to the cloud.

Anuj et al. [18] introduced another hybrid model for cloud storage security consisting of RSA and DES ciphers. As shown in Fig. 11, data is double encrypted using RSA then DES cipher.

Another cloud storage security model was implemented by Shivam et al. [19] consisted of AES, RC4 [6], and DES ciphers. As shown in Fig. 12, the original file

Fig. 10 Hybrid cryptographic model proposed by Huili et al.

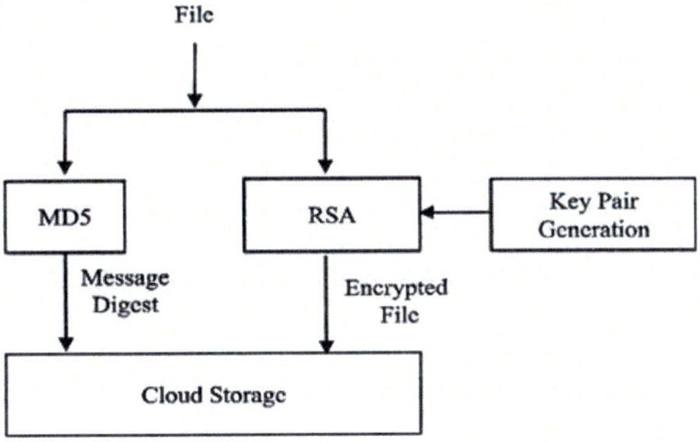

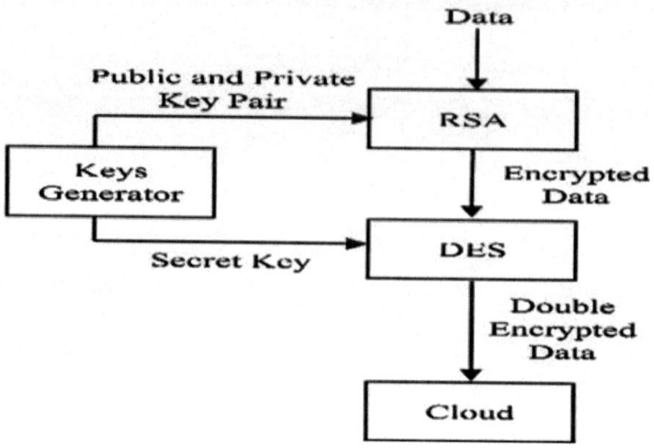

Fig. 11 Hybrid cryptographic model proposed by Anuj et al.

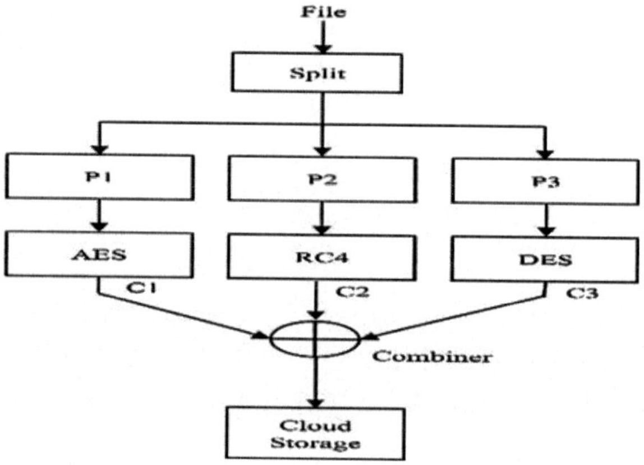

Fig. 12 Hybrid cryptographic model proposed by Shivam et al.

is broken into three equal chunks, and each will be encrypted using a distinct cipher. Finally, encrypted parts are merge and uploaded to the cloud storage server.

A hybrid model consists of AES, 3DES [6], RSA ciphers combined with LSB steganography was presented by Vinay et al. [20]. As shown in Fig. 13, two-layer encryption is achieved via AES then 3DES ciphers. RSA cipher is employed to encrypt secret keys. Finally, both encrypted file and keys are hidden into an image using the LSB insertion method.

4 Comparison Between Proposed Models

A chronological summary of the studied hybrid models proposed from the oldest to the most recent within the period from 2013 to 2020. In Table 1, we have presented a comparison between the proposed models based on hybrid cryptography in terms

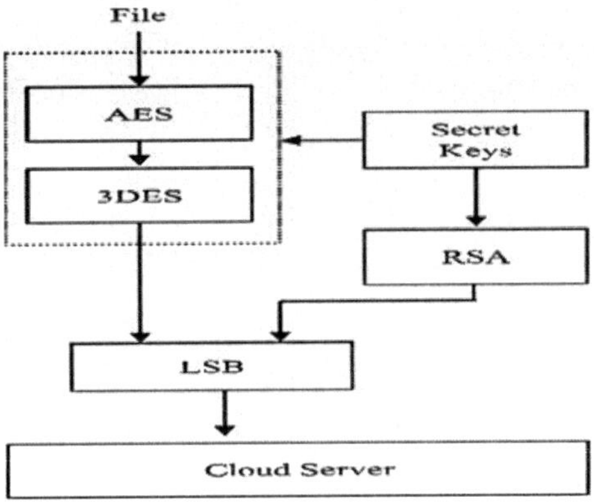

Fig. 13 Hybrid cryptographic model proposed by Vinay et al.

of elements composing the model, features achieved, limitations, and suggested applications for each proposal.

5 Conclusions

After studying all of the trending hybrid cryptographic models, we came up with the conclusion that data security is the most considerable topic related to cloud computing technology. To overcome security limitations, the integration between symmetric and asymmetric cryptosystems is employed. By applying the hybrid of different encryption algorithms such as DES, 3DES, AES, Blowfish, RSA, and SHA, we would try to secure sensitive data on the cloud. Our study also concludes that hybrid cryptography enhances the performance and adds more security levels to the data compared to applying these algorithms individually. All of the examined studies have benefits and some drawbacks. As a future direction, we aim to overcome these drawbacks to enhance security and performance.

Table 1 Summary of the trending studies based on hybrid cryptography

References	Elements	Features achieved	Limitations found	Applications
[7]	Caesar and Vigenere	• More security	• Vigenere key and message must be equal in length • Caesar is not secure	Cloud infrastructure
[8]	AES and RSA	• Confidentiality, integrity, and reliability	• RSA has more computations cost	Cloud storage
[9]	AES and FHE	• Confidentiality, integrity • Malware attack protection	• Not compatible with all cloud services	Cloud storage
[10]	AES, Blowfish, RC6, BRA	• Less delay (multithreading) • Confidentiality, integrity, and authentication	• Secret keys could be compromised or altered	Cloud storage
[11]	Blowfish, MD5	• Confidentiality and integrity	• MD5 is adequately secure • Key distribution issue	Cloud storage
[12]	RSA and HMAC	• Confidentiality and integrity	• RSA takes a long time and a large memory size	Cloud storage
[13]	DAES, and Blowfish	• Brute force and algebraic attack protection	• Key distribution issue • Encryption is slow	File and web servers—social media programs
[14]	Blowfish-Visual Cryptography, LSB	• Very secure • Fingerprint authentication • Robust against steganalysis	• Physical keys distribution • Almost used for documents files	Law enforcement
[16]	Permutation-substitution	• More fast and efficient than asymmetric-Brute-force attack protection	• Key distribution issue • No authentication	Cloud storage

(continued)

Table 1 (continued)

References	Elements	Features achieved	Limitations found	Applications
[17]	RSA and MD5	• Confidentiality, integrity, and authentication	• RSA computations cost • MD5 is adequately secure • RSA is not for large files	Cloud intelligent robot
[18]	RSA and DES	• Provide security to IoT data	• DES is slow and less secure • RSA computations cost • RSA is not for IoT	IoT applications on the cloud
[19]	AES, RC4, and DES	• It is better for bigger file sizes • More efficient compared to AES alone	• RC4 is adequately secure • DES is not secure enough • Key distribution issue	Cloud storage
[20]	AES, 3DES, RSA, and LSB	• Higher security due to double encryption	• Long encryption time • 3DES is very slow	Banking and private sectors

References

1. Badger ML, Grance T, Patt-Corner R, Voas JM (2012) Cloud computing synopsis and recommendations. National Institute of Standards & Technology
2. Kyriazis D, Voulodimos A, Gogouvitis SV, Varvarigou TA (2013) Data intensive storage services for cloud environments. Business Science Reference
3. Singh V, Pandey SK (2020) Cloud computing: vulnerability and threat indications. In: Performance management of integrated systems and its applications in software engineering. Springer, Singapore, pp 11–20
4. Sinchana MK, Savithramma RM (2020) Survey on cloud computing security. In: Innovations in computer science and engineering. Springer, Singapore, pp 1–6
5. Forouzan BA (2007) Cryptography and network security. McGraw-Hill, Inc.
6. Schneier B (2007) Applied cryptography: protocols, algorithms, and source code. C. John Wiley & Sons
7. Sengupta N, Holmes J (2013) Designing of cryptography based security system for cloud computing. In: 2013 international conference on cloud & ubiquitous computing & emerging technologies. IEEE, pp 52–57

8. Mahalle VS, Shahade AK (2014) Enhancing the data security in Cloud by implementing hybrid (Rsa & Aes) encryption algorithm. In: 2014 International Conference on Power, Automation and Communication (INPAC). IEEE, pp 146–149
9. Olumide A, Alsadoon A, Prasad PWC, Pham L (2015) A hybrid encryption model for secure cloud computing. In: 2015 13th International Conference on ICT and Knowledge Engineering (ICT & Knowledge Engineering 2015). IEEE, pp 24–32
10. Maitri PV, Verma A (2016) Secure file storage in cloud computing using hybrid cryptography algorithm. In: 2016 international conference on wireless communications, signal processing and networking (WiSPNET). IEEE, pp 1635–1638
11. Chauhan A, Gupta J (2017) A novel technique of cloud security based on hybrid encryption by Blowfish and MD5. In: 2017 4th International conference on signal processing, computing and control (ISPCC). IEEE, pp 349–355
12. Sharma T (2018) Proposed hybrid RSA algorithm for cloud computing. In: 2018 2nd international conference on inventive systems and control (ICISC), pp 60–64. IEEE
13. Salma RF, Khaizuran Abdullah R, Darwis H (2018) Enhancing cloud data security using hybrid of advanced encryption standard and blowfish encryption algorithms. In: 2018 2nd East Indonesia Conference on Computer and Information Technology (EIConCIT). IEEE, pp 18–23
14. Murad SH, Gody AM, Barakat TM (2019) Enhanced security of symmetric encryption using combination of steganography with visual cryptography. arXiv preprint arXiv: 1902.11167
15. Naor M, Shamir A (1994) Visual cryptography. In: Workshop on the theory and application of cryptographic techniques. Springer, Berlin, Heidelberg, pp 1–12
16. Kaushik S, Patel A (2019) Secure cloud data using hybrid cryptographic scheme. In: 2019 4th international conference on internet of things: smart innovation and usages (IoT-SIU), pp 1–6. IEEE
17. Cai H, Liu X, Cangelosi A (2019) Security of cloud intelligent robot based on RSA algorithm and digital signature. In: 2019 IEEE symposium series on computational intelligence (SSCI). IEEE, pp 1453–1456
18. Kumar A, Jain V, Yadav A (2020) A new approach for security in cloud data storage for IOT applications using hybrid cryptography technique. In: 2020 international conference on power electronics & IoT applications in renewable energy and its control (PARC), pp 514–517. IEEE
19. Sharma S, Singla K, Rathee G, Saini H (2020) A hybrid cryptographic technique for file storage mechanism over cloud. In: First international conference on sustainable technologies for computational intelligence, pp 241–256. Springer, Singapore
20. Poduval V, Koul A, Rebello D, Bhat K, Wahul RM (2020) Cloud based secure storage of files using hybrid cryptography and image steganography. (IJRTE) Int J Recent Technol Eng 8(6)
21. Nigoti R, Jhuria M, Singh S (2013) A survey of cryptographic algorithms for cloud computing
22. Sajjan RS, Ghorpade V, Dalimbkar V (2016) A survey paper on data security in cloud computing. Int J Comput Sci Eng (IJCSE) 4(4):9–13

Big Data and AI in Digital Transformation

A Study of Sentiment Analysis Approaches in Short Text

Ahmed F. Ibrahim, M. Hassaballah, Abdelmgeid A. Ali, and Ibrahim A. Ibrahim

Abstract Recently, the remarkable growth of Internet technology, particularly on social media networking sites, enables gathering data for analyzing and gaining insights. It is challenging to analyze such a huge amount of information that causes time-consuming. So, it is necessary to make an intelligent system that automatically analyzes a great amount of data. Sentiment analysis methods appear to analyze sentiments and opinions of people through what they write on social networking sites. Different sentiment analysis approaches have been proposed to understand the sentiments and opinions expressed by the individuals in the text. However, some methods produce an improper result when applied to short text due to text briefness and sparsity. In this paper, we present sentiment analysis models that analyzing people feelings and opinions in the short text such as tweets and instant messages. Then, we illustrate the evaluation metrics used to assess the quality of the generated hierarchical structure in extracting an ideal tree.

Keywords Sentiment analysis · Hierarchical structure · Evaluation methods · Short text

A. F. Ibrahim (✉)
Department of Computer Science, Faculty of Computer Science, Nahda University, Banisuef, Egypt
e-mail: i.ibrahim@minia.edu.eg; ahmed.farouk@nub.edu.eg

M. Hassaballah
Department of Computer Science, Faculty of Computers and Information, South Valley University, Qena, Egypt
e-mail: m.hassaballah@svu.edu.eg

A. A. Ali · I. A. Ibrahim
Department of Computer Science, Faculty of Computers and Information, Minia University, AL Minia, Egypt
e-mail: dr_abdelmgeid@mu.edu.eg

D. A. Magdi et al. (eds.), *Digital Transformation Technology*, Lecture Notes in Networks and Systems 224, https://doi.org/10.1007/978-981-16-2275-5_8

1 Introduction

The notable progress of Internet technology, especially on social media networking sites, has made people live in glass houses. It contributes to discovering a massive amount of information created by users like tweets, customer feeling, opinions, comments, and reviews. This information may be about events, products, people, etc., which is very useful for organizations and individuals. Analyzing such information enables organizations to measure their customer's satisfaction with a certain product (service) faster, more comprehensive, and more reliable feedback than traditional questionnaires. Further, it provides governments with important information to help in developing future strategies and policies. It permits an individual to make a decision efficiently by checking people's attitudes about certain products.

However, analyzing this huge amount of information is challenging and time-consuming. Consequently, it is necessary to create an intelligent system that analyzes a large amount of information automatically. Sentiment analysis is widely used for analyzing people's feelings and opinions to show the causes of why they favor certain products through what they write using natural language processing (NLP). Many methods have been introduced to understand the sentiments and people's opinions in text. However, applying such methods on short text in social networking sites such as Twitter and Weiboa usually produce understandable and inefficient results due to text briefness and sparsity.

Other methods attempt to extract and capture the hierarchical structure from a collection of documents and comprehend basic information. Some researchers studied the problem by proposing a model that discovers a hierarchy from review data collection. Discovering hidden relations from short text makes many applications easier to use such as summarization [1, 2], sentiment analysis applications [3], and user [4] recommendation. Some methods make a flat sentiment analysis on each obtained aspect separately and disregard the concept hierarchy. Evaluating the quality and correctness of the extracted hierarchical structure attracted many researchers to apply subjective methods. Such that, arises the need to evaluate these proposed techniques.

In this paper, we illustrate sentiment analysis approaches in text and classify based on the topic structure: (1) approaches based on flat model and (2) other methods which depend on the hierarchical topic model. We will show what happens when applying these methods in a short text. We discuss metrics are used to evaluate generated hierarchical structure.

The rest of the paper is organized as follows: In Sect. 2: illustrate the sentiment analysis approaches. Then, we present what happens when applying these methods to short text. In Sect. 3: we introduce the evaluation methods of hierarchical models. In Sect. 4: we present the analysis and discussions of sentiment analysis techniques. In Sect. 5: a conclusion.

2 Sentiment Analysis Approaches

Sentiment analysis methods have been implemented to various types of texts which include news [5–7], tweets [8–11], and customer reviews [12–14]. Topic modeling and sentiment analysis are becoming a spread research topic among researchers' community. Researchers proposed several approaches to perform sentiment analysis. It can be divided into two main categories; the first type is flat sentiment analysis and the second type is hierarchical sentiment analysis. In this section, we introduce the approaches used to learn the structural relations between concepts in text. Then, we illustrate the results described of the presented approaches.

2.1 Flat Sentiment Analysis Approaches

Different approaches have been introduced to solve the problem of extracting features from the text. For example, Hu and Liu [15] proposed a set of techniques for mining and summarizing product reviews based on data mining and natural language processing methods. The main objective is to provide a feature-based summary of a large number of customer reviews of a product sold online. These techniques extract nouns and noun phrases by using part-of-speech (POS) tagging. Then, they catch highly frequent words by used association rules. However, the proposed techniques are unable to determining the strength or the influence of opinions. Moreover, it fails to inspect opinions expressed with adverbs, verbs, and nouns.

Popescu and Etzioni [16] proposed an unsupervised information extraction system, namely OPINE, which extracts fine-grained features, and associated opinions from reviews based on a novel relaxation-labeling technique. Although, the system can identify opinions expressed by the customer and their polarity with high precision and recall, but it is not able to procedure a hierarchal structure. Moreover, topic model tools have produced excellent results in discovering the hidden patterns within texts in various domains [17–19].

Weng et al. [9] focus on the problem of identifying influential users of micro-blogging services particularly twitter. TwitterRank algorithm [9] measures the topic-sensitive influence by examining the topical similarity between users and the link structure into account. Xu et al. [20] proposed an improved word representation approach that performs sentiment analysis of comment texts. They can better get the text representation of the comments. Wang et al. [21] proposed a new model that improves the ability of feature extraction.

Nevertheless, all of the above approaches do not consider the hierarchical structure of the aspects and their polarity. So, when applied these methods to short text produce incomplete results. In natural language processing, text sentiment analysis [22, 23] or classify short text is a difficult task [24]. Hierarchical sentiment analysis approaches have been suggested to effectively solve the problem of the existent flat topic models.

2.2 Hierarchical Sentiment Analysis Methods

Based on the aforementioned drawbacks flat sentiment analysis approaches, the researchers have proposed new methods to extract the hierarchical structure from text effectively. Titov and McDonald [1, 25] proposed a multi-grain topic framework for extracting the assessable characteristics of objects from online user reviews. These multi-grain models are extensions to standard topic modeling methods such as LDA and PLSA to produce topics that correspond to global properties of objects. Kim et al. [26] applied a hierarchical aspect-sentiment model (HASM) that discover hierarchical tree structure from online reviews. (HASM) [26] automatically extracts both the structure and parameters of the tree based on a Bayesian nonparametric model, and it is based on recursive Chinese Restaurant Process (rCRP), rCRP to discover aspect-sentiment topics over multiple granularities from an unlabeled review corpus. Although HASM is flexible to discover aspects with more than two sentiments, but it cannot discover a set of topics with shared features in a hierarchical tree structure.

Wang et al. [27] presented the problem of constructing a topical hierarchy from short and content-representative texts, where topics are represented by ranked lists of phrases. Wang et al. proposed a recursive method for building a hierarchy of topics from a set of content-representative documents. The proposed approach is focusing on a phrase-centric assessment rather than a unigram-centric to produce high-quality topics over multiple levels. Moreover, some research papers proposed methods to discover the hierarchical of topics in review data, such as [28–30]. Chen et al. [31] proposed a new approach for hierarchical topic extraction by using a class of graphical models, namely hierarchical latent tree models (HLTMs), which used to model document collection. HLTMs structure patterns of word co-occurrence and co-occurrence in a hierarchical latent tree model.

Li and McCallum [32] proposed a method to detect the correlations among topics using the hierarchical Pachinko allocation model (hPAM) uses a directed acyclic graph (DAG). This model creates many levels of super- and sub-topics. Blei et al. [33] attempt to learn latent structures from data hierarchically, namely the nested Chinese restaurant process (nCRP). It is a process that assigns probability distributions to ensembles of infinitely deep, infinitely branching trees. However, (hPAM) provides a flexible alternative to recent work by Blei and Lafferty, which captures correlations only between pairs of topics. Jain et al. [34] proposed a hybrid system that shows great results when applied to classify large-sized text.

However, all of these methods mentioned above have only been developed to apply to traditionally long text. They produce an incomplete and incomprehensible result when applied to short text.

There are three main difficulties when extracting sentiment tree hierarchically from short texts. First, due to the shortness of text, that make it difficult to discover data more deeply and unsuccessful in learning relations between concept. Second, it is challenging to create comprehensible summaries. People want to show the result of sentiment tree in an understandable way. Third, existing methods extract aspects then perform independently flat sentiment analysis and disregard the concept hierarchy. The sentiment polarity of an aspect must additionally cover polarity of its all children.

To effectively extract the hierarchical structure from short text, Zhao and Li [4] have developed an algorithm to construct a hierarchical tree of aspects of a certain product. This algorithm depends on the term frequency-inverse document frequency (TFIDF) and formal concept analysis (FCA). When applying this method to short texts, it cannot discover semantics and hidden pattern in text and not study the evolution of the tree's quality. To solve the problems that appeared in this method, Almars et al. [1] proposed an LDA-based method, namely structured sentiment analysis (SSA) approach, which automatically extracts the product's hot aspects from short text and organizes them hierarchically in a tree. Hot aspects are the most aspects persons speak about. This model has three advantages; first, they introduced a hierarchical method to extract products hot aspects and organized them hierarchically in the tree.

This hierarchical method succeeded in extracting weak relationships between aspects. Second, they analyze opinions expressed by the people on those aspects by using the hierarchical sentiment method. By using this method, the sentiment polarity of an aspect must cover the polarity of aspects itself and its offspring in the hierarchical tree. Third, they generate a structured summary of the result. This summary helps people to know the total polarity of the product, the reasons why people prefer or avoid those aspects and the most appropriate and inappropriate aspects of the product. There are two main drawbacks; the first, it cannot discover the optimal number of depth and width of the tree automatically which must be identified manually. The second, in this method to evaluate the quality of extracted hierarchical tree, they are applying subjective method.

To solve the first drawback, Almars et al. [35] propose a top-down recursive model called context coherence-based model (CCM). This model analyzes the relations between words to discover a concept hierarchy from short text automatically without a pre-defended hierarchy depth and width. It is simple and easy to implement which can discover the data more deeply. The main task of this model is the extracting concept. They extract a concept by using context coherence. It analyzes the relations between the words in a whole text to measure the coverage of a given word. In this model, word coverage refers to the importance of this word in the entire text. The coverage is measured by the number of words that are related to it. They use two functions to extract the concept hierarchically. First, splitting operation which applying to create a hierarchical tree in which concepts near to the root are general and near to the leaf is specific. Another task of this operation is to build a hierarchical tree in which the concept defined as a parent must be semantically related to its children rather than to its non-children. Second, merging operation which tried to discover similar words and collected them under a new concept.

To solve the second problem, there are other methods proposed to evaluate the quality of the extracted hierarchical tree as shown in Sect. 3.

3 Evaluation Methods of Generated Hierarchical Structure

There is a need for evaluating the quality of the hierarchical tree extracted from unstructured text using sentiment analysis hierarchical model, which helps to know the ability of these methods in extracting an optimal tree. Some existing hierarchical models such as hierarchical Pachinko allocation model (hPAM) [32] and model based on a formal concept analysis (FCA) [4], not objectivity measure the goodness of extracted hierarchical structure. Other models such as a recursive method proposed by Wang et al. [27] and an LDA-based method proposed by Almars et al. [1] use subjective evaluation (e.g., surveys). Some approaches such as hierarchical aspect-sentiment model (HASM) proposed by Kim et al. [26] and knowledge-based hierarchical topic model (KHTM) proposed by Xu et al. [36] evaluate the quality of topics only. Existing evaluation methods cannot apply to all hierarchical models. These methods do not measure other characteristics of an ideal tree. The ideal tree has three important properties. First, the concept that appears near to the root, at a high level, must cover a more deeply area than concepts that appears near to leaf node. Second, all words represent a topic in the tree that must be related to the other words semantically. Third, the parent concept in the tree must be related semantically to its children rather than to its non-children.

Almars et al. [37] proposed three methods of parent–child relatedness, coverage, and coherence to evaluate the quality of the hierarchical tree extracted from the unstructured text. These methods are used to reflect the three main properties of an ideal tree, which can be used to evaluate any hierarchical models. Moreover, to evaluate the quality of the hierarchical tree extracted from structured data such as relational data, they proposed a new measuring method named interest-based coherent. They consider the importance of attributes in the database reflects the quality of the tree. They measure the importance of each node on the tree using the utility score added to the frequency score. The result of applying these methods shows that it can reflect the three properties of an ideal tree and can be used to evaluate the quality of any hierarchical models.

4 Analysis and Discussion

Several researchers evaluate the quality of their models by calculating the precision [38]. Pang et al. [39] apply three machine learning classifiers (Naïve Bayes, maximum entropy, and support vector machines) to the sentiment classification problem. They achieved an accuracy of 81.5% when applying Naïve Bayes, 81% when applying maximum entropy and 82.9% when applying support vector machines. Turney [40] able to obtain 66% accuracy for the movie review domain. Dave et al. [41] achieved from 85.8 to 87.2% when applying support vector machines and achieved from 81.9 to 87% when applying Naïve Bayes. Hu and Liu [15] achieved 84% when applying lexicon-based techniques.

Table 1 Summary of precision and hierarchical relatedness of sentiment analysis approaches

Paper	Dataset	Technique (Precision, %)	Hierarchical relatedness
Pang et al. [39]	IMDB	NB (81.5), ME (81.0), SVM (82.9)	–
Turney [40]	Epinions	PMI (66)	–
Dave et al. [41]	Amazon, CNET	SVM (85.8–87.2), NB (81.9–87.0)	–
Hu and Liu [15]	Amazon, CNET	Lexicon (84.0)	–
Almars et al. [1]	Smartphone (IPhone–Galaxy–HTC)	–	IPhone (82%) Galaxy (75%) HTC (75%)
Almars et al. [35]	Smartphone, DBLP	–	Smartphone(89%), DBLP (83%)

However, Almars et al. [1] introduced a new metric, hierarchical relatedness, to evaluate the quality of their model. Almars et al. [1] measure the usefulness of extracted tree. For IPhone customers, the result shows that 82% of the participants decided the aspects of the product were relevant. For the Galaxy and HTC, the result shows 75% of the participants viewed the quality of the structure was satisfactory. Table 1 presents summary of precision and hierarchical relatedness of sentiment analysis.

5 Conclusion

We present in this paper sentiment analysis approaches which proposed to analyze people's opinion and feeling of a particular product. We divided them into two key categories based on the output produced, namely flat sentiment analysis and hierarchical sentiment analysis. When applying flat sentiment analysis models to short text, the results are confusing and inappropriate. Some hierarchical sentiment analysis models succeeded in extracting hierarchical structure from short text. Then, we discussed the evaluation of generated hierarchical structure which is very critical to measure the quality and correctness, and the ability of used model in extracting an ideal tree. We discussed the evaluation of the discussed approaches across various measurements such as classical precision, coverage, and relatedness which evaluate the quality of hierarchical models in extracting hierarchical structure.

References

1. Almars A, et al (2017) Structured sentiment analysis. In: International conference on advanced data mining and applications. Springer, Berlin, pp 695–707
2. Gerani S, Carenini G, Ng RT (2019) Modeling content and structure for abstractive review summarization. In: Computer Speech & Language
3. Kontopoulos E et al (2013) Ontology-based sentiment analysis of twitter posts. Expert Syst Appl 40(10):4065–4074
4. Zhao P, Li X, Wang K (2013) Feature extraction from micro-blogs for comparison of products and services. In: International conference on web information systems engineering. Springer, Berlin
5. Godbole N, Srinivasaiah M, Skiena S (2007) Large-scale sentiment analysis for news and blogs. ICWSM 7(21):219–222
6. Li J et al (2014) Hierarchical classification in text mining for sentiment analysis. In: 2014 international conference on soft computing and machine intelligence. IEEE
7. Li J et al (2016) Hierarchical classification in text mining for sentiment analysis of online news. Soft Comput 20(9):3411–3420
8. Ali F et al (2019) Transportation sentiment analysis using word embedding and ontology-based topic modeling. Knowl-Based Syst 174:27–42
9. Grandin P, Adan JM (2016) Piegas: a systems for sentiment analysis of tweets in portuguese. IEEE Lat Am Trans 14(7):3467–3473
10. Korkontzelos I et al (2016) Analysis of the effect of sentiment analysis on extracting adverse drug reactions from tweets and forum posts. J Biomed Inform 62:148–158
11. Morente-Molinera JA et al (2019) Carrying out consensual group decision making processes under social networks using sentiment analysis over comparative expressions. Knowl-Based Syst 165:335–345
12. Ji P, Zhang H-Y, Wang J-Q (2018) A fuzzy decision support model with sentiment analysis for items comparison in e-commerce: the case study of http://PConline.com. IEEE Trans Syst Man Cybern Syst 49(10):1993–2004
13. Morente-Molinera JA et al (2019) An automatic procedure to create fuzzy ontologies from users' opinions using sentiment analysis procedures and multi-granular fuzzy linguistic modelling methods. Inf Sci 476:222–238
14. Wang L et al (2020) The differences in hotel selection among various types of travellers: a comparative analysis with a useful bounded rationality behavioural decision support model. Tourism Manage 76:103961
15. Hu M, Liu B (2004) Mining and summarizing customer reviews. In: 10th ACM SIGKDD international conference on Knowledge discovery and data mining
16. Popescu A-M, Etzioni O (2007) Extracting product features and opinions from reviews. In: Natural language processing and text mining. Springer, Berlin, pp 9–28
17. Lin C, He Y (2009) Joint sentiment/topic model for sentiment analysis. In: 18th ACM conference on Information and knowledge management. ACM
18. Ramage D, Dumais S, Liebling D (2010) Characterizing microblogs with topic models. In: 4th international AAAI conference on weblogs and social media
19. Weng J et al (2010) Twitterrank: finding topic-sensitive influential twitterers. In: Proceedings of the third ACM international conference on Web search and data mining
20. Xu G et al (2019) Sentiment analysis of comment texts based on BiLSTM. IEEE Access 7:51522–51532
21. Wang H et al (2020) Chinese implicit sentiment analysis based on hierarchical knowledge enhancement and multi-pooling. IEEE Access 8:126051–126065
22. Le Q, Mikolov T (2014) Distributed representations of sentences and documents. In: International conference on machine learning
23. Lipton ZC, Berkowitz J, Elkan C (2015) A critical review of recurrent neural networks for sequence learning. arXiv preprint arXiv:1506.00019

24. Sun X, He J (2020) A novel approach to generate a large scale of supervised data for short text sentiment analysis. Multimed Tools Appl 79(9):5439–5459
25. Titov I, McDonald R (2008) Modeling online reviews with multi-grain topic models. In: 17th international conference on World Wide Web
26. Kim S et al (2013) A hierarchical aspect-sentiment model for online reviews. In: Twenty-seventh AAAI conference on artificial intelligence
27. Wang C et al (2013) A phrase mining framework for recursive construction of a topical hierarchy. In: 19th ACM SIGKDD international conference on knowledge discovery and data mining
28. Blei DM, Griffiths TL, Jordan MI (2007) The nested chinese restaurant process and bayesian nonparametric inference of topic hierarchies. arXiv preprint arXiv:0710.0845
29. Kim JH et al (2012) Modeling topic hierarchies with the recursive chinese restaurant process. In: 21st ACM international conference on information and knowledge management
30. Paisley J et al (2014) Nested hierarchical Dirichlet processes. IEEE Trans Pattern Anal Mach Intell 37(2):256–270
31. Chen P et al (2017) Latent tree models for hierarchical topic detection. Artif Intell 250:105–124
32. Li W, McCallum A (2006) Pachinko allocation: DAG-structured mixture models of topic correlations. In: 23rd international conference on machine learning
33. Blei DM, Griffiths TL, Jordan MI (2010) The nested chinese restaurant process and bayesian nonparametric inference of topic hierarchies. J ACM (JACM) 57(2):7
34. Jain A et al (2020) Senti-NSetPSO: large-sized document-level sentiment analysis using neutrosophic set and particle swarm optimization. Soft Comput 24(1):3–15
35. Almars A et al (2018) Learning concept hierarchy from short texts using context coherence. In: International conference on web information systems engineering. Springer, Berlin
36. Xu Y et al (2018) Hierarchical topic modeling with automatic knowledge mining. Expert Syst Appl 103:106–117
37. Almars A et al (2018) Evaluation methods of hierarchical models. In: 14th international conference, ADMA 2018, Nanjing, China, November 16–18, 2018, Proceedings, pp 455–464
38. Vohra S, Teraiya J (2013) A comparative study of sentiment analysis techniques. J JIKRCE 2(2):313–317
39. Pang B, Lee L, Vaithyanathan S (2002) Thumbs up?: sentiment classification using machine learning techniques. In: ACL-02 conference on Empirical methods in natural language processing.
40. Turney PD (2002) Thumbs up or thumbs down?: semantic orientation applied to unsupervised classification of reviews. In: 40th annual meeting on association for computational linguistics.
41. Dave K, Lawrence S, Pennock DM (2003) Mining the peanut gallery: opinion extraction and semantic classification of product reviews. In: 12th international conference on World Wide Web

Teleportation; Next Leap to Outer Space

Khaled Elbehiery and Hussam Elbehiery

Abstract Going to space and in particular to the moon was the distinguished accomplishment of mankind for the twentieth century; however, teleportation and in particular, human teleportation is going to be the greatest achievement of all human being knowledge combined for the twenty-first century. The simple scientific definition of teleportation is when an object or a person be transported through a machine or a device and ends up reappearing at a different place. It all starts from the human brain which is very complex organ and in particular the human consciousness which is considered mathematically complicated. Consequently, many questions arise, could the human consciousness be decoded and mapped, could it be transported to an android or an avatar along with human's mind thoughts, is the quantum mechanics and computing is the methodology to achieve teleportation. On the other hand, the visionaries who changed our technological history might have been humans have crossed to other universes or they could be teleported humans from other universes. Lastly, if the human teleportation is a possibility, will it face any physical or moral challenges. Significant questions indeed and fortunately, this research paper intend to answer all of them. It is worth it to mention that Albert Einstein claimed that humans are theoretically capable of time travel or being teleported.

Keywords Teleportation · Multi-verse · Consciousness mapping · Avatar · Mind reading · Quantum · Teleported humans

1 Introduction

Former United States President John Kennedy on May 25th, 1961 had set the goal to land on the moon and when Neil Armstrong stepped down from the Apollo Lunar Module on July 21st, 1969 was the reason that has driven many great men to accomplish the impossible forward. NASA Mission Control Center 1969, the massive IBM

K. Elbehiery
Park University, Denver, CO 80126, USA

H. Elbehiery (✉)
October 6 University (O6U), Giza 12572, Egypt

D. A. Magdi et al. (eds.), *Digital Transformation Technology*, Lecture Notes in Networks and Systems 224, https://doi.org/10.1007/978-981-16-2275-5_9

System/360 Model 75 computer, with processing power 16.6 million instructions per second and up to 8 megabytes of main memory, with the goal to accomplish the greatest leap in human history which is putting a man on the moon, people across the world marveled at this technological achievement. But incredibly, only six decades later, a handheld device weighing less than half a pound dwarfs the total technology NASA possessed in 1969. Today's smartphone contains a staggering one million times the computing power used to carry out the moon landing. What we had when they went to the moon is like nothing compared to what an average teenager carries around now, the kind of computing power, the ability to access information, the ability to reach people, and it is an astonishing technological achievement. Teleportation in its simplistic meaning is just like when we scan an object, and it converts it to a digitized form then email it or send it as a normal data transmission to a destination. In turn, if the capability to scan a human being or just their mind exists and send the digitized human copy some other place, we have achieved human teleportation.

The research paper will cover few important angles of teleportation; first, mapping the human brain and consciousness is the initial step of teleportation to digitize every organ, every tiny bit of the human being, followed by transporting it to an android or even an organic form like an avatar. Second, it explores what the quantum mechanics and quantum computing has accomplished when it comes to transporting particles. Third, the journey of teleportation will face challenges, and the research paper will also shade lights on that topic. In addition, we could not ignore the evidences over history and the clues that teleportation might have already happened but sometimes in different format, but through technological visions from geniuses around us or they could be teleported from what is called multi-verse telepathically, or it could have happened with normal individuals that defy all law of physics and could go around the globe in just a blink of eye. The research paper is able to cover different views between science and evidences.

2 Human Consciousness Mapping

The technology of transferring someone's consciousness into another human's body is similar to the process of downloading a file from one computer and uploading it to another [1]. Over the centuries, we have compared our brain to clocks, telephone switchboards, and now of course computers; but really, the brain is none of these things, and it is its own thing. Stephen Hawking thinks so, and futurist Ray Kurzweil believes humans are only forty years away from achieving brain backups [2]. To read from the source which is the brain, scientists will have to learn how we store those memories [3]. Optogenetics technology is how memories are written, erased, and reactivated, and it could be taken further to "implant" a false memory into another genetically modified human.

Assuming we get to a point where we can read and write to the brain, we need to copy this information; but, the human's brain is not a finite storage system like a computer hard drive, and the estimate is 2.5 petabytes which is about a million

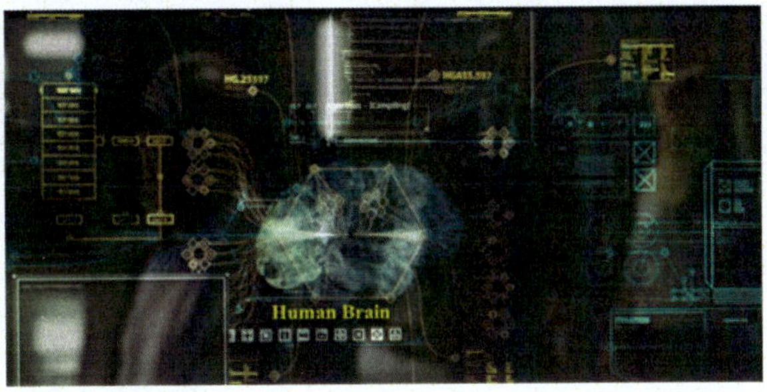

Fig. 1 Human consciousness mapping

gigabytes; if we want to copy 2.5 petabytes over a USB 3.0 connection, it would have to run continuously for more than 80 days, and even over Thunderbolt 3, it would take almost a week [4].

The biggest issue in consciousness transfer is going to be mapping the brain accurately. Computer scientists from giant companies, major universities, and world governments have been working on mapping and translating the brain network to a computer network for decades [5]. Billions of dollars are being spent on this, and according to top minds in the field, it is only a matter of time until we learn to emulate a brain in a computer and map a consciousness on top of it as depicted in Fig. 1 [6].

3 Human Consciousness Teleportation to an Android

This type of teleportation would end up involving an android in a very advanced robotic and bionics form, and it will look like the replicated person with a digital copy of the mind uploaded to it; basically, it is transferring human consciousness into a machine, as long as the neurological data is accessible, we are going to take the biological brain and imprint it into that synthetic brain to replicate the human mind. Then, the android goes and does whatever, exactly as the real person while the real person either sleeps or work or does anything else. Periodically, an updated copy of the mind could be sent to the android with new objectives or else or vice versa. Whatever the android experience and interpreted as memories get integrated back in to the original mind in the real person. The problem with mind uploading to an android that we get the brain scanned into a machine, and the digital copy ends up waving back from a monitor, we can say it is a live computerized copy of a human. Much money is already spent on this kind of teleportation and already has a high value especially in military application such as saving mortally of wounded soldiers, the best pilot put into a thousand individual drones, and the mind of a hacker loaded into the virus [7].

Google and Boston Dynamics latest robotic android is now going through basic trainings with the US Marine Corps which recently recruited the robotic dog for field trials with a name "spot" as shown in Fig. 2. The four-legged robot can trot on a smooth surface, navigate a variety of terrain with ease, and maintain its balance

Fig. 2 Android with human consciousness

even when been kicked. DARPA Robotics Challenge (DRC) engineers accompanied the Boston Dynamics Rover to the Marine Corps base in Virginia with a rabbit worked alongside soldiers in several tests combat situations, "spot" was required to scramble over rocky hills, walk through woods, and make its way around an urban environment, "spot" also acted as a scout; a role often filled by a working dog, in tests "spot" entered a building before the Marines with a goal of identifying any potential threats besides its uncanny ability to scramble over obstacles. "spot" also is safe to operate as it is wirelessly controlled using a game controller connected to a laptop computer; while the robot is exploring a dangerous area, the operator can remain at a safe distance controlling the rivet up to 500 feet away, and the operator also does not need a line of sight for the controls to work a useful feature that allows the robot to move along the optimal trajectory necessary to complete its mission [8].

At Zurich fly machinery in Switzerland, quadcopters are shown flying around and putting together a very sturdy road bridge that could support the weight of a human being. One of the big selling points of drones is that they can get to areas that are not exactly safe or accessible by humans such as assembling a road bridge that is sturdy enough for a person to walk across that is done entirely by flying machines, and every knot and braid was tied by the UAV using Dyneema rope as described in Fig. 3. The researchers claimed this demonstration as the first effort showing that tiny airborne drones are able to build a load bearing structures at full scale [9].

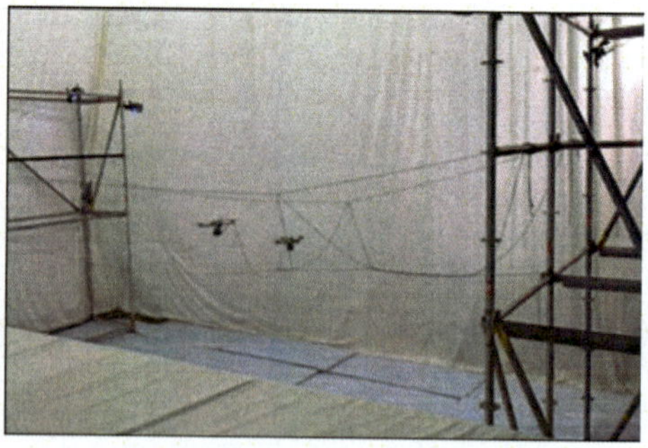

Fig. 3 Drones with human consciousness

4 Human Consciousness Teleportation to an Avatar

The android teleportation methodology has an obvious drawback which is the communication's update delay between the biological brain in the real person and the synthetic brain in the android. As known, technology does not know the impossible, and the real-time teleportation uploading becomes crucial. Consciousness comes from the biological activity of the brain; the brain is this incredibly complex organ composed of about 88 billion neurons; each neuron has a connection with about 10,000 other neurons. If the technology already succeeded to make a robotic hand move by your thoughts of a human being, so why not to control a whole robotic body [10].

Researchers at medical cybernetics and artificial intelligence have succeeded together to create a bionic exoskeleton that allowed a paralyzed man with a complete spinal cord to walk again using his own brainwaves. The paralyzed man was hooked up to a device that would receive electrical signals directly sent from the brain and transmit them to his knees given him the ability to move once again, and the man learned to operate the system by first wearing an electric AG cap that reads his brainwaves as he visualizes moving an avatar in a virtual reality environment. Scientists say that this is the first time a person with a complete paralysis in both legs has been able to walk without manually controlled robotic limbs even after years of paralysis, and the brain can still generate robust brain waves that can be harnessed to enable basic walking. This non-invasive system for leg muscle stimulation is a promising method and is an astonishing advance of current brain control systems that use virtual reality of a robotic exoskeleton as demonstrated in Fig. 4 [11].

Mind reading is possible, and researchers from the University of Washington successfully sent thoughts over the Internet in an unprecedented mind-reading experiment that shows two brains can be linked directly to allow one person to guess what is on the other person's mind as depicted in Fig. 5.

It is becoming the new normal that human being could be in fact a cyborg mixture between biological body parts and machines, and from the point of view of consciousness, it is about examining whether we can induce new conscious experiences that people have not experienced before. The goal at end is to remotely control a genetically engineered body telepathically to do what no human could do, to sustain an environment, tolerate extreme cold temperature, or high temperature that no human could afford, this called "Avatar." Avatar creates a believability for human on Earth,

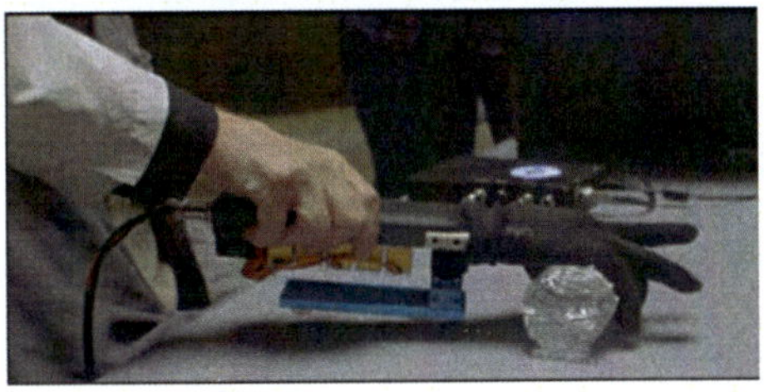

Fig. 4 Synthetic robotic telepathy

Fig. 5 Human telepathy

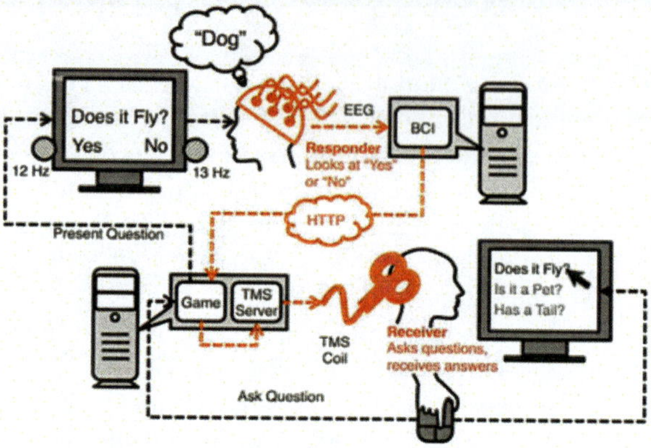

for example, to believe they are on Mars in a whole different environment, and it is already part of Elon Musk plan for the initial phase of Mars City. The Avatar program is working progress as shown in Fig. 6 [12].

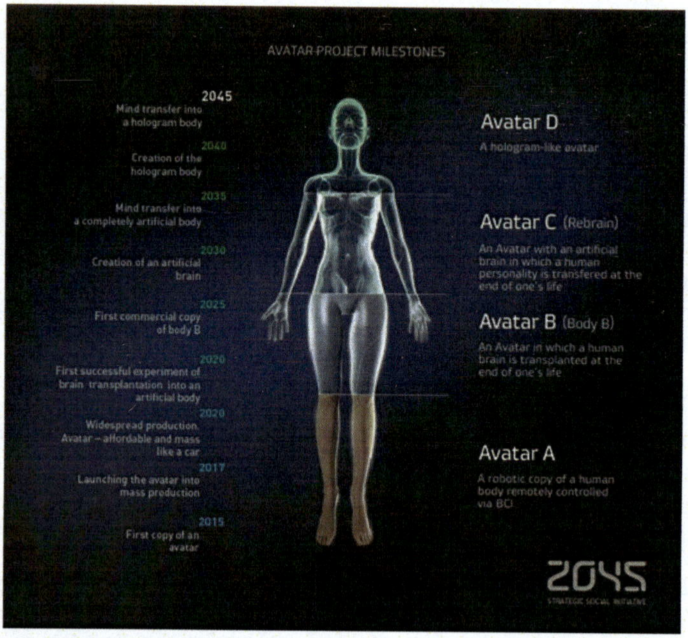

Fig. 6 Avatar program progress

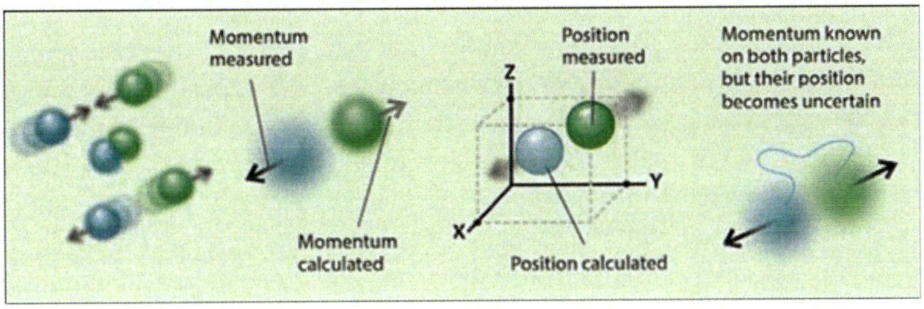

Fig. 7 Quantum entanglement

5 Teleportation Science

It all starts with the simple concept of quantum entanglement also referred to as the spooky action at a distance by Albert Einstein, and it is when two particles as twins and despite how much distance is between them, they will always be twins. The particles share an unexplainable bond as long as we know the information about one, and we can know the information about the other regardless of the amount of distance between the two. Remember that the twins are more like archenemies; whatever one is doing, the other is doing the exact opposite, and scientists measure this in terms of spin; if one particle is spinning up to the right, then we automatically know that the other particle is spinning down and to the left [13].

Quantum entanglement goes along with another phenomena, when observed quantum particles change, and when unobserved, they are simultaneously in all possible states which is referred to as a superposition [14]. Scientists found a way to get around this problem by adding a third particle to the mix and indirectly measuring them together. In simple terms, the sender adds particle one into a box with a third particle; then the sender can indirectly observe these particles and learn some information but not all the information about the particles, and this is called a bell measurement; while the sender still does not have all the information because they cannot look directly at the particles, they will still have enough information to achieve some very complicated algebra, and then the sender can send this information to the receiver; once the receiver has this information, they two can prepare some complicated algebra, but they still need one vital piece of information; the actual state of the first particle which means the sender will now look still indirectly at the particle and send that piece of information to the receiver as described in Fig. 7.

6 Teleportation Transport

The first successful teleportation experiment was in 1998, physicists at the California Institute of Technology partnered with two European groups and approved the teleportation theory as described by IBM, and they managed to successfully teleport a photon that represents the particle of energy that carries light [15]. This experiment teleported the photon just one meter in distance, and the experiment had given the hope that the theory should work regardless of the distance between the entangled particles. In 2004, the distance of a successful teleportation increased to 600 m, and then a few years later to 16 km, then 97 km, and as of 2017; the most impressive accomplishment for quantum teleportation, the Chinese scientists managed to teleport the quantum state of a photon from Tibet to especially designed quantum satellite in orbit around the earth for a distance of 1400 km or about 870 miles [16].

The Chinese scientists fired a laser beam carrying one end of the entangled particles into space; when sending the particle all the way into orbit, we risk interacting with objects and particles and therefore breaking the entanglement. To get around this issue, the scientists actually sent out millions of photons, and out of that only 911 made it to the satellite undisturbed. All in all, the experiment ended up being a success, and there are many examples of successful quantum teleportation that followed [17].

The Furusawa group at the University of Tokyo has succeeded in demonstrating complete quantum teleportation of photonic quantum bits by a hybrid technique for the first time worldwide. In 1997, quantum teleportation of photonic quantum bits was achieved by a research team at Innsbruck University in Austria; however, such quantum teleportation could not be used for information processing because measurement was required after transport, and the transport efficiency was low; it was only achieved in a probabilistic sense; so, quantum teleportation was still a long way from practical use in quantum communication and quantum computing [18].

The demonstration of quantum teleportation of photonic quantum bits by Furusawa group shows that transport efficiency can be over 100 times higher than before. Also, because no measurement is needed after transport, this result constitutes a major advance toward quantum information processing technology as architected in Fig. 8 [19]. The hybrid technique was developed by combining technology for transporting light waves with a broad frequency range and technology for reducing the

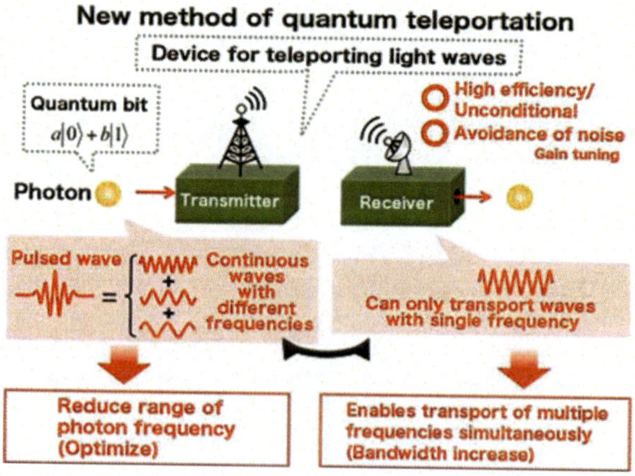

Fig. 8 Quantum transport methodology

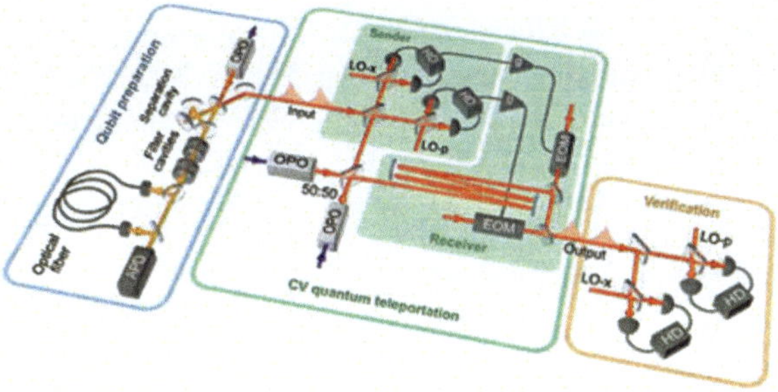

Fig. 9 Quantum gates fundamentals

frequency range of photonic quantum bits. This has made it possible to incorporate photonic quantum bit information into light waves without disruption by noise [20].

Teleportation can be thought of as quantum gate where input and output are the same; consequently, Furusawa group aims to increase the transport efficiency and make the device smaller by using photonic chips. In this way, the researchers plan to achieve further advances toward quantum computing, Fig. 9 [21].

7 Visionaries and Teleported Humans

At the Mind Research Network at the University of New Mexico, neuropsychologist Rex Young uses an imaging machine to measure the brainwave activity of a test subject. Dr. Young believes this test may show how creative inspiration strikes the brain. As the test subject performs these everyday tasks, the image machine measures normal brain activity, but then, Dr. Young has the test subject clear his mind, as this subject relaxes his mind, his brain activity decreases. But then suddenly, the image machine detects something new, alpha waves. Scientists say that these brainwaves indicate the unconscious mind is working behind the scenes, outside our conscious thoughts. The alpha wave has found to be associated with divergent thinking, the manifestation of creative cognition, and it is also associated with relaxing the brain, and this is something very important; it is the manifestation of genius, this ability to think of new and useful ideas that have not been thought of before [22]. Those incredible individuals always know farther than most intelligent minds in their time would ever know, and they had the vision to see what was important, and they had so many beautiful ideas that some of which are just waiting to be developed. In many cases with no formal training, repeatedly stuns the academic world with innovative theorems, even some of the world's leading mathematicians are confounded by their remarkable formulas [23].

On the other hand, because those visionaries seem to access knowledge outside the brain, and those true human geniuses could be the product of a universal mind or collective unconsciousness of another world or another universe that ended up teleported into our universe, the world we know. Albert Einstein received the inspiration for his ground-breaking theory of relativity in a dream, and Friedrich August Kekul discovered the elusive shape of the benzene molecule during a daydream in which he saw a snake chasing its tail, and the brilliant Russian chemist, Dimitri Mendeleev, literally dreamed up the periodic table of the elements, he saw exactly where all of the elements lined up [24]. The following are just a glimpse on three visionaries or believed to be teleported humans that no doubt they changed the world:

Srinivasan Ramanujan: Emory University, Atlanta, Georgia, December 2012, after years of work, mathematician Ken Ono, and two of his former students come up with a ground-breaking mathematical formula that will allow scientists to study black holes in an entirely new way. Incredibly, they achieved this by studying a single paragraph written by an Indian mathematician over nine decades earlier, Srinivasan Ramanujan as shown in Fig. 10. Srinivasan Ramanujan was an Indian mathematician who is unlike any other genius in world history, and Ramanujan's work has now formed the basis for superstring theory and multi-dimensional physics, and some of the most advanced math that all the high-end scientists are still using today is called modular functions, which could lead to time travel, anti-gravity, limitless free energy, and it is used by physicists dealing with relativity and quantum mechanics, basically all of this futuristic technology. Ramanujan made breakthroughs in integral calculus which can be used to determine the drag force buffeting a wing as it slides through the air or the gravitational effects of the Earth on a man-made satellite. But perhaps what is most noteworthy is that Ramanujan insisted these baffling theorems were not simply the product of his own genius. He claimed they were communicated to him by Hindu goddess known as "Namagiri," transmitted these theorems to him. Was it really his goddess or actually was another mind somewhere else or some other time or some other universe communicating with his mind on a certain frequency with a special brain capability.

Nikola Tesla: Colorado Springs, July 1899, while testing a magnifying transmitter built to track storms, Nikola Tesla claims he received some sort of transmission from an unknown source. One night he was tracking thunderstorms 600 miles away; but all of a sudden, he heard these beeps, and it was three beeps in sequence, and it means it was mathematical, and it did not make any sense to him, the more he thought about it, he thought that they came from outer space. These communications went for a while, and it led him to the point where he was actually able to receive useful information that was helping him build his inventions. Tesla had direct contact through the thing he had

Fig. 10 Srinivasan Ramanujan

Fig. 11 Nikola Tesla

invented to the world with teleportation intelligence as shown in Fig. 11. At Nikola Tesla Museum, Belgrade, Serbia, it is housing more than 160,000 documents that offer rare insight into some of Tesla's future plans, including what some believe to be drawings of spaceships. Tesla began work on his flying machine in 1910, focusing on the use of field propulsion, or anti-gravity and during this period, Tesla had discovered that high amounts of electricity could actually create lift in an object. Throughout the 1920s and 1930s, Tesla continually talked about anti-gravity ships that could derive power from his Wardenclyffe Towers that were going to be broadcasting power. He claimed these ships did not have wings or fuel; they were completely electric. The everlasting conclusion is that Tesla had received otherworldly knowledge in order to design anti-gravity spaceships.

Steve Jobs: It is worth it to point to the visionary who jump-started the micro-computer revolution, Steve Jobs. Steve Jobs was one of the greatest visionaries in Silicon Valley, the idea of what he was doing is how you popularize computing, he managed to deliver into the hands of consumers a device that was usable, it was intuitive, it was easy to use, and it was easy to understand and that is not a small thing in the simplicity and the beauty of it. Steve Jobs and his team of engineers at Apple harnessed technology that connected society digitally and put all the world's knowledge literally at mankind's fingertips as shown in Fig. 12. But the seeds of this technological revolution were planted in 1973 when Jobs was 19-year-old college student dropped out of school, and traveled to India, he discovered a Hindu guru known as Haidakhan Babaji. Haidakhan Babaji claimed that he had no mother or

Fig. 12 Steve Jobs

father, and Steve Jobs did spend some time with him, and he was being influenced telepathically by an Haidakhan Babaji that claimed be a celestial being who had come to earth to help enlighten Earth planet and to advance humanity forward.

8 Multi-verse Teleportation

Scientists have achieved quantum teleportation in the laboratory, however, is human teleportation even possible, and if so, how often does it happen. Human history is full of examples of real-life teleportation that were written, verified for their trustworthy, and should be mentioned [25]:

The Germans during World War II were testing a highly advanced machine known as "Die Glocke" or the "Bell" as depicted in Fig. 13. The German scientists Hermann Oberth and Wernher von Braun were working on Bell shaped machine with electro-magnetic and propulsion system, and it was suggested to be "Wunder Waffe" or the "Wonder Weapon," and a weapon is so powerful that even in the last minute of the war as the soviet army closed on Poland, it could have changed the whole course of the war that could only happen if the German army were teleported behind the enemy lines without actually seen, and the bell might have been a time or a teleporting machine in this sense [26].

At Tokyo International Airport, Japan, 1954, an ordinary passenger was handing over his passport to the customs' officer, the authorities noticed something suspicious that his passport looked authentic enough but it listed his country of origin as a mysterious place called Taured, unfortunately this place does not exit, interestingly enough he claimed he was flying to Japan for business as it has been for the last five years and the visa stands from his travels, he also carried legal currency from various countries and even a driver's license that was issued by the Taured government. After a prolonged interrogation, the man from Taured was sent to a nearby hotel where he was guarded by two immigration officers, and mysteriously, the following morning, the officers went into the hotel room and discovered they had vanished without a trace.

Fig. 13 Bell (Die Glocke)

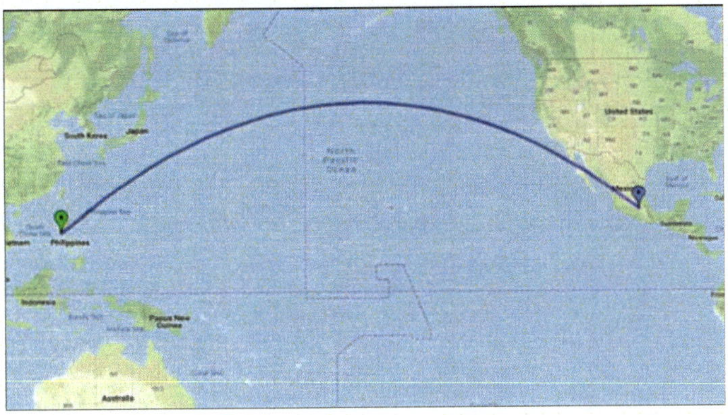

Fig. 14 Manila, Philippines to Mexico City, Mexico (9000 Miles)

Mexico City, Mexico, October 25, 1593, from the country's official records, Gil Perez was wearing a foreign uniform and wielding a different type of musket to the rest of the soldiers outside the governor's palace in Manila, Philippines, but in fact he was in Mexico City, Mexico. Manila and Mexico City are nearly 9000 miles apart on opposite sides of the Pacific Ocean as shown in Fig. 14. Gil Perez was later jailed for two months until the ship from Manila docked in Mexico City, and the soldiers on board the ship knew Perez and were able to confirm that he was in Manila the day before his ill-fated visit to Mexico City. In a time before powered flight, how did Perez travel so far so quickly and why did he have no memory of the journey.

Buenos Aires, Argentina, 1968, Dr. Geraldo Vidal and his wife Bravo David Alma were driving their Peugeot 403 along a remote road; when they were suddenly surrounded by a thick fog, everything went black, and the Vidals were rendered unconscious, a full 48 h later Dr. Geraldo Vidal called his family to let them know that he was safe but that he and his wife were in Mexico City around 4000 miles away, and they had no idea how they got there, and they both had sore necks and felt like they had spent a long time asleep the car had been badly burned (see Fig. 15).

9 Human Teleportation Challenges

The application of the technology so far exists just in the quantum world, weirdly enough, the quantum world seems to have different laws than that are macroscopic world, the macroscopic world being that the objects that can be viewed by the naked eye, the body of laws that apply to the quantum world is referred to as quantum mechanics, this means the first challenge of teleportation would be to imitate a similar situation that would work outside of the world of quantum mechanics, as of now that seems pretty unattainable for physicists [27].

Fig. 15 Buenos Aires, Argentina to Mexico City, Mexico (4000 Miles)

But for the sake of the argument, assume that we are not sending the object itself but just the information of the state of the object while destroying the original; so, we could teleport a suitcase full of all belongings to the vacation destination; but in reality, it would just be an exact replica but not the original. This should work perfectly fine but the real ethical challenge comes when we talk about human teleportation; because if we were to teleport a human using the same rules, humans themselves would not be teleported but the information about the exact makeup of the human would be transported to a set of particles, ending up with a destroyed human on the other end [28].

Let's also assume for a minute the teleportation was successful, and then, there is an important thing to think about at this point; the soul, how do we really know when something is actually alive, theoretically, we would assume active brain circuits qualify it as a quality to life, and perhaps, a pumping heart would mean alive in the physical sense but we have never tested the idea of transferring one life to another body, does that even mean would the body on the other side actually be alive [29].

Assume the scientists succeed, then every human teleportation would be a substantial risk, and the human body is very complex and extremely deliberate even the slightest mix-up of a single molecule could lead to severe neurological or physiological damage; there could be no margin of error at all if we ever wanted to really use the technology, that in itself would be a difficult challenge.

10 Conclusion

If cloning of embryonic stem cell has changed not only the course of human life, but ultimately the course of humanity, scientists have actually made some significant strides when it comes to teleportation to make it physically possible, and teleportation will allow humanity to make major leaps forward. At the present time, we can only teleport photons (particles of light) and atoms like Cesium and Rubidium; however, in the coming years, scientists expect to teleport molecules such as water and carbon dioxide, perhaps afterward DNA and organic molecules. Now to teleport a human, atoms will fall apart then to rebuilt again with the same memories and personality of the original human; basically, the original dies first then a new one somewhere else a carbon copy of the original comes to life, and this logic raises an important question about the human soul, could it be copied, could it be altered, could it be multiplied, could it be hacked, is it just information. Physicists say it is physically possible to teleport an entire human being across the room or maybe to Mars, but the concern about soul will always remain until actually human teleportation happens.

With regard to transmitting the human digitized scan copy, there will be much bandwidth involved because human body have about 1026 atoms in every kilogram of the body, and we are going to need a kilobyte of data just to accurately record each atom type and isotope as it was, where it was positioned, a person might weigh 100 kg, then we are likely be looking at as much as 1031 bytes of data, or simply said 10 billion, billion terabytes. The current technology with a 400-nm wavelength blue laser could do 200 Tb/s, even a shorter wavelength UV laser might be able to send 1031 bytes in a 1017 s, or simply said a few billion years, not a very fast way to travel.

All in all, there is undeniable progress with quantum physics and quantum computing, this progress is measured in results, and the results are obvious and promising; teleportation is feasible if we look how far the computers have come just in the past fifty years, teleportation is possible.

Acknowledgements The authors would like to thank ***Park University, USA*** and ***October 6 University (O6U), Egypt*** for their support to introduce this research paper in a suitable view and as a useful material for researchers. The authors also would like to thank their colleagues who provided insight and expertise that greatly assisted the research.

References

1. MIT Technology Review (2019) Available: https://www.technologyreview.com/2019/07/19/238809/nathan-copeland-man-with-brain-implant-on-elon-musk-neuralink-brain-computer-interface/
2. The Institution of Engineering and Technology (2019) Musk's Neuralink aims to implant devices in human brains in 2020. The Institution of Engineering and Technology, E&T editorial staff
3. Bloomberg; ©2020 Bloomberg L.P., 2020. Available: https://www.bloomberg.com
4. International Business Times; ©Copyright 2020 IBTimes Co., Ltd., 2020. Available: https://www.ibtimes.co.uk
5. Tong F, Pratte MS (2011) Decoding patterns of human brain activity. Ann Rev Psychol 63(1):483–509. https://doi.org/10.1146/annurev-psych-120710-100412
6. Martin S, Mikutta C, Knight RT, Pasley BN (2016) Understanding and decoding thoughts in the human brain. NeuroScience Published. https://doi.org/10.3389/frym.2016.00004
7. Kurzweil Network (2020) Accelerating Intelligence. Available: https://www.kurzweilai.net

8. MIT Media Lab Developing Lighter (2020) More Powerful Bionic Ankle; IEEE Spectrum, © Copyright 2020 IEEE Spectrum. Available: https://spectrum.ieee.org
9. Elbehiery K, Elbehiery H (2020) Artificial intelligent bionics; Disability to exceptional ability. Int J Eng Sci (IJES) 9(04):23–33, ISSN (e): 2319-1813 ISSN (p): 20-24-1805. https://doi.org/10.9790/1813-0904012333.http://www.theijes.com/Vol9-Issue4.html
10. Borghino D (2012) "Avatar" project aims for human immortality by 2045, July 2012. Available: https://newatlas.com/avatar-project-2045/23454/
11. Deutsch D (2003) Science and ultimate reality: from quantum to cosmos. In: Barrow JD, Davies PCW, Harper CL (eds) Cambridge University Press, Cambridge
12. Elbehiery K, Elbehiery H (2020) Space industry; future giant leap investment. Seventh Sense Research Group. Int J Indus Eng (IJIE), 7(2):23–32. ISSN: 2349-9362. https://doi.org/10.14445/23499362/IJIE-V7I2P105
13. Davis EW (2004) Warp Drive Metrics, AFRL-PR-ED-TR-2003-0034. Teleportation Physics Study
14. Aczel AD (2002) Entanglement: the greatest mystery in physics. Four Walls Eight Windows Press, New York
15. Moeller C (1972) The theory of relativity, 2nd edn. Clarendon Press, Oxford
16. Langley P, Simon H, Bradshaw G, Zytkow J (1987) Scientific discovery. MIT Press, Cambridge
17. Heitler W (1954) The quantum theory of radiation, 3rd edn. Clarendon Press, Oxford
18. University of Tokyo (2020) Large scale quantum computing with quantum teleportation. Available: http://www.alice.t.u-tokyo.ac.jp/index-en.php
19. Takeda S, Mizuta T, Fuwa M, van Loock P, Furusawa A (2013) Deterministic quantum teleportation of photonic quantum bits by a hybrid technique. Nature 500:315–318. News & Views: T. C. Ralph, Nature 500:282
20. Bennett GL, Forward RL, Frisbee RH (1995) Report on the NASA/JPL workshop on advanced quantum/relativity theory propulsion. AIAA-95-2599, 31st AIAA/ASME/ASE/ASEE Joint Propulsion Conference and Exhibit, San Diego, CA
21. Tsurumoto K, Kuroiwa R, Kano H, Sekiguchi Y, Kosaka H (2019) Quantum teleportation-based state transfer of photon polarization into a carbon spin in diamond. Available: https://www.nature.com/articles/s42005-019-0158-0
22. Howell E (2017) Time travel: theories, paradoxes & possibilities, © Future US, Inc. Space is part of Future US Inc.
23. Northrop Grumman Co.; Where is creativity in the brain?, Neuroscience scientific discoveries, @2020 Northrop Grumman Co., August 2020. Available: https://now.northropgrumman.com/where-is-creativity-in-the-brain/
24. Miller AI (1986) Secrets of Genius. Review of Imagery in Scientific Thought, MIT Press, Cambridge. New Ideas in Psychol 8(2):231–246, 1990 0732-118X/9O Printed in Great Britain Pergamon Press
25. Moskowitz C (2012) 5 reasons we may live in a multiverse. December 2012. Available: https://www.space.com/18811-multiple-universes-5-theories.html
26. Tegmark M (2003) Parallel Universes. Cambridge University Press, Cambridge
27. Bolton D (2016) Russian billionaire Dmitry Itskov seeks 'immortality' by uploading his brain to a computer, March 2016. Available: https://www.independent.co.uk/news/science/dmitry-itskov-2045-initiative-immortality-brain-uploading-a6930416.html
28. Birrell ND, Davies PCW (1982) Quantum fields in curved space. Cambridge University Press, Cambridge
29. Elbehiery K, Elbehiery H (2020) Quantum cloud. Seventh Sense Research Group®, Int J Comput Trends Technol (IJCTT) 68(7):8–16, Louisiana, USA, P-ISSN: 2349-0829, E-ISSN: 2231-2803, July 2020. https://doi.org/10.14445/22312803/IJCTT-V68I7P102

Practical Comparison Between the LR(1) Bottom-Up and LL(1) Top-Down Methodology

Nabil Amein Ali

Abstract The syntax analysis or parser is the essential stage of designing compilers, and the role of the parser is parsing strings according to a definite number of rules. Researchers often use either LL(1) top-down technique or LR bottom-up technique to parse strings to decide whether all these strings or some of them are accepted by some defined language. Theoretically, many papers illustrated that the LR method is more suitable for LL(1). The current paper treats the problem of parsing practically. It proves that the LR technique is more suitable of LL(1) in terms of the computing time of parsing in each technique.

Keywords Parser · Top-down · Bottom-up · LL(1) · LR(1) · String

1 Introduction

The aim of the parser is to verify the correctness of programs in terms of the syntax of the programming language which means that the purpose of parser is to know whether the source program (strings) can be derived or not from the distinguished nonterminal symbol using definite given productions rules [1]. It is known that every programming language has precise production rules that prescribe the syntactic structure of efficient formed programs. The syntax of programming language statements can be expressed by context-free grammars (CFGs). CFGs offer significant benefits for compiler designers. It gives a precise, yet easy-to-understand, syntactic specification of a programming language [2]. The CFGs are used in the paper as suitable production rules. The left-to-right top-down (LL(1)) and left-to-right bottom-up (LR(1)) are the two essential parsing approaches. These approaches are deterministic parsing methods and the term "deterministic" means that no searching is involved during the process of parsing strings. The deterministic parsing methods have the advantage that they require an amount of time that is a linear function of the length of the input: They are linear time methods [3]. The paper develops two programs, top-down LL(1) and bottom-up LR(1) programs. The two programs will be implemented

N. A. Ali (✉)
Suez Institute of Management Information Systems, Suez, Egypt

D. A. Magdi et al. (eds.), *Digital Transformation Technology*, Lecture Notes in Networks and Systems 224, https://doi.org/10.1007/978-981-16-2275-5_10

practically to parse a number of string using definite CFGs. The parsing of strings verifies the syntactic correctness [4]. In all practical implementation processes of the two programs, the computing time will be calculated for comparing the performance of the two approaches.

2 Theoretical Comparison Between LL(1) and LR(1)

LL(1) grammar is a grammar that can be parsed by an LL parser with one lookahead symbol. That is, if a nonterminal has more than one right-hand side, the parser can decide which one to apply by looking at the next input symbol. It is possible to have an LL(k) grammar, where k is some number other than 1, and which requires k symbols of lookahead, but such grammars are not very practical [5]. The LL(1) technique is different from LR(1) technique. It has to check whether the CFG is qualified to use the LL(1) method or not. There are many problems concerned with the used CFGs such as left recursion, left factoring and ambiguous. If these problems exist, then the LL(1) cannot be applied immediately until they are resolved, i.e., not all grammars represent LL grammars. However, some non-LL grammars can be converted to equivalent LL grammars by a transformation that removes the undesirable grammatical phenomenon. Unfortunately, the transformation cannot be used for all non-LL grammars, for some grammars; there exists no equivalent LL [6]. Also, the resolving of some of these problems by transformation will exceed the number of grammars, and so the computing time. Since LR(1) doesn't have problems, it can be used immediately to a wider category of grammars and languages. Furthermore, usually there is no need for the transformation of the grammar. Therefore, theoretically LR(1) method is more superior than LL(1). This makes the LL(1) as a property not enough for obtaining proper parsing [7]. Furthermore, the comparison between the two approaches in terms of the sizes of syntactical analysis tables illustrates that the elements of LL syntactical analysis that can be stored in single word may reduce the size of the typical LL(1) syntactical analysis table to be less than the LR(1) syntactical analysis table. However, in some cases, comparison is not always in favor of the LR(1) method [1]. The best way for parsers is to take advantages of both methods and combine them to carry out the analysis process.

Some languages have been designed with specific parsing methods in mind: Pascal was designed for LL(1) parsing while C was originally designed to fit LALR(1) parsing which belongs to the LR class, but this property was lost in later versions of the language, which has more complex grammars. Most parser generators are based on LALR(1) parsing, but a few use LL parsing [8]. The C compiler on PDP-11 adopts recursively descending method, i.e., the top-down syntactical analysis method to handle most of the expressions, while for some other expressions; it uses the simple bottom-up method [1]. It is obvious that the theoretical comparison is not in favor to one against the other; the next section illustrates how the practical comparison can be effective to favor one approach over the other approaches.

3 Parsing Strings Using Parsers

The paper suggests two programs, parsing strings by LL(1) program (PSLL1) and parsing strings by LR(1) program (PSLR1). The two programs have been developed for parsing strings, and so, the paper supposes that LL(1) and LR(1) tables are given in each case. Now, the purpose of the paper is to calculate the computing time for parsing strings using PSLL(1) and PSLR(1). Then the resulted computing time can be used for holding a practical comparison between the LL(1) and The LR(1) methods. To guarantee that this comparison is fair, the running of PSLL(1) and PSLR(1) requires multiple definite criteria. These criteria are concerned with the hardware ability, the available programming software, source code, the used production rules and the string itself. These criteria can be stated by the next section.

3.1 The Criteria of Parsing Strings Using PSLL(1) and PSLR(1)

In this section, the paper explains the criteria for running the two programs, in general.

- The two programs must be executed by the same device with the same operating system.
- The source codes of the two programs must be designed using the same software.
- The input string must be the same length for each execution.
- All the input string must be accepted.
- The two sets of grammar rules for the two programs are identically used in the two programs.

If the problems appear during the LL(1), it is solved on spot according to the kind of the appearing problem. This is apparent in Fig. 3, where the left recursion problem is solved first before passing the program to the next step of compilation.

The computing time is calculated by executing PSLL(1) and PSLR(1), of course, the time depends on the abilities of the used device, and therefore, the paper introduces the used device's specifications, as follow:

- Intel Core i3 CPU, 2.4 GHz.
- RAM: 3 GBs.
- Operating System: 32-bit, Windows 7.
- The source program is designed using Microsoft Visual C++ 2005.
- The production rules can be expressed by the CFGs.

E=E+T|T

T=T*F|F

F=(E)|i

Fig. 1 CFGs for LR(1)

3.2 *The LR(1) Grammars*

The CFGs are used here for expressing the production rules of the LR(1) method Fig. 1.

The CFGs of Fig. 1 can be used directly through the PSLR(1) program without any transformation, where E, T and F are called nonterminals while the symbols (,), i are called terminals.

It should be noted that the CFGs of Fig. 1 are a subset of the production rules of the C language [9]. Since, the C language constructs can be divided into different syntactic categories. A syntactic category is a sub-language that embodies a particular concept. For example, statements may contain expressions, so some of the productions for statements use the main nonterminal for expressions E, T represents terms, F represents factors and i represents identifiers [10]. This means that the designed programs in our paper deal with a sub-language of the C language not with the whole language.

3.3 *The LL(1) Grammars*

The CFGs of Fig. 1 cannot be used here with the same form by LL(1) method, since; the grammars in Fig. 2, are left recursion.

Now, a transformation is needed to eliminate left recursion, by rewriting the left recursion grammars of Fig. 1, we get the CFGs of Fig. 3.

Now, the CFGs of Fig. 3, can be used by The PSLL(1) program.

E=E+T

E=T

T=T*F|F

F=(E)|i

Fig. 2 Left recursion CFGs

E=TA

A=+TA

A=e

T=FB

B=*FB

B=e

F=(E)

F=i

Fig. 3 CFGs for LL(1)

4 Parsing Strings Using PSLR(1)

In bottom-up parsing technique, first, the input string is considered, and then by using the grammars, the string is reduced to obtain the start symbol. The process of parsing halts successfully as soon as the start symbol is obtained (acceptance state) [11]. The PSLR(1) uses this concept to calculate the computing time for parsing the input string.

4.1 The Data Structure of the PSLR(1) Program

A number of data structure are used by the PSLR(1) program. These data structure may be C++ struct data structure, arrays and functions. Table 1 illustrates the data structures of the program.

Also, the program has the following given data:

Table 1 Interested data structures of the PSLR(1) program

Data structure	Type	Purpose
struct action	Data structure	Representing the action part of the parsing table
struct goto	Data structure	Representing the goto part of the parsing table
struct grammar	Data structure	Representing the cfgs
ter[]	Array	Storing the terminals of PSLR(1)
nter[]	Array	Storing the nonterminals of PSLR(1)
states[]	Array	Storing states ('a', 'b', 'c', 'd', 'e', 'f', 'g', 'h',
stack[]	Array	'm', 'j', 'k', 'l')
char push(char *,int *,char)	Function	Storing the pushed elements of the input string
char pop(char *,int *);	Function	Moving of the input symbol from the input buffer to the stack
		Popping the RHS of the appropriate rule

- The production rules or CFGs as in Fig. 1.
- The states ('a', 'b', 'c', 'd', 'e', 'f', 'g', 'h', 'm', 'j', 'k', 'l') which corresponding to ('I0', 'I1', 'I2', 'I3', 'I4', 'I5', 'I6', 'I7', 'I8', 'I9', 'I10', 'I11').
- The LR(1) parsing table, which represented by the data structure action and goto.

4.2 The Main Function Code of PSLR(1)

The main code function of the PSLR(1) can be stated as follow:

```
int main(){
char inp[80],x,p,dl[80],y,bl='a';
int i=0,j,k,l,n,m,c,len;
float Start,End,Diff;
printf(" Enter the input :");
scanf("%s",inp);  /* reading the input string*/
Start= omp_get_wtime();* calculating the initial time*/
len=strlen(inp);/* calculating the length of the input string*/
inp[len]='$';
inp[len+1]='\0';
push(stack,&top,bl);   /* pushing "a" to the stack*/
do{
x=inp[i];
p=stacktop(stack);
isproduct(x,p);
if(strcmp(temp," ")==0) error();
if(strcmp(temp,"acc")==0)
break;
else{ if(temp[0]=='s'){
push(stack,&top,inp[i]);
/* pushing input string elements to the stack*/
push(stack,&top,temp[1]);
i++;}
else{
if(temp[0]=='r'){
j=isstate(temp[1]);
strcpy(temp,rl[j-2].right);
dl[0]=rl[j-2].left;
dl[1]='\0';
n=strlen(temp);
for(k=0;k<2*n;k++)
pop(stack,&top);/*popping the top element from the stack*/
for(m=0;dl[m]!='\0';m++)
push(stack,&top,dl[m]);
l=top;
y=stack[l-1];
isreduce(y,dl[0]);
for(m=0;temp[m]!='\0';m++)
push(stack,&top,temp[m]);
}}}
printt(stack,&top,inp,i);}
while(inp[i]!='\0');
```

```
if(strcmp(temp,"acc")==0)
printf("\n accept the input ");
else
printf("\n do not accept the input ");
End= omp_get_wtime();/*calculating the final time*/
Diff=End-Start;/* calculating the actual computing time*/
cout<<"Diff="<<Diff*1000<<endl;
}
```

4.3 The Subroutine Codes of PSLR(1)

The subroutine codes of PSLR(1) can be stated as follow:

```
void push(char *s,int *sp,char item)
{
if(*sp==100)
printf(" stack is full ");
else
{
*sp=*sp+1;
s[*sp]=item;
}}
char stacktop(char *s)
{
char i;
i=s[top];
return i;}
void isproduct(char x,char p)
{
int k,l;
k=ister(x);
l=isstate(p);
strcpy(temp,A[l-1].row[k-1]);
}
int ister(char x){
int i;
for(i=0;i<6;i++)
if(x==ter[i])
return
i+1;
return 0;
}
int isnter(char x)
{int i;
for(i=0;i<3;i++)
if(x==nter[i])
return i+1;
return 0;
}
int isstate(char p)
```

```
{
int i;
for(i=0;i<12;i++)
if(p==states[i])
return i+1;
return 0;}
void error()
{
printf(" error in the input ");
exit(0);
}
void isreduce(char x,char p)
{
int k,l;
k=isstate(x);
l=isnter(p);
strcpy(temp,G[k-1].r[l-1]);
}
char pop(char *s,int *sp)
{
char item;
if(*sp==-1)
printf(" stack is empty ");
else
{
item=s[*sp];
*sp=*sp-1;
}
return item;
}
void printt(char *t,int *p,char inp[],int i)
{
int r;
printf("\n");
for(r=0;r<=*p;r++)
rep(t,r);
printf("\t\t\t");
for(r=i;inp[r]!='\0';r++)
printf("%c",inp[r]);
}
```

4.4 Running PSLR(1) Program

A number of strings must be passed to PSLR(1). The passed input strings must be accepted. For example, the following input strings can be selected:

- The 1st string: "i + i" (identifier + identifier).
- The 2nd string: "i + i * i" (identifier + identifier * identifier).
- The 3rd string: "i * i * i" (identifier * identifier * identifier).

Table 2 Parsing the first string using PSLR(1) program

Input string	i + i	
	Stack	Input
Parsing steps	0	i + i$
	0i5	+ i$
	0F3	+ i$
	0T2	+ i$
	0E1	+ i$
	0E1 + 6	i$
	0E1 + 6i5	$
	0E1 + 6F3	$
	0E1 + 6T9	$
	0E1	$
Result of parsing	Accepted	
Computing time	3.39 ms	

The following tables illustrate the results of parsing input string and the corresponding computing times in each time of running PSLR(1) program.

Table 2 illustrates the parsing of "i + i" input string and the computing time.

Table 3 illustrates the parsing of "i + i * i" input string and the computing time.

Table 4 illustrates the parsing of "i * i * i" input string and the computing time.

The computing time of parsing of the three input strings using PSLR(1) program can be illustrated by Table 5.

Table 3 Parsing the second string using PSLR(1) program

The input string	i + i * i	
	Stack	Input
Parsing steps	0	i + i * i$
	0i5	+ i * i$
	0F3	+ i * i$
	0T2	+ i * i$
	0E1	+ i * i$
	0E1 + 6	i * i$
	0E1 + 6i5	* i$
	0E1 + 6F3	* i$
	0E1 + 6T9	* i$
	0E1 + 6T9 * 7	i$
	0E1 + 6T9 * 7i5	$
	0E1 + 6T9 * 7F10	$
	0E1 + 6T9	$
	0E1	$
Result of parsing	Accepted	
Computing time	8.31 ms	

Table 4 Parsing the third string using PSLR(1) program

The input string	i * i * i	
	Stack	Input
Parsing steps	0	i * i * i$
	0i5	* i * i$
	0F3	* i * i$
	0T2	* i * i$
	0T2 * 7	i * i$
	0T2 * 7i5	* i$
	0T2 * 7F10	* i$
	0T2	* i$
	0T2 * 7	i$
	0T2 * 7i5	$
	0T2 * 7F10	$
	0T2	$
	0E1	
Result of parsing	Accepted	
Computing time	5.98 ms	

Table 5 Computing time of PSLR(1) program

Input string	Computing time (in ms)
1st string	3.39
2nd string	8.31
3rd string	5.98

5 Parsing Strings Using PSLL(1)

The top-down method scans the input string from left to right, one symbol at a time. It attempts to find a leftmost or rightmost derivation for an input string. A parse tree for the input string will be constructed from the root and creating the nodes of it. The input string is accepted if and only if, at the moment when the blank symbol is read and the top symbol on the stack is $ [12]. The PSLL(1) uses this technique to calculate the computing time for parsing the input string.

The program has the following given data:

- The production rules or CFGs as in Fig. 3.
- The values of the first and follow functions of each nonterminal.
- The LL(1) parsing table, which has been constructed using the first and follow functions.

5.1 The Main Function Code of PSLL(1)

/*The main code function of the PSLL(1) can be shown as follow:*/

```
int main(){
printf("Enter the input:");
cin>>input;
Start= omp_get_wtime();
for(j=0;j<ttt;j++)
top++;
stack[top]='$';
top++;
stack[top]=Nonter[0];
cout<<"\nStack Input Action \n "<<endl;
i=0;
cout<<"$"<<stack[top]<<"\t ";
cout<<input<<"\n"<<endl;
while(stack[top]!='$'){
int x=0;
r=NonTerIndex(stack[top],tnt);
s=TerIndex(input[i],Tter,ttt);
L=LL1[r+1][s+1];
int Le=L.length();
if(L=="r")
{cout<<"\nString is rejected "<<endl;break;}
else
if(L=="e"){pop();}
else{
pop(); for(j=Le-1;j>=0;j--)
{ch=L.at(j);
push(ch);}}
for(k=0;k<=top;k++)
cout<<stack[k];
for(k=i;input[k]!='\0';k++)
cout<<input[k];
cout<<"\t ";
if((stack[top]==input[i])&&(stack[top]!='$'))
{ pop();
  i++;}}
if((stack[top]=='$')&&(top==0))
cout<<"\nThe string is accepted";
End= omp_get_wtime();
Diff=End-Start;
cout<<"Diff="<<Diff*1000<<endl;
return 0;
}
```

5.2 The Subroutine Functions Code

The subroutine codes of PSLL(1) can be stated as follow:

```
/*Function to return Nonterminal index*/
int NonTerIndex(char c,int tnt)
{
int i;
//cout<<"c ="<<c<<endl;
for(i=0;i<tnt;i++)
if(c==Nonter[i])
return i;
}
/*Function to return terminal index*/
int TerIndex(char c, char Tter[m],int ttt)
{
int j;
//cout<<"c ="<<c<<endl;
for(j=0;j<ttt;j++){
//cout<<"Tter ="<<Tter[j]<<endl;
if(c==Tter[j])
return j;
}
}
/*Function to add symbol to stack*/
void push(char c){
top++;
stack[top]=c;
return;
}
/*Function to pop symbol from stack*/
void pop()
{
top--;
return;
}
```

5.3 Running PSLL(1) Program

Depending on the criteria of parsing strings using PSLL(1) and PSLR(1) programs, the passed input strings to PSLL(1) program must have the same length as in the case of the PSLR(1) program.

The following tables illustrate the results of parsing of the input string and the corresponding computing times each time of running PSLL(1) program.

Table 6 illustrates the parsing of "i + i" input string and the computing time using PSLL(1).

Table 7 illustrates the parsing of "i + i * i" input string and the computing time.

Table 8 illustrates the parsing of "i * i * i" input string and the computing time.

Now, from Tables 6, 7 and 8, The computing time of the parsing of the three input strings using PSLL(1) program can be stated by Table 9.

Table 6 Parsing of The first string using PSLL(1)

Input string	i + i		Action
	Stack	Input	
Parsing steps	$E $AT $ABF $ABi $A $AT + $ABF $ABi $A $	i + i$ + i$ + i$ + i$ + i$ i$ $ $ $ $	E -> TA T -> FB F -> i B -> e A -> +TA T -> FB F -> i B -> e A -> e
Result of parsing	Accepted		
Computing time	52.98 ms		

Table 7 Parsing of the second string using PSLL(1)

Input string	i + i * i		Action
	Stack	Input	
Parsing steps	$E $AT $ABF $ABi $A $AT + $ABF $ABi $ABF * $ABi $A $	i + i * i$ i + i * i$ i + i * i$ i + i * i$ + i * i$ + i * i$ i * i$ i * i$ * i$ i$ $ $	E -> TA T -> FB F -> i B -> e A -> +TA T -> FB F -> i B -> *FB F -> i B -> e A -> e
Result of parsing	Accepted		
Computing time	62.26 ms		

Table 8 The parsing of the third string using PSLL(1)

Input string	i * i * i		Action
	Stack	Input	
Parsing steps	$E $AT $ABF $ABi $ABF * $ABi $ABF * $ABi $A $	i * i * i$ i * i * i$ i * i * i$ i * i * i$ * i * i$ i * i$ * i$ i$ $ $	E -> TA T -> FB F -> i B -> *FB F -> i B -> *FB F -> i B -> e A -> e
Result of parsing	Accepted		
Computing time	55.18 ms		

Table 9 The computing time of PSLL(1) program

Input string	Computing time (in ms)
1st string	52.98
2nd string	62.26
3rd string	55.18

6 Practical Comparison Between the PSLR(1) and PSLL(1)

From Tables 5 and 9, the following table can be concluded.

The results of Table 10 can be represented by Fig. 4.

NP: It is taken into consideration that the printing time is included in all of the displayed results, which could be excluded from each case to give the net computing time and then the net used printing time can be added, whatever it might be.

Table 10 The computing time of PSLR(1) and PSLL(1) programs

Input string	The computing time (in ms)		Almost ratio
	PSLR(1)	PSLL(1)	
1st string	3.39	52.98	1/15
2nd string	8.31	62.26	1/8
3rd string	5.98	55.18	1/9

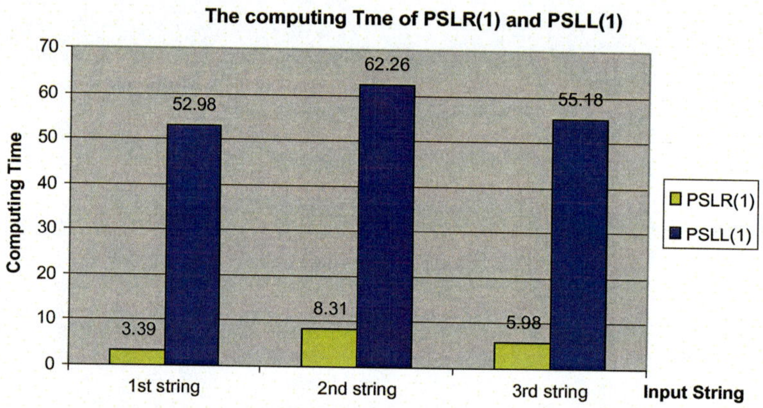

Fig. 4 Practical comparison between the PSLR(1) and PSLL(1)

7 Conclusion

Practically, it looks apparent from the above results that the LR(1) technique give a smaller time than the LL(1) method. This makes us adopt this method for compilation since it looks superior to the LL(1).

References

1. Su Y, Yan SY (2011) Principles of compilers—a new approach to compilers including the algebraic method. Higher Education Press, Beijing and Springer Heidelberg Dordrecht, London, New York
2. Ah AV, Lam MS, Sethi R, Ullman JD (2007) Compilers—principles, techniques, and tools. Pearson Addison Wesley, New York
3. Jacobs CJH, Kees DG, Bal HE, van Reeuwijk (2012) Modern compiler design. Springer, New York, Heidelberg Dordrecht London
4. Waite WM, Goos G (1994) Compiler construction. Springer-Verlag, Berlin, New York
5. Holub AI (1990) Compiler design in C. Prentice Hall, New Jersey
6. Meduna A (2008) Elements of compiler design. Auerbach Publications, New York
7. Ismail IA, Ali NA (2019) LL(1) as a property is not enough for obtaining proper parsing. In: The first world conference on internet of things: applications & future. French University-Cairo, SPRINGER Peer-Reviewed Proceedings, Cairo
8. Ægidius Mogensen T (2011) An introduction to compiler design. Springer-Verlag, London
9. Ismail IA, Ali NA (2017) Parsing strings using a subset of the production rules of the C-language. Egypt Comput Sci J 41:30–35
10. Ægidius Mogensen T (2010) Basics of compiler design. Published through lulu.com. University of Copenhagen, Copenhagen
11. Puntambekar AA (2008) Compiler design. Technical Publication
12. Maheshwari A, Smid M (2017) Introduction to theory of computation. Carleton University, Canada

A Solution for Handling Big Data Heterogeneity Problem

Ahmad M. Gamal El-Din and M. B. Senousy

Abstract In the recent years, Big Data became a prominent tendency in many different fields. It refers to any data that are intricate and gigantic in volume, with tremendous velocity and vast variety, which transcend the processing ability of conventional tools. Heterogeneity is one of the eminent problems facing Big Data. Most of the data that are gathered from diverse resources are semi-structured and unstructured, which makes information extraction a difficult task. In this paper, a framework and algorithms are proposed to provide a solution that can help in handling the heterogeneity problem of big data through increasing the homogeneity and decreasing the heterogeneity of data to a certain degree or level for text formats. The proposed solution is implemented using R language, R AnalyticFlow software environment and Apache Spark which is one of the big data processing tools.

Keywords Big data · Heterogeneity · Heterogeneity solutions

1 Introduction

Big Data has earned a great stature in the world of business and technology, leaving behind many challenges and issues that need a lot of research and giving a different view for the nature of data and how to deal with it. Big Data is huge and unstructured. Structuring it as a whole is obviously infeasible and impractical for many reasons such as:

- Big Data comes from heterogeneous diverse sources with different formats (Heterogeneity) [1].
- Lack of analytical tools that can analyze such large data in a timely manner fashion (Timeliness) [1].
- The difficulty to build 100% reliable fault tolerant machines or software (Fault tolerance) [1].

A. M. G. El-Din (✉) · M. B. Senousy
Computer and Information Systems Department, Sadat Academy for Management Sciences, Cairo, Egypt

D. A. Magdi et al. (eds.), *Digital Transformation Technology*, Lecture Notes in Networks and Systems 224, https://doi.org/10.1007/978-981-16-2275-5_11

- Big data collected from different sources may include incorrect, untrusted and missing value data (Inconsistency, Incompleteness and Quality of Data) [1].
- Conventional database tools cannot deal with such huge data (Storage) [2].
- Big Data needs huge processing capabilities and scalable database tools to deal with such large data (Scalability) [3].
- Limited capabilities of data visualization tools that can visualize this huge data (Visualization) [4].
- The existing hardware and techniques are not able to handle that large size of data in real time (Processing) [4].
- Security issues concerning the storage of such huge data (Security) [5].
- Privacy issues of storing personal data with that massive size (Privacy) [5].

Heterogeneity is one of the paramount features of big data. As almost all of the data that are collected from various resources are semi-structured and unstructured, which makes information extraction a complex task. Also, heterogeneity corresponds to one of the main big data characteristics (also known as the V's), which is the variety. As variety and heterogeneity have common properties in terms of the diversity of formats and sources. There are some researches that discuss solutions for handling big data heterogeneity. For more details please see [6–9]. The objective of this paper is to provide a solution that can handle the big data heterogeneity problem, through increasing the homogeneity and decreasing the heterogeneity of data to a certain degree or level.

2 Proposed Solution

A framework and algorithms are proposed as a solution for handling big data heterogeneity problem written in R language. The framework consists of four phases including two sub-phases in the second phase. As for the algorithms, there are three algorithms, which correspond to the phases of the framework. Then some clustering techniques are applied with certain and specific measurements. After that, accuracy is calculated, the clusters are visualized and the clustering results are evaluated.

2.1 Proposed Framework

The framework consists of four phases including two sub-phases in the second phase. Each phase includes a series of steps. The framework is implemented using R Language, R AnalyticFlow [10] and Apache Spark. The phases of the framework are explained briefly as follows and shown in Fig. 1:

- **Phase (1) Big Data Preprocessing**: the first phase preprocesses the data before analyzing, beginning from importing the data and the required R libraries till scaling the data and converting it to a data frame.

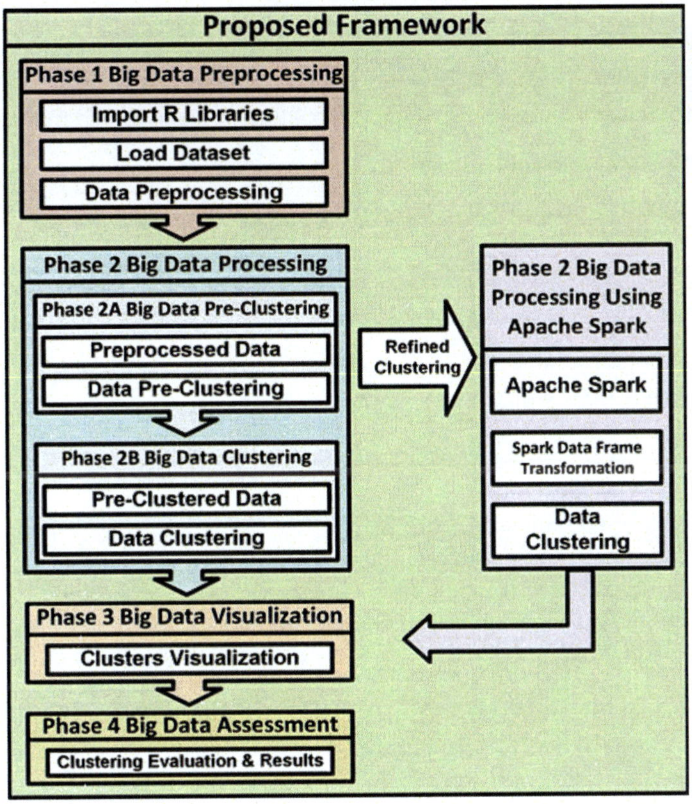

Fig. 1 Proposed framework

- **Phase (2) Big Data Processing**: the second phase processes the data through two sub-phases, Big Data Pre-Clustering and Big Data Clustering.
  - **Phase (2A) Big Data Pre-clustering**: in the first sub-phase, the preprocessed data is checked to see if it contains meaningful clusters, through visual inspection using Principle Component Analysis (PCA) [11] and statistically using Hopkins method [12, 13].
  - **Phase (2B) Big Data Clustering**: in the second sub-phase, the pre-clustered data is clustered using some clustering techniques such as K-Means or clustered by parallel processing in Apache Spark.
- **Phase (3) Big Data Visualization**: in the third phase, the created clusters are visualized using PCA to see how far the clusters are separated from each other.
- **Phase (4) Big Data Assessment**: in the last phase, the clustering results are evaluated in order to check the goodness of clustering and if the degree or level of heterogeneity decreased and the homogeneity increased.

2.2 Proposed Algorithms

The following algorithms correspond to each phase of the proposed framework:

- **Big Data Preprocessing Algorithm**

The first algorithm describes the steps of big data preprocessing.

1: Input: Crimes Dataset.
2: Output: Preprocessed Crimes Dataset.
3: Load dataset.
4: Delete records with missing values.
5: ***if*** Any missing value still exists ***then***
6:    Delete records with missing values.
7: ***else***
8:    Sort values in ascending or descending.
9: ***end if***
10: Search for non-numeric columns.
11: Add factors to non-numeric columns.
12: Convert non-numeric columns to numeric.
13: ***if*** Column is redundant and/or unnecessary ***then***
14:    column name = NULL.
15: ***end if***
16: Scale the data.
17: Transform the data to a data frame.

- **Big Data Processing Algorithm**

The second algorithm describes the steps of big data processing. This algorithm is divided into two sub-algorithms for big data pre-clustering and big data clustering, respectively.

– **Big Data Pre-clustering Algorithm**

1: Input: Preprocessed Crimes Dataset.
2: Output: Pre-Clustered Crimes Dataset.
3: Load preprocessed dataset.
4: Check for meaningful clusters.
5: ***if*** No meaningful clusters exist ***then***
6:    End the process.
7: ***else***
8:    Check for cluster tendency.
9: ***end if***
10: Visualize the distance matrix.
11: Determine the optimal number of clusters.
12: Validate the clusters.
13: Determine the optimal clustering algorithm.

– **Big Data Clustering Algorithm**

1: Input: Pre-clustered Crimes Dataset.
2: Output: Clustered Crimes Dataset.
3: Load pre-clustered dataset.
4: Apply the optimal clustering algorithm.
5: Visualize the clusters.
6: Calculate the clustering accuracy.
7: Validate the clustering results.

- **Big Data Clustering Using Apache Spark Algorithm**

The third algorithm describes the steps of big data clustering using Apache Spark.

1: Input: Pre-clustered Crimes Dataset.
2: Output: Clustered Crimes Dataset Using Apache Spark.
3: Load pre-clustered dataset.
4: Transform the pre-clustered dataset data frame to spark data frame.
5: Apply the optimal clustering algorithm.
6: Visualize the clusters.
7: Calculate the clustering accuracy.
8: Validate the clustering results.

3 Results and Discussion

The proposed algorithms were applied on Chicago Crimes dataset [14], which is a real data with real time updates (weekly updates), that includes a large number of records (6,882,009 records) and attributes with heterogeneous data types. Before applying the proposed framework and algorithms, the heterogeneity degree of the dataset was measured by applying some data quality metrics which correspond to these degrees as [15] follows (Table 1):

$$\text{Completeness} = \frac{\text{Total Non Missing Values}}{\text{Number of Expected Values}} \quad (1)$$

Table 1 Degree of heterogeneity measurement

Data quality metric	Number of records and percentage	Corresponding degree of heterogeneity
Completeness	6,179,427 records, 89.79%	Intra-dataset heterogeneity or attribute heterogeneity
Semantic consistency	6,179,427 records, 89.79%	Intra-dataset heterogeneity
Uniqueness	6,882,009 records, 100%	Attribute homogeneity

$$\text{Semantic Consistency} = \frac{\text{Total Semantically Consistent Rows}}{\text{Total Rows}} \tag{2}$$

$$\text{Uniqueness} = \frac{\text{Total Unique Rows}}{\text{Total Rows}} \tag{3}$$

The total number of records after preprocessing is 6,179,427 records.

According to the achieved results of clustering, the optimal number of clusters [16] is 2 clusters. So, the Chicago Crimes Dataset can be clustered according to the columns of AM PM or Arrest or Domestic. The AM PM column denotes if the crimes were committed day or night, while the Arrest and Domestic columns denote if it is true or false. Three clustering algorithms were applied on the dataset after preprocessing as follows:

- **K-Means Algorithm**

In Fig. 2 the dataset was clustered with K-Means Algorithm according to the column of Domestic which has 3,211,952 correctly classified records from 6,179,724 records which is the total number of records and with accuracy of 51.98%. Below is the code that checks if each class is located in a cluster. This low accuracy is due to the unscalable K-Means algorithm, which was not able to collect that large number of points.

```
table(KMeansHW$cluster,CrimesCompleteCasesXDF$Domestic)
      false    true
  1  2643939   498725
  2  2713227   325041
```

Fig. 2 K-Means clustering

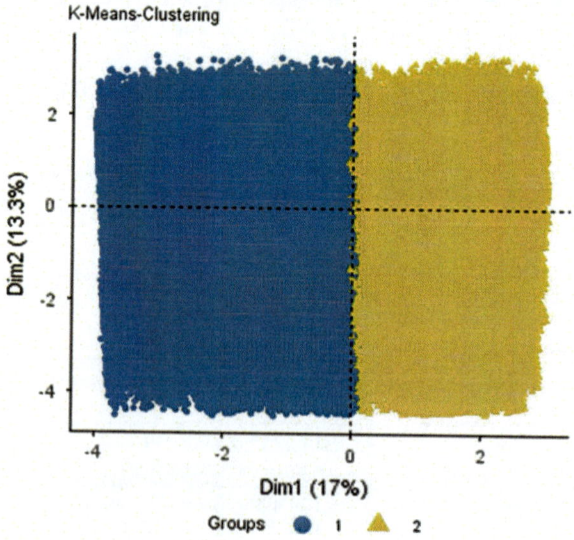

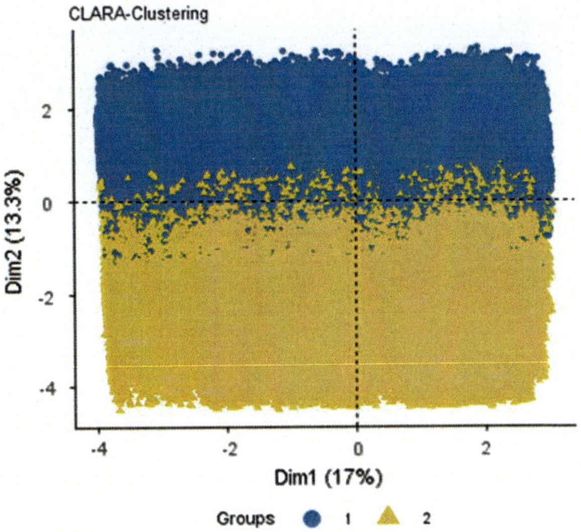

Fig. 3 CLARA clustering

From the code, it is shown that most of the domestic (false) records are located in the second cluster and the domestic (true) records in the first cluster. So, the total correctly classified records are 3,211,952.

- **Clustering Large Applications (CLARA) Algorithm**

In Fig. 3 the dataset was clustered with CLARA Algorithm according to the column of Arrest which has 5,246,065 correctly classified records from 6,179,724 rows which is the total number of records and with accuracy 84.90%. Below is the code that checks if each class is located in a cluster.

```
table(CLARA$clustering,CrimesCompleteCasesXDF$Arrest)
        false    true
  1  3940655  390407
  2   542955 1305410
```

From the code, it is shown that most of the arrest (false) records are located in the first cluster and the arrest (true) records in the second cluster. So, the total correctly classified records are 5,246,065.

- **Tree Clustering Using CLARA Algorithm**

In Fig. 4 the dataset was clustered with Tree Clustering Using CLARA Algorithm [17] according to the column of Arrest which has 4,377,040 correctly classified records from 6,179,724 records which is the total number of records and with accuracy 70.83%. Below is the code that checks if each class is located in a cluster.

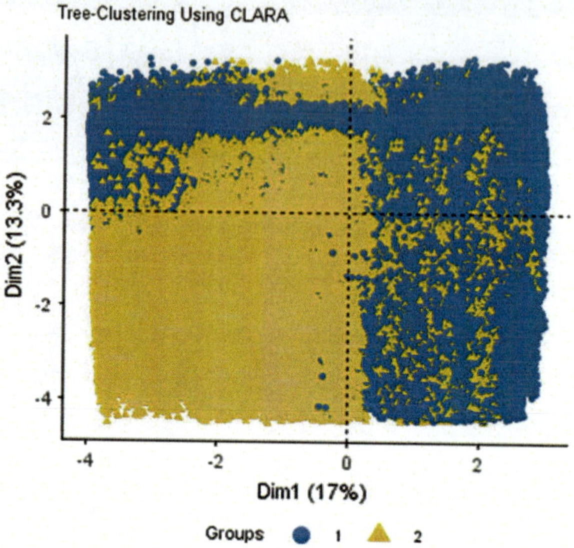

Fig. 4 Tree clustering using CLARA

```
table(TreeClusteringCLARA$final.clust$cluster,CrimesCom
pleteCasesXDF$Arrest)
      false    true
  1 3227710  546487
  2 1255900 1149330
```

From the code, it is shown that most of the arrest (false) records are located in the first cluster and the arrest (true) records in the second cluster. So, the total correctly classified records are 4,377,040.

- **K-Means Algorithm Using Apache Spark**

In Fig. 5 the dataset was clustered with K-Means Algorithm using Apache Spark [18, 19] according to the column of Domestic which has 5,149,140 correctly classified records from 6,179,724 records which is the total number of records and with accuracy 83.33%. The accuracy is 31.35% higher than the K-Means Algorithm that was not processed using Apache Spark. In addition, the K-Means algorithm is scalable to deal with huge data. Below is the code that checks if each class is located in a cluster.

```
KMeansPredict <- ml_predict(spark_kmeans, Crimes_tbl) %>% collect
table(KMeansPredict$prediction,KMeansPredict$Arrest)
              false        true
 0          4040111      586788
 1           443499     1109029
```

From the code, it is shown that most of the arrest (false) records are located in the zero cluster and the arrest (true) records in the first cluster. So, the total correctly classified records are 5,149,140.

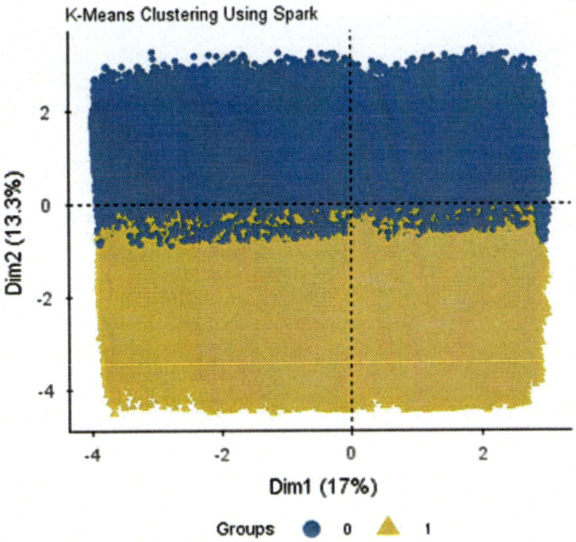

Fig. 5 K-Means algorithm using Apache Spark

4 Conclusion

There are many challenges and obstacles concerning the extraction of information from big data, beginning with the issues related to scalability of resources till the heterogeneity of the big data itself. The framework and algorithms were implemented using R language to preprocess and prepare the data before doing analytics using R AnalyticFlow software environment. Then, some clustering algorithms were applied to check the achieved results from preprocessing of data after using the proposed algorithms. Also, one of the clustering algorithms was applied on Apache Spark, which is a big data processing tool and a part of the collection of modules included in the big data platforms. The accomplished results show that the proposed framework and algorithms are capable to handle the heterogeneity of big data in text format. Moreover, the clustering results reached high accuracy percentage up to 84.90%, which is considered a great achievement to prove the capability of the algorithms in preprocessing and preparing the data before doing analytics.

References

1. Desai PV (2018) A survey on big data applications and challenges. In: 2018 second international conference on inventive communication and computational technologies (ICICCT), Coimbatore
2. Zhang Y, Ren J, Liu J, Xu C, Guo H, Liu Y (2017) A survey on emerging computing paradigms for big data. Chin J Electron 26(1):1–12

3. Athmaja S, Hanumanthappa M, Kavitha V (2017) A survey of machine learning algorithms for big data analytics. In: 2017 international conference on innovations in information, embedded and communication systems (ICIIECS), Coimbatore
4. Wani MA, Jabin S (2018) Big data: issues, challenges, and techniques in business intelligence. In: Proceedings of CSI 2015, New Delhi
5. Lee I (2017) Big data: dimensions, evolution, impacts, and challenges. Bus Horiz 60(3):293–303
6. Sindhu CS, Hegde NP (2015) A framework to handle data heterogeneity contextual to medical big data. In: 2015 IEEE international conference on computational intelligence and computing research (ICCIC), Madurai
7. Kang L, Yi L, Dong L (2014) Research on construction methods of big data semantic model. In: Proceedings of the world congress on engineering (WCE 2014), London
8. Gaffar A, Darwish EM, Tridane A (2014) Structuring heterogeneous big data for scalability and accuracy. Int J Digit Inf Wirel Commun (IJDIWC) 4(1):10–23
9. Kumar TKA, Liu H, Thomas JP (2014) Efficient structuring of data in big data. In: 2014 international conference on data science & engineering (ICDSE), Kochi
10. Ef-prime Incorporation. R AnalyticFlow. https://r.analyticflow.com/en/, 7 Nov 2020
11. Lê S, Josse J, Husson F (2008) FactoMineR: an R package for multivariate analysis. J Stat Softw Am Stat Assoc 25(1):1–18
12. Banerjee A, Dave RN (2004) Validating clusters using the Hopkins statistic. In: 2004 IEEE international conference on fuzzy systems, vol 1, pp 149–153
13. YiLan L, RuTong Z (2015) R. l package version 1.4. https://cran.r-project.org/package=clustertend, 7 Nov 2020
14. Chicago Police Department. https://data.cityofchicago.org/Public-Safety/Crimes-2001-to-Present/ijzp-q8t2, 7 Nov 2020
15. Micic N, Neagu D, Campean F, Zadeh EH (2017) Towards a data quality framework for heterogeneous data. In: 2017 IEEE international conference on internet of things (iThings) and IEEE green computing and communications (GreenCom) and IEEE cyber, physical and social computing (CPSCom) and IEEE smart data (SmartData), Exeter
16. Malika C, Ghazzali N, Boiteau V, Niknafs A (2014) NbClust: an R package for determining the relevant number of clusters in a data set. J Stat Softw Am Stat Assoc 61(6):1–36
17. Buttrey SE, Whitaker LR (2015) treeClust: an R package for tree-based clustering dissimilarities. R J 7(2):227–236
18. The Apache Software Foundation. Apache Spark. https://spark.apache.org/, 7 Nov 2020
19. Javier L, Kuo K, Ruiz E (2019) k-Means clustering. In: Mastering spark with R. O'Reilly Media, Inc.

Recognizing Clothing Patterns and Colors for BVI People Using Different Techniques

Mahmoud Allam, Islam ElShaarawy, and Sara Ahmed Farghal

Abstract For blind and visually impaired (BVI) people, it is an arduous task to recognize clothing patterns and colors. It is especially complex to recognize them automatically. This is a highly researched area in image processing. This paper provides BVI with the ability to detect patterns and colors without depending on personal assistance, leading to increasing their confidence. The user first captures an image; then, our system detects colors and patterns. The color model used here is the hue saturation value (HSV). The main colors detected by our system are magenta, cyan, blue, green, red and yellow. The saturation of each color is also taken into consideration. We propose various methods that include multiple feature extraction and classification techniques on the City College of New York (CCNY) clothing pattern dataset. The combinations of the feature extraction and classification techniques exceeded 200 methods. The proposed methods were performed using MATLAB and Weka under different preprocessing and compression methods, such as converting the dataset to grayscale, using raw red, green and blue (RGB) images, in addition to grayscale and RGB scalograms from continuous wavelets. Using a combination of histogram of oriented gradients (HOG), feature extraction and a sequential minimal optimization (SMO) classifier resulted in the highest recognition rate, achieving 91.8% accuracy.

Keywords Clothing pattern recognition · CCNY · Color detection · BVI

M. Allam
The Knowledge Hub Universities, Cairo, Egypt
e-mail: allam@tkh.edu.eg

I. ElShaarawy
Department of Electrical Engineering, Faculty of Engineering at Shoubra, Benha University, Cairo, Egypt
e-mail: islam.elshaarawy@feng.bu.edu.eg

S. A. Farghal (✉)
Department of Software Engineering, Faculty of Computer Science, Nile University, Giza, Egypt
e-mail: s.farghal@nu.edu.eg

D. A. Magdi et al. (eds.), *Digital Transformation Technology*, Lecture Notes in Networks and Systems 224, https://doi.org/10.1007/978-981-16-2275-5_12

1 Background of Study

Sometimes, matching colors of the outfits with each other and attempting to reach a suitable style are hard; however, human's eyes can observe, detect and recognize the surroundings easily. Therefore, detecting patterns and colors is not a comfortable task for the BVI. Although there are enormous types of applications that can recognize the patterns and colors, it is hard for blind users to imagine the patterns and colors. Besides, fitting suitable colors with body features is another troublesome task.

Based on [1], BVI persons tend to use smartphones that enable digital services for being easy to use. Although these devices are considered one of their significant assistive tools, the information communication technology projects are fewer for BVI comparing with other projects [2]. But according to the study [3], it is expected that the assistive technologies (AT) field will be pushed further across challenges and the reality of developing success. Our proposed methods can be a smart application that enables BVI users to make shopping, detect patterns and colors and have a convenient style based on their body shape. It also enables them to recognize their clothing without memorizing, which fortunately can be daily use.

Our model needs only a mobile camera to capture the hanging cloth while making shopping or before outings. Every user has his own profile which is customized according to his/her body characteristics such as height and weight. After capturing the image, the model will analyze this image to identify the pattern and its colors. Finally, the user will receive feedback about this piece, and therefore, they can make their own decision to wear it or not.

2 Literature Review

Nowadays, detecting patterns and colors of clothing has attracted the attention of many researchers. Among all the astonishing works, various researches have been picked to be discussed, in order to reach the proposed solution. These researches are classified generally, under AT and specifically to apparel.

2.1 Apparel

Classifying and detecting apparel's color and clothing patterns have been presented by varied methods. For most of them, they follow a skeleton which is image acquisition, feature extraction, classification and feedback. So this subsection can be categorized using machine learning, CCNY dataset and wavelets.

2.1.1 Detecting Colors and Clothing Patterns Using Machine Learning Techniques

For instance, in [4], two snapshots for clothing are needed as input; then, after processing, the match of a pair of clothing is identified whether they match together or not. A tie and a shirt were their case study, and it has been developed using a neural network. Pattern and color recognition using the neural network has been presented in [5–7], and the patterns are classified into four classes which are irregular, plaid, stripes and patternless. The systems in [5, 6] have the ability to discover 11 clothing colors, but in [7], it identifies another additional five more colors. The systems start with capturing an image as an input, then providing the color name and the pattern recognition category. They do not offer any matching colors; in addition, the identified colors are limited.

The online shopping challenge for the BVI has been presented in [8]. The system focuses on speech recognition; it enables the user to get the desired product which includes color, material, pattern and other features through voice input. Then, after processing, the speech recognition using the deep learning method, the speech is classified into local and global features to display the proper results from the database of clothing on the computer. Clothing coordination has been proposed in [9], which is an android application that classifies the apparels into several categories, in addition to recognizing the color and matching it with the saved apparel images using KNN algorithm. Meanwhile, it does not focus on pattern recognition.

Classification pipeline is discussed in [10]. The distinctive decisions of fashion categories have been taken by the multiclass learner and achieved by the transfer forest algorithm. It showed enhancements among the other baseline algorithms of random forest and SVM. This pipeline has the ability to identify and classify just upper apparel. Yang [11] proposed a system for the BVI that uses the Canny edge detector. It can detect the match regarding the color and pattern. The system consists of a camera, a microphone, a computer and a speaker. Then, the result is displayed verbally to the user via the speaker.

Pattanavichai also suggests a system [12] that can detect the color of clothing by capturing an image and present the color verbally via a speaker in the Thai language. It does not focus on matching patterns and colors. This system [13] is an android application that does not focus on matching patterns and colors, but it can just detect colors using hue, saturation, lightness (HSL) color space after capturing an image. Another application that focuses on clothing recommendations has been presented in [14]. It recommends the user to wear what is suitable for them based on the event and their saved photo album using latent SVM.

Based on classifying apparel into distinct categories of shoes, clothing, accessories in addition to their colors, styles, fabrics and patterns, recommendations and feedback for clothing have been discussed in [15, 16]. Kalantidis et al. [16] have considered an outdated version of clothing feedback that had misclassification. Thus, i-Stylist has been proposed to enable clothing recommendations to the users based on their styles from their social images, and the recommendations can be enhanced through users' feedback [15]. The core of i-Stylist model is based on merging state-of-the-art fine-tuned CNN features besides the acquired metadata from SVM models.

Another clothing problem has been discussed in [17] which is a challenge to match clothing with multiple colors and complex patterns, to aid visually impaired and blind people to realize both patterns and colors information. In addition to Radon transform, it has been combined with other techniques which are wavelet features and co-occurrence matrix for pattern matching. The matching color algorithm is based on normalized color in HSI color space. It can identify various colors; for instance, red, orange, yellow, green, cyan, blue, purple, pink, black, gray and white. A simple edge-based pattern detection has been developed to boost the efficiency of the algorithm. The system gives audio feedback after capturing a clothing image. It also has speech commands in case the user needs it. In [18], also an important fashion model is presented that depends on determining the main points on a cloth, for example, the corners of the neckline, hemline and cuff using CNN. Deep fashion the alignment showed efficacy among the other methods which are ground truth, image dependent pairwise relations (IDPR) and DeepPose.

2.1.2 Machine Learning Techniques Applied to CCNY Dataset

Numerous research papers have applied their methods using CCNY dataset for training and testing their models. In [19], they collected a dataset that includes 627 images with four different pattern designs. Figure 1 represents sample of images for each class. The pattern designs consist of the following:

- Patternless pattern includes 156 images
- Striped pattern includes 157 images
- Plaid pattern includes 156 images
- Irregular includes 158 images

Radon signature is considered a substantial feature extraction for pattern recognition, which has been represented in plenty of research papers such as in [17, 20–24]. It has been used with different combinations of techniques. In [21], Radon signature is used with statistical descriptor (STA) to extract the local features and scale-invariant feature transform (SIFT) for global features. The result has been evaluated on the CCNY dataset. This system has three main units which are a camera, a data analysis and finally a speaker; therefore, the clothing pattern is recognized after capturing an image for it; then, the data analysis unit can recognize the pattern, and finally, the user can realize the pattern from the audio feedback.

The Radon signature technique is used for several types of patterns. It is used for clothing, food and traffic signal recognition. All these challenges have been presented in [22]. CCNY dataset is used for recognizing clothing patterns. Three descriptors are used to extract features of clothing pattern image in [20] which are as follows: Radon signature for statistical properties, the wavelet subbands for global features and local features SIFT. All these descriptors are used in addition to the support vector machine to classify the patterns, and CCNY is the dataset that was used. The

Fig. 1 Four classes of the dataset [19]

proposed method is used to help visually impaired people to detect clothing patterns after capturing a clothing image. SIFT, statistical features and Radon transform are used as feature extraction in [23]. They are used to extract local, global and global directionality features, respectively. For classification, a support vector machine with radial basis function (RBF) kernel in Weka has been used to achieve better accuracy. As a result, this proposed method was trained and validated on the CCNY dataset.

Multiple combinations have been used in [25]. They are divided into three feature extraction techniques and three classifiers. The feature extraction techniques are Radon feature extraction, projection of rotated gradient (PRG) and quantized histogram of gradients (QHOG). The classifiers are KNN, neural network (NN) and SVM classifier. Amongst these combinations, the highest accuracy has been accomplished after using Radon signature and KNN.

2.1.3 Wavelets Combined with Other Techniques

Another method is discussed in [26] which enables visually impaired people to recognize clothing patterns through applying three different techniques which are speedup robust features (SURF), gray level co-occurrence matrix (GLCM) and DWT. These techniques are used to recognize the patterns; in addition to detecting the colors of the patterns using HSV color space. Matching complex patterns of two images has been represented also in [24]. Several techniques have been integrated which are Radon transform, wavelet transform and co-occurrence matrix. Radon transform is applied for extracting directional properties, and Haar wavelet is applied for extracting texture features based on three directions which are horizontal, vertical and diagonal.

3 Proposed Method

In this section, our proposed methods will be illustrated in different environments. We have used various combinations that consist of multiple feature extraction techniques with various classifiers. The proposed methods are divided into three main subsections. The first one is classifying user type. The second elaborates on the image acquisition and preprocessing that we have used in the proposed methods, besides the feature extraction techniques and the main classifiers that we have applied. The last one illustrates the color detection and the feedback that the user receives as a result.

Figure 2 represents the pipeline of our proposed model. It is divided into seven modules. The first one is classifying the user type based on calculating the body mass index (BMI) of the user. The second one is image acquisition which is the first stage of any computer vision system that is responsible for obtaining the image. We have used CCNY dataset [19] that is obtained from our desktop. Then, the third module is preprocessing over the dataset; we have used the original images and grayscale, as well. The fourth is feature extraction; in this module, we have used several techniques of feature extraction. The fifth is classification or machine learning techniques that are responsible for categorizing the image to a specific class. The sixth part is color

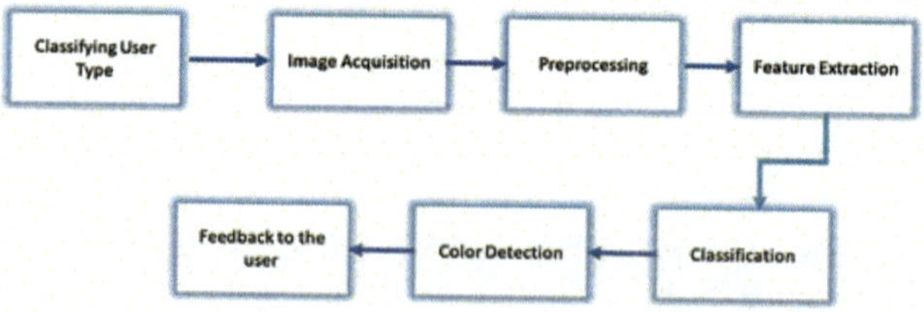

Fig. 2 Block diagram

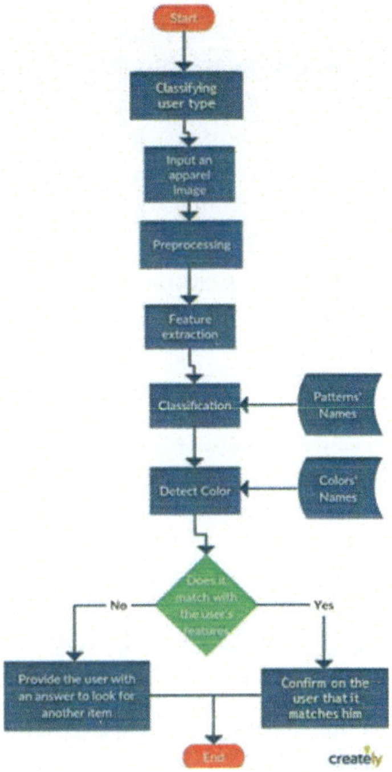

Fig. 3 Our model flowchart diagram

detection; it is responsible for detecting the colors of the classified pattern. Finally, the feedback module is responsible for providing the user with proper feedback that matches his characteristics.

Figure 3 represents the flowchart for our model. It starts with entering the user's weight and height, then the BMI formula is executed, and this attribute is saved. Then, the user captures an image of a clothing pattern as an input, and then, the system makes a preprocessing method on the image such as converting the RGB image to grayscale. Then, extracting the features from the image and classify it into one of the four categories which are irregular, patternless, plaid and striped. After that the model will detect the color of the pattern. Finally, providing the user with feedback if it matches his features or not.

3.1 Classifying User Type

It is based on calculating the BMI formula which can be calculated as

$$\text{Weight}/\text{Height}^2.$$

Then, from this formula, we get the BMI that enables the system to categorize the user either as heightened, averaged or low. Based on the user's category, the feedback of the colors will be provided to him [27]. Somatotype is a term that is used for classifying humans according to their body shape and physique type. This term is developed by US psychologist William Sheldon. In Sheldon's system, humans are classified into three types based on their body shape.

The endomorphic person should wear dark colors, to hide the overweight size. The ectomorphic person should wear light colors, to accentuate the form and that can help them to look a little more filled out. Finally, the muscular person can wear any color.

They are categorized as the following:

- Endomorphic, round or fat type
- Mesomorphic or muscular type
- Ectomorphic, slim or linear type

3.2 The Methods

We have downloaded the CCNY dataset that is imported for the training and testing phase. We have split it to 70–30% for training and validation, respectively. Before the training phase, we have used both the original images and a preprocessed one that is converted to grayscale.

The fourth and fifth modules include enormous combinations of feature extraction and classification techniques, to reach our model. We have used several descriptors that can be listed as the following:

- Wavelets
- Histogram of oriented gradient
- Bag of words
- Radon transform

3.2.1 Method 1: CWT

We have structured our dataset as a structure array called PatternData that includes two fields: data and labels. The data field is a 162-by-65536 matrix where each row is considered as image recording sampled at 128 Hz. Labels vector is a 162-by-1 cell array, one label, respectively, for each row of the data. The four classes are as follows: "irregular," "patternless," "plaid" and "stripped." We have used a function called "CreatePatternsDirectories," to store the new pre-processed data of each class. This function creates a data directory that creates four sub-directories for each class; this directory is created inside a parent directory. So this function accepts two parameters the first one is the directory of the patterns data, and the second is

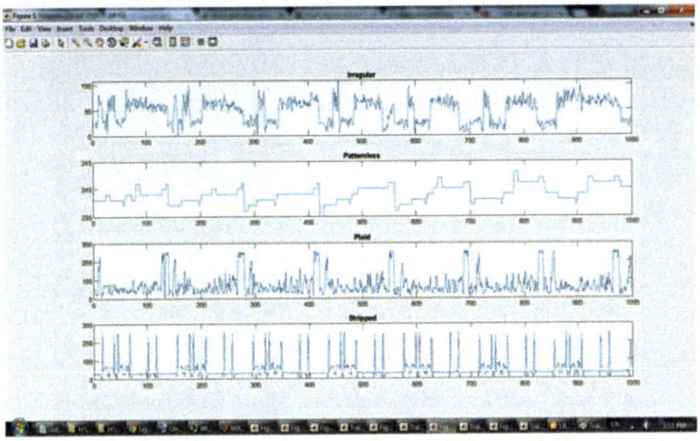

Fig. 4 Signals for the four classes

the parent directory. Figure 4 represents the patterns' signals for each class. There is another function that is responsible for that which is "PlotScalograms" it accepts the patterns data as input and plots them as categories.

Scalogram is defined as "the absolute value of the CWT of a signal" [28]. To create scalograms, the CWT filter bank should be precomputed. When getting the CWT of many signals using the same parameters, precomputing the CWT filter bank is considered as the appropriate method. Figure 5 depicts the scalograms after creating time-frequency illustrations of the patterns' signals. All the scalograms are created using a function called CreateRGBfromTF, and it is responsible for creating the scalograms as RGB images. The scalograms are generated from the wavelet transform using cwtfilterbank to obtain the continuous wavelet transform of the data signals. After that, it stores the classes of the scalogram as sub-directories in the data directory. Each image is 140-by-140-by-3.

Then, we divided the dataset into training and validation, the scalogram images are loaded as an image datastore which enables the data to be stored as ImageDataStore objects and to be labeled according to the folders' names. The training set is 70%, and the testing is 30% for validation. The training and validation images are randomly set. These scalograms will be used in the next proposed methods with several descriptors and classifiers such as:

- Bag of visual words and SVM
- Histogram of oriented gradients and SVM
- Bag of visual words and decision tree
- Discrete wavelets and SVM

Fig. 5 Scalograms classes

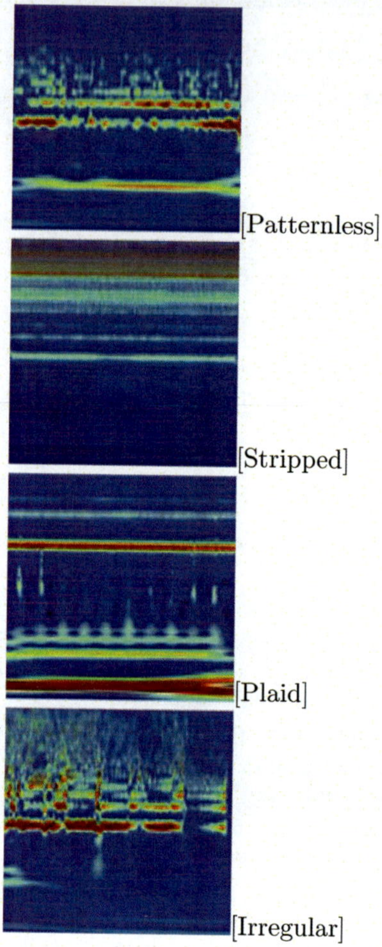

3.2.2 Method 2: Bag of Visual Words and SVM

In this method, BoVW or Bag of features is used to extract the features, and SVM is the classifier that is used to classify the results into four categories. The following list represents the MATLAB built-in functions that are used:

1. imageDatastore
2. bagOfFeatures (ImgesTrain2)
3. encode (bag, ImgesTrain2)
4. fitcecoc(trainFeatures, ImgesTrain2.Labels)

The first function is used for managing a set of image files. An object can be created from "imageDatastore" function which enables the importing and processing of the data through the object's functions. The features are extracted from the second

Table 1 Bag of features and SVM results

ID	Dataset (RGB)	Dataset (scalograms B&W)	Dataset (B&W)	Dataset (scalograms RGB)	Accuracy (%)
1	No	Yes	No	No	83
2	Yes	No	No	No	88
3	No	No	Yes	No	86
4	No	No	No	Yes	80

function which is "bagOfFeatures." These features will classify the images. And it needs parallel computing toolbox^{textTM} as a prerequisite. After extracting the features, we need to have a histogram for the visual word occurrences. And "encode" is the built-in function which is responsible for this. After that extracting features from 404 images, 523584 features could be extracted and finally, using K-Means clustering to create a 500-word visual vocabulary.

- Number of features: 418,868
- Number of clusters (K): 500

Table 1 shows the accuracy after applying our model on the CCNY dataset under different conditions, the data is preprocessed to be converted to grayscale, RGB scalograms and grayscale scalograms.

3.2.3 Method 3: Histogram of Oriented Gradients and SVM

In this method, HOG and SVM are used. HOG is used as a descriptor and SVM as a classifier. The main MATLAB built-in functions that are used can be listed as the following:

1. imageDatastore ("Location," "LabelSource," "foldernames")
2. extractHOGFeatures (image, "CellSize,");
3. fitcecoc (trainingFeatures, trainingLabels)

As described earlier "imageDatastore" for managing a collection of images, the "extractHOGFeatures" function is responsible for returning the HOG features either from an RGB or a gray input image. And "fitcecoc" is for classifying the output using SVM. This model starts with selecting feature point locations using the grid method. Then, extracting SURF features from the selected feature point locations. Table 2 represents the results after applying our model to the CCNY dataset with different preprocessed methods.

Table 2 Hog and SVM results

ID	Dataset (RGB)	Dataset (scalograms B&W)	Dataset (B&W)	Dataset (scalograms RGB)	Accuracy (%)
1	No	Yes	No	No	76.70
2	Yes	No	No	No	67.61
3	No	No	Yes	No	69.80
4	No	No	No	Yes	80

3.2.4 Method 4: Bag of Features and Decision Tree

In method 4, the combination that is used is bag of features for feature extraction and decision tree for classification. The built-in functions which are used:

1. fitctree(trainFeatures,ImgesTrain2.Labels)
2. predict(Mdl, featureMatrix)
3. crossval(Mdl)

The first function fit binary classification decision tree for multiclass classification. The second function is to predict the class from two parameters the model that is built from the first function and from the features that have been extracted. The third function is to estimate loss using cross-validation. Tables 3 and 4 present the results.

Table 3 Decision tree results

ID	Dataset (RGB)	Dataset (B&W)	Dataset (scalograms RGB)	Dataset (scalograms B&W)	Accuracy (%)	Loss (%)
1	Yes	No	No	No	59	49.8
2	No	Yes	No	No	54.2	42.6
3	No	No	No	Yes	69.3	31.5
4	No	No	Yes	No	71.6	30

Table 4 Bag of features and decision tree results

ID	Dataset (RGB)	Dataset (B&W)	Dataset (scalograms RGB)	Dataset (scalograms B&W)	Accuracy (%)	Loss (%)
1	Yes	No	No	No	74.42	28
2	No	Yes	No	No	74.42	29.4
3	No	No	No	Yes	76.74	31.19
4	No	No	Yes	No	65	36.39

Table 5 Discrete wavelets and SVM results

ID	Dataset (RGB)	Dataset (scalograms B&W)	Dataset (B&W)	Dataset (scalograms RGB)	Accuracy (%)	Loss (%)
1	No	No	No	Yes	76.87	19
2	Yes	No	No	No	84.9	18.4
3	No	No	Yes	No	80.34	21.3
4	No	Yes	No	No	82.65	16.66

3.2.5 Method 5: Discrete Wavelets and SVM

Wavelet-based feature extraction is used, in addition to the SVM classifier. The idea is to extract a smaller set of features that serve in aggregation, to distinguish the four classes.

1. Wavelet toolbox™
2. Signal processing toolbox™
3. Statistics and machine learning toolbox™

Features have been extracted for each signal in this method on eight blocks of each signal. We have used the following:

- Autoregressive model (AR) coefficients of order 4
- Shannon entropy (SE) values for the maximal overlap discrete wavelet packet transform (MODPWT) at level 4
- Multifractal wavelet leader estimates

The AR coefficients are assessed using the Burg method "arburg" for each window. The number of the equivalent coefficients in the j-th node defines the Shannon entropy for the undecimated wavelet packet transform. The singularity spectrum and cumulant of the scaling exponents are two fractal measures that are used as features. The singularity spectrum width is used from "dwtleader" as a measure of the multifractal nature of the signal. The scaling exponents depicting power-law behavior in the signal at various resolutions, and the other cumulant illustrates the departure of the scaling exponents from linearity.

The wavelet variance for the entire signal is obtained using a built-in function called "modwtvar." The benefit of wavelet variance is to measure the variability in a signal by scale. Table 5 represents the results for the discrete wavelets and SVM combination.

3.2.6 Method 6: Radon Transform

We have applied Radon transform as a feature extraction technique, to use it in the following proposed methods. We have used a built-in function called "Radon."

It accepts two parameters the image and theta. It gets back the Radon transform of 2D grayscale image for the angles given by theta, and theta in our case was in the range [0, 180]. It is applied to the CCNY dataset as RGB and RGB scalograms as it will be shown in the next subsection.

3.2.7 Other Proposed Methods

Waikato environment for knowledge analysis (Weka) is open-source software that is written in java, and it includes a set of machine learning algorithms. These algorithms are used for data mining tasks. Weka includes tools for data preparation, classification, regression, clustering, association rules mining and visualization. We have performed our proposed methods on Weka for time consumption as it saves time, and it allows the user to perform many proposed methods. We have exported our datasets as CSV files from MATLAB for each feature extraction such as BoVW, Radon, Wavelets and HOG. Each feature extraction technique is applied on the CCNY dataset as being RGB, grayscale, grayscale scalograms and RGB scalograms.

There are a huge number of classifiers that Weka provides. These classifiers are categorized as Bayes, Trees, Meta, Misc, Rules, Mi and Functions. Each category includes multiple numbers of classifiers. These classifiers combined with feature extraction techniques have shown astounding results that will be shown in the next chapter.

3.3 Color Detection and Feedback

In the color detection module, the colors of the patterns are detected using HSV color space. The main colors that our model detects are as follows:

- Light cyan
- Light green
- Light blue
- Light yellow
- Light magenta
- Light red
- Dark cyan
- Dark green
- Dark blue
- Dark yellow
- Dark magenta
- Dark red.

The four figures in Fig. 6 show the original images, the histogram of the hue channel and the colors which are included in the four different patterns.

Fig. 6 Colors in patterns

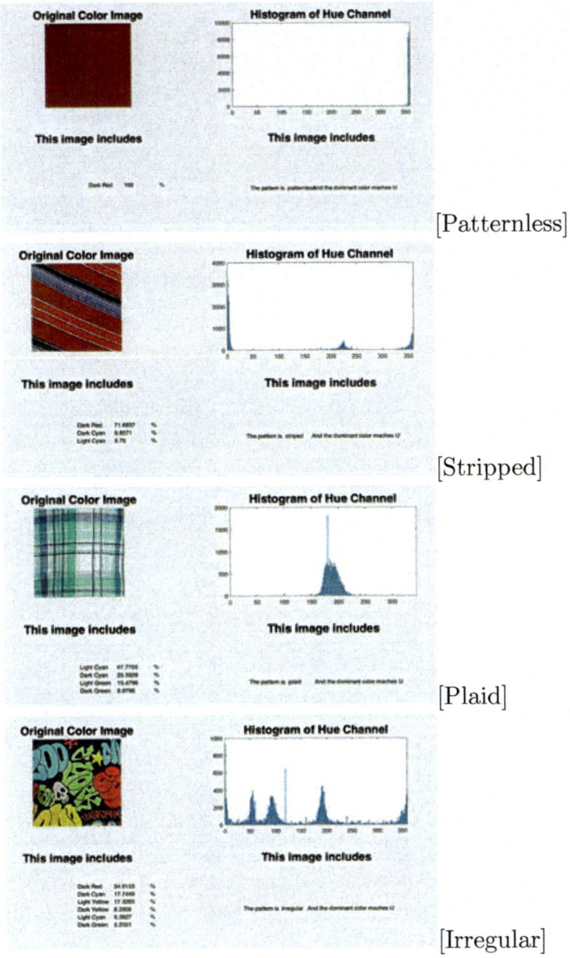

In the feedback module, we provide the user with feedback that matches the dominant color with his BMI. We have two use cases that include an irregular pattern, and the dominant color is dark cyan. For the first use case, the user entered weight as 40 kg and Height 180 cm, so his BMI is low. The first figure in Fig. 7 shows the feedback that is appeared to the user. In the second use case, the user entered weight as 100 kg and Height 180 cm, so his BMI is Heightened. The second figure in Fig. 7 represents the feedback which is appeared to the user.

Fig. 7 BMI Results

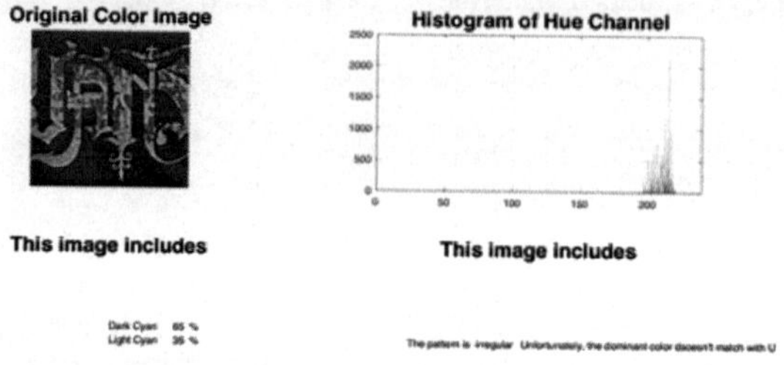

[First use case]

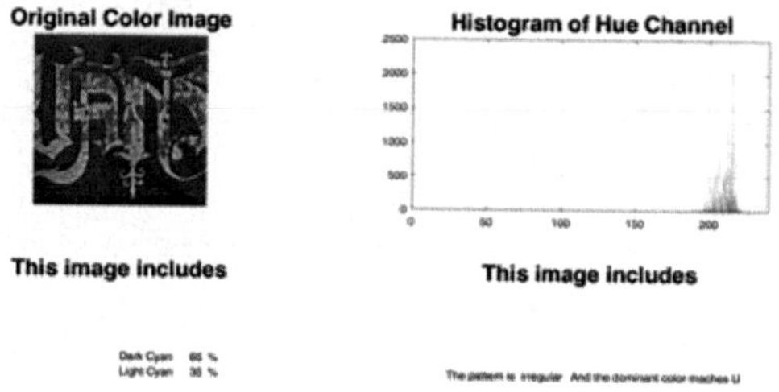

[Second use case]

4 Experimental Results and Discussion

As explained before our system consists of seven phases, the first phase is calculating the BMI of the user. The second is obtaining the image. The third is a preprocessing of the dataset. The fourth is extracting the features from the image. The fifth is classifying the pattern into four classes. The sixth is to detect the colors of the pattern. Moreover, the final phase is to provide the user with suitable feedback. So in this chapter, we assess the performance of our proposed methods using CCNY dataset as it is used in [6, 19, 21–23, 25]. Our dataset is split into 70–30% for training Data-Test Data, respectively. We have performed the proposed methods in two different environments. These environments are either MATLAB or Weka. The following lists will describe each environment.

For MAATLAB, the environment is as follows:

- Windows10
- MATLAB R2018
- Wavelet toolbox™
- Image processing toolbox™
- Neural network toolbox™
- Parallel computing toolbox™

Table 6 Related work results

Method	Percentage of training data (%)	Accuracy (%)
STA+SFT+Radon signature [6]	70	92.55
Radon signature [22]	70	92.55
STA+SFT+Radon signature [21]	70	88.68
SIFT+Radon signature+Statistical features [23]	70	92
WSIFT*WSTA [19]	70	91.86
Radon signature +KNN [25]	70	93.7

And for Weka, the environment is as follows:

- Windows7
- Weka-3-8

The descriptors that are compared in recognition experiments are Radon transform, bag of visual words, wavelets and histogram of oriented gradients combined with multiple classifiers using MATLAB and Weka. Some of the descriptors are used by [6, 19, 21–23, 25] that is why we have chosen them, in order to have a comparative work. And for the classifiers, we have expanded them to be more than the others used. The recognition results of the related work using MATLAB are demonstrated in Table 6.

Figures 8, 9, 10, 11 and 12 depict the results as 3D histograms for all the heatmaps. The combination of BoVW and SVM on the RGB dataset yields better recognition results comparing to all other results that we have performed using MATLAB. From our observation, while performing our experiments on the RGB dataset, it takes more time than on the grayscale dataset. However, the accuracies of the RGB dataset are better. The recognition for all BoVW with all the classifiers combinations using Weka shows consistent growth. HOG shows the growth of RGB and grayscale scalograms rather than on the original or grayscale datasets. Radon transform combinations with the classifiers do not show that much consistent growth.

Although random forest tree performs a consistent growth with the feature extraction techniques, SMO showed the best result with HOG on RGB Scalograms. Multilayer perceptron took the longest time while classifying the dataset, but SMO and random forest were the fastest classifiers. Clothing patterns are not complex patterns; furthermore, extracting features and image compression before classifying the patterns have shown effective results. The scalograms which are generated from CWT presented effectiveness at detecting specific features in the images. Thus, in our experiments, HOG, BoWV, wavelets and Radon transform are considered as an advantage before classifying the datasets.

Radon transform could not provide us good accuracies, because it has the least number of extracted features which are just 180 features compared with the others which are 1296, 542 and 500 for HOG, wavelets and BoVW, respectively. So, it

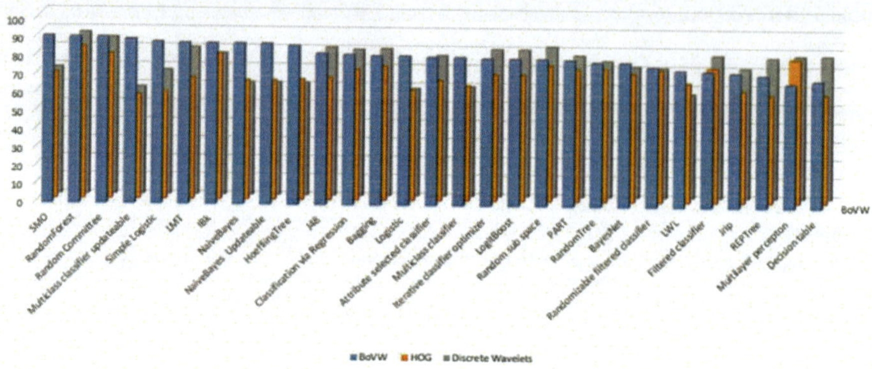

Fig. 8 Experimental results histogram using Weka for grayscale dataset

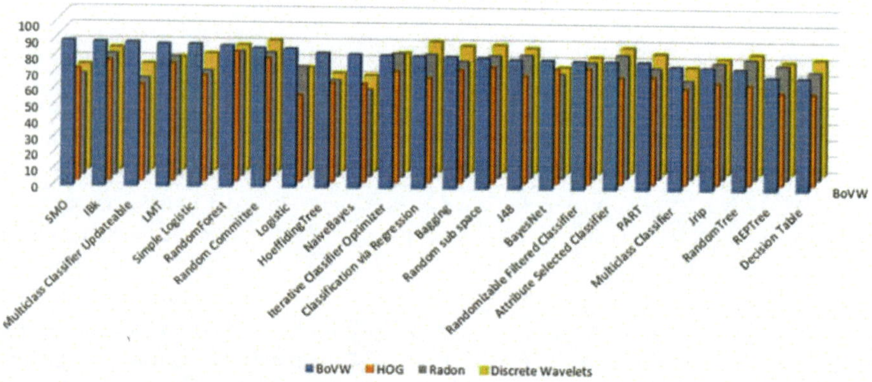

Fig. 9 Experimental results histogram using Weka for RGB dataset

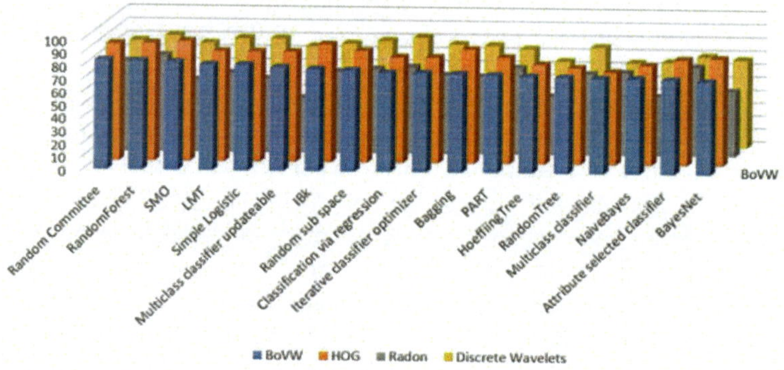

Fig. 10 Experimental results histogram using Weka for RGB scalograms dataset

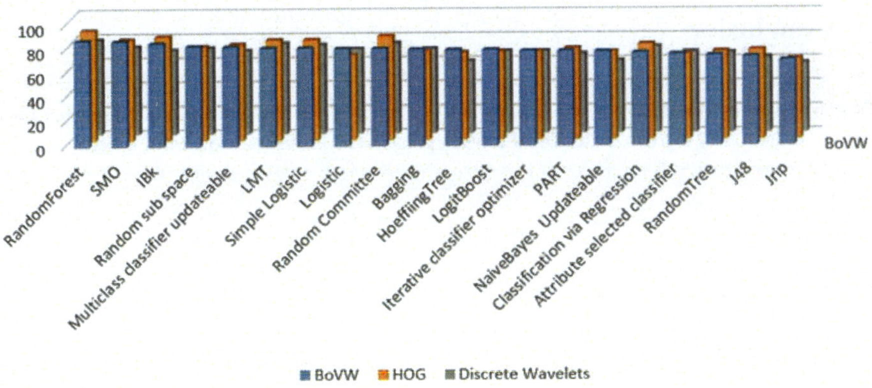

Fig. 11 Experimental results histogram using Weka for grayscale scalograms dataset

Fig. 12 Experimental results histogram using MATLAB

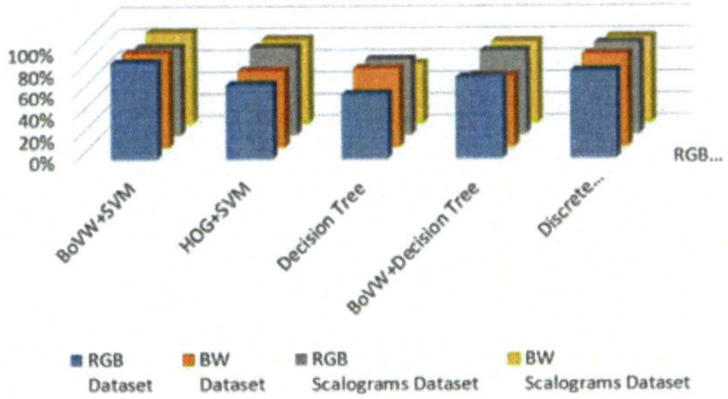

seems that the extracted features of Radon are not the required ones. BoVW is considered as the best feature extraction for our problem as it was in [6, 21]. HOG was affected by the rotation of the objects in the images, but due to our problem, the rotation was a minor issue that is why it worked well. RGB scalograms have better accuracies than BW scalograms; the RGB scalograms provide us with the desired features. The combinations that include random forest as a classifier showed better accuracies than the other combinations that include the decision tree. The reason behind this is that random forest is considered as an ensemble classifier that consists of multiple numbers of decision trees. The better predictive power of random forest is due to exhibiting the lower variance. The proposed methods of detecting clothing patterns can handle large illumination variances, diverse colors and several patterns. Our paper has the following impact:

1. It enriches the study of detecting patterns and colors. It pushes to substantial enhancements over the existing methods.
2. The methods that are developed in this paper provide new combinations of descriptors and classifiers to enhance the clothing pattern recognition.

5 Conclusion

Although there are various AT applications for BVI people, the aesthetics remain vague. It needs to be more detailed and expressive to boost confidence and do away with the need to have an assistant during shopping or over the day. We performed a tremendous number of experiments on clothing pattern recognition using different feature extractions combined with various classifiers using MATLAB and Weka. Extracting features had a useful impact on the accuracy, fine-tuning the results. Our proposed methods were executed over the CCNY dataset. The highest result 91.86%, obtained at 70% of the training data, was achieved by using HOG feature extraction and SMO classifier on RGB scalogram images. While using MATLAB, BoVW combined with SVM achieved the highest accuracy on RGB images. We thus observed that BoVW consistently performed better than all other feature extraction techniques in all combinations.

Multilayer perceptron took the longest time during training in terms of operation speed; SMO took the shortest time while classifying, meaning that it does not require special training. As a result, SMO is considered to be the fastest classifier. In MATLAB, discrete wavelets took the longest time for feature training while HOG was the fastest. All considerations resulted in the conclusion that HOG and SMO are the best feature extraction classifier combination for our specified problem.

6 Future Work

Our future work on pattern detection will include complex combinations of feature extraction techniques and ensemble classifiers and more informative description of detected patterns. Describing the patterns in more detail, such as whether the stripes are horizontal or vertical, and the size of the plaid is small or big. In addition to identifying more colors and considering their effect on the user's characteristics, such as skin tone and gender, will allow the user to receive coherently feedback.

References

1. Pal J, Viswanathan A, Chandra P, Nazareth A, Kameswaran V, Subramonyam H, Johri A, Ackerman MS, O'Modhrain S (2017) Agency in assistive technology adoption: visual impairment and smartphone use in bangalore. In: Proceedings of the 2017 CHI conference on human factors in computing systems. ACM, pp 5929–5940
2. Ashraf MM, Hasan N, Lewis L, Hasan MR, Ray P (2016) A systematic literature review of the application of information communication technology for visually impaired people. Int J Disab Manage 11
3. Bhowmick A, Hazarika SM (2017) An insight into assistive technology for the visually impaired and blind people: state-of-the-art and future trends. J Multimodal User Interf 11(2):149–172

4. Paisios N, Subramanian L, Rubinsteyn (2012) Choosing which clothes to wear confidently: a tool for pattern matching. In: Workshop on frontiers in accessibility for pervasive computing. ACM
5. Thilagavathi B et al (2015) Recognizing clothes patterns and colours for blind people using neural network. In: Innovations in information, embedded and communication systems (ICIIECS). IEEE, pp 1–5. IEEE
6. Yang X, Yuan S, Tian Y (2014) Assistive clothing pattern recognition for visually impaired people. IEEE Trans Human-Mach Syst 44(2):234–243
7. Bharadwaj S, Sharath HK, Praveena MB, Shetty A, Shivarudraiah B Clothing color and pattern recognition for visually impaired people
8. Mohadikar K, Navkhare R (2017) Review on online shopping for visually impaired people
9. Hamdy A, Kareem N, Nagaty K (2016) Apparel coordination based on contour and dominant colours matching. In: SAI computing conference (SAI), 2016. IEEE, pp 394–402
10. Bossard L, Dantone M, Leistner C, Wengert C, Quack T, Gool LV (2012) Apparel classification with style. In: Asian conference on computer vision. Springer, pp 321–335
11. Yuan S (2010) A system of clothes matching for visually impaired persons. In: Proceedings of the 12th international ACM SIGACCESS conference on Computers and accessibility. ACM, pp 303–304
12. Pattanavichai S, Pattamavorakun S (2015) Creating detectors clothes' colors and presentations for the visually impaired. In: 2015 international conference on science and technology (TICST). IEEE, pp 342–347
13. Al-Doweesh SA, Al-Hamed FA, Al-Khalifa HS (2014) What color? a real-time color identification mobile application for visually impaired people. In: International conference on human-computer interaction. Springer, pp 203–208
14. Liu S, Feng J, Song Z, Zhang T, Lu H, Xu C, Yan S (2012) Hi, magic closet, tell me what to wear! In: Proceedings of the 20th ACM international conference on multimedia. ACM, pp 619–628
15. Sanchez-Riera J, Lin J-M, Hua K-L, Cheng W-H, Tsui AW (2017) i-stylist: finding the right dress through your social networks. In: International conference on multimedia modeling. Springer, pp 662–673
16. Kalantidis Y, Kennedy L, Li L-J (2013) Getting the look: clothing recognition and segmentation for automatic product suggestions in everyday photos. In: Proceedings of the 3rd ACM conference on international conference on multimedia retrieval. ACM, pp 105–112
17. Shuai Y, YingLi T, Aries A (2011) Clothing matching for visually impaired persons. Technol Disab 23(2):75–85
18. Liu Z, Yan S, Luo P, Wang X, Tang X (2016) Fashion landmark detection in the wild. In: European conference on computer vision. Springer, pp 229–245
19. Yang X, Yuan S, Tian Y (2011) Recognizing clothes patterns for blind people by confidence margin based feature combination. In: Proceedings of the 19th ACM international conference on multimedia. ACM, pp 1097–1100
20. Shweta J, Aabhas M, Hemant D (2017) A review: recognizing clothes patterns and colours for blind people using neural network. Int J Adv Res Comput Sci 8(8)
21. Susithra K, Sujaritha M (2016) Clothing pattern recognition based on local and global features. Int J Sci Eng Res 7(3):106–110
22. Savitha R, Ramya R Portable camera based assistive pattern recognition for visually challenged persons
23. Chaitra GN, Khare N (2015) Feature extraction of clothing texture patterns for classification. In: Recent and emerging trends in computer and computational sciences (RETCOMP), 2015. IEEE, pp 6–9
24. Yuan S, Tian Y (2010) Rotation and illumination invariant texture analysis: matching clothes with complex patterns for blind people. In: 2010 3rd international congress on image and signal processing (CISP), vol 6. IEEE, pp 2643–2647
25. Shinde A, Shinde A (2018) Analysis of multiple features and classifier techniques combination for image pattern recognition. In: Intelligent computing and information and communication. Springer, pp 237–245

26. Dhongade M Clothing pattern recognition for blind using surf and combined glcm, wavelet. Int J Sci Res (IJSR), pp 2319–7064. ISSN (Online)
27. Sheldon WH, Stevens SS, Tucker WB (1940) The varieties of human physique
28. URL https://mathworks.com/help/signal/ug/scalogram-computation-in-signal-analyzer.html

Fuzzy Logic in Control Theory

Michael Gr. Voskoglou

Abstract A control system regulates the operation of a device or another system with the help of a set of IF–THEN logical rules. In particular, a fuzzy control system analyzes the input data in terms of variables taking continuous values in the interval [0, 1]. Fuzzy logic has the advantage of expressing the solution of a problem in the natural language, which enables the mechanization of tasks that have been already performed successfully by humans. The present paper studies the operation of a fuzzy control system and illustrates it by an application to a building's central heating boiler.

Keywords Fuzzy set (FS) · Fuzzy logic (FL) · Control system (CS) · Triangular fuzzy number (TFN) · Center of gravity (COG) defuzzification technique · Fuzzy control system (FCS) · Fuzzy approximation theorem (FAT) · Fuzzy associative memory (FAM)

1 Introduction

Fuzzy Set (FS) theory and the based on it *Fuzzy Logic (FL)*, initiated by Zadeh approximately 50 years ago [1, 2], despite the fact that have initially received tough critiques from several directions [3], have enabled the scientists to model under conditions of imprecision and/or vagueness. FL has found nowadays many and important applications to almost all sectors of human activity, becoming a complement and support of probability theory, which is suitable for modeling only situations of uncertainty which are due to randomness [4].

The traditional supporters of the bivalent logic, based on a culture of centuries, argue that, since this logic works effectively in science, functions the computers, and explains satisfactorily the phenomena of the real world, except perhaps those that happen in the boundaries, there is no reason to make things more complicated by introducing the unstable principles of FL, an infinite-valued in the interval [0, 1] logic.

M. Gr. Voskoglou (✉)
Graduate Technological Educational Institute, 26334 Patras, Greece

D. A. Magdi et al. (eds.), *Digital Transformation Technology*, Lecture Notes in Networks and Systems 224, https://doi.org/10.1007/978-981-16-2275-5_13

FL, however, aims exactly at smoothing the situation in the boundaries! Look, for example, at the graph of Fig. 1 corresponding to the FS T of "tall people." People with heights less than 1.50 m have membership degree 0 in T. The membership degree is continuously increasing for heights greater than 1.50 m, taking its maximal value 1 for heights equal or greater than 1.80 m. Therefore, the "fuzzy part" of the graph—which is conventionally represented in Fig. 1 by the straight line segment AC, but its exact form depends upon the way in which the membership function has been defined—lies in the area of the rectangle *ABCD* defined by the *OX* axis, its parallel through the point *E,* and the two perpendicular to it lines at the points *A* and *B*. In fact, the way of perceiving a concept (e.g., "tall") is different from person to person, depending on the "signals" that each one receives from the real world about it. Mathematically speaking, that means that the definition of the membership function of a FS is not unique, depending on the observer's personal criteria. The only restriction is that this definition must be compatible to the common logic, because otherwise the corresponding FS does not give a reliable description of the corresponding real situation. On the contrary, bivalent logic defines a bound, e.g., 1.80 m, above which people are considered to be tall and under which are considered to be short. Consequently, one with height 1.79 m is considered to be short, whether another with height 1.81 m is considered to be tall!

Zadeh realized that FSs are connected to words (adjectives and adverbs) of the natural language; e.g., the adjective "tall" indicates the FS of the tall people, since "how tall is everyone" is a matter of degree. A grammatical sentence may contain many adjectives and/or adverbs, therefore it correlates a number of FSs. A synthesis of grammatical sentences, i.e., a group of FSs related to each other, forms what we call a *fuzzy system*. A fuzzy system provides empirical advices, mnemonic rules, and common logic in general. It is not only able to use its own knowledge to represent and explain the real world, but can also increase it with the help of the new data; in other words, it learns from the experience. This is actually the way in which humans think. Such applications started to appear in industry during the 1970s, the first one being

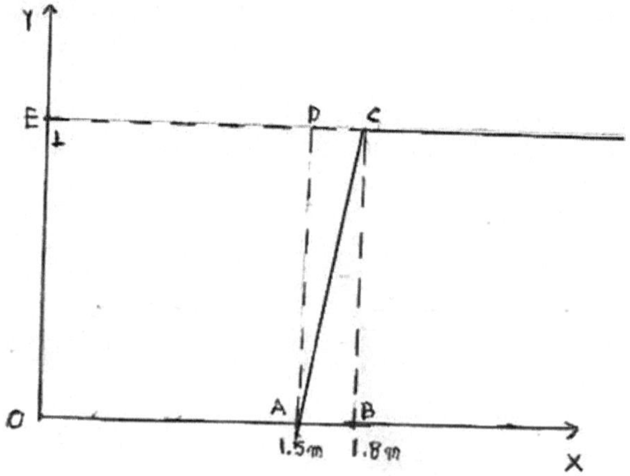

Fig. 1 Fuzzy set of "tall people"

in the area of cement kiln control [5]. This is an operation demanding the control of a highly complex set of chemical interactions by dynamically managing 40–50 "rules of thumb." This was followed by Mamdani's work in the Queen Mary College of London [6], who designed the first fuzzy system for controlling a steam engine and later the operation of traffic lights. Another type of fuzzy inference systems was developed later in Japan by Takagi-Sugeno-Kang [7]. Nowadays, a fuzzy system can control, for example, the function of an electric washing-machine or send signals for purchasing shares from the stock exchange, etc.

The target of the present paper is the study of the fuzzy control systems, which are the most important category of fuzzy systems. The rest of the paper is organized as follows: Sect. 2 studies the structure of the traditional control systems of bivalent logic. Section 3 is devoted to the theoretical study of fuzzy control, which is highly based on the fuzzy approximation theorem. The structure of a fuzzy control system is fully illustrated by an example of controlling a building's central heating boiler presented in Sect. 4, and the article closes with the general conclusions, which are stated in Sect. 5.

2 Control Systems

The notion of a *system* has a very broad context in general. It can be defined as a set of interacting components forming an integrated whole. We distinguish among physical, social, economic, biological, abstract (mathematical, philosophical, etc.) systems, and many others.

The study of a system is usually performed through the modeling process. A *model* is a simplified representation of the basic characteristics of a real system, which identifies and simplifies the relationships among those characteristics in a form amenable to analysis. Several kinds of models are used according to the form of the corresponding problem, their most common types being the mathematical and simulation models. In the former case, mathematical symbols and methods are used for representing the system. In the latter case, the system is represented by a sequence of logical orders, usually functioning with the help of computers (for more details see [8], Chap. 1).

A *control system (CS)* regulates the behavior of other devices or systems by using a feedback controller. We distinguish between two main types of CSs, the *open-loop* and the *closed-loop* CSs. In the former case, the action of the controller is independent of the *process variable (PV)* to be controlled. This is, for example, the case of a central heating boiler controlled by a timer switching on or off the boiler after a certain period of time, regardless the building's temperature (which is the PV). In the latter case, on the contrary, the controller's action depends on the PV and the desired value of it, which is usually referred as the *set point (SP)*. More explicitly, the difference SP-VP, called the *SP-VP error*, is applied as a control signal on the purpose of equalizing the values of SP and VP. In case of the boiler analogy, for example, this is achieved with the help of a thermostat, which, working as a sensor,

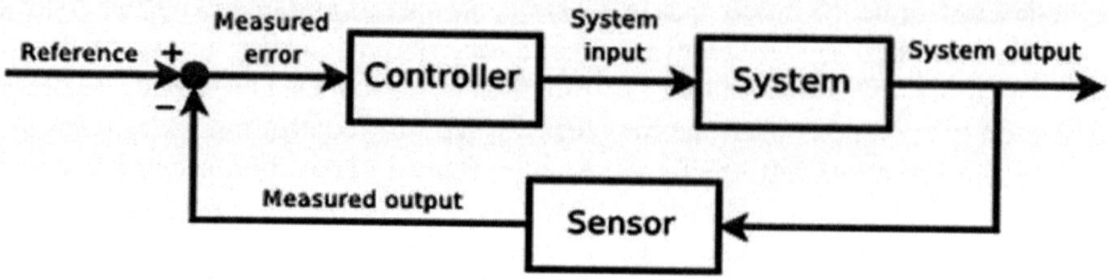

Fig. 2 Structure of a closed-loop control system

monitors the building's temperature to ensure that the controller's output maintains it close to the temperature set by the thermostat (the SP). The structure of a closed-loop CS is graphically represented in Fig. 2, retrieved from the Web.

The operation of a controller is regulated by a set of *IF–THEN logical rules. A* rule connects a process or event *A* to another process or event *B*. For example, IF the temperature of the building is lower than 18 °C (*A*), then the thermostat sends a signal, which, with the help of one or more mechanical valves, turns on the boiler (*B*). On the contrary, if the building's temperature exceeds 27 °C (*A*), then the thermostat sends a signal to the controller to turn off the boiler (*B*). Knowledge, therefore, could be seen as a combination of such rules, an idea originated at least from the Aristotle's time.

CSs are used to achieve, through automatic operation, increased productivity, high quality products, and improved performance of a device or another system. Control theory dates from 1868, when J. C. Maxwell described the theoretical basis for the operation of the governors, i.e., mechanical devices for regulating the operation of industrial engines (Fig. 3). Some years later E. Routh, C. Sturm, and A. Hurwitz established the control stability criteria.

The traditional CSs are described by mathematical models using one or more differential equations that define the CS's response to its inputs. Such CSs, initiated by N. Minorsky in 1922, are usually referred as *proportional—integral—derivative (PID) controllers.* Although the main application of mathematical control theory is the design of CSs for industry (CSs engineering), other important applications can also be found in life sciences, computer engineering, sociology, and operation research [9].

3 Fuzzy Control

In many cases, the mathematical model of a classical controller may not exist, or may be too "expensive" in terms of computer processing power and memory. The classical expert systems of Artificial Intelligence (AI) [10], for example, work by using 100–1000 bivalent IF–THEN rules. It has been estimated, however, that the human intelligence could be approached by a machine using at least 10^5 such rules! In such cases, therefore, a system based on empirical rules could be more effective.

Fig. 3 Centrifugal governor in a Boulton and Watt engine of 1788

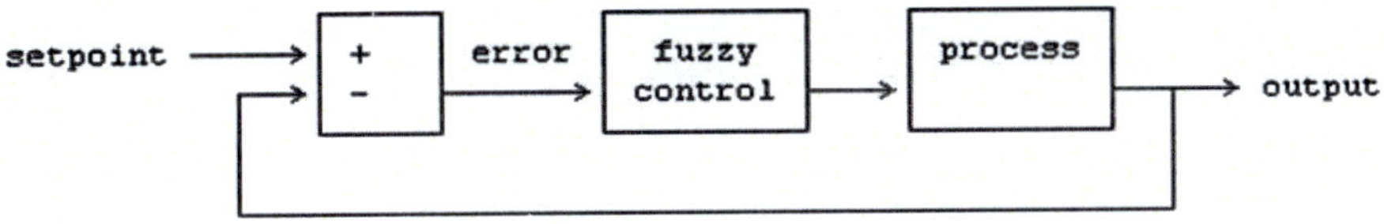

Fig. 4 Structure of a fuzzy control system

This was a good reason to introduce principles of FL for constructing "clever" CSs. It is of worth noting that, although in many cases, genetic algorithms and neural networks can perform as well as FL, the latter has the advantage of expressing the solution of the problems in the natural language. In this way, the human experience can be used for designing the controller, which makes easier the mechanization of tasks that have been already performed successfully by humans. FL needs also logical rules but, in contrast to bivalent logic, does not need too many rules.

A *fuzzy control system (FCS)* is a CS that analyzes the input values in terms of variables which take on continuous values in the interval [0, 1]. On the contrary the traditional CSs operate on the basis of the discrete values 0 (false) and 1 (true) only. The structure of a FCS, which is graphically represented in Fig. 4, is relatively simple and similar to the structure of a traditional CS (Fig. 2).

The interest on FCSs sparked by the development of a system for accelerating, breaking, and stopping the vehicles of the Sedai Subway, a rapid transit network in Japan. S. Yasunobu and S. Miyamoto of Hitachi provided in 1985 simulations that demonstrated the feasibility of FCSs in that case. Since then, a wide range of FCSs was developed by Japanese engineers concerning vacuum cleaners, refrigerators, air conditions, autofocusing cameras, voice controlled helicopters, robots, etc. In 1988, Japan established the Laboratory of Intelligent Fuzzy Engineering (LIFE), a cooperative of 48 companies, on the purpose of pursuing applied fuzzy research. Development of FCSs is also proceeding in USA and Europe, though in a smaller scale than in Japan. NASA, the US Environmental Protection Agency, big firms as Boeing, General Motors, Chrysler, Whirlpool, etc., have worked on FL for improved automatic transmissions, energy-efficient electric motors, and several other applications.

An important advantage of the FCSs with respect to the traditional CSs is that they do not need the existence of a model for studying the corresponding real world system. FCS design is an empirical approach based on the *trial-and-error method* characterized by repeated attempts, which are continued until success or until the subject stops trying ([11], Sect. 3). The general process involves the following steps:

- Determine the system's operational specifications, inputs, and outputs.
- Choose the FSs for the inputs and outputs.
- Document the set of IF–THEN fuzzy rules.
- Choose the suitable defuzzification method.
- Run tests to validate the system's behavior.
- Complete document by adjusting details as required and release to production.

An *IF–THEN fuzzy rule* is of the form: If X is A, then Y is B, where A, B are FS on the crisp sets of the input X and the output Y, respectively. For example: If the temperature (X) is very low (A), then the boiler (Y) works in full rate (B).

The operation of a FCS is based on the *Fuzzy Approximation Theorem (FAT)* [12]. Roughly speaking, the FAT theorem, called so because the abbreviation "FAT theorem" sounds better than "FA theorem," states that any system can always be "covered" or approximated by a fuzzy system with a finite number of IF–THEN fuzzy rules and is geometrically represented in Fig. 5 [3].

Fig. 5 FAT theorem

The existing in Fig. 5 curve represents the original system and each of the *patches* covering it corresponds to an IF–THEN fuzzy rule. The covering fuzzy system is the set of all those fuzzy rules, i.e., the set of all patches. The more uncertain an IF–THEN rule, the bigger the corresponding patch. If a rule is too accurate, then it is not fuzzy and the corresponding patch degenerates to a point on the curve. More knowledge means more rules, i.e., more patches and better covering. The better the curve is covered by the patches, the cleverer the covering fuzzy system.

The design of a FCS highlighted above will be fully illustrated by the example that follows in the next section.

4 An Application: Fuzzy Control of a Central Heating Boiler

4.1 Triangular Fuzzy Numbers

For the better understating of the following application, the reader is assumed to be familiar with the basics of the theory of *Fuzzy Numbers (FNs)* and in particular of *Triangular FNs (TFNs)*, being the simplest form of FNs [13]. It is recalled that a TFN $A = (a, b, c)$, with a, b, and c real numbers such that $a < b < c$ is a FS on the set R of real numbers with membership function defined by

$$y = m(x) = \begin{cases} \frac{x-a}{b-a}, & x \in [a, b] \\ \frac{c-x}{c-b}, & x \in [b, c] \\ 0, & x < a \text{ or } x > c \end{cases}$$

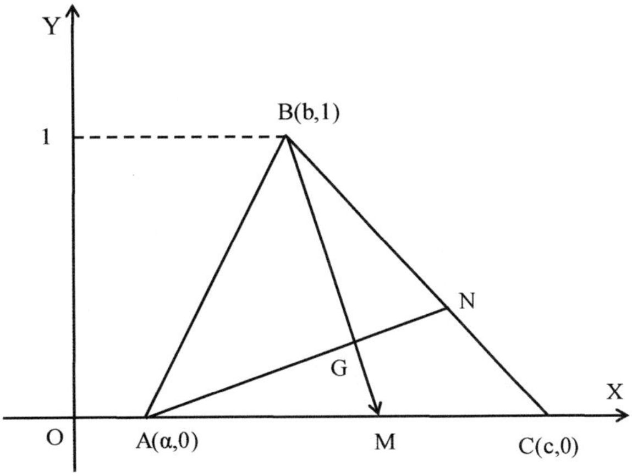

Fig. 6 Graph and COG of the TFN (a, b, c)

The graph of the TFN $A = (a, b, c)$ is the triangle ABC of Fig. 6. It is straightforward to check that the center of gravity (COG) of this triangle, being the intersection point G of its medians, has coordinates $(x, y) = (\frac{a+b+c}{3}, \frac{1}{3})$ ([14], Sect. 5).

According to the *COG defuzzification technique* ([14], Sect. 4), the x-coordinate of G can be used as a ranking function R for defufuzzifying the TFNs. Therefore, we can write

$$R(A) = \frac{a+b+c}{3} \tag{1}$$

It is noted that there are several other defuzzification approaches in use, the most commonly used being perhaps the *height method*, which takes the value of the biggest contributor (greatest output value).

4.2 Fuzzy Control of the Boiler

Reconsider the case of controlling a building's central heating boiler, where the input X is the building's temperature in degrees of Celsius and the output Y is the rate of the boiler's operation expressed in a climax from 0 to 100.

Without loss of generality and for reasons of simplicity, let us consider only three FSs on X, namely:

- A_1 = low temperature (≤18 °C)
- A_2 = normal temperature (between 15 and 25 °C), and
- A_3 = high temperature (>22 °C).

Assume also that the corresponding FSs on Y are:

- B_1 = high rate of the boiler's operation (≥70)
- B_2 = low rate of the boiler's operation (between 70 and 20), and
- B_3 = boiler's turn off (rate between 30 and 0).

The simplest way of representing geometrically the FSs defined above is by using triangles (TFNs), although one could use trapezoids (trapezoidal FNs) as well, or any other figure representing the membership functions of the defined FSs. Using triangles, the A_i's are represented on the coordinate system XOY and the B_i's on the coordinate system $X'O'\Upsilon'$ of Fig. 7. The triangles representing the FSs A_1 and A_2 are not complete, because we could have temperatures lower than 0 °C or higher than 27 °C. Those triangles can, however, be always completed in case of temperatures being outside the bounds of Fig. 7.

In Fig. 7, the perpendicular lines to the OX and $O'X'$ axes from the vertices of the triangles A_i and B_i, respectively, lying on those axes define the rectangles $A_i \times B_i$, $i = 1, 2, 3$. In this way, the system obtains the required FAT, with the rectangles being the patches of the following fuzzy rules:

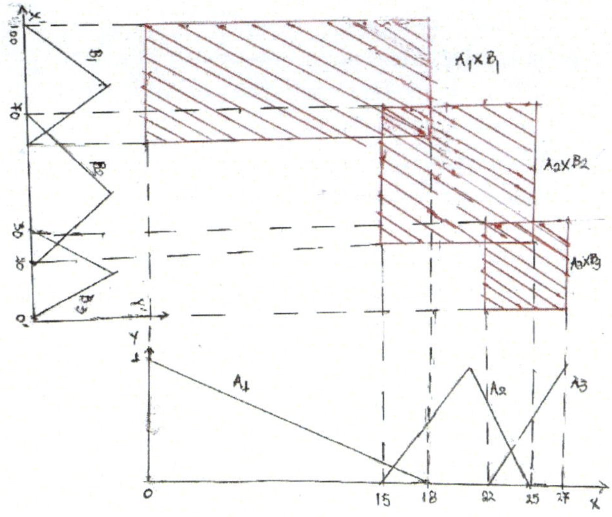

Fig. 7 Fuzzy control system of the boiler

- $A_1 \times B_1$: IF the building's temperature is low, THEN the boiler operates in full rate.
- $A_2 \times B_2$: IF the building's temperature is normal, THEN the boiler operates in low rate.
- $A_3 \times B_3$: IF the building's temperature is high, THEN the boiler turns off.

Those three patches cover a curve which represents the traditional control system of the boiler. That curve starts at the top of the patches on the left and ends at the bottom on the right. In case of a dynamic system changing by the time, the corresponding curve is moving. However, if the curve's movement is not too fast, the patches can be extended or moved in order to cover always the curve. In other words, a fuzzy system has usually the ability to cover or approximate an existing dynamic system.

The boiler, however, does not "understand" the fuzzy rules. In order to respond, one must give to it a number instead, representing the rate of its operation. This is succeeded by defuzzifying the corresponding FSs. Assume, for example, that the building's temperature is 19 °C. This temperature has membership degrees zero in the FSs A_1 and A_3 on X and nonzero in A_2. This means that the fuzzy rule $A_2 \times B_2$ will be fully activated (100%). Since the triangle B_2 represents the TFN B_2 (20, 45, 70), using the COG defuzzification technique one finds by Eq. (1) that $R(B_2) = 45$, i.e., the boiler receives the order to operate in a low rate.

Assume now that the building's temperature drops to 16 °C, which has membership degrees zero in A_3 and nonzero in A_1 and A_2 (Fig. 7). In this case, therefore, both the rules $A_2 \times B_2$ and $A_1 \times B_1$ will be activated.

In Fig. 8, BAC and HOZ correspond to the triangles A_2 and A_1 of Fig. 7, respectively, while the point E corresponds to the temperature of 16 °C. Applying Thales' theorem in the triangle ALB, one finds that $\frac{AB}{DB}=\frac{AL}{KL}$ and $\frac{BA}{BD}=\frac{BL}{BE}=\frac{5}{1}$, which gives that KL $= \frac{1}{5}$AL $= \frac{1}{5}$ = DE. Therefore, (DEB) $= \frac{1}{10}$. We also have that (BAC) = 5 and (ADEC) = (BAC) − (DEB) = $5 - \frac{1}{10} = \frac{49}{10}$, which gives that $\frac{(ADEC)}{(ABC)} = \frac{49}{50} =$ 0.98. Therefore, the input 16 °C belongs by 98% to the FS A_2, i.e., the fuzzy rule $A_2 \times B_2$ will be activated by 98%. As a result, the corresponding triangle B_2 in the output must be reduced to the 0.98 of its height (Fig. 9).

Also, in the triangle HOZ of Fig. 8, we have that $\frac{HO}{IO}=\frac{HZ}{TZ}$ and $\frac{ZH}{ZT}=\frac{ZO}{ZE}=\frac{18}{2}$, which gives that IO $= \frac{1}{9}$ = TE. Then (TEZ) $= \frac{1}{9}$, (HOZ) = 9 and $\frac{(TEZ)}{(HOZ)} = \frac{1}{81} \simeq 0.012$. Therefore, the fuzzy rule $A_1 \times B_1$ will be activated by 1.2%, which means that the corresponding triangle B_1 in the output must be reduced to the 0.12 of its height (Fig. 9).

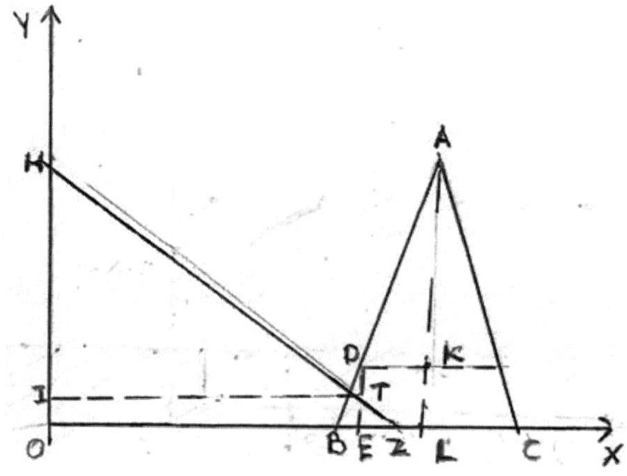

Fig. 8 Triangles A_1 and A_2 (inputs) of Fig. 7

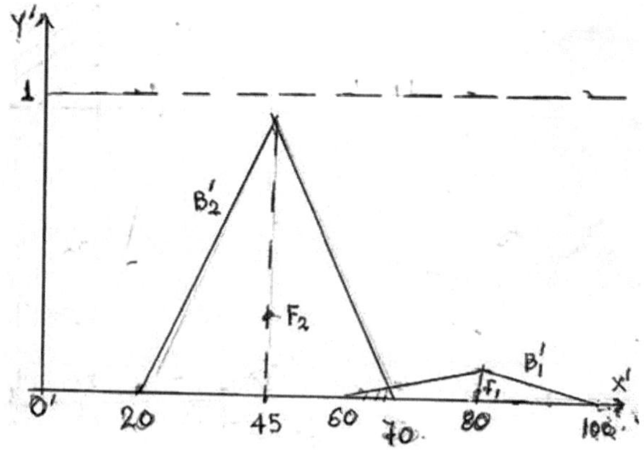

Fig. 9 Reduced triangles $B_1{}'$ and $B_2{}'$ (outputs) from Fig. 6

Let F_1 and F_2 be in Fig. 9 the COGs of the triangles B_1' and B_2' with x-coordinates $x_1 = 80$ and $x_2 = 45$, respectively. Then, the x-coordinate of the COG of the whole area defined by the two triangles, in which their common part is included twice (since it belongs to both outputs), being the system of the COGs F_1 and F_2 [15], is calculated by the equation

$$x = \frac{1}{S}(x_1 S_1 + x_2 S_2) \tag{2}$$

In Eq. (2), S_i denotes the area of the triangle B_i', $i = 1, 2$ and $S = S_1 + S_2$. But $S_1 = (0.12 \times 40){:}2 = 2.4$, $S_2 = (0.98 \times 50){:}2 = 24.5$, and replacing to (2), one finds that $x \approx 48$. Therefore, the rate of the boiler's operation increases and the temperature rises up again.

The process described above is performed by of a fuzzy chip, which, receives a number from the input, compares it to all the input's FSs, provides output FSs, and turns them to a number. Those steps are repeated continuously in millions of flips per second. That process is usually referred as *Fuzzy Associative Memory (FAM).* Note that computers have a direct memory, since they save an acquired information in an address—i.e., a set of digital characters determining a place in their memory—and, whenever is needed, they search for this information in that particular address. On the contrary, FAM activates the whole memory of the FCS in a way analogous to what humans do when they search all their past memory in order to recognize a familiar person among the crowd. Thus, FAM provides a reliable model of the way that humans think and decide.

5 Conclusions

The discussion performed in this paper leads to the following general conclusions:

- The traditional CSs are used to achieve, through automatic operation, increased productivity, high-quality products, and improved performance of a device or another system in terms of a set of bivalent IF–THEN rules. A model is needed in order to describe the operation of a traditional CS. The operation of the PID controllers, initiated by Minorsky in 1922, can be described by one or more differential equations that define the CS's response to its outputs. Apart from industry, CSs have found important applications to life sciences, computer engineering, and operations' research.
- The operation of a FCS is based on the FAT theorem, according to which a system can be "covered" or approximated by a fuzzy system. An advantage of FCs with respect to the traditional CSs is that they do not need a model to describe their operation. On the contrary, they do need IF–THEN fuzzy rules, which, however, could not be too many as in a traditional controller. A FCS operates on the basis of its FAM, which simulates the way that humans think and decide. Starting from

Japan during the 1980s, fuzzy controllers have found many industrial applications, and not only, while the research about them is nowadays in a phase of continuous development.

References

1. Zadeh LA (1965) Fuzzy sets. Inf Control 8:338–353
2. Zadeh LA (1973) Outline of a new approach to the analysis of complex systems and decision processes. IEEE Trans Syst Man Cybern 3:28–44
3. Kosko B (1993) Fuzzy thinking: the new science of fuzzy logic. Hyperion, New York
4. Kosko B (1990) Fuzziness vs. probability. Int J Gen Syst 17(2–3):211–240
5. Umbers IG, King PJ (1980) An analysis of human decision-making in cement kiln control and the implications for automation. Int J Man Mach Stud 12:11–23
6. Mamdani EH, Assilian S (1975) An experiment in linguistic synthesis with a fuzzy logic controller. Int J Man Mach Stud 7(1):1–13
7. Sugeno M (1985) Industrial applications of fuzzy control. Elsevier Science Pub. Co.
8. Taha HA (1967) Operations research—an introduction, 2nd edn. Collier Macmillan, New York, London
9. Portal.dnb.de. GND, Katalog der Deutschen Nationalbibliothek (authority control). https://portal.dnb.de/opac.htm?method=simpleSearch&cqlMode=true&query=nid=4032317-1. Retrieved 26 Apr 2020
10. Voskoglou MGr, Salem A-B (2020) Benefits and limitations of the artificial with respect to the traditional learning of mathematics. Mathematics 8:611
11. Athanassopoulos E, Voskoglou MGr (2020) A philosophical treatise on the connection of scientific reasoning with fuzzy logic. Mathematics 8:875
12. Kosko B (1992) Fuzzy systems as universal approximators. In: 1st IEEE conference on fuzzy systems, pp 1153–1162
13. Kaufmann A, Gupta M (1991) Introduction to fuzzy arithmetic. Van Nostrand Reinhold Company, New York, NY
14. Voskoglou MGr (2019) Methods for assessing human-machine performance under fuzzy conditions. Mathematics 7:230
15. Wikipedia. Center of mass: a system of particles. http://en.wikipedia.org/wiki/Center_of_mass#A_system_of_particles. Retrieved 25 Oct 2014

Text-to-Image Synthesis: A Comparative Study

Razan Bayoumi, Marco Alfonse, and Abdel-Badeeh M. Salem

Abstract Text-to-image synthesis or conditional image generation is referring to converting natural language text descriptions into realistic images, where both are semantically consistence. Deep convolutional generative adversarial networks (GANs) are rapidly changing field that have shown great revolution in the generative models. GANs are the recent widely used networks in many different frameworks to conditionally and unconditionally generate realistic images. In this paper, we aim to explore and review the majority of the work in this domain and provide a comparative study to get a full overview image about this domain.

Keywords Text-to-image · Generative models · GAN · Deep learning · Computer vision · Conditional generation

1 Introduction

Generating images based on input text, known as text-to-image synthesis, is a new field that comes up a few years ago. Converting textual features into pixels and understanding the relation between visual content and natural language is considered a challenge because of the difference between the two latent spaces. There are a number of researches in this point that start achieving promising results which opens the door to more researchers to start working in this direction. They can be used in many useful aspects such as facilitating literacy, assisting communication tool for people with disabilities, helping in generating prototypes, sketches, architectures,

R. Bayoumi (✉) · M. Alfonse · A.-B. M. Salem
Computer Science Department, Faculty of Computer and Information Sciences, Ain Shams University, Cairo, Egypt
e-mail: razan.bayoumi@cis.asu.edu.eg

M. Alfonse
e-mail: marco_alfonse@cis.asu.edu.eg

A.-B. M. Salem
e-mail: abSalem@cis.asu.edu.eg

D. A. Magdi et al. (eds.), *Digital Transformation Technology*, Lecture Notes in Networks and Systems 224, https://doi.org/10.1007/978-981-16-2275-5_14

and realistic images of faces can also help in Crime Scene investigation (CSI) to recognize criminals.

Deep learning is an artificial intelligence, which was inspired from the structure and how the human brain works. It is a subfield from machine learning that consists of networks aim to learn and understand pattern of unstructured data. GANs are deep networks that are used as generative models. GANs are the main models that used in text-to-image synthesis task for their abilities to learn the regularities or patterns in input data and generate output that looks realistic such as input data.

The GANs consists of two convolutional networks; generator and discriminator. The generator networks work to produce realistic output images with distribution as close as to the dataset distribution to fool the discriminator, while the discriminator role is to determine whether the image is real (from dataset) or not (generated). The two models are trained together as a two players' minimax game. In this paper, we will review different GAN architectures that is used in in different models for text-to-image synthesis task.

The performance of GANs is assessed by evaluating the quality and diversity of the generated images qualitatively and/or quantitatively. The qualitative measures are non-numerical evaluation that usually depend on comparison, while the quantitative measures refer to calculated score values that reflect the quality of the output images as inception score (IS) [1], Frechet Inception Distance (FID) [2], visual-semantic similarity (VS) [3], R-precision [4], and Structural Similarity Index Measure (SSIM) [5]. IS measures the quality and diversity of the generated images by using a pre-trained model that classifies the generated images and predicts the probability of the image belonging to each class. The probabilities are then summarized into score to measure how much the image belongs to class and how diverse are the generated images over the classes. FID is an improvement of IS as it compares the statistics (mean and covariance) of the generated images with the real images and calculate the distance between them, instead of depending only on the generated images. VS measures the alignment between the input text and the generated images by calculating the distance between feature vector of the input text and the generated image. A feature vector is extracted from the synthesis images and mapped with the embedding text vector to a common space. R-precision is used for ranking retrieval results as a complementary method for evaluation. SSIM is used to evaluate the quality of images by measuring the similarity between the generated and the original images.

This paper is organized as follow, Sect. 2 presents text-to-image methodologies grouped into seven subsections according to the idea and the architecture of the models, and Sect. 3 discusses the conclusion and future work.

2 Text-to-Image Methodologies

This section presents the most recent work concerning the text conditional image generation methodologies. These works are presented through years from 2016 to 2019.

2.1 Baseline Generation Method

The baseline architecture for generating images from text using GANs, is built using simple deep convolutional generative adversarial network DCGAN that consists of a generator and discriminator as shown in Fig. 1. The both GAN networks; the generator and the discriminator perform feed-forward inference conditioned on the text, where the text is encoded previously into a vector which fed into the generator as input.

Reed and colleagues [6] considered the baseline for generating images from an input text description. The deep convolutional generative adversarial network (DCGAN) aims to generate 64 × 64 images based on a single-sentence human-written description. The text description is encoded by a hybrid character-level convolutional recurrent neural network before entering the GAN network as an input vector. The input text is encoded into a vector and then concatenated with noise vector and fed as input into the generator network. There are three proposed models: GAN-CLS, GAN-INT, and GAN-INT-CLS. In GAN-CLS, instead of doing binary discrimination (real images with matching text, and synthetic images with arbitrary text) like Naive GAN, the discriminator distinguishes three different cases; real images with matched text, synthetic images with random text, and real images with mismatched text. In GAN-INT, by applying interpolation between the embedding vectors of the images' descriptions, many extra embedding texts are generated without additional labeling cost as they do not have actual natural language corresponding. GAN-INT-CLS combines both foregoing. Reed et al. also introduced the idea of inverting the generator for style transfer, while the input text description captures the main image objects, the generated noise vector z should capture the remaining factors such as the

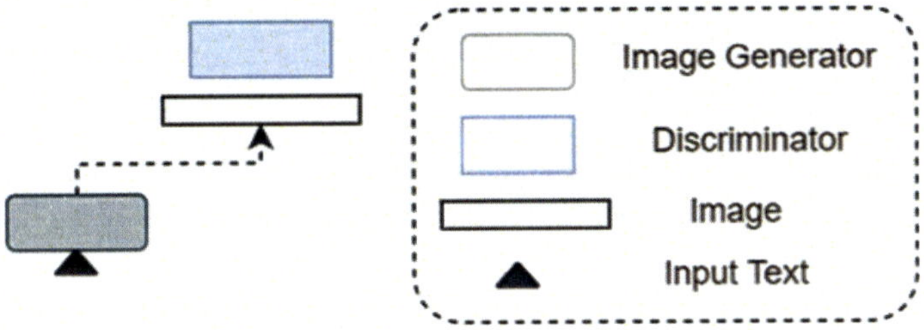

Fig. 1 Baseline framework

pose and background colors in order to generate a realistic images. Caltech-UCSD Birds dataset, the Oxford-102 Flowers dataset, and the MS-COCO dataset are used in the training process.

Cha and colleagues [7] built a model upon the GAN-INT-CLS architecture of Reed et al. [6] aiming to extend this work to improve the quality and consistency of the generated images by introducing two main loss terms: contextual loss term and perceptual loss terms. Contextual loss term guarantees good cross-modal translation, which means the consistency between the generated images and the input text description will be higher. Perceptual loss terms, which corresponding to pixel, feature activation, and texture reconstruction losses, are for improving the quality and generating more realistic images. The input text description is encoded and used as input for both networks: generator and discriminator. The text-encoded vector is concatenated with the noise vector z and fed into the generator network, the same encoded vector is concatenated with the synthesized generated image and fed into the discriminator network. The three proposed models: (GAN-INT-CLS-Pixel), (GAN-INT-CLS-VGG), and (GAN-INT-CLS-Gram), are evaluated with perceptual losses on Caltech-UCSD Bird (CUB) and Oxford-102 flower datasets. For quantitative evaluation, GoogLeNet is used to evaluate how obvious the objects are in the generated images and can be correctly classified as birds or flowers.

2.2 *Multistage Generation*

Instead of using a single generative adversarial network to synthesize an image based on input text description at once, a multistage generative adversarial network architecture is proposed as shown in Fig. 2. It consists of multiple generators and multiple discriminators, each of them works on a different scale in an incremental fashion to generate a high-resolution image with fine-grained details. Some models are built based on this network, which show improvement in the quality and resolution of the generated images.

Zhang and colleagues [8] proposed a model called Stacked Generative Adversarial Networks (StackGAN) to generate photo-realistic images conditioned on input text captions. This method synthesizes images from text as a novel sketch-refinement process in wo stages, each stage is considered a separate network. The Stage-I GAN outlines the basic shape and colors of the image object based on the input text description and the noise vector, resulting in low-resolution images. The Stage-II GAN generates high-resolution images with more realistic details by taking Stage-I outputs and text descriptions as inputs. A novel conditioning augmentation technique is introduced to increase the diversity of the generated images and help to stabilize the training process because of its ability to improve smoothness and continuity in the latent conditioning manifold. The datasets used are the Caltech-UCSD Birds dataset, the Oxford-102 Flowers dataset, and the MS-COCO dataset. To verify the superiority of StackGAN, the inception score approach is used for quantitative evaluation as well

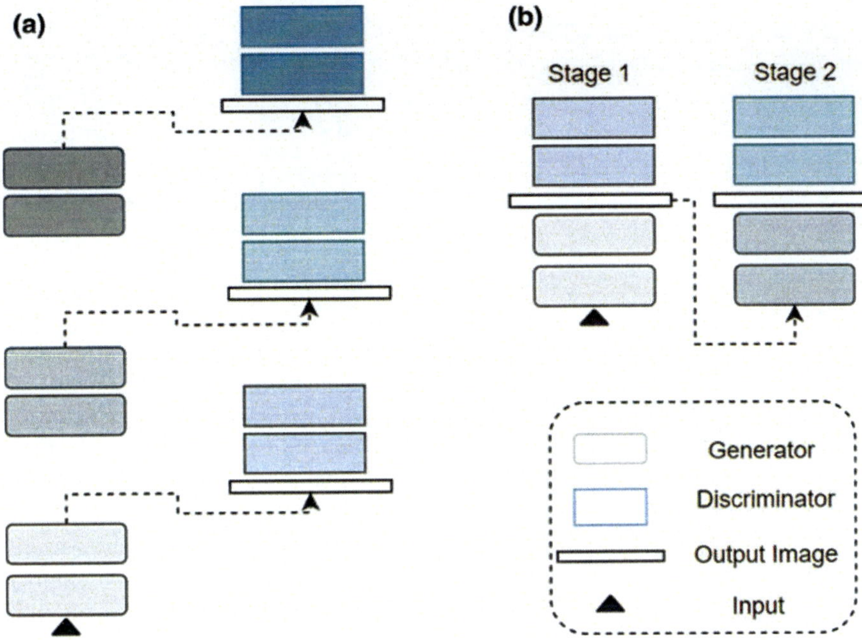

Fig. 2 Multistage GAN frameworks. From left to right, **a** uses multiple generators and multiple discriminators, **b** uses multiple stage GANs where the output from one GAN is fed to the next GAN as input

as human evaluation to reflect if the generated images are well-conditioned on the input text descriptions or not.

Then, Zhang and colleagues [9] proposed a new version of their model called Multi-Distribution Generative Adversarial Network (StackGAN-v2), which is proposed to enhance the nature of generated images and balance the GANs' training process. StackGAN-v2 is composed of several generators and discriminators in a "tree-like" structure. Images are generated from various tree branches, from low-resolution to high-resolution. At each branch, the generator generates a synthesis image at that scale, and the discriminator determines whether the image is real or not by estimating the probability that a sample originated from training images of that scale rather than the generator. The conditioning variable vector for StackGAN-v2 is also generated by conditioning augmentation. The conditioning variable is the embedding of the input text concatenated with a noise vector. The generators and discriminators have trained alternately. StackGAN-v2 discriminators are trained to recognize genuine images from fake ones for unconditional image generation. As for conditional image generation, images and their corresponding conditioning variables (i.e., class or text description) enter the discriminator to determine whether the pair matches or not, which motivates the generator to approximate the conditional image distribution. A color-consistency regularization term is introduced to keep the generated samples more consistent in color and therefore improves the quality of the generated images. StackGAN-v2 is trained like the first version on the CUB, Oxford-102, and COCO datasets. IS and Fréchet Inception Distance (FID) are used for quantitative evaluation. To better evaluate the proposed methods, especially to see whether the generated images are well-conditioned on the given text descriptions, user studies are also conducted.

Xu and colleagues [4] proposed a model called Attentional Generative Adversarial Network (AttnGAN) that aims to generate fine-grained synthesis images in a multistage fashion by paying attention to the relevant between the input description words and subregions of the generated images. AttnGAN consists of two novel components; attentional generative network and the deep attentional multimodal similarity model (DAMSM). The attentional generative network allows each subregion of the generated images to be drawn based on the most relevant words of the input text description. DAMSM measures the similarity between the subregions and input words by mapping both to a common semantic space to compute fine-grained loss for image generation. This model is composed of multiple generator and discriminator networks. The input natural language description is encoded as a whole sentence into a global vector and each word of it separately into a word vector. Conditioning augmentation converts the sentence vector to the conditioning vector and concatenated it to a noise vector. At the first stage, the generator network generates a low-resolution image using the global sentence vector. In the next stages, an attention layer is used to form a word-context vector by using the image vector of each subregion to query word vectors. Then, the subregion vector and the corresponding word vector are integrated into a multimodal context vector, which is used to generate more features and details of this subregion. This process leads to generate a higher resolution image with more realistic details in a gradual fashion. Discriminators are structurally separated; therefore, they can be trained in parallel and each of them works on a single image scale. CUB and MS-COCO datasets are used for evaluating and training this model. The IS approach as well as R-precision are used to verify the effectiveness of the AttnGAN model. IS is used as the quantitative evaluation measure. R-precision is a complimentary evaluation as IS cannot measure if the generated images and the input text are consistent.

2.3 *Single Stream Generator and Multiple Discriminators*

New designs have been proposed to make headway in the field rather complicated designs such as a different architecture that consists of a single stream generator and multiple discriminators as shown in Fig. 3. The multiple discriminators may be hierarchically nested discriminators, multi-purpose discriminators in which the discriminator focuses in multiple aspects in the generated images or independent multi-scale discriminators.

Zhang and colleagues [3] proposed a model defined as HDGAN, referring to high definition generated images and the idea of hierarchically nested discriminators.

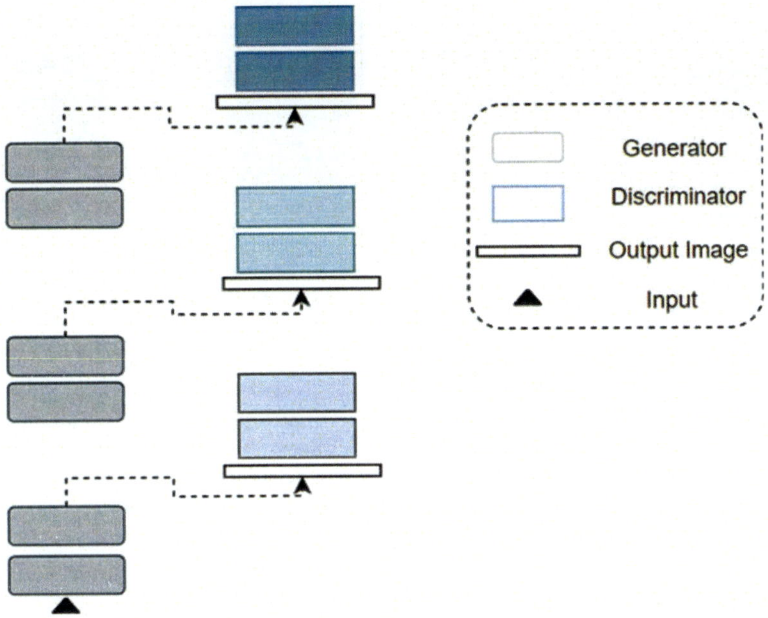

Fig. 3 A single stream generator with multiple discriminators

HDGAN consists of a single stream generator and hierarchically nested discriminators at multi-scale intermediate layers. First the input text description is encoded using pre-trained char-RNN text encoder then added to a random noise vector. The generator consists of three modules; termed K-repeat res-blocks, stretching layers, and linear compression layers. A single res-block in the K-repeat res-block is a modified residual block. The stretching layer is used to change the size and dimension of the feature map. The linear compression layer directly compresses feature maps to the RGB space. The discriminators are used to evaluate the output of the intermediate layers of the generator to assure the diversity and semantic consistency between the input text and its corresponding generated image. There are two branches that are added to the upper layer of the discriminator. One of them classifies each patch of the generated image as real or fake, and the other distinguishes real and fake image-text pair. The model is evaluated on three datasets; the CUB dataset, the Oxford-102 dataset, and the MS-COCO dataset. IS, multi-scale structural similarity (MS-SSIM), and visual-semantic similarity are the three quantitative metrics that are used to evaluate HDGAN.

Chen and colleagues [10] proposed Fully-trained Generative Adversarial Network (FTGAN) for fine-grained text-to-face generation, in which a new technique is used for the training process that includes training both the text encoder and image decoder at the same time, unlike the previous frameworks that train each of them separately as shown in Fig. 4.

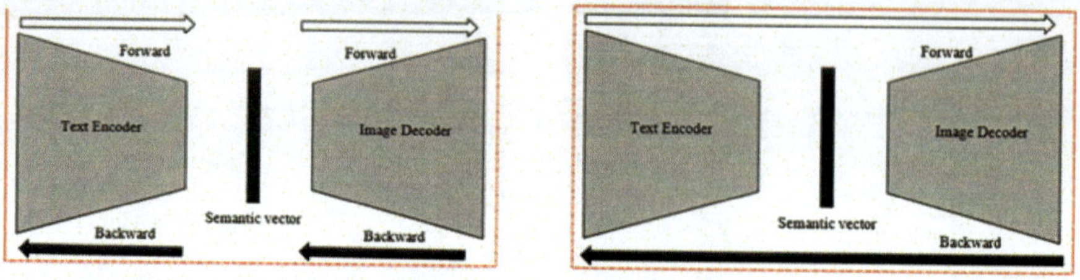

Fig. 4 Framework comparison between the proposed FTGAN (right) and previous frameworks (left) [10]

The GAN network consists of a generator and 3 discriminators for different scales. The generator, which does the main job, can be divided into 2 parts; text encoder and image decoder. The text encoder part extracts the semantic features from the input text description using a bi-directional long short-term memory (BiLSTM) which is then concatenated with a noise vector sampled from normal distribution. The image decoder part takes the text encoder's output and maps it to nature realistic image by a three-stages convolutional neural network (CNN). The discriminators' inputs are the input text embedding and its corresponding generated images (64 × 64, 128 × 128, 256 × 256, respectively), and they output true or false, true for the real image (ground-truth) with the matched semantic vector and false otherwise. CUB dataset is utilized to compare FTGAN with the previous state-of-the-art text-to-image GAN models. A new dataset called SCU-Text2face is built for text-to-face synthesis and used to compare the generated face images of FTGAN with the results of AttnGAN. To evaluate the results of the network in CUB, a pre-trained Inception-V3 network is used. Though, for the face dataset, there is no pre-trained Inception-V3 model. So, the FID score is used instead, which is considered another common measure for evaluating synthesized images. FaceNet extracts the feature vector of generated faces and computes the average face semantic distance (FSD) and average face semantic similarity (FSS) between the generated face and real one (ground-truth).

Gao and colleagues [11] proposed Perceptual Pyramid Adversarial Network (PPAN) for synthesis multi-scale realistic images based on input text description in an adversarial way. The model is composed of three main components: conditioning augmentation, a generator, and three discriminators. Conditioning augmentation is usually applied to overcome the problem of the limited training data, which causes a discontinuity in the latent space in higher dimensions, by taking the text embedding and generating more training pairs which is then concatenated to noise vector and used as input to the generator. The generator network consists of two components: Residual Generative Blocks (RG-Blocks) and Cumulative Blocks (C-Blocks). C-Blocks are developed to generate visual features based on the textual features captured by the RG-Blocks. The perceptual loss is defined on the generator to apply semantic similarity between the generated and ground-truth images which guarantee diversity of generated images. The three multi-purpose discriminators are designed to encourage semantic consistency between the input texts and generated images, guarantee image fidelity, and ensure class invariance at different pyramids by calculating matching-aware pair loss, local image loss, and class information loss, respectively. PPAN is evaluated on two datasets; CUB and Oxford-102. Two quantitative metrics are used to evaluate this model; IS, which measures the quality and diversity of the generated images, while visual-semantic similarity (VS) is used to measure the semantic consistency between the generated images and the input text.

2.4 *Bi-directional Generation Method*

To improve the performance of generating images from a text description, an idea of learning text-to-image generation by re-description or in other words by integrating these two models (text-to-image and image-to-text) came up.

Dong and colleagues [12] used it as a base to propose the Image-Text-Image (I2T2I) model which consists of three modules: image captioning module, image-text mapping module, and conditional generative adversarial network for text-to-image synthesis (GAN-CLS). The image captioning module encodes the input image and generates its corresponding natural language description. The image-text mapping module generates vector representation that embeds images and sentences into one common space. The conditional generative adversarial network for text-to-image synthesis, which is used in this model is GAN-CLS the same as Reed et al. comparing between both results, I2T2I performs better for certain classes like "people." Two datasets were used in the experiments, the MS-COCO dataset and the MPII Human Pose dataset (MHP).

Qiao and colleagues [13] used the re-description idea and proposed the MirrorGAN model to enhance the diversity and semantic consistency of the generated images. MirrorGAN consists of three modules; a semantic text embedding module (STEM), a global–local collaborative attentive module (GLAM), and a semantic text regeneration and alignment module (STREAM). STEM is used for generating word- and sentence-level embeddings of the input text description using recurrent neural network (RNN). The conditioning augmentation is then applied to augment the text descriptions and fed with the noise vector to GLAM. GLAM has a multistage stream generator for generating images from coarse to fine scales using local and global attention. STREAM tries to regenerate the image captions from the generated image that aligns with the input text description semantically. Datasets used are the CUB bird dataset and the MS-COCO dataset. The IS method is used to measure both the quality and diversity of the generated images, while R-precision is used to measure how similar is the generated images and their corresponding text description semantically. Human perceptual evaluation is used to conduct two tests: the Image Authenticity Test and the Semantic Consistency Test to compare the generated images of AttnGAN and MirrorGAN.

2.5 Layout Image in Generation Process

While most of the previous works at this point turned to apply direct mapping from text to image, some researchers tended to decompose this process by introducing an intermediate layout stage. The input text is encoded and used to generate a semantic layout image, which is then converted to an image as shown in Fig. 5. This method improves the semantic consistency between the input text and the generated images.

Hong and colleagues [14] proposed a novel hierarchical approach for text-to-image synthesis by inferring semantic layout to generate semantically meaningful images. The suggested idea in this work is to decompose the generation process into multiple phases. The model mainly consists of three parts; text encoder, layout generator, and image generator. Text encoder extracts the feature vector of the input text description. Layout generator, which consists of two generator networks: box generator and shape generator, constructs a semantic layout from the input text. The image generator transforms the layout into an image. To generate an image using a refining semantic structure, the box generator takes the text embedding (the output of the text encoder) as an input and constructs a layout of the image which contains a set of bounding boxes that determine the position and size of the objects in the image. The shape generator then took the box generator's output and refine the bounding boxes with the objects' shape. At this step, the user can modify the generated layout image, by changing the number, size, or position of objects in the image. Using the constructed layout and the input embedding concatenated with a noise vector, the image generator generates a pixel-wise image. This model was trained using the MS-COCO dataset and evaluated using three evaluation metrics: IS, caption generation, and human evaluation.

Li and colleagues [15] proposed Object-driven Attentive Generative Adversarial Networks (Obj-GANs) to generate high quality, semantically consistent, and realistic multi-object images from input text descriptions. The text-to-image synthesis process in Obj-GAN is done in two main steps: semantic layout generation and image

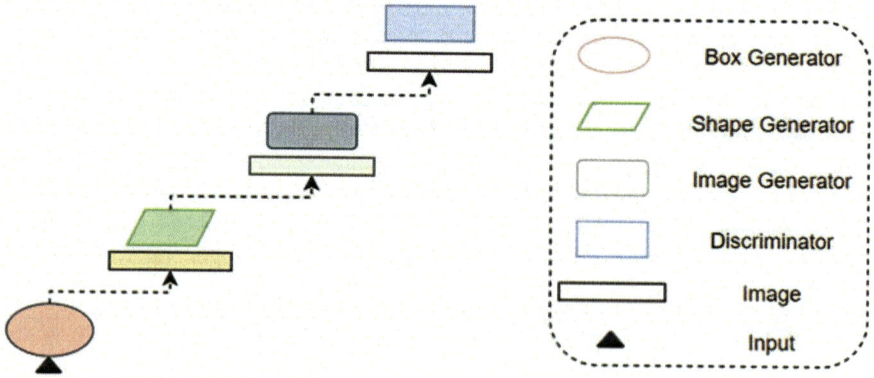

Fig. 5 Generating images based on text using layout stage framework

generation. Pretrained biLSTM model is used for encoding the input's description words as word vectors and the entire sentence as a global sentence vector. The layout generation phase consists of two generators: box generator and shape generator. The box generator first generates a sequence of bounding boxes with class labels, and then, a shape generator predicts and generates the shapes of each object in the bounding boxes. The image generation phase is composed of attentive multi-stage image generator and discriminators. Attentive multistage image generator has two generator networks: the *base generator* that generates low-resolution images based on the global input sentence and the layout generated image, and then, the *refiner generator* that improves the details of the subregions in the images based on the most related words and pre-generated class labels to generate high-resolution images. There are two types of discriminators: patch-wise and object-wise. The patch-wise discriminator determines both the realism of the patch (unconditional) and the consistency between the patch and the input text description (conditional). The object-wise discriminator checks every bounding box to ensure that the generated object matches the pre-generated semantic layout. The MS-COCO dataset is used for evaluation. The IS and FID score are used as the quantitative evaluation metrics. The R-precision is also used for evaluation but to estimate how accurate the generated image is conditioned and consistent with the input text description.

2.6 *Human-Inspired Generation*

The main idea behind the human-inspired generation models is imitating the way the human follow in order to generate or draw an image.

Singh and colleagues [16] proposed a model called CanvasGAN for generating semantically consistent images with input description, and the model is built based on the idea of how painters deal with any painting, they paint images incrementally by focusing on a particular part at once. For CanvasGAN, the model focuses on a specific part of the text for creating a new patch at each timestep instead of dealing with the whole sentence at once. A new visual-semantic embedding generation mechanism that uses self-attention is introduced to enable concentrating on the critical hidden states of RNN rather than taking hidden state at last timestep to generate a sentence embedding. GloVe is used to generate the embedding of the words in the input description and then to encode them using the visual-semantic network which provides encoded word embeddings and a sentence embedding. The sentence embedding is resampled by the conditioning augmentation and used with a noise vector to generate the first hidden state. Initially, the canvas is empty, then at each timestep, the patches start to be generated iteratively based on attended features from the text description. The discriminator is then used to measure the semantic consistent between the generated image and text description. The model is evaluated by conducting experiments on the Oxford flowers-102 dataset. IS is used to calculate the model generative capabilities using pretrained ImageNet.

Qiao and colleagues [17] proposed LeicaGAN, a novel text-to-image model that aims to generate realistic and semantically consistent images. LeicaGAN is built based on the idea of how humans generate an image from a written description. First people should have prior knowledge about colors, textures, shapes, etc. Then, they imagine an overall visual impression and progressively draw an image by adding more and more details. According to this, LeicaGAN is consisted of three phases: textual-visual co-embedding (TVE), multiple priors aggregation (MPA), and cascaded attentive generator (CAG). TVE consists of two encoders: *text-mask encoder* for learning shape and layout and a *text-image encoder* for learning texture, colors, and semantic. MPA, in which complementary priors learned in the previous phase, is combined together, and noise is added for diversity. At the final phase, CAG, the images are progressively drawn from coarse to fine. LeicaGAN is evaluated by conducting experiments on CUB and Oxford-102 datasets. The performance of the model is measured quantitatively and qualitatively. inception score, R-precision, and human evaluation is the used quantitative measures.

2.7 Others

Reed and colleagues [18] proposed a new model called Generative Adversarial What-Where Network (GAWWN) to generate 128 × 128 images based on a given instructions which determine the specifications and the place of the drawn object. There are two ways to encode the spatial constraints that have been introduced. The first way depends on the rough location of the bird, which is applied using spatial transformers, by including cropping and masking modules in the GAN model. The second way depends on location of some parts of the birds and humans structured as a set of normalized (x, y) coordinates. The GAWWN model, consists of two networks; generator and discriminator. The generator takes the text embedding vector (encoded using pre-trained encoder) as input and concatenate it with noise vector, then use one of the previous mentioned way to localize the object position and generate a realistic image, which then fed into the discriminator to determine whether it is real or not by producing the scalar discriminator score. To train the model, datasets need to have both global annotations of the images like class label or text description as well as localized annotations as bird part keypoints in Caltech-USCD birds (CUB) and human joint locations in the MPII Human Pose dataset (MHP).

Dash and colleagues [19] proposed Text-Conditioned Auxiliary Classifier Generative Adversarial Network (TAC-GAN), a text-to-image GAN for synthesizing images of size 128 × 128 from input text descriptions. TAC-GAN can be considered a new version of AC-GAN [20] as shown in Fig. 6. In the AC-GAN model, the images are generated based on a class label, and the discriminator as well as distinguishing between the synthesized images and real ones also classifies those images to their classes. While in TAC-GAN, the generator output is conditioned on a text description and the discriminator does the same two tasks as in AC-GAN. The input of the generative network is the text description encoded using skip-thought vectors

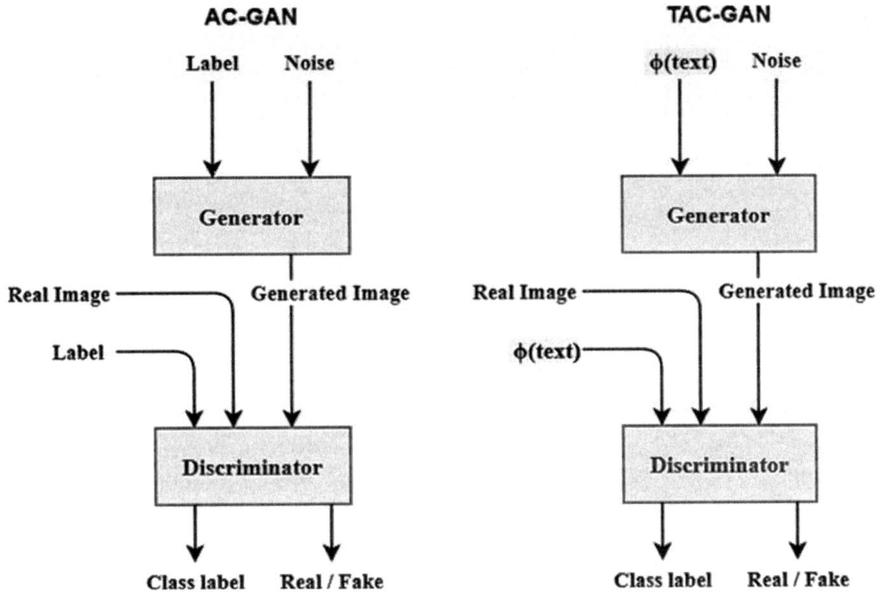

Fig. 6 A simple architecture comparison between AC-GAN and TAC-GAN

to generate the text embedding vector in addition to a noise vector z. The Dataset used is Oxford-102 flowers dataset that has a class label and five text descriptions for each image. As an evaluation metric, the IS is used to estimate the diversity and the quality of the generated images.

Cha and colleagues [21] proposed Text-Conditioned Semantic Classifier GAN (Text-SeGAN), a new GAN approach that aims to increase the diversity of the generated images and improve the training process by introducing different strategies for choosing the negative samples instead of the default random selection way. Text-SeGAN is built upon the GAN-INT-CLS architecture, both GAN networks; generator and discriminator are deep convolutional neural networks. The generator's input is the embedded vector of the text description concatenated with the noise vector that is sampled from a normal distribution. In addition to the normal task of the discriminator, distinguishing the real and fake images, it also has an auxiliary task of measuring a semantic relevance between the input description and image. There are six different ways: Random negatives (RN), Easy negatives (EN), Hard negatives (HN), Semi-easy negatives (SEN), Semi-hard negatives (SHN), Easy-to-hard negatives (EHN), for sampling negative images based on the semantic distance to its reference text using Euclidean distance. In random negatives, which is used in the previous approaches, the negative sample is chose randomly from any class outside the reference class. In easy negatives, choosing the furthest text embedding from the reference text, which is classed outside the reference class, while on the other hand, hard negatives is finding the closest embedded text to the reference text from an outer class. In semi-easy negatives, apply easy negative in a randomly selected M outer samples. In semi-hard negatives, choose M random samples that are classified outside the reference class and then apply hard negative over them. In easy-to-hard negatives, select negative text from M outer samples randomly. Generate a histogram of cosine similarity values between positive and M negative text, select 100β-th percentile of the histogram ($0 < \beta \leq 1$) and increase β gradually. Oxford-102 flower dataset is used for evaluation. The IS is used for quantitative evaluation, which measures the quality and diversity of the generated images.

Zhu and colleagues [22] proposed Dynamic Memory Generative Adversarial Network (DM-GAN) model to generate accurate high-quality images based on input description. DM-GAN model consists of two phases; initial image generation and dynamic memory-based image refinement. At the initial image generation phase, the input text is encoded as a whole sentence into a global vector and each word of it separately into a word vector. The sentence vector is resampled by the conditioning augmentation to avoid overfitting. Using the global vector concatenated with a noise vector sampled from a random distribution, an initial low-resolution image is generated with basic objects and primitive details. At the dynamic memory-based image refinement phase, more details are added and a higher resolution image is generated. This phase is composed of four modules: Memory Writing, Key Addressing, Value Reading, and Response. In Memory Writing, the text embedded data is stored in a key-value structured memory, which enables accurately generated images from the text description. Key Addressing and Value Reading are then used to retrieve features from the memory module to improve and enhance the visual features of the low-resolution images. The Response operation is added to dynamically control the fusion of the image features and the memory reads. Experiments on the CUB and the MS-COCO datasets are run to show the ability of the model. The quantitative performance of the DM-GAN is measured using three methods: IS, FID, and R-precision.

Yin and colleagues [23] proposed Semantics Disentangling Generative Adversarial Network (SD-GAN) for text-to-image synthesis. It aims to distill the semantic commons from text descriptions to generate consistent images under description variants and meanwhile keep the low-level semantic details and diversities for a fine-grained generation. The model is built upon Siamese structure, which consists of two identical neural networks (branches), each taking one of the two inputs, and learns the similarity between them. The purpose of the Siamese structure is to improve the generation consistency despite the variation of the input descriptions expressions. The intra-class pair of input descriptions should generate a close distance apart images in the feature space, while the inter-class pairs should generate larger distances. Semantic-conditioned batch normalization (SCBN) considered a complement to the Siamese structure, which keeps the detailed and diverse semantics of the input texts. There are two input descriptions, one for each branch, first the text passes through a text encoder to extract the semantic feature vector and then concatenates it with a noise vector. Then, there is the Hierarchical Generative Adversarial Networks, which consists of multiple generators to generate realistic images from low to high resolutions, each of them followed by a discriminator which determines whether the generated images at each stage are real or fake. SD-GAN is evaluated on two datasets, the CUB and the MSCOCO datasets. IS is used to measure the quantitative performance, as well as human evaluation.

Li and colleagues [24] proposed a novel Controllable Text-To-Image Generative Adversarial Network (ControlGAN) that generates high-quality images with controlling parts of them based on the input text description. ControlGAN consists of three main parts: word-level spatial and channel-wise attention-driven generator, word-level discriminator and adoption of the perceptual loss. The word-level spatial and channel-wise attention-driven generator allows to synthesize subregions of the images corresponding to the most relevant words. The word-level discriminator explores the correlation between the input words and image subregions to distinguish between different visual attributes and provides the generator with accurate training signals. The adoption of the perceptual loss helps the generator to achieve the goal of "controlling" the generated image by reducing the randomness in the generation process by only changing the subregion of the image related to the modified text while preserving the rest of the visual appearance. ControlGAN was evaluated on CUB and MS-COCO datasets qualitatively and quantitatively. The used quantitative performance metrics are IS, R-precision, and L2 reconstruction error. The L2 error reconstruction scored 0.18 for CUB database and 0.17 for MS-COCO.

Liang and colleagues [25] proposed Content-Parsing Generative Adversarial Networks (CPGAN), which aims to improve the semantic alignment between the input text and the generated images by performing full-spectrum content parsing in a deep semantic level. CPGAN model is composed three components: memory-attended text encoder, object-aware image encoder, and fine-grained conditional discriminator. The memory-attended text encoder is designed to capture semantic correspondence between a word and its different visual contexts from all the related images across the training dataset. The object-aware image encoder is proposed to parse the synthesized images in the semantic level. The fine-grained conditional discriminator is used to decide whether the input text and the generated images are semantically consistent for optimizing the model. CPGAN is evaluated quantitatively and qualitatively on MS-COCO dataset. IS and R-precision are the used quantitative measures.

Table 1 presents text-to-image generation models with respect to objective, datasets, quantitative performance metric; IS, HE, FID, VS, R-precision, and SSIM.

Most of the previous mentioned models are trained and validated on one or more of those three benchmarks datasets: CUB, MS-COCO and Oxford-102. In FTGAN, SCU-Text2face dataset was built to allow to the GAN to train on human faces. The IS is the main evaluation metric that almost all the researches used to evaluate the GAN output, which is used to measure both the diversity and quality of the generated images. The CPGAN has achieved the highest IS value over the MS-COCO dataset, while DM-GAN has been the highest on CUB dataset. StackGAN-v2 scored the lowest (best) FID value on CUB dataset, while Obj-GANs scored the best on MS-COCO dataset.

Table 1 A summary of different methods with respect to objective, datasets, performance metrics: inception score (IS), human evaluation (HE), Fréchet Inception Distance (FID), visual-semantic similarity (VS), R-precision, and SSIM scores

References	Method	Objective	Dataset	Evaluation					
				IS	HE	FID	VS	R-precision	SSIM
Baseline generation method									
Reed et al. [6]	GAN-INT-CLS	Generate 64 × 64 images	CUB Oxford 102 MS-COCO	2.88 ± 0.04 2.66 ± 0.03 7.88 ± 0.07 [8]	2.81 ± 0.03 1.87 ± 0.03 1.89 ± 0.04 [8]	68.79 79.55 60.62 [9]	–	–	–
Cha et al. [7]	GAN-INT-CLS-Pixel GAN-INT-CLS-VGG GAN-INT-CLS-Gram	Improve the quality and consistency of the generated images	CUB Oxford 102	–	0.83 0.80 0.85 0.66 0.76 0.89	–	–	–	–
Multistage generation									
Zhang et al. [8]	StackGAN	Generate 256 × 256 photo-realistic images	CUB Oxford 102 MS-COCO	3.70 ± 0.04 3.20 ± 0.01 8.45 ± 0.03	1.37 ± 0.02 1.13 ± 0.03 1.11 ± 0.03	51.89 55.28 74.05 [9]	–	–	0.234

(continued)

Table 1 (continued)

References	Method	Objective	Dataset	Evaluation					
				IS	HE	FID	VS	R-precision	SSIM
Zhang et al. [9]	StackGAN-v2	Enhance the nature of generated images and balance the GANs' training process	CUB Oxford 102 MS-COCO	4.04 ± 0.05 3.26 ± 0.01 8.30 ± 0.10	1.19 ± 0.02 1.30 ± 0.03 1.55 ± 0.05	**15.30** 48.68 81.59	–	–	–
Xu et al. [4]	AttnGAN	Generate fine-grained synthesis images in a multistage fashion	CUB MS-COCO SCU-Text2face [10]	4.36 ± 0.03 25.89 ± 0.47	–	23.98 35.49 [22] 45.56 [10]	–	67.82 ± 4.43 85.47 ± 3.69	–
Single stream generator and multiple discriminator									
Zhang et al. [3]	HDGAN	Generating photographic images	CUB Oxford 102 MS-COCO	4.15 ± 0.05 3.45 ± 0.07 11.86 ± 0.18	–	–	0.246 ± 0.157 0.296 ± 0.131 0.199 ± 0.183	–	0.215
Chen et al. [10]	FTGAN	Fine-grained text-to-face generation	CUB SCU-Text2face	4.63 ± 0.05	–	44.49	–	–	–
Gao et al. [11]	PPAN	Synthesis multi-scale realistic images	CUB Oxford 102	4.38 ± 0.05 3.52 ± 0.02	–	–	0.290 ± 0.149 0.297 ± 0.136	–	–

(continued)

Table 1 (continued)

References	Method	Objective	Dataset	Evaluation					
				IS	HE	FID	VS	R-precision	SSIM
Bi-directional generation method									
Dong et al. [12]	I2T2I	Enhance the quality of multi-categories images	MS-COCO MHP	–	–	–	–	–	–
Qiao et al. [13]	MirrorGAN	Enhance the diversity and semantic consistency of the generated images	CUB MS-COCO	4.56 ± 0.05 26.47 ± 0.41	–	–	–	$K = 3$ 60.42% 80.21%	–
Layout image in generation process									
Hong et al. [14]		Generate semantically meaningful images	MS-COCO	11.46 ± 0.09	58.3%	–	–	–	–
Li et al. [24]	Obj-GANs	Generate high-quality, semantically consistent, and realistic multi-object images	MS-COCO	29.89 ± 0.22	–	**20.75**	–	–	–

(continued)

Table 1 (continued)

References	Method	Objective	Dataset	Evaluation					
				IS	HE	FID	VS	R-precision	SSIM
Human-inspired generation									
Singh et al. [16]	CanvasGAN	Generating semantically consistent images with input description	Oxford 102	2.94 ± 0.18	–	–	–	–	–
Qiao et al. [17]	LeicaGAN	Generate realistic and semantically consistent images	CUB Oxford-102	5.69 ± 0.06 3.80 ± 0.01	✓	–	–	85.28 85.81	–
Others									
Reed et al. [18]	GAWWN	Generate 128 × 128 images based on a given	CUB	3.62 ± 0.07 [8]	1.99 ± 0.04 [8]	67.22 [9]	–	–	–
Dash et al. [19]	TAC-GAN	Synthesizing images of size 128 × 128	Oxford 102	3.45 ± 0.05	–	–	–	–	–

(continued)

Table 1 (continued)

References	Method	Objective	Dataset	Evaluation					
				IS	HE	FID	VS	R-precision	SSIM
Cha et al. [21]	Text-SeGAN	Increase the diversity of the generated images and improve the training process	Oxford 102	4.03 ± 0.07	–	–	–	–	–
Zhu et al. [22]	DM-GAN	Generate accurate high-quality images	CUB MS-COCO	**4.75 ± 0.07** 30.49 ± 0.57	–	16.09 32.64	–	72.31 ± 0.91 88.56 ± 0.28	–
Yin et al. [23]	SD-GAN	Distill the semantic commons from text descriptions to generate consistent images under description variants	CUB MS-COCO	4.67 ± 0.09 35.69 ± 0.50	–	–	–	–	–
Li et al. [24]	ControlGAN	Generate high-quality images with controlling image parts	CUB MS-COCO	4.58 ± 0.09 24.06 ± 0.60	–	–	–	69.33 ± 3.23 82.43 ± 2.43	–

(continued)

Table 1 (continued)

References	Method	Objective	Dataset	Evaluation					
				IS	HE	FID	VS	R-precision	SSIM
Liang et al. [25]	CPGAN	Improve the semantic consistence between the input text and generated images	MS-COCO	**52.73 ± 0.61**	63.73%	–	–	93.59%	–

Bold indicates to match the mentioned results after the table as the best results

3 Conclusion and Future Work

In this paper, we reviewed the most recent GAN based text-to-image synthesis work that contribute in the extension and improvement of this domain. Generating diverse, semantically consistent, and high-resolution realistic images are the main base objectives that the previously mentioned work in this paper tried to achieve one or more of them. The [6] reviewed work was grouped in classes according to the idea and the architecture of the model into seven classes. First class includes the models that uses the baseline architecture. Second class consists of the multistage generation models. In the third class, the models were built on using a single stream generator with multiple discriminators. Fourth class's models used I2T (image-to-text) and T2I (text-to-image) in the training process. In the fifth class, the generation process includes an intermediate layout layer. In the sixth class, the output images are generated based on input text and image. The seventh class includes different ideas and architecture. All methods were evaluated on benchmark datasets that included birds, flowers, humans, and other multi-object images. There are still some limitations in the proposed works, such as the disability of fully evaluate the performance of the models and the limitation of the categories of generated images, so in future studies more work may be directed to measure both the quality and the consistency between the input textual data and output images which minimize the human intervention as well as collecting and building new datasets to expand the benefits of applying this field in more different categories.

References

1. Salimans T, Goodfellow I, Zaremba W, Cheung V, Radford A, Chen X (2016) Improved techniques for training gans. Adv Neural Inform Process Syst 2234–2242
2. Heusel M, Ramsauer H, Unterthiner T, Nessler B, Hochreiter S (2017) Gans trained by a two time-scale update rule converge to a local nash equilibrium. Adv Neural Inform Process Syst 6626–6637
3. Zhang Z, Xie Y, Yang L (2018b) Photographic text-to-image synthesis with a hierarchically-nested adversarial network. In: Proceedings of the IEEE conference on computer vision and pattern recognition, pp 6199–6208
4. Xu T, Zhang P, Huang Q, Zhang H, Gan Z, Huang X, He X (2018) Attngan: fine-grained text to image generation with attentional generative adversarial networks. In: Proceedings of the IEEE conference on computer vision and pattern recognition, pp 1316–1324
5. Wang Z, Simoncelli E, Bovik AC (2003) Multi-scale structural similarity for image quality assessment. In: Signals, systems, and computers, pp 1398–1402
6. Reed S, Akata Z, Yan X, Logeswaran L, Schiele B, Lee H (2016a) Generative adversarial text to image synthesis. arXiv preprint arXiv:1605.05396
7. Cha M, Gwon Y, Kung HT (2017) Adversarial nets with perceptual losses for text-to-image synthesis. In: 2017 IEEE 27th international workshop on machine learning for signal processing (MLSP), pp 1–6
8. Zhang H, Xu T, Li H, Zhang S, Wang X, Huang X, Metaxas DN (2017) Stackgan: text to photo-realistic image synthesis with stacked generative adversarial networks. In: Proceedings of the IEEE international conference on computer vision, pp 5907–5915

9. Zhang H, Xu T, Li H, Zhang S, Wang X, Huang X, Metaxas DN (2018a) Stackgan++: realistic image synthesis with stacked generative adversarial networks. IEEE Trans Pattern Anal Mach Intell 41(8)
10. Chen X, Qing L, He X, Luo X, Xu Y (2020) FTGAN: a fully-trained generative adversarial networks for text to face generation. IEEE Access
11. Gao L, Chen D, Song J, Xu X, Zhang D, Shen HT (2019) Perceptual pyramid adversarial networks for text-to-image synthesis. Proc AAAI Conf Artif Intell 33:8312–8319
12. Dong H, Zhang J, McIlwraith D, Guo Y (2017) I2t2i: learning text to image synthesis with textual data augmentation. In: 2017 IEEE international conference on image processing (ICIP), pp 2015–2019
13. Qiao T, Zhang J, Xu D, Tao D (2019a) Mirrorgan: learning text-to-image generation by redescription. In: Proceedings of the IEEE conference on computer vision and pattern recognition, pp 1505–1514
14. Hong S, Yang D, Choi J, Lee H (2018) Inferring semantic layout for hierarchical text-to-image synthesis. In: Proceedings of the IEEE conference on computer vision and pattern recognition, pp 7986–7994
15. Li W, Zhang P, Zhang L, Huang Q, He X, Lyu S, Gao J (2019a) Object-driven text-to-image synthesis via adversarial training. In: Proceedings of the IEEE conference on computer vision and pattern recognition, pp 12174–12182
16. Singh A, Agrawal S (2019) CanvasGAN: a simple baseline for text to image generation by incrementally patching a canvas. In: Science and information conference. Springer, pp 86–98
17. Qiao T, Zhang J, Xu D, Tao D (2019b) Learn, imagine and create: text-to-image generation from prior knowledge. Adv Neural Inform Process Syst 887–897
18. Reed SE, Akata Z, Mohan S, Tenka S, Schiele B, Lee H (2016b) Learning what and where to draw. Adv Neural Inform Process Syst 217–225
19. Dash A, Gamboa JCB, Ahmed S, Liwicki M, Afzal MZ (2018) Tac-gan-text conditioned auxiliary classifier generative adversarial network. arXiv preprint arXiv:1703.06412
20. Odena A, Olah C, Shlens J (2017) Conditional image synthesis with auxiliary classifier gans. In: Proceedings of the 34th international conference on machine learning-volume 70, pp 2642–2651
21. Cha M, Gwon YL, Kung HT (2019) Adversarial learning of semantic relevance in text to image synthesis. Proc AAAI Conf Artif Intell 33:3272–3279
22. Zhu M, Pan P, Chen W, Yang Y (2019) Dm-gan: dynamic memory generative adversarial networks for text-to-image synthesis. In: Proceedings of the IEEE conference on computer vision and pattern recognition, pp 5802–5810
23. Yin G, Liu B, Sheng L, Yu N, Wang X, Shao J (2019) Semantics disentangling for text-to-image generation. In: Proceedings of the IEEE conference on computer vision and pattern recognition, pp 2327–2336
24. Li B, Qi X, Lukasiewicz T, Torr P (2019b) Controllable text-to-image generation. Adv Neural Inform Process Syst 2065–2075
25. Liang J, Pei W, Lu F (2019) CPGAN: full-spectrum content-parsing generative adversarial networks for text-to-image synthesis. arXiv preprint arXiv:1912.08562

COVID-19 Pandemic's Impact on E-learning Platforms: A Survey

Jeanne Georges and Dalia Magdi

Abstract After the announcement of the World Health Organization (WHO) COVID-19 (Corona Virus Disease) as a global pandemic, many governments start following some limitations by following a lockdown or applying a curfew according to the number of cases recorded in each country. Many industries and organizations start to come up with alternative ways to accomplish tasks after the sudden lockdown either by splitting teams into smaller groups or by working from home (if this is applicable) according to the organization's policy. Like many other aspects of everyday life, COVID-19 had a serious impact on students, instructors, and all educational organizations worldwide. The pandemic caused schools, colleges and universities across the globe to shut down their campuses so that students could follow social distancing measures (Muhammad Adnan1and Kainat Anwar1, Online learning amid the COVID-19 pandemic: Students perspectives, Journal of Pedagogical Research - June 2020). Because of the pandemic, many possible solutions and alternatives were suggested to solve the dilemma of the education system amidst the global pandemic (Christian Paul Chua, Lawrence Dale Valencia, The Role of Artificial Intelligence in Education Amidst of the COVID-19 Pandemic, conference paper, August 2020.). Schools and universities start to work online using open tools technologies to be able to deliver the course contents including the course objectives and the intended learning outcomes of each course. Many online educational platforms were available to use by universities. This paper discusses some of the educational platforms and the impact of COVID-19 on these platforms.

Keywords COVID-19 · Online learning · Educational platforms

J. Georges (✉)
French University in Egypt, Cairo, Egypt

D. Magdi
Canadian International College, Cairo, Egypt

Sadat Academy for Management Sciences, Cairo, Egypt

D. A. Magdi et al. (eds.), *Digital Transformation Technology*, Lecture Notes in Networks and Systems 224, https://doi.org/10.1007/978-981-16-2275-5_15

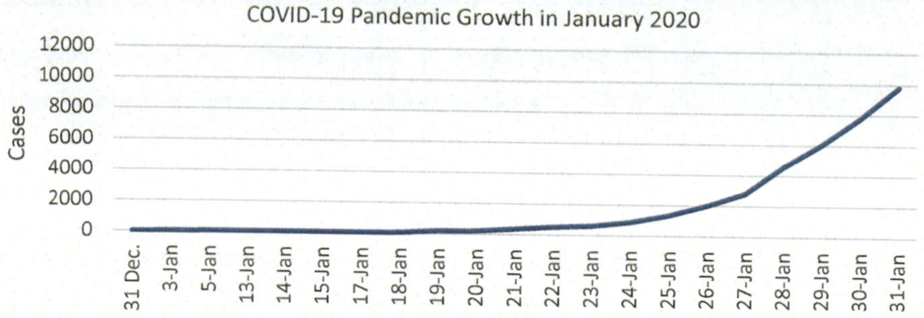

Fig. 1 COVID-19 Pandemic growth in January 2020 [2]

1 Introduction

1.1 COVID-19 Pandemic

As COVID-19 starts spreading with an exponential growth all over the world forcing countries either to make a total lockdown or to follow a curfew based on the total number of cases recorded specifically in each country and according to the government policy, Technology has played an essential role either in the educational part or in the organizations.

Technological development and the Internet have changed the lives of people immensely during this period and have brought a huge change in various fields including the educational organizations such as schools, universities and so on [1].

As of September 20, 2020, there are about 30 million reported cases of COVID-19 around the world. After many countries start their lockdown, Schools and universities begin working on online open sources to be able to deliver and finish course contents of the semester.

The two figures below show the total number of cases reported in January and in September jumping from 10,000 cases by the end of January to 2,500,000 by the beginning of September, it is quite clear the huge difference in the number of cases in less than one year [2] (Figs. 1 and 2).

1.2 Online Learning

Learning courses are interactive which means that courses are available by instructors who exist in different locations for all age groups everywhere and anytime, and presented in the Internet following varieties of learning methods. Virtual classroom, simulation, collaboration, assembly, and even physical classroom include all educational stages namely enrollment, education, evaluation, and presentation of certification [3].

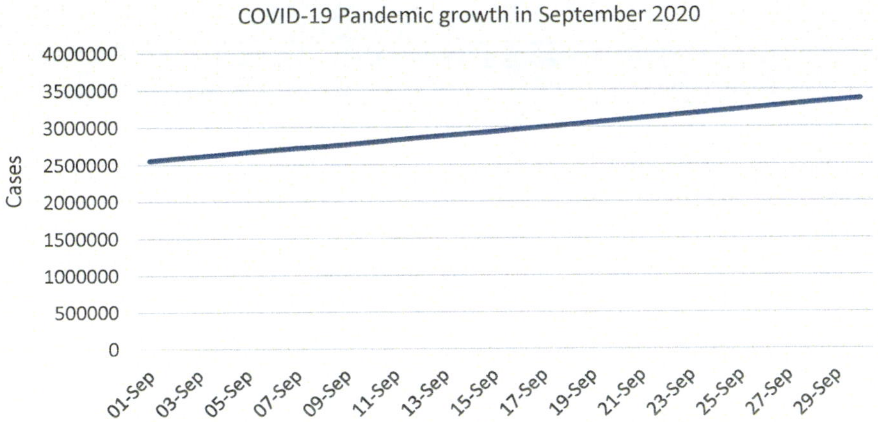

Fig. 2 COVID-19 Pandemic growth in September 2020 [2]

When talking about online learning, we must focus on open education platform that allows free access to the course learning materials and other resources [4]. As we mentioned above that the technology has played an essential role during this crisis, e-learning has become an important part of our educational life. The way to develop any web-based system or applications depends on the learning culture of each country. The key issues are to facilitate new learning modalities for the younger generations to cope with the e-learning process [5].

E-learning system is composed of learning management system (LMS) and learning content management system (LCMS).

Learning management system (LMS) is a software that takes over the learning management of an organization. The system is a mixture of a set of learning resources; this tool makes LMS automated and makes available new features.

While the learning content management system (LCMS) is a knowledge management system that provides the opportunity for collection of information in various forms and formats, their storage in storehouse, sharing, reusing, modularity management, the packing available resources to present them as a learning course. The system takes over the management of learning things available in learning storehouses [3] (Fig. 3).

We are going to discuss alternative ways followed by schools and universities to be able to cope with the education and deliver the required syllabus, there exist many available resources for online learning, we are going to discuss some here.

2 Materials and Methods

Before the COVID-19 Pandemic, Educators were using technology to engage students in the learning process, and numerous studies have shown evidence of

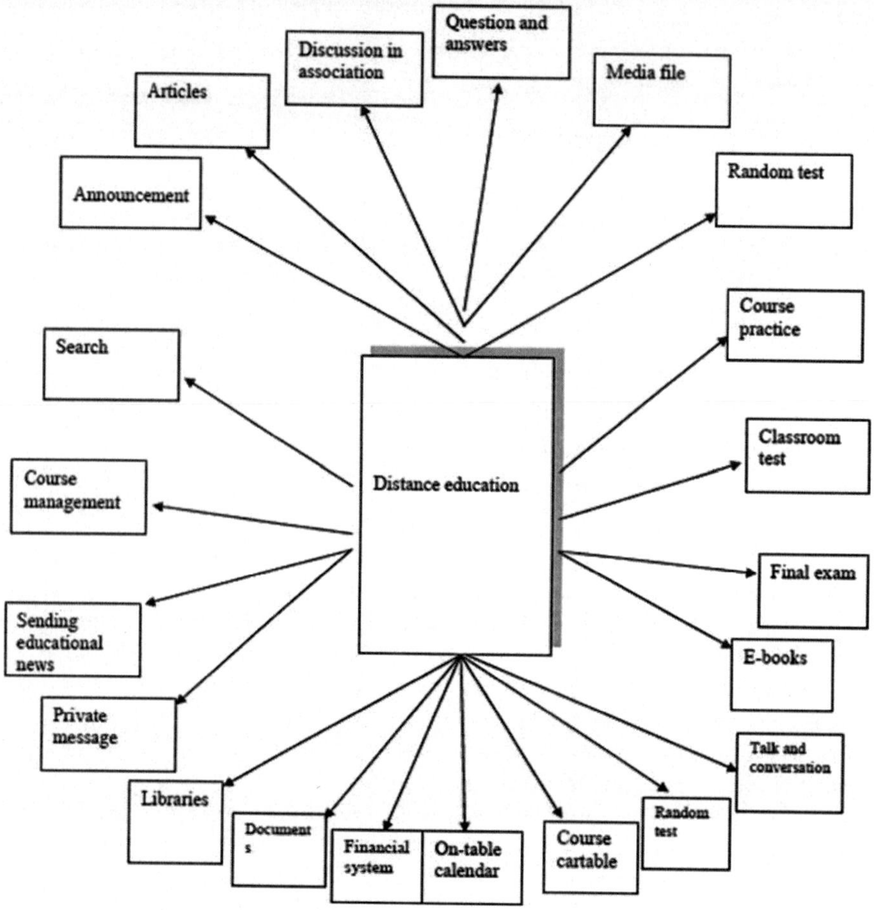

Fig. 3 A sample example of LMS components [3]

increased interest to learn when digital devices are incorporated with the learning environment [6].

When COVID-19 arises, it has been a very important and essential way to use educational platforms to limit the contact between people to minimize the spreading of the virus. This paper presents a research on the impact of COVID-19 on academic achievement and motivation of digital learning with an open education platform for higher education students [6].

Google Classroom (GC) and Microsoft (MS) Teams are defined as "learning management systems" designed to facilitate the creation, distribution and grading of assignments electronically [7].

There are many platforms used during the pandemic.

The most popular and commonly used platforms were Google Classroom, Moodle and Microsoft teams.

We are going to discuss below some of the strengths and weaknesses of every platform from many points of view (Figs. 4, 5 and 6).

Comparing the three platforms from their User Interface (Fig. 7):

Point of Comparison		Google Classroom	Moodle	Microsoft teams
Plans and Pricing	Pros	• Google provides everyone with free access to tools like Gmail and Google Drive.	• Install on your server (self-hosted) • Run it on our server (Moodle Cloud), or get it set up with the help of a certified Moodle Partner	• The integration with Share point and Office 365 tools is really awesome, works very well, and also you can create and built-in Apps from Power Platform inside your team's Group
	Cons	• Individuals can also access Google Classroom free, although some organizations may need to pay for the tool as a part of Google's G Suite service.	The costs of implementing and hosting the solution on your own include: • The hardware costs of a server (an estimated $4,000 up front cost) • The pay for IT employees, including their overtime hours for the project and daily hours for maintaining a server • Any implementation, training or other ongoing services required through a third-party consultant or other contracted specialists	• The price is not so good to medium/small companies that probably will use a less professional tool, or a freeware with limitations.

Fig. 4 Comparison of the three platforms from the plans and pricing point of view

Point of Comparison		Google Classroom	Moodle	Microsoft teams
Features & Utilities	Pros	• Simple and organized Learning Management System • It has other options besides Classroom. Google Meet and Google Hangouts are also very useful. • Compatible with many online education platforms like other Google Suite products • Organizes class announcements, assignments, and materials • Built in grading management system, including auto grading on quiz assignments • Ability to add parents of students who will receive weekly summary of all of child's Google Classroom assignments and materials • Automatically saves all materials and assignments posted to the classroom's Google Drive • Email notifications for students when materials posted, all posts saved under Stream tab in Google Classroom, able to serve as a class email list-serve • Ability to create assignments with teacher-owned documents shared with students to view or edit • App available on smartphones	• Free and effective open source learning. • Create easy assignments and learning management as per subject or class. • Integrated with many plugin - more professional for higher education • It has many features 'right out of the box', setting it up once installed is relatively easy and it has a great community where you can find answers for almost every issue or 'how to setup X' you could find. In addition, there are many modules made for the platform that integrate really well.	• Within one application, Microsoft Teams enables users to schedule video or audio meetings with a single person or a team, Administrators can also organize webinars and large meetings with up to 10,000 participants. • Microsoft Teams is great for one Team to stay in touch for instance or to conduct daily tasks. • Full integration with Microsoft 365 means that calls easily scheduled and invites shared among the organization. • Microsoft Teams also provides the full range of features expected from a leading video conferencing software provider. • Sharing files across teams, having record of sent messages.

Fig. 5 Comparison about the features advantages of the three platforms

Point of Comparison		Google Classroom	Moodle	Microsoft Teams
Features & Utilities	Cons	• No notification for date due assignments • For non-profit organizations, not being able to stream live Google meets for the students with simply click on the live stream link. • Lack of communication like chat option • No option to set time for post • Difficult account management. • Lacks some features associated with a traditional learning management system, limited integration options. • No automated updates. • Difficult learner sharing. • Editing problems. • No automated quizzes and tests. • Not integrated Google Hangouts, which creates a problem; online interaction between teachers and learners is only possible through Google documents	• It is a little confusing; you have to focus a lot. It can be simpler and everything will be wonderful; sometimes needs technical support. • Poor navigation and not user friendly	• Sometimes all of my Teammates appear as offline, even though we hear seconds ago and my internet was working normally • Sometimes we are unable to share screenshots with my colleagues, as my pictures corrupt during the upload.

Fig. 6 Comparison about the features disadvantages of the three platforms

Point of Comparison	Google Classroom	Moodle	Microsoft Teams
User Interface	• Free, easy to use and accessible from all devices. • Clean and user-friendly interface. • Effective communication and makes it easy to share materials and announcements	• Excellent open source LMS for SMBS (Small and Medium Business Solution) and larger organizations • Works very effectively and is regulated • It is easy to learn and use and is quite versatile. • Good Tool to organize the Training Session, save a lot of time and work • Moodle is a veritable and excellent LMS with a cloud and mobile friendly platform • <u>Cons:</u> There are many competing applications, so not all students prefer to use it. • <u>Cons:</u> It can be added more notification and integration methods such as text messages and Google calendar synchronization.	• Easy of making a new meeting and Making sure that those who are in need to attend are notified. Blur background feature in video conferencing. • Trouble teaching others to find their schedule and the apps are different on some people phones, which makes consistency hard. • The structure and the setup of the software is confusing and people tend to confuse which version of a file is the latest.

Fig. 7 Comparison of the user interface of the three platforms

2.1 Microsoft Teams and Google Classroom

Microsoft teams is a cloud app digital hub that brings channels, conversations, meetings, files and apps to Microsoft 365. Teams are composed of channels, which are used to divide teams into topic, or in this case undergraduate modules, one of its

	Microsoft Teams (MS Teams)	Google Classroom
Points of Strength	• Instructor-led instantaneous student registration • **Offline functionality available** after syncing of associated apps (e.g., OneNote, Class Notebook, OneDrive) • **Compatible with all 3 major platforms** (PC, Android and iOS), finger/ stylus functionality on any device with MS apps	• Admin-led mass student registration • Easy access to submissions via app or browser • Variety of 3rd party associated apps (rubrics, appending peer/ instructor feedback to assignments) • Assignment filename auto-renamed with students' names

Fig. 8 Unique Strength of both systems [10]

advantages is the use of the assignment tab to facilitate PowerPoint presentations in-group workshops, allocate required reading and grade formative assignments [8].

Besides, Google Classroom is a free web service developed by Google for schools that aims to simplify creating, distributing, and grading assignments. The primary purpose of Google Classroom is to streamline the process of sharing files between teachers and students [9] (Fig. 8).

2.2 *Moodle*

Modular object-oriented dynamic learning environment abbreviated as Moodle is a course management system (CMS) designed and developed by an Australian teacher. It is a free access platform for the construction of open source code of courses and Web sites based on the Internet. It is a "learning management system" (LMS) that combines educational theory taken with practice and supports teachers and students to organize and implement teaching activities to engage the students more in the learning process.

Let us see some of the Moodle platform advantages:

2.3 *Advanced Concept and Comprehensive Functions*

Moodle especially emphasizes its educational philosophy. It emphasizes that the basic idea of programming is social constructivism education. Moodle is a learning center system, while most CMS is just a collection of commercial tools. The main functions of Moodle can be divided randomly into three parts: Web site management, learning management and course management.

2.4 High Compatibility, Low Technical Threshold, Easy to Use

Moodle's online teaching module adopts dynamic modular design, which can be freely combined. When teachers build online courses, it is as simple and interesting as building blocks.

2.5 Open Source and Free

Moodle, as free software downloaded with open source code, anyone is allowed to use and modify the source code free of charge, but the copyright license notice of the original author must be retained. It is the open source nature of Moodle that promotes its continuous improvement and development [11].

Using this platform improved professor-student communication and increased the student's satisfaction for courses. Another encouraging aspect was that this approach (with the e-learning platform) changed a lot of student's perception regarding homework and their importance in the educational process [12].

2.6 Challenges

The e-learning is a way of learning by utilizing electronic technologies to access educational curriculum outside of a traditional classroom. In most cases, it refers to a course, program or degree delivered completely online, many students worldwide prefers to take online courses that may not be available in their countries to improve their knowledge and gain more skills. Others prefer accredited universities to take courses there so that this will add credibility to their profile.

Besides, there are many challenges facing the e-learning or distant learning. One of the most common challenges is the stability of Internet connection in some countries.

One other challenge is that the students have not worked before in such online platforms. They haven't any knowledge about how to use the system or log in by their educational mail, see if there are any announcements in any course, view the course contents, upload and submit their assignments, participate in discussion forums if there exists, joining online meetings via Zoom, Microsoft teams or Moodle.

We can overcome this challenge by giving the students a small tutorial video which explain the usability of everything in the platform so they can be able to deal with the courses, assignments online meetings and so on.

3 Discussion

Since COVID-19 hit, there has been a huge shift in the way people teach, learn and work. Access to online learning is now more important than ever, as we witness the dramatic changes in the use of online platforms from the start of COVID.

From the comparative study done Figs. 4, 5,6,7,8 for specific users such as school students, we recommend that they use Google Classroom for its simple user interface and the students will have the ability to deal with it without having to go through some sessions for training about how to use the platform.

For higher education students, we recommend using Moodle as it has many features useful from downloading course contents; its interface is a bit more complex than Google Classroom, so it will fit higher education students.

About organizations and large institutions, we recommend using Microsoft teams as it has many available tools like video conferencing, which is somehow a very needed tool for achieving work online. Using Teams in education is also an added value for its integration with office 365, some students prefer using Microsoft teams.

The figure below illustrates the total number of cases starting from the middle of March to the middle of September (Fig. 9).

As shown in Fig. 3, the huge growth of the total number of cases within eight months forces many countries to follow some limitations.

The educational part was directly affected by this pandemic. Many solutions were already available as discussed but they need some improvements as many students and universities used them during the pandemic discovering many things that needs to be added or updated (Fig. 10).

Figure 4 represents the COVID-19's terrible impact on global education; the *X*-axis shows the months at the beginning of the crises, the *Y*-axis shows the number of learners impacted by national school closures worldwide.

Educational platform's importance was a great evidence during this pandemic and will continue as they proved more engagement from the students to their courses; it helped them to facilitate distant learning by videos, tutorials, online meetings

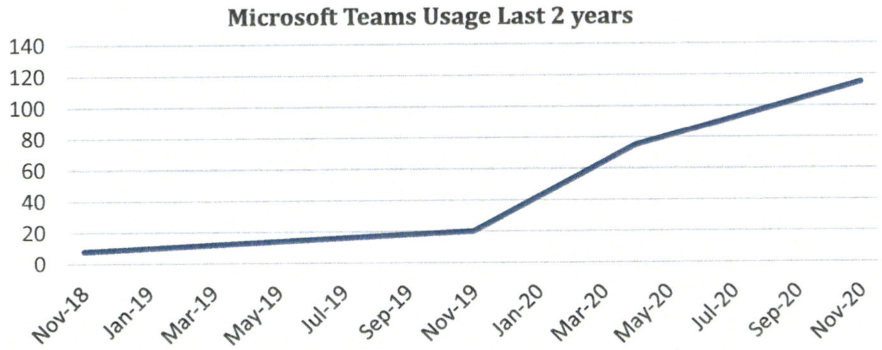

Fig. 9 COVID-19 Cases worldwide from 15 March to 15 September [2]

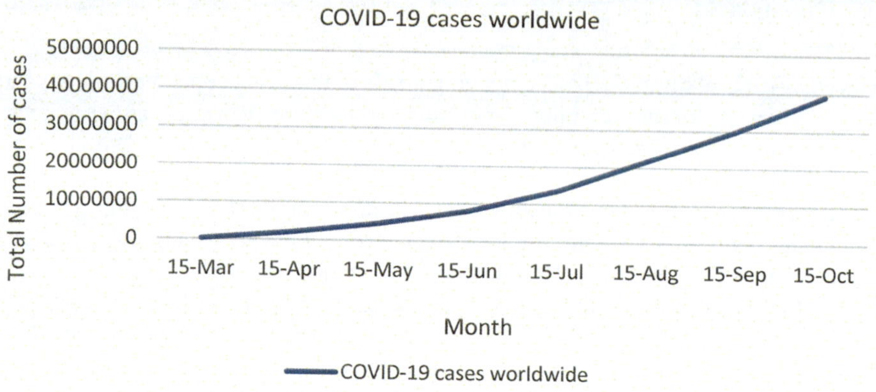

Fig. 10 COVID-19's terrible impact on global education [13]

and sharing screen, submitting their assignments, joining online discussion forums, which is very useful containing many opinions and sharing different thoughts.

The implementation of lockdown and social distancing has been one of the measures to prevent the spreading of the COVID infection. The education system which was completely shut and to proceed with the academic curriculum, there is great shift from the regular learning process to electronic learning. This can show an increased number of online classes, conferences, meetings, discussion forums, etc. In other words, the world is completely dependent on information technologies during this crisis [14].

From Fig. 11 we notice the huge increase in the number of Microsoft teams users since the start of the COVID Pandemic (Figs. 10 and 11).

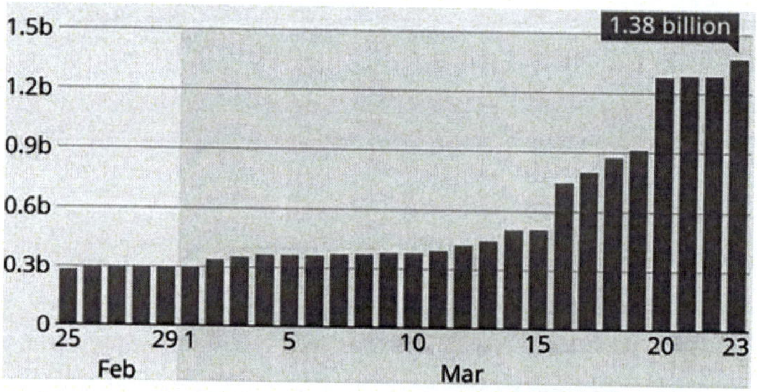

Fig. 11 Increased number of Microsoft teams during the last two years

4 Conclusion and Future Work

E-Learning is the globalization of education system. Due to the improvement in technology, it is unbelievable to imagine where the world will be after another 20 years. Many Applications of IOT & big data analytics in e-learning will achieve the revolutionary development. This combination also reshapes customized learning environments to the learners [15].

During this pandemic, different countries through-out the globe have introduced several solutions to carry forward the educational process. TV broadcast, Online libraries, resources, guidelines, online channels, video lectures, are highly introduced in around 96 countries. The quality of e-learning needs much improvement. As discussed the Google Classroom and Microsoft teams had played an important role in this sudden out-break of COVID-19, there was not sufficient time to assure the quality of the e-learning process because the focus was on saving and continuing the education process at the midst of a semester [7].

Following the limitations of lockdown or social distancing, many universities or educational institutions were obliged to use online platforms not to affect the learning process.

Many online educational platforms were discussed comparing each other, coming up with what can be improved, it is quite clear that many European countries have started another lockdown due to the 2nd wave of COVID-19. There is no possible solution except to continue improving our educational platforms to be able to cope with any sudden lockdown as it seems that there is no solution to this pandemic until coming up with a vaccine. This may take much longer time as it is a new virus so they must finish all the clinical trials to produce a powerful vaccine against COVID-19.

Acknowledgements I would like to express my deep gratitude to my supervisor, Dr. Dalia Magdi, for providing guidance and feedback throughout this paper, for her patient guidance, enthusiastic encouragement and useful critiques of this survey.

References

1. Chua CP, Valencia LD (2020) The role of artificial intelligence in education amidst of the COVID-19 pandemic, conference paper
2. Statistics of COVID-19 Cases worldwide from: https://ourworldindata.org/covid-cases
3. Fasihfar Z, Rokhsati H Study of application of intelligent agents in e-learning systems, Buletin de la société Royale des sciences de liege
4. Sousa MJ, Rocha Á (2018) Digital learning in an open education platform for higher education students, conference paper
5. Kristensen t, Bech Ø, Dyngeland M (2013) Towards a multi-agent e-learning platform. J Comput Eng Inf
6. Sousa MJ, Rocha A (2018)Digital learning in an open education platform for higher education students, research gate
7. Google (2017) About classroom—classroom Help (2017). Support.google.com. Retrieved, 2 December 2017 rom: https://support.google.com/edu/classroom/answer/6020279?hl=en

8. Martin L, Tapp D (2019) Teaching with Teams: An introduction to teaching an undergraduate law module using Microsoft teams. Innovative Practice in Higher Education
9. Larry M (2017) Google classroom offers assignment center for students and teachers. Forbes, Retrieved April 28, 2017
10. Davidson P, Molnar A, Ting CY (2018) MS teams and google classroom: preliminary qualitative comparisons & user feedback, Conference Paper
11. Using open-source learning platform (moodle) in university teachers' professional development. J Phys Conf Series
12. Bentaa D, Bologaa G, Dzitaca I (2014) e-learning platforms in higher education. Case Study, Elsevier
13. UNESCO
14. Soni VD (2020) Global impact of e-learning during COVID 19. SSRN Electron J
15. Kusuma S, Kasi Viswanath D (2018) IOT and big data analytics in e-learning: a technological perspective and review. Int J Eng Technol
16. Adnan M, Anwar K (2020) Online learning amid the COVID-19 pandemic: students perspectives. J Pedagog Res

Digital Transformation in Business

A Comparative Analysis of Credit Card Fraud Detection Using Machine Learning and Deep Learning Techniques

Mohamed Ashraf, Mohamed A. Abourezka, and Fahima A. Maghraby

Abstract E-commerce is growing rapidly around the world and this causes a significant increase in credit card transactions, both normal and fraud transactions. Financial institutions throughout the world lose billions because of credit card fraud. Fraudsters have no fixed styles; they always change their behavior and try to learn new technologies that allow them to commit frauds through online transactions. Moreover, they assume that the regular behavior of consumers and fraud patterns change fast. Fraud detection systems have become necessary for banks and financial institutions to minimize their losses. However, not much research has been conducted on credit card fraud detection methodologies, due to the unavailability of credit card transaction dataset for researchers. Also, most of the researchers apply either machine learning or deep learning techniques without comparing the results of applying both techniques in the same dataset. This research aims to apply different machine learning (Logistic Regression, K-Nearest Neighbor, Random Forest) and deep learning (Deep Neural Network, Convolutional Neural Network) techniques in a real-life credit card dataset to choose the most efficient algorithm for detecting fraud transactions. After applying different experiments with different parameters using all the algorithms mentioned before, the Random Forest Algorithm gives slightly better accuracy than Deep Neural Networks.

Keywords Fraud detection · Machine learning · Deep Neural Networks · SMOTE · Convolutional neural network · Credit card fraud

M. Ashraf (✉) · M. A. Abourezka · F. A. Maghraby
Faculty of Computing and Information Technology, Arab Academy for Science, Technology & Maritime Transport, Cairo, Egypt
e-mail: m__ashraf@live.com

M. A. Abourezka
e-mail: M.Aborizka@aast.edu

F. A. Maghraby
e-mail: fahima@aast.edu

D. A. Magdi et al. (eds.), *Digital Transformation Technology*, Lecture Notes in Networks and Systems 224, https://doi.org/10.1007/978-981-16-2275-5_16

1 Introduction

Banking fraud is defined as the unauthorized use of an individual's confidential information to make purchases, or to take away funds from the user's account. The use of online shopping, digital payments, net banking, and transactions through payment cards is increasing day by day. Fraud detection system should distinguish between fraud transactions and normal transactions and make the detection possible in real-time transactions. Credit card fraud increases daily and fraudsters' actions rarely follow one predictable pattern [1].

E-banking popularity is increasing rapidly, which results in increasing online transactions owing to heavy online shopping and immediate online bill payment. Consequently, the fraud cases have also increased, putting great burdens on the economy and affecting both customers and financial institutions [2]. Moreover, fixing the harm done is a money consuming and time-consuming task. It is difficult to build an efficient fraud detection and prevention system that works in real-time, as the number of fraudulent card transactions is very small compared to the total volume (VISA alone processes over 65,000 + transactions every second). Hence, establishing a training set is difficult due to the highly unbalanced data [3].

Traditional rule-based systems are insufficient, often resulting in false positive rates that exceed 90%. Further, this gives rise to a huge number of false positive alerts that need to be manually cleared by human intervention. Repetitive analyst action can lead to a significant number of "false positives", raising the operational risk and subsequently regulatory risk in the process.

Merchants, Acquirers, and Issuers have started to create solutions to bring down fraud transactions to a minimum and decrease merchant charge-back rates. The long-time delay between fraud transaction time and the time taken to detect it is one of the main challenges of fraud prevention. This means that if there is no fraud prevention technique, fraudsters can easily inflict huge damage on a business before any stakeholder realizes the problem [4].

According to Shift Processing article on credit card theft statistics in 2018, $24.26 billion was lost due to the payment card frauds committed worldwide. The USA alone suffered 38.6% of reported card fraud losses in 2018. Credit card fraud increased by 18.4% in 2018 and it is on the rise. Card-Not-Present (CNP) fraud is now 81% of the cases, more than Point-of-Sale (POS) fraud. Identity theft was the 3rd largest cause of fraud in the USA in 2018, 14.8% of the reported frauds. Credit card fraud was on top of the list of Identity theft fraud. In 2018, credit card fraud was almost 35% of all identity theft frauds. Yahoo's breach in 2013 resulted in the exposure of 3 billion victims, the biggest single informational breach. Also, in 2018, the data breaches in Business Sector were 46%, including the Marriott data breach. 69% of frauds started with a consumer being contacted by telephone or email, for overdue loans or prize scams, for example. Moreover, the report focused on the top 7 data breaches of all Time: (1) Due to Yahoo data breach in 2013, 3 billion consumers in total were exposed. It is considered the biggest single data breach; (2) First American Financial Corp. had 900 million customer files exposed in 2019; (3) 540 million users' information was exposed by Facebook in 2019; (4) The biggest breach was by Marriott International in 2018 which exposed as many as 500 million records; (5) Adult Friend Finder had 412 million records compromised in 2016; (6) Equifax had 143 million records exposed in 2017, and (7) In 2019, Capital One had more than 100 million records which could have been exposed [5].

This research aims to detect fraud transactions by applying different techniques of machine learning and deep learning. In the analysis phase, the account usage patterns are monitored and compared to the history related to this account in order to identify any suspicious transaction which would be flagged as fraud. The major contribution made through this research is applying different algorithms of machine learning and deep learning to a real credit card dataset extracted from one of the biggest banks in the Middle East then, and comparing the results in terms of accuracy, recall, precision, and area under curve (AUC) so as to detect the best algorithm to be used in the fraud detection process

2 Background

In this research different types of credit card fraud are defined. There are two types which are Card-Not-Present (CNP) fraud, and Card-Present fraud. Card-Not-Present (CNP) fraud includes online transactions in which physical card is not required. About Card-Present fraud, physical card is mandatory. It is classified into Counterfeit Card fraud, and Lost and Stolen Card fraud, Card ID theft, ATM fraud, and Point-of-Sale (POS) fraud. Counterfeit Card fraud occurs when the card details are copied through its magnetic strip and then the card is recreated. In Lost and Stolen card fraud, the card is either lost or stolen and used by others to make purchases. In Card ID theft, a new account with the owner's details is created as the criminal is aware of the card's details. As for ATM fraud, this usually includes instances where a card is charged but cash is not delivered, or where a physical card is skimmed. Point-of-Sale (POS) fraud occurs as a result of an employee defrauding his/her employer at the point of sale [5].

2.1 Random Forest

Random Forest is a supervised classification algorithm that consists of a collection of tree-structured classifiers where independent random vectors are distributed identically, and each tree casts a unit vote for the most popular class at input x. A random vector is generated. It is independent of the past random vectors with the same distribution and a tree is generated by using the training test [6]. Equation (1) below shows the formula for the random forest algorithm. In general, having more trees in the forest means more robust the forest looks like.

$$\hat{y} = \frac{1}{m}\sum_{j=1}^{m}\sum_{i=1}^{n} W_j(x_{i,}x')y_i = \sum_{i=1}^{n}\left(\frac{1}{m}\sum_{j=1}^{m} W_j(x_{i,}x')\right)y_i \quad (1)$$

Equation (1) shows that the whole forest is again a weighted neighborhood scheme, with weights that average those of the individual trees. The neighbors' of x' in this interpretation are the points x_i sharing the same leaf in any tree j. In this way, the neighborhood of x' builds on a complex way in the structure of the trees, and thus in the structure of the training set.

2.2 K-Nearest Neighbor

K-Nearest Neighbor is a supervised learning technique being used in solving classification and regression problems. However, it is commonly used in classification problems. The main objective of K-Nearest Neighbor algorithm is to identify the new points based on similarity measure where the new points are being classified based on the distance between them and the points in the training set. K-Nearest Neighbor algorithm takes into account Lazy Learning Algorithm as it uses all the data for training during the prediction phase. Also, this algorithm takes Non-Parametric Learning Algorithm into consideration as it has nothing to assume about the data. Equation (2) shows the K-Nearest Neighbor mathematical formula where $Vd = \pi d/2$, $\Gamma(d2 + 1)$ is the volume of a unit d-dimensional ball and $\Gamma(x)$ is the Gamma function [7].

$$\begin{aligned}\hat{P}_{Knn}(x) &= \frac{K}{n}.\frac{1}{V_d.R_k^d(x)} \\ &= \frac{K}{n}.\frac{1}{\text{Volume of a d} - \text{dimensional ball with radius being } R_k(x)}\end{aligned} \quad (2)$$

Here are the results when $d = 1, 2,$ and 3

- $d = 1, V1 = 2 : b \quad \hat{P}_{Knn}(x) = kn12Rk(x).$
- $d = 2, V1 = \pi : b \quad \hat{P}_{Knn}(x) = kn1\pi R2k(x).$
- $d = 3, V1 = 43\pi : b \quad \hat{P}_{Knn}(x) = kn34\pi R3k(x).$

2.3 Logistic Regression

Logistic Regression algorithm is regarded as one of the supervised learning techniques being used in classifying problems with two possible outcomes. It is used to predict a categorical dependent variable by using independent variables. Logistic Regression uses complex cost function which is a sigmoid function. This function scales any real number to be a number between 0 and 1 [8]. The simple logistic model is shown below in Eq. (3) where π is the probability of the outcome of interest or event. Equation (4) shows the formula of calculating π),α is the Y intercept, β is the regression coefficient, and $e = 2.71828$ is the base of the system of natural logarithms.

$$\log\text{it}(Y) = \text{natural } \log(\text{odds}) = \ln\left(\frac{\pi}{1-\pi}\right) = \alpha + \beta X \tag{3}$$

$$\pi = \frac{e^{\alpha+\beta X}}{1 + e^{\alpha+\beta X}} \tag{4}$$

2.4 *Deep Neural Networks (DNN)*

Deep Neural Network (DNN) is a multiple layer Artificial Neural Network (ANN) which consists of an input layer, an output layer, and multiple hidden layers between them. Neural Network is a technology built to simulate the human brain. The most basic element of the human brain is the neuron which helps us in remembering, thinking, and reacting to each action based on one's experience. The basic unit of neural networks is the artificial neurons which simulate the basic functions of brain neurons. It receives input and produces output. Each input is assigned a random weight to make the output of each neuron random. After applying feeds errors back algorithm, these weights are adjusted to produce better results. DNN has the solution to complex problems that machine learning cannot overcome, such as computer vision and natural language processing. However, the training in DNN requires huge computational power [9]. Equation (5) shows the DNN formula where x is the layer number and w is the weight associated with this layer. Sigmoid is also used as an activation function between layers. Equation (6) shows the formula of the sigmoid function.

$$z = x_1 * w_1 + x_2 * w_2 + \cdots + x_n * w_n + b * 1 \tag{5}$$

$$\text{sigmoid}(z) = \frac{1}{1 + e^{-z}} \tag{6}$$

2.5 *Convolutional Neural Networks (CNN)*

Convolutional Neural Networks are a class of artificial neural networks which consists of neurons. These neurons take in a weighted sum of inputs and produce an activity level which is frequently a nonlinear function of the input and a rectified linear unit where the activity contains all positive inputs and all non-positive inputs are converted to 0.

CNN are feedforward networks with units organized into 2D Grids (Feature Maps) which contain spatial arrangements. These feature maps resulted from a performed convolution on the layer below which indicates that the same convolutional filter is applied at each location in the layer below. Therefore, a unit at a certain location on the 2D grid can only receive input from units at a similar location in the layer below. Furthermore, the weights attached to the inputs are the same for each unit in a feature map. Also, there is a common operation, which is called pooling, that refers to the use of maximum activity in a small spatial area of each 2D feature map grid to represent this area. It reduces the feature maps size [9]. Equation (7) shows CNN formula where net (i, j) is the output in the next layer, x is the input image, w is the kernel or filter matrix and $*$ is the convolution operation.

$$\text{net}(i, j) = (x * w)[i, j] = \sum_{m} \sum_{n} x[m, n]w[i - m, j - n] \tag{7}$$

2.6 *Synthetic Minority Oversampling Technique (SMOTE)*

SMOTE is an oversampling technique to increase the number of minority cases in the dataset without changing the number of majority cases. Neighbors from the K-Nearest Neighbors are randomly chosen based on the required amount of oversampling. If the needed oversample cases, for example, are 200%, then only two neighbors are chosen, and one sample is generated in each direction. Synthetic samples are generated by taking the difference between the feature vector under consideration and its nearest neighbor, multiplying it by a random number between 0 and 1 and then adding it to the feature vector. This results in the selection of a random point along the line segment between two specific features. This approach effectively forces the decision region of the minority class to become more general [10].

3 Related Work

Credit card is a popular payment method in online shopping for goods and services. Fraudsters try to falsely simulate the normal behavior of users to make their payments. As a result of these problems, the studies conducted into credit card fraud detection have focused on matching patterns in which fraud patterns are identified as distinct from normal transactions. Recently, researchers have used different techniques for credit card fraud detection that were applied in the last few years. Some of the studies carried out by those researchers are discussed below and a summary of these researches is shown in Table 1.

Table 1 Summary of Literature Survey

Research author	Objectives	Approach	Dataset	Accuracy
Alae Chouiekha, EL Hassane Ibn EL Hajb., 2018	Fraud detection in CDRs	SVM RF GBC CNN	Artificial images describing the profile history of different subscribers during two months of behavior	SVM:77% RF: 72% GBC: 79% CNN: 82%
Reshma R S, 2018	Fraud Detection in Banking Dataset	Autoencoder Variational Autoencoder Restricted Boltzmann Machine Deep Belief Network	European credit card dataset	Autoencoder: 96.26% Variational Autoencoder: 96.55% Restricted Boltzmann Machine: 96% Deep Belief Network: 96.31%
Krishna Modi, 2017	Credit card fraud detection using Neural Network and CNN	NN NN with SMOTE NN with added features CNN CNN with SMOTE	Kaggle credit card fraud data	NN: 0.84 NN with SMOTE: 0.87 NN with added features: 0.877 CNN: 0.862 CNN with SMOTE: 0.89
Maja Puh, Ljiljana Brkic, 2019	Credit card fraud detection using machine learning techniques	Random Forest SVM Logistic Regression	European credit card dataset	Random Forest: 90.13% SVM: 86.78% Logistic Regression: 91.07%
Shiv Shankar Singh, 2019	Credit card fraud detection	Logistic Regression SVM Decision Tree Random Forest	European electronic cardholders	Logistic Regression: 97.7% SVM: 97.5% Decision Tree: 95.5% Random Forest: 98.6%
Kaithekuzhical Leena Kurien, Dr. Ajeet Chikkamannur, 2019	Credit card fraud detection in Financial Transactions	Logistic Regression Random Forest	Kaggle credit card dataset	Logistic Regression: 97.8% Random Forest: 97.5%

(continued)

Table 1 (continued)

Research author	Objectives	Approach	Dataset	Accuracy
Our Research	Credit card fraud detection in Banking Industry	Logistic Regression KNN Random Forest DNN CNN	Real credit card dataset from one of the biggest banks in the private sector	Logistic Regression: 94.8% KNN: 85.8% Random Forest: 98.8% DNN: 98.2% CNN: 91.7%

In [11] Deep Convolutional Neural Network architecture was introduced to detect fraud transactions in a dataset containing profile history of different subscribers in two months and compared them using various machine learning algorithms. The accuracy level was 82%, the highest degree in all the modules (SVM, RF, and GBC algorithms). This model has also given the best performance in training duration.

In [12] Unsupervised deep learning was used to detect fraud detection in the European dataset. One with higher AUC value was considered a better model. Four unsupervised deep learning models have been used in this study (Autoencoder Model—Restricted Boltzmann Machine (RBM)—Variational Autoencoder—Deep Belief Network (DBN)). The Variational Autoencoder model created a greater AUC of 0.9655 which was greater than the other model used. There was also an assumption of improving the value of AUC when real-time data is used.

In [1] Neural Network and Convolutional Neural Network were applied with Synthetic Minority Oversampling Technique (SMOTE). The experiment results showed that the precision of CNN was poorer than NN in CNN, and the ratio of detecting fraudulent transaction is more than NN which improves the performance of recall and F1 score. However, the limitation raised was a fraudulent transaction in which behavior is the same as legitimate transactions cannot be detected by this model.

In [13] Machine learning algorithms techniques [Random Forest, Support Vector Machine, and Logistic Regression] were used in the credit card fraud detection dataset. The performance was measured, and experiments were performed using two approaches (A) static (B) Incremental learning of algorithms. The results of experiments showed that the SVM was the poorest performance in both static and incremental setups. The performance of Logistic Regression and Random Forest was slightly different, and Logistic regression showed better results in incremental setup.

In [4] Logistic Regression, Support Vector Machine, and Decision Tree were applied to a dataset of electronic payment card which was taken from European electronic cardholders. The machine learning techniques were applied to the unstructured data without processing them. The results of this study showed that Random Forest was highly accurate and gave better results in credit card fraud detection.

In [14] Logistic Regression and Random Forest in credit card Dataset was published on Kaggle, collected and analyzed during a research collaboration of worldline and the machine learning group of ULB (Université Libre de Bruxelles) after applying SMOTE oversampling technique to balance the dataset. The result of the experiments showed that Random Forest produced better performance in terms of precision, recall, and F1 Score. However, Logistic Regression provided slightly better results after using ROC-AUC Curve

4 Proposed Framework

The framework of detecting credit card fraud transactions is divided into 4 phases as shown in Fig. 1. They are data preparation phase, training phase, testing Phase and Scoring phase:

4.1 Data Preparation

In this phase, the data are prepared to be used in the training process. This is done by:

1. Handling missing values by replacing them with the mean value.
2. Removing noisy data.
3. Feature selection to choose the best columns with the highest impact on the target variable.
4. Feature scaling to the numeric variables on the dataset.
5. Dividing the data into two groups, Training (80%) and Testing (20%).
6. Balancing the dataset by using the SMOTE technique.

4.2 Training Phase

In this phase, the algorithms used in our experiment started to learn from the data to classify the normal transactions and fraud transactions by learning the behavior of fraudsters from the historical transactions provided in the dataset.

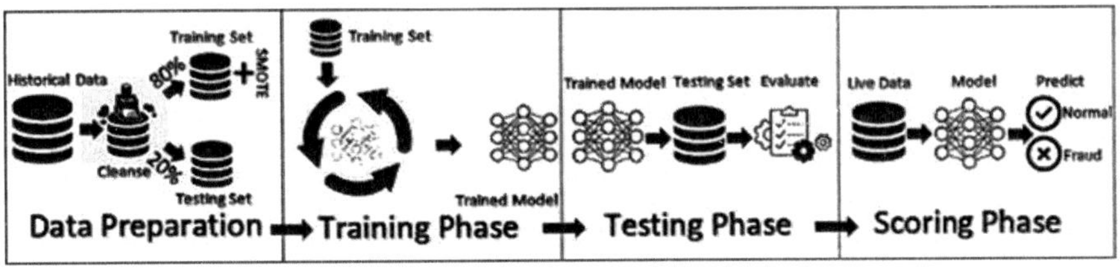

Fig. 1 Proposed framework

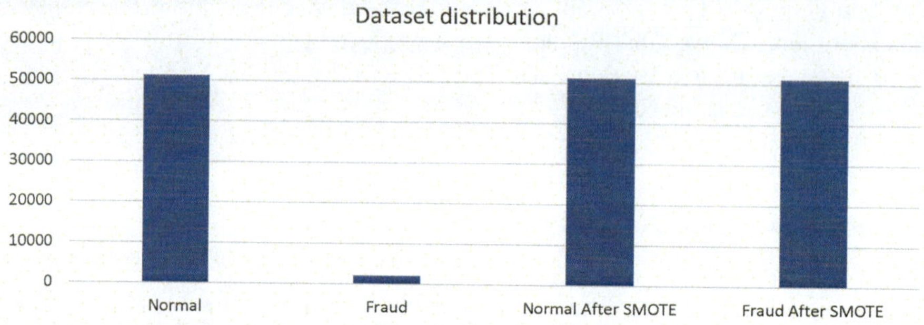

Fig. 2 Dataset distribution before and after applying SMOTE

4.3 Testing Phase

In this phase, the models generated from the algorithms used in our experiment are being evaluated by using the test data to check the true positive, true negative, false positive and false negative which are used in calculating accuracy, recall, precision.

4.4 Scoring Phase

In this phase, the model with highest ranking is used with the live data to classify the transactions into normal and fraud and this process is carried out by matching the patterns being learned from the training data.

5 Dataset

The dataset used is real historical transactions from one of the biggest banks in the private sector in the Middle East. The dataset contains 60,428 rows and 12 columns. It is highly unbalanced as it contains 58,179 normal transactions and only 2249 fraud transactions as shown in Fig. 2. It also contains the data distribution before and after applying the SMOTE technique to balance the dataset. Table 2 contains a description of the features used in the dataset.

6 Environment Setup

The hardware specs used in this experiment is CUDA Cores:768 Graphic Clock (MHz):1290, Processor Clock: 1392, Graphics Performance: high-6747, Memory

Table 2 Features and description

Feature name	Description
Type	It contains flag whether the transaction is rejected or not
Card number	It contains the masked credit card number
Date	It is the date of transaction
Time	It is the time used for transaction
Amount	It is the amount of transaction
Currency	It is the credit card currency
POS	Point-of-Sale type
MCC	Merchant category code
Transaction currency	The currency used by the merchant in transactio
Acceptor ID	Merchant ID number
Rejection code	Code of rejection if transaction is rejecte
Fraud flag	It is normal transaction or fraud

Clock:7 Gbps, Standard Memory Config: 4 GB, Memory Interface: GDDR5, Memory Interface Width: 128-bit and Memory Bandwidth (GB/s): 112.

7 Evaluation Criteria

The evaluation of each algorithm is conducted using some measures that are calculated and then the performance of different algorithms will be compared. The various methods for evaluation are: (i) accuracy, the ratio of true positive and true negative observation to the total observations, (ii) precision, the ratio of the true positive observations to the true positive and false positive observations, (iii) recall, the ratio of correctly predicted positive observations to the true positive and false negative observations, (iv) AUC-ROC curve, a probability curve and AUC represents degree or measure of separability. It tells how much the model is capable of distinguishing between classes. The higher the AUC is, the better the model is at predicting 0 s as 0 s and 1 s as 1 s. True positive (TP) is the number of fraud transactions that were committed and detected by the algorithm. True negative (TN) is the number of non-fraud transactions that were not committed and classified as normal transactions by the algorithm. False positive (FP) is the number of normal transactions that were classified as frauds by the algorithm. False negative (FN) is the number of fraud transactions that were classified by the algorithm as normal transactions.

8 Experiment Results

In this research, experiments were performed using a real dataset for one of the biggest banks in the private sector. Five algorithms were applied to these data to classify a transaction as fraudulent or legitimate, two of them are deep learning techniques (Deep Neural Network, Convolutional Neural Network) and three of them are machine learning techniques (Random Forest, Logistic Regression and K-Nearest Neighbor). The performance of each algorithm was evaluated by using five metrics which are accuracy, recall, precision, ROC–AUC as shown in Fig. 3 and Time as shown in Fig. 4.

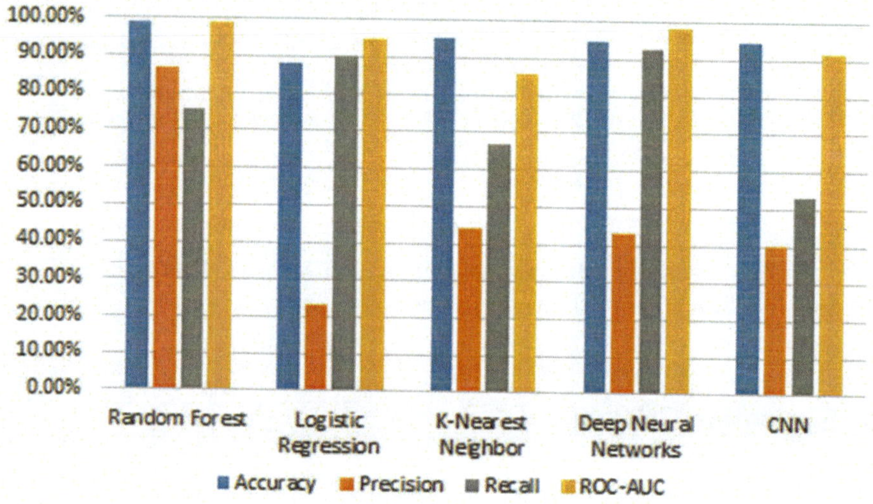

Fig. 3 Accuracy, precision, recall and ROC-AUC comparison

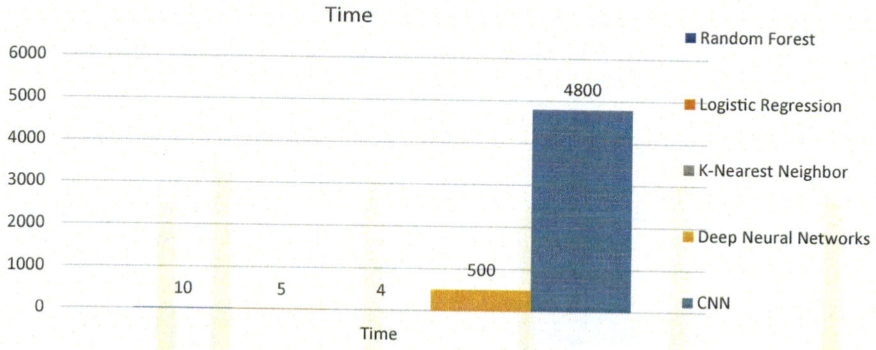

Fig. 4 Time comparison

The highest accuracy was achieved by Random Forest algorithm at 98.6%, the second highest accuracy was by K-Nearest Neighbor at 95.4%, the third highest accuracy was by CNN at 95% the fourth highest accuracy was by DNN at 94.9%, and the least accuracy was by Logistic Regression at 88.1%. The highest precision was achieved by Random Forest at 86.5%, the second highest precision was by K- Nearest Neighbor at 44%, the third highest precision was by DNN at 42.8%, the fourth highest precision was by CNN at 39.8% and the least precision was by Logistic Regression at 23.2%. The highest recall was achieved by DNN at 92.5%, the second highest recall was by Logistic Regression at 90.4%, the third highest recall was by Random Forest at 75.7%, the fourth highest recall was by K-Nearest Neighbor at 66.8%, and the Fifth highest recall was by CNN at 53.2%. The highest AUC-ROC score was by RF at 98.8%, the second highest AUC-ROC score was by DNN at 98.2%, the third highest score was by Logistic Regression at 94.8%, the fourth highest score was by CNN at 91.7% and the least score was by K-Nearest Neighbor at 85.8%. The longest consumption time was by CNN for 4800 s, the second longest consumption time was by DNN for 500 s, the third longest consumption time was by Random Forest for 10 s, the fourth longest consumption time was by Logistic Regression and the least consumption time was by K-Nearest Neighbor for 4 s.

9 Results Discussion

The experiment results show that Random Forest algorithm (Machine Learning) gives better results than Deep Neural Networks and CNN algorithm (deep learning) when applied to the dataset as Random Forest has the highest accuracy (98.6%) in classifying credit card fraud transactions and the highest accuracy in ROC (98.8%) with slightly different percentage compared to Deep Neural Networks (98.2%). Also, it comes third in training running time with 10 s after Logistic Regression and K-Nearest Neighbor. This may be changed if the number of rows in the dataset is increased. However, in the current situation with the available dataset, the most efficient way is to use machine learning models as its training time is short compared to the training time of deep learning models. Moreover, it gives better results than deep learning models. Figures 5, 6, 7, 8, and 9 show a detailed comparison between algorithms applied in our experiment. Table 3 also shows the consolidated comparison between all algorithms used.

10 Conclusion

Credit card fraud is a critical business problem. This results in huge losses in terms of money and reputation. As a result, financial institutions invest a lot of money in developing new methodologies for fraud detection and prevention. The main goal of this research is to compare machine learning techniques and deep learning techniques in

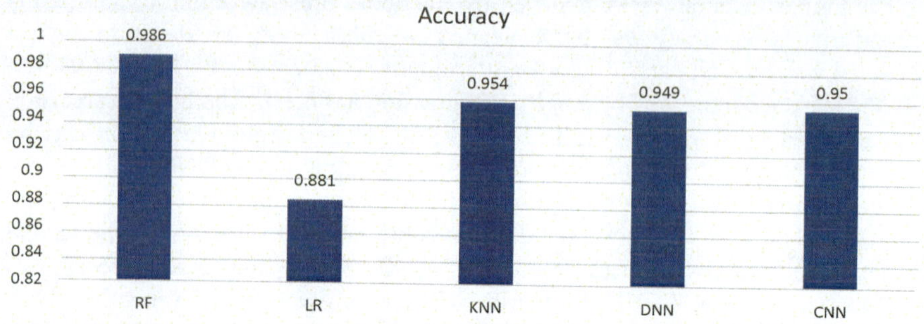

Fig. 5 Accuracy comparison between RF, LR, K-Neighbor, DNN

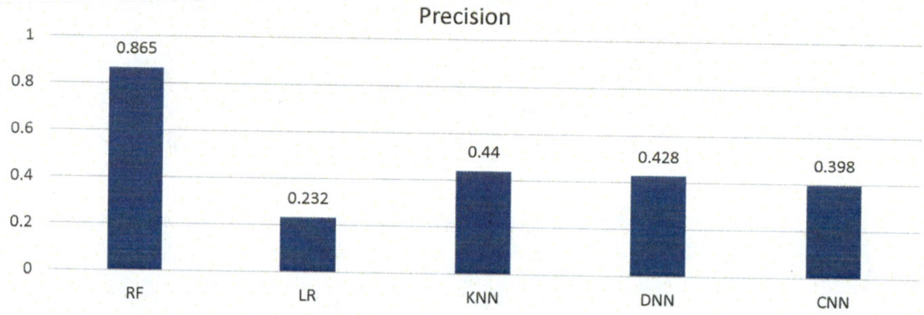

Fig. 6 Precision comparison between RF, LR, K-Neighbor, DNN

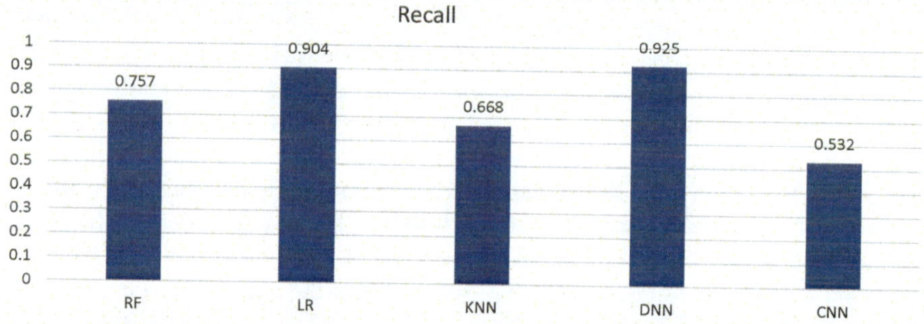

Fig. 7 Recall comparison between RF, LR, K-Neighbor, DNN

fraud detection and choose the best algorithm from among them. After conducting the experiments, comparison was made, and it showed that Random Forest algorithm produces the best results in knowing whether a transaction is fraud or not. This was concluded after using different measurements which are recall, accuracy, precision, AUC-ROC, and time. For fraud detection problem, it is important to have Recall with high value. Feature selection and balancing of the dataset have proved

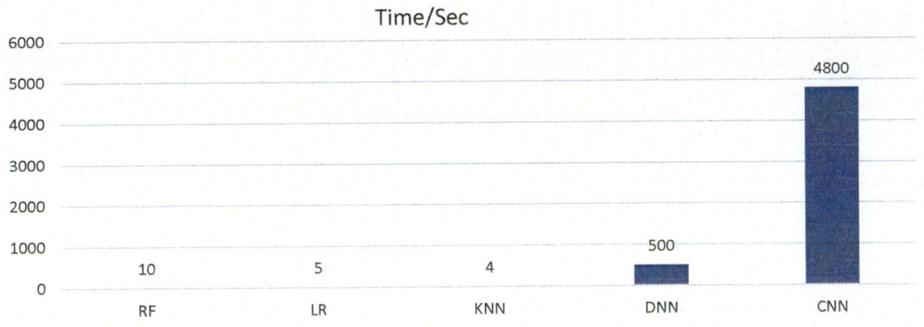

Fig. 8 Running time comparison between RF, LR, K-Neighbor, DNN

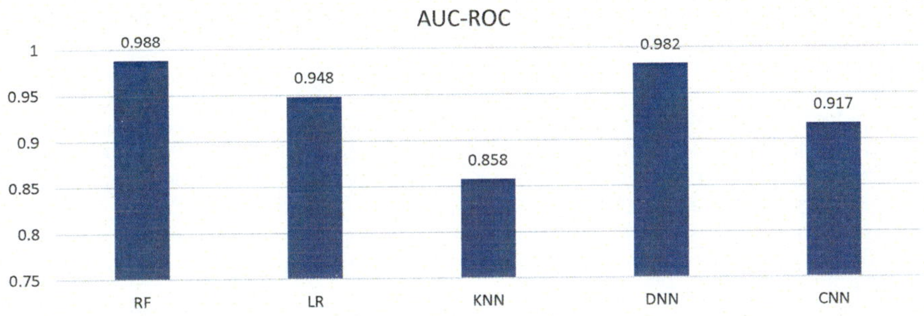

Fig. 9 AUC-ROC comparison between RF, LR, K-Neighbor, DNN

Table 3 Summary of results

	RF	LR	KNN	DNN	CNN
Accuracy	0.986	0.881	0.954	0.949	0.95
Precision	0.865	0.232	0.44	0.428	0.398
Recall	0.757	0.904	0.668	0.925	0.532
ROC	0.988	0.948	0.858	0.982	0.917
Time/s	10	5	4	500	4800

to be extremely important in achieving significant results, especially because most fraud datasets show a small number of fraud transactions. Further research should focus on increasing the number of rows in the dataset and monitoring the results of deep learning techniques which might provide better accuracy than machine learning techniques.

References

1. Modi K (2017) Fraud detection technique in credit card transactions using convolutional neural network. Int J Adv Res Eng Sci Technol 4(8)
2. Singh M, Singh P, Kumar R (2014) Fraud detection by monitoring user behavior and activities. In: 2nd international conference on computer and intelligent systems (ICCIS'2014) & 2nd international conference of electrical, electronics, instrumentation and biomedical engineering (ICEEIB 2014) May 30–31, 2014. Bangkok
3. Visa website,https://usa.visa.com/dam/VCOM/download/corporate/media/visanet-technology/aboutvisafactsheet.pdf, Jun 19,2020
4. Singh SS (2019) Electronic credit card fraud detection system by collaboration of machine learning models. Int J Innov Technol Exploring Eng (IJITEE) 8(12S)
5. Shift Processing website,https://shiftprocessing.com/credit-card-fraud-statistics/, Jun 19,2020
6. Goel E, Abhilasha (2017) Random forest: a review. Int J Adv Res Comput Sci Softw Eng 7(1)
7. Govindarajan M, Chandrasekaran RLVL (2009) Intrusion detection using k-nearest neighbor. In: 2009 first international conference on advanced computing. Chennai, pp 13–20
8. Pengkuk CYJ, Lee KL, Ingersoll GM (2002) An introduction to logistic regression analysis and reporting. J Educ Res
9. Sze V, Chen Y-H, Yang T-J, Emer J (2017) Efficient Processing of deep neural networks: a tutorial and survey. In: Proceedings of the IEEE, 105(12)
10. Chawla NV, Bowyer KW, Hall LO, Philip Kegelmeyer W (2002) SMOTE: synthetic minority over-sampling technique. J Artif Intell Res 321–357
11. Chouiekha A, Hajb EHIE (2018) ConvNets for fraud detection analysis. In: Procedia computer science, the first international conference on intelligent computing in data sciences 2018, vol 127. Elsevier, pp 133–138
12. Reshma RS (2018) Deep learning enabled fraud detection in credit card transactions. Int J Res Sci Inn (IJRSI) V(VII)
13. Puh M, Brkic L (2019) Detecting credit card fraud using selected machine learning algorithms. University of Zagreb Faculty of Electrical Engineering and Computing
14. Kurien KL, Chikkamannur A (2019) Detection and prediction of credit card fraud transactions using machine learning. Int J Eng Sci Res Technol IJESS7. ISSN 2277–9655

The Mediating Role of Customer Experience Management in the Relationship Between E-Commerce and Supply Chain Management Practices

Dalia Youniss

Abstract As markets become more and more connected on a global scale, the manner customer demands are being fulfilled is getting redefined. Labor, raw materials, and production can be acquired from a dynamic range of sources as big and small businesses look for the lowest cost. This is why supply chain management is extremely important to the growing e-commerce sector. E-commerce is prospering with the development of new business model and will be continuously developed in the following years. E-commerce is considered one of the main factors in economy development. E-commerce becomes a worldwide phenomenon that affords flexibility, convenience, and accessibility. E-commerce affords many benefits to customers like lower cost, availability, and reduce time. The impact of e-commerce appears in various business aspects from customer service to product design and delivery. A growing number of customers order products through the internet because of low prices, a wider range of products, convenience, etc. They assume these products will arrive at their destination, as soon as possible, with the potentiality of various forms of payment, free returns. Survival in today's economic competitive climate requires more than just price discount and innovative products. To achieve competitive advantage, companies must focus and understand customer experience. We are living in the age where fulfilling customer's needs is the main differentiator between Companies. Hence, companies need to switch their focus to customer experience management (CEM). It includes all touchpoints customers interact with the company, its products, and service while customer experience management represents the strategy followed by the company to manage the customer experience. This paper analyses the interaction of two topics: Supply Chain Management practices (SCMP) and e-commerce This paper will address the concepts related to impact of e-commerce on supply chain and future impact of e-commerce by answering different questions. And due to the vital role customer experience management play in differentiating between companies, the researcher introduces conceptually customer experience management as a mediator that impacts the relationship between e-commerce and supply chain management practices.

D. Youniss (✉)
Arab Academy of Science Technology and Maritime Transport (AASTMT), Alexandria, Egypt

D. A. Magdi et al. (eds.), *Digital Transformation Technology*, Lecture Notes in Networks and Systems 224, https://doi.org/10.1007/978-981-16-2275-5_17

Keywords E-commerce · Supply chain management practices · Customer experience management

1 Introduction

With the intensive competition and technological development, businesses should be capable to perform and use strategies, tools, and techniques that are unique to achieve competitive advantage and channels of distributions that are not complicated. The emergence of E-commerce technologies has created numerous opportunities for businesses to develop and streamline their supply chains. Hence, coordinating e-commerce with supply chain activities, a company can be capable to get raw materials and other components then manufacture and distribute the products to retailers or customers on time and at the right place lacking defections which will decrease the cost and increases customers' satisfaction. Furthermore, using e-commerce makes it easy for businesses to reach wide range of customers through the internet. Thus, integrating the ecommerce with the supply chain practices will lead to the elimination of retailers from the channel of distribution which permits the business to directly sell and deliver the products to the customers.

The combination and coordination of supply chain management practices with e-commerce focus on customers, integrates the whole process of supply chain, use of external resources, recognizes fast reaction, immensely reduces the level of stock. Supply chain management practices become a critical way for enterprise to improve competitiveness and adaptability. E-commerce is a segment of business model that allows individuals or firms to perform business through internet-commerce includes selling, buying and transferring funds. Customers use the internet to compare between products in terms of quality, price, durability etc. and to make online shopping experience and to get shopping experience. Internet affords a platform for customers and firms to meet as potential buyers and sellers. Companies by getting aware of the importance to cut costs to gain competitive advantage over competitors through omitting intermediaries by using the internet as a tool to reach customers and fulfilling their needs.

The emergence of customer experience management has considerably changed the whole game play. Previously the focus was on service or product quality, but customer experience is about elements which are out of control—like the impact of customers on each other, impact of devices like smart phones on the customers. Customer experience is receiving more and more attention as a business driver and a main strategic objective for companies as highlighted by [1]. Yet, in the following years, online customer experience became the emphasis of service marketing researchers [2]. The trigger for the change is largely accredited to the evolution of e-commerce platforms.

2 Literature Review

2.1 Supply Chain

The supply chain origin began in the 1980s and it has developed and became one of the main important business concepts. Smart [3] clarified that supply chain has been evolved from an operational activity focusing on distribution to a strategic concept which cover various functions and crosses interorganizational boundaries. Many researchers defined supply chain like Pires [4] who defined it" as a network of independent or semi-independent companies responsible for acquiring, producing and delivering a product or service to the end customer. In addition to Slack et al. [5] who affirmed that no network exists independently; all operations are part of a bigger network, interconnected with other operations. A network consists of a set of relationships between customers and suppliers.

Laud on and Laud on [6] stated that the supply chain is a network of organizations and business processes to choose raw materials to transform them into intermediate and finished products and distribute the finished products to customers. The supply chain connects suppliers, distribution centers, industrial plants, means of transportation, retail stores, and information by means of processes such as, inventory control, selection of raw materials, distribution, and delivery, to provide products and services from the source till the point of consumption.

2.2 Supply Chain Management

Turban et al. [7] stated that SCM is a complex process that involves the coordination of many activities in a manner that facilitates the transfer of goods and services from the supplier to the customer in the most efficient and effective way. While Chaffey [8] limited his description of SCM as the whole activities of the organization from suppliers to customers. Moreover, he classified SCM to:

- Upstream supply chain: transactions among an organization and its suppliers and intermediaries, corresponding to buy-side e-commerce.
- Downstream supply chain: transactions among an organization and its customers and intermediaries, corresponding to sell-side e-commerce.

Likewise, SCM is management of upstream and downstream relationships between suppliers and customers to achieve superior value at the lowest cost for the whole supply chain [9]. As asserted by Ninlawan et al. [10], SCM is the integration and management of a complex network of activities related to convey a finished product to the end-user or customer. Supply chain management includes not only cost management since it influences other aspects, such as performance, reliability of deliveries and speed, the flexibility with which the network can adapt, and finally the quality of the products [11].

Therefore, SCM involves all activities related to the flow and transformation of goods from the raw material stage up to the end user. Thus, relates to the integration of these activities, it involves coordination of these activities by establishing improved relationships in the supply chain to gain sustainable competitive advantage. [12]. Supply chain management was initially established as a way to reduce costs. It emphasized on precise elements in the supply chain and attempted to detect opportunities for process efficiency. Nowadays, supply chain management is used mainly to add value in the form of benefits to the final consumer at the end of the supply chain. As mentioned by [7] SCM includes E-procurement, Inventory management using wireless devices, Supply chain replenishment, Collaborative planning, Collaborative design and product development, E-logistics and Supply chain monitoring, and control.

2.3 Supply Chain Management Practices

"SCM practices were defined as a set of activities carried out by an organization to promote effective management of its supply chain" [13]. He suggested SCM practices as a multi-dimensional construct that consists of both upstream and downstream sides of the supply chain. Donlon [14] assumed outsourcing, information sharing, supplier partnership, continuous process flow and cycle time compression, as SCM practices. Tan [15] considered quality, purchasing, and customer relations as SCM practices, in their empirical study. Through usage of factor analysis, Tan et al. [16] confirmed that: supply chain integration, customer service management, geographic proximity, information sharing, and JIT capability, as the key aspects of SCM practice. Lee [17] identified outsourcing, strategic supplier partnerships, information sharing, customer relationships, and product modularity as SCM practices.

Chen and Paulraj [18] mentioned that cross-functional teams, long-term relationships, supplier involvement and supplier base reduction are the basic SCM practices. While Min and Mentzer [19] suggested that information sharing, long-term relationship, vision, goals, risk, award sharing, cooperation, process integration, and supply chain leadership are the main SCM. Practices.Li et al. [13] specified strategic supplier partnership, customer relationship, and information sharing are the main SCM practices. Li et al. [20] developed a reliable and valid instrument to measure SCM practices. The same instrument has been adopted in this study.

SCM practices impact overall organizational performance as well as the competitive advantage of any organization through price/cost, delivery dependability, time to market, quality, and product innovation. Previous studies have shown that various components of SCM practices (such as strategic supplier partnership) have an impact on achieving competitive advantage (such as price/cost). For example, strategic supplier partnership can improve supplier performance, reduce time to market [12], and increase the level of customer responsiveness and satisfaction [21].

2.4 E-Commerce

Hoffman and Novak [22] described e-commerce as an Internet technology that affords the capability to buy and sell online—includes market creation, ordering, supply chain management (SCM), and transfers by opening protocol. E-commerce was defined to be useful in recognizing possible exchange partners and responding shortly to customer changing demands, and relates more to buying and selling products on the internet. E-commerce is connected to individual consumers, where e-procurement usually deals with companies [23]. E-commerce is a vital concept that defines the process of buying, selling, or exchanging products, services, and information, through the Internet. E-commerce was defined by Turban [24] from different perspectives:

- **Communication perspective**: concerned with e-commerce in terms of goods delivery, services, information, or e-payments through internet.
- **Business process perspective**: concerned with e-commerce in terms of digital technology application toward the automation of business transactions and workflow
- **Service perspective**: concerned with e-commerce as a tool that addresses the need of companies, consumers, and management to reduce service costs while refining the quality of goods and increasing the speed of service delivery
- **Online perspective:** concerned with e-commerce in terms of its capability to buy and sell products and information on the Internet and other online services
- **Collaborations perspective**: concerned with e-commerce as a framework for inter- and intra-organizational collaboration
- **Community perspective**: concerned with e-commerce as a gathering place for community members, to learn, transact, and collaborate.

Turban et al. [7] interpreted e-commerce as the process of buying, selling, or trading products, services, or information via computer. Currently, e-commerce has the ability to revolutionize the manner a business is run [25]. Actually, e-commerce developed in the last few years into a wide range of relationships known as e-business, afford the potential to increases the speed of the order cycles. As a consequence, customers can operate business more easily than ever, mostly when a mix of price, quality, and speed of response represents an important requirement.

Further, businesses become more agile in managing the information between suppliers and customers besides the movement of products which will lead to low operating costs and better customer service. With e-commerce, customers can buy products, track delivery orders, and pay for the product immediately. It similarly allows companies to record details of products being shipped. E- commerce can assist companies to know what is needed to be ordered ahead of time. The implementation of technology in the supply chain will shrink the need for manual intervention and paperwork leading to high efficiency. The Growth of e-commerce relies on effective IT security systems for which vital technological and legal provisions need to be reinforced regularly. It is noticed mobile commerce (M-commerce) is growing quickly as a stable and secure addition to e-commerce industry. the new opportunities that will be afforded by e-commerce, will be available to both large corporations and small companies.

2.5 Customer Experience

The term "customer experience" was first introduced by Pine and Gilmore [26] in their article "the experience economy" than many academics and researchers have given attention to that topic. Many articles related to that topic concentrate either on experience measurement or quality measurement or analysis of the dimensions that impact customer experience. Puccinelli et al. [27] focus their research on seven consumer behavior research fields that impact the customer experience: (1) memory; (2) attitudes; (3) involvement; (4) goals, schemas, and information processing; (5) atmospherics; (6) affect; and (7) consumer attributions and choices

Although a service is the process or activity, the customers experience is their own interpretation of the service process and their involvement during their journey or interaction through a series of touch points [28, 1] while the experience is the customer own point of view existing only in his mind as mentioned by [29]. Stuart and Tax [30] claimed that customer experience can be enhanced by planning the service system to encourage greater customer participation. Bate and Robert [31] presented an approach which comprises customers in the design of the experience-based design. Pickles et al. [32] established this methodology to prove how three theoretical components of good design: engineering, functionality, and aesthetics can be utilized as a framework to enhance performance, safety, and governance.

2.6 Definition of Customer Experience

Schmitt [33] defined customer experience as "the process of strategically managing a customer's entire experience with a product or company." Gentile et al. [34] provide a definition of customer experience: "The customer experience stems from a set of interactions between a customer and a product, a company, which lead to a reaction. This experience is completely personal and involves the customer's involvement at different levels (sensorial, emotional, rational, physical, and spiritual)". Meyer and Schwager [35] stated that the interaction can either be direct and done by the customer, such as an intended purchase, or the interaction can be indirect, which involves unplanned activities in the course of purchase or service. Shaw et al. [36] asserted that the experience is measured through the customer's expectations across all these touch points. Rich [37] defined customer experience as the outcome of perceptions, observations, feelings, and thoughts arising from interactions and relationships between a customer and their provider over a period of time. Customer experience is not established only on a single situation with the customer, but rather on the entire journey with the company.

2.7 Customer Experience Management

The emergence of customer experience management has considerably changed the whole game play. Previously the focus was on service or product quality, but customer experience is about elements which are out of control—like the impact of customers on each other, impact of devices like smart phones on the customers. Customer experience is receiving more and more attention as a business driver and a main strategic objective for companies as highlighted by [1]. Yet, in the following years, online customer experience became the emphasis of service marketing researchers [2]. The trigger for the change is largely accredited to the evolution of e-commerce platforms. The portals became more dynamic and interactive affording a better interface for the customers to explore the offerings. Online customer experiences influenced customers' online buying behavior as stated by Bridges and Florsheim [38].

Customer experience management as a concept was introduced by Holbrooke and Hirschman [22] in the Journal of Consumer Research. Creating satisfactory customer experience is the most favorable way to create customer loyalty and competitive advantage for the company [39]. Experiences are about the observed value which occurred by the individual from such interactions as mentioned by [40].

As customers' potentials to choose between products, services, and preferred supply chain channels increase, customer service excellence becomes ever more important to all companies that provide a product or a service to their customers. The service-profit chain as mentioned by Heskett et al. [41] is one of the first concepts that address the significance of quality customer service and its relationships with other variables in the enterprise. [42] argued that since 2010 the age of the customer started. All previous manufacturing strength, information capability, distribution power has begun to disappear as competitive boundaries and customer choice became the main differentiator, hence enterprises had to shift their focus on customer experience management.

Meyer and Schwager [43] mentioned that customer experience in order to improve it should become the top priority of the enterprise including its structure, systems, and processes to reflect customer-centric priority. While Rae [44] accepted that companies started to focus on customer experience as a critical success factor combined with strategy, integration of technology, brand management, and orchestrating business models. More and more studies attempted to explore online customer experiences in different contexts [45] calling it as a relatively new construct. Hahn and Kim [22] highlighted that consumer's trust can be kept by ensuring offline to online.

As stated by [46] Traditional marketing communication channels are not anymore sufficient to influence nowadays consumers. Moreover, consumers influenced by their experience attitude in selecting goods and services as mentioned by Sharma and Sharma [47]. Experience is the main focus of future marketing according to Schmitt, 1999 while according to [46] it's all about customer perceptions, thoughts, feelings, and reviews about either a product or service. Usually enhancing satisfaction by focusing on specific touch points doesn't reflect the actual picture, the whole customer journey should be entirely of prime importance.

CEM represents the process used to manage a customer's cross-channel exposure, product, brand, interaction, and transaction with a company, or service comprehensively as claimed by Schmitt [33]. It involves every aspect of a company's offering; the quality of customer care, as well as packaging, advertising, product and service features, reliability, and ease of use as asserted by Meyer and Schwager [43]. These definitions show that CEM is a broad field that tries intentionally to align the enterprise and its various activities to deliver unique customer experiences that satisfy the enterprise's customers.

3 Through a Literature Review, Five Research Questions Were Identified

1. What is the relationship between e-commerce and supply chain management practices?
2. What are the changes caused by e-commerce in supply chain management practices?
3. To what extent does the relationship between e-commerce and supply chain management practices vary according to customer experience management?
4. What are the main benefits of the application of customer experience management in improving the performance of supply?
5. Does e-commerce change the competitive position in a supply network?

4 Research Methodology

This section is mainly concerned with the philosophy, approach, and design through which the data would be analyzed. As this paper depends on the philosophy of "Positivism" as it illustrates the data to find out the effect and cause relationship between the independent variable which is the E-Commerce and the dependent variable which is the SCM Practices with the mediation role of CEM and decide whether the phenomenon is described in a correct manner or not, in order to generalize these findings. Therefore, this current paper would be adopting the "Deduction" method to logically infer the quantitative data.

Unit of Analysis: This section refers to what is analyzed through this paper, therefore, the unit of analysis regarding this paper is the Egyptian corporales regardless of their size, as the small, medium, and large corporales are included in this analysis.

Population and Sample: The population in this paper is the employees in the Egyptian corporales, however, it is not available to include all the employees in every Egyptian corporate in the sample, therefore, a sample of 543 employees of the Egyptian corporate was conducted, out of this number, we have received 363 respondents, the valid respondents upon which we can depend were 317 respondents.

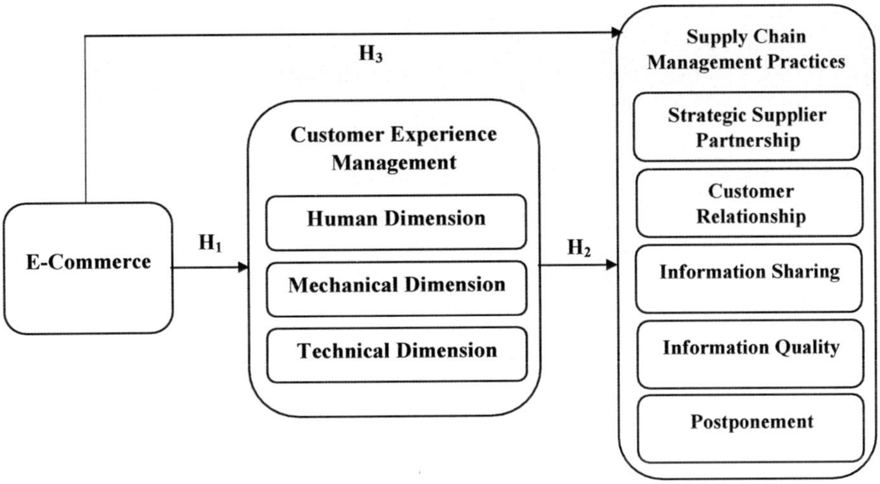

Fig. 1 Conceptual framework

Data Collection: The data is collected by a questionnaire, through the questions of this questionnaire the relation between the E-Commerce, CEM, and SCM can be measured.

Research Framework and Hypotheses: The framework of this current paper is illustrated in Fig. 1 which shows the E-Commerce is the independent variable and the Supply Chain Management Practices is the dependent variable which includes five dimensions, while, the Customer Experience Management represents the mediator variable which consists of three dimensions.

From this, the research hypotheses can be constructed as follows:

H_1: there is a significant relationship between E-Commerce and Customer Experience Management

H_2: there is a significant relationship between Customer Experience Management and Supply Chain Management Practices

H_3: there is a significant relationship between E-Commerce and Supply Chain Management Practices

Research Variables Measurement: Table 1 illustrates the variables included in this paper, identifying their type, whether dependent or independent, and exhibiting also the questions through which each variable is measured.

Table 1 Research variables measurement

Variable	Measurement scale
E-Commerce (Al-Alawi and Al-Al 2015)	Interacting with e-commerce is flexible
	Learning to operate e-commerce is easy
	It would be easy for me to become skillful at using e-commerce
	My interaction with e-commerce is clear and understandable
Human dimension (employees' reputation) (San-Martín et al. 2019)	Most consumers think that this m-site has good employees
	This m-site/vendor is acknowledged as a leader in its sector
	This m-site/vendor shows concern about its customers
	This m-site/vendor has a good reputation because of its employees
Mechanical dimension (m-site design) (San-Martín et al. 2019)	This m-site/vendor contains images that make a better experience of shopping
	This m-site/vendor has an attractive, modern, and professional design
	The advertising I receive from this m-site/vendor adapts to my situation and interests
	This m-site/vendor allows me to ask for personalized products/services
Technical dimension (m-site security) (San-Martín et al. 2019)	This m-site/vendor assures that the risk of abusing my personal information is low when I pay via a mobile
	This m-site/vendor makes sure that the risk of abusing my banking information is low
	This m-site/vendor makes sure that the risk of an unauthorized person accessing my m-payment process is low
	I think that the m-payment services of this m-site/vendor are secure
Strategic supplier partnership [13]	We consider quality as our number one criterion in selecting suppliers
	We regularly solve problems jointly with our suppliers
	We have helped our suppliers to improve their product quality
	We have continuous improvement programs that include our key supplier

(continued)

Table 1 (continued)

Variable	Measurement scale
Customer relationship [13]	We frequently interact with customers to set reliability, responsiveness, and other standards for us
	We frequently measure and evaluate customer satisfaction
	We frequently determine future customer expectations
	We facilitate customers' ability to seek assistance from us
Level of information sharing [13]	We inform trading partners in advance of changing needs
	Our trading partners share proprietary information with us
	Our trading partners keep us fully informed about issues that affect our business
	Our trading partners share business knowledge of core business processes with us
Level of information quality [13]	Information exchange between our trading partners and us is timely
	Information exchange between our trading partners and us is accurate
	Information exchange between our trading partners and us is complete
	Information exchange between our trading partners and us is adequate
Postponement [13]	Our products are designed for modular assembly
	We delay final product assembly activities until customer orders have actually been received
	We delay final product assembly activities until the last possible position (or nearest to customers) in the supply chain

5 Results and Findings

5.1 Data Testing

A validity and reliability analysis are conducted to be sure the questionnaire statements are phrased in a good format. Table 2 shows the results of the validity and reliability of the research questionnaire. Validity is measured by the two main factors. First, the Average Variance Extracted (AVE); it represents the average community for each latent factor. The (AVE) result should be greater than 0.5 to imply adequate

Table 2 Data validation

Variables	KMO	AVE	Cronbach's alpha	Items	Factor loading
E-Commerce	0.817	79.207	0.912	EC1	0.769
				EC2	0.755
				EC3	0.847
				EC4	0.797
Human dimension	0.801	69.962	0.851	HD1	0.758
				HD2	0.786
				HD3	0.523
				HD4	0.732
Mechanical dimension	0.777	65.539	0.817	MD1	0.678
				MD2	0.785
				MD3	0.737
				MD4	0.422
Technical dimension	0.651	63.066	0.705	TD1	0.668
				TD2	0.697
				TD3	0.527
				TD4	–
Strategic supplier partnership	0.839	74.949	0.887	SSP1	0.773
				SSP2	0.793
				SSP3	0.766
				SSP4	0.666
Customer relationship	0.726	77.851	0.857	CR1	–
				CR2	0.816
				CR3	0.757
				CR4	0.763
Information sharing	0.698	69.996	0.785	IS1	0.706
				IS2	0.736
				IS3	0.658
				IS4	–
Information quality	0.766	74.165	0.882	IQ1	0.638
				IQ2	0.779
				IQ3	0.829
				IQ4	0.72
Postponement	0.759	86.5	0.922	POS1	0.846
				POS2	0.881
				POS3	0.868

validity. Second is the factor loading for each item (statement) which should be greater than or equal to 0.4. To examine reliability, each factor is measured using a group of statements, indicates how stable and consistently the instrument taps the variable which can be examined by Cronbach's Alpha, the most commonly used test of reliability. The range of Alpha coefficient comes between 0 and 1, the higher the score the higher the reliability. If Alpha coefficients are greater than or equal to 0.7, it implies adequate reliability. By studying the variables in the model, it was found that all variables exceed 50% in the AVE indicator and also exceeding 0.4-factor loading for each item after deleting statements with factor loading problem.

Table 3 shows the discriminant validity of the research variables, where it could be observed that all square roots of AVE value are greater than the correlations between the corresponding construct and other constructs. This means that the research variables have adequate discriminant validity.

5.2 Descriptive Analysis

Table 4 shows the Mean and Standard Deviation for Research variables. It could be observed that the mean and the frequencies of most responses are in the agreement zone, as the mean values for the research variables: E-Commerce, Human Dimension, Mechanical Dimension, Technical Dimension, Strategic Supplier Partnership, Customer Relationship, Information Sharing, Information Quality, and Postponement are 3.8297, 3.9022, 3.7729, 3.7003, 3.9905, 4.0757, 3.8707, 3.9590, and 4.1167, respectively.

5.3 Normality Testing for the Research Variables

In order to check the normality for the data, two types of tests are conducted; formal and informal. Table 5 shows the formal testing of normality assumption for the research variables using the Kolmogorov-Smirnov test of normality. It could be observed that the research variables are not normally distributed, as the corresponding P-values are less than 0.05.

As the formal test shows that the values are not normally distributed, an informal test is used to detect the approximate normality. Table 6 shows the informal test of normality, where it could be shown that the skewness and kurtosis values are in the acceptance level of ± 1.5, which means that the data under study are approximately normal.

Table 3 Discriminant validity of the research variables

Variables	1.	2.	3.	4.	5.	6.	7.	8.	9.
1. E-commerce	(0.890)								
	317								
2. Human dimension	0.116*	(0.836)							
	0.038								
	317	317							
3. Mechanical dimension	0.397**	0.317**	(0.810)						
	0.000	0.000							
	317	317	317						
4. Technical dimension	0.299**	0.305**	0.404**	(0.794)					
	0.000	0.000	0.000						
	317	317	317	317					
5. Strategic supplier partnership	0.192**	0.249**	0.331**	0.314**	(0.866)				
	0.001	0.000	0.000	0.000					
	317	317	317	317	317				
6. Customer relationship	0.224**	0.250**	0.376**	0.327**	0.451**	(0.882)			
	0.000	0.000	0.000	0.000	0.000				
	317	317	317	317	317	317			
7. Information sharing	0.296**	0.375**	0.457**	0.426**	0.305**	0.364**	(0.837)		
	0.000	0.000	0.000	0.000	00.000	0.000			
	317	317	317	317	317	317	317		
8. Information quality	0.394**	0.386**	0.429**	0.438**	0.363**	0.415**	0.485**	(0.861)	

(continued)

Table 3 (continued)

Variables	1.	2.	3.	4.	5.	6.	7.	8.	9.
	00.000	0.000	0.000	00.000	0.000	0.000	0.000		
	317	317	317	317	317	317	317	317	
9. Postponement	0.385**	0.304**	0.406**	0.466**	0.352**	0.367**	0.392**	0.530**	(0.930)
	0.000	0.000	0.000	0.000	0.000	0.000	0.000	0.000	
	317	317	317	317	317	317	317	317	317

Table 4 Descriptive analysis for the research variables

	N	Mean	Std. deviation	Frequency				
				1	2	3	4	5
E-Commerce	317	3.8297	0.67693	0	0	104	163	50
Human dimension	317	3.9022	0.55109	0	0	65	218	34
Mechanical dimension	317	3.7729	0.56155	0	0	94	201	22
Technical dimension	317	3.7003	0.46569	0	0	96	220	1
Strategic supplier partnership	317	3.9905	0.49991	0	0	41	238	38
Customer relationship	317	4.0757	0.60636	0	0	47	199	71
Information sharing	317	3.8707	0.48944	0	0	61	236	20
Information quality	317	3.9590	0.59125	0	0	62	206	49
Postponement	317	4.1167	0.67195	0	0	55	170	92

Table 5 Formal testing of normality

	Kolmogorov-Smirnov[a]			Shapiro-Wilk		
	Statistic	df	Sig.	Statistic	df	Sig.
E-commerce	0.271	317	0.000	0.794	317	0.000
Human dimension	0.365	317	0.000	0.726	317	0.000
Mechanical dimension	0.361	317	0.000	0.730	317	0.000
Technical dimension	0.437	317	0.000	0.596	317	0.000
Strategic supplier partnership	0.378	317	0.000	0.677	317	0.000
Customer relationship	0.326	317	0.000	0.767	317	0.000
Information sharing	0.412	317	0.000	0.661	317	0.000
Information quality	0.332	317	0.000	0.758	317	0.000
Postponement	0.279	317	0.000	0.796	317	0.000

5.4 Testing Hypotheses

Table 7 shows the SEM analysis of the effect of E-Commerce on Customer Experience Management Dimensions. It could be observed that there is a significant positive effect of E-Commerce on Human Dimension as the estimate value is 0.118 and the *P*-value is less than 0.05. Further, the R^2 is 0.101, which means that the E-Commerce explains 10.1% of the variation of the Human Dimension. Moreover, there is a significant positive effect of E-Commerce on Mechanical Dimension as the estimate value is 0.302 and the *P*-value is less than 0.05. Further, the R^2 is 0.151, which means

Table 6 Informal testing of normality

	N	Skewness		Kurtosis	
	Statistic	Statistic	Std. error	Statistic	Std. Error
E-commerce	317	0.221	0.137	−0.829	0.273
Human dimension	317	−0.048	0.137	0.205	0.273
Mechanical dimension	317	−0.007	0.137	−0.305	0.273
Technical dimension	317	−0.784	0.137	−1.128	0.273
Strategic supplier partnership	317	−0.019	0.137	1.047	0.273
Customer relationship	317	−0.036	0.137	−0.291	0.273
Information sharing	317	−0.297	0.137	0.774	0.273
Information quality	317	0.009	0.137	−0.126	0.273
Postponement	317	−0.141	0.137	−0.789	0.273

Table 7 SEM analysis of the effect of E-commerce on customer experience management dimensions

			Estimate	P	R^{lm}
Human dimension	←	E-commerce	0.118	0.019	0.101
Mechanical dimension	←	E-commerce	0.302	***	0.151
Technical dimension	←	E-commerce	0.188	***	0.021

that the E-Commerce explains 15.1% of the variation of the Mechanical Dimension. Also, there is a significant positive effect of E-Commerce on Technical Dimension as the estimate value is 0.188 and the *P*-value is more than 0.05. Further, the R^2 is 0.021, which means that the E-Commerce explains 2.1% of the variation of the Technical Dimension.

The model fit indices are presented in Table 8 which could be observed that are all within their acceptable levels. The SEM model conducted for the effect of E-Commerce on Customer Experience Management Dimensions is illustrated in Fig. 2.

Table 9 shows the SEM analysis of the effect of Customer Experience Management dimensions on Supply Chain Management Practices.

At First, it could be observed that there is a significant positive effect of Mechanical Dimension, and Technical Dimension on Strategic Supplier Partnership as the estimate values are 0.461, and 0.425, respectively and the *P*-values are less than 0.05, while, there is an insignificant effect of Human Dimension on Strategic Supplier Partnership as the *P*-value is more than 0.05. However, the R^2 is 0.242, which means

Table 8 The model fit indices

CMIN/DF	GFI	AGFI	CFI	RMSEA
2.482	0.914	0.882	0.945	0.068

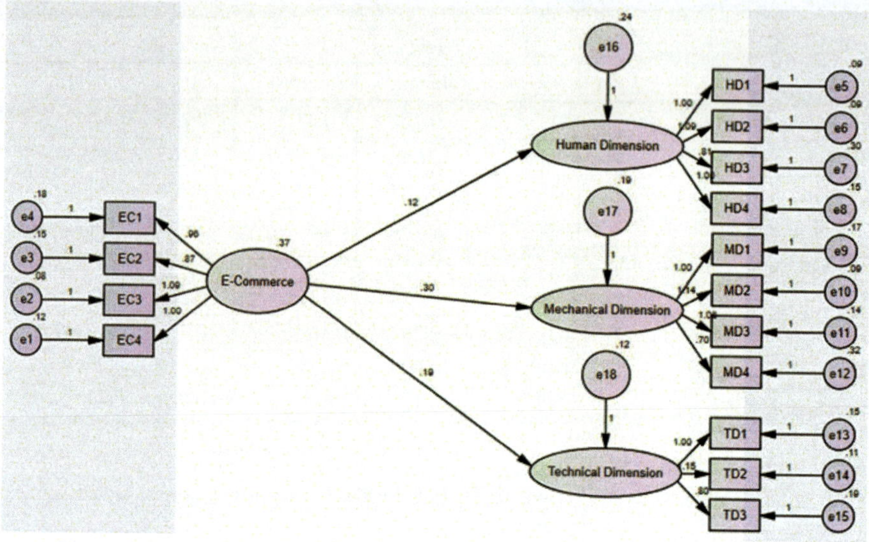

Fig. 2 SEM analysis of the effect of e-commerce on customer experience management dimensions

Table 9 SEM analysis of the effect of customer experience management dimensions on supply chain management practices

			Estimate	P	R square
Supplier Partnership	<---	Human Dimension	.039	.557	
Supplier Partnership	<---	Mechanical Dimension	.461	***	.242
Supplier Partnership	<---	Technical Dimension	.425	.002	
Customer Relationship	<---	Human Dimension	.037	.644	
Customer Relationship	<---	Mechanical Dimension	.504	***	.262
Customer Relationship	<---	Technical Dimension	.622	***	
Information Sharing	<---	Human Dimension	.126	.034	
Information Sharing	<---	Mechanical Dimension	.330	.001	.356
Information Sharing	<---	Technical Dimension	.518	***	
Information Quality	<---	Human Dimension	.142	.006	
Information Quality	<---	Mechanical Dimension	.275	.002	.383
Information Quality	<---	Technical Dimension	.537	***	
Postponement	<---	Human Dimension	.098	.203	
Postponement	<---	Mechanical Dimension	.482	***	.388
Postponement	<---	Technical Dimension	.914	***	

that the model explains 24.2% of the variation of the Strategic Supplier Partnership. Secondly, there is a significant positive effect of Mechanical Dimension, and Technical Dimension on Customer Relationship as the estimate values are 0.506, and 0.622 respectively and the *P*-values are less than 0.05, while, there is an insignificant effect of Human Dimension on Customer Relationship as the *P*-value is more than

Table 10 The model fit indices

CMIN/DF	GFI	AGFI	CFI	RMSEA
1.512	0.902	0.880	0.965	0.040

0.05. However, the R^2 is 0.262, which means that the model explains 26.2% of the variation of the Customer Relationship.

Thirdly, there is a significant positive effect of Human Dimension, Mechanical Dimension, and Technical Dimension on Information Sharing as the estimate values are 0.126, 0.330, and 0.518 respectively and the P-values are less than 0.05. Moreover, the R^2 is 0.356, which means that the model explains 35.6% of the variation of the Information Sharing. Fourthly, there is a significant positive effect of Human Dimension, Mechanical Dimension, and Technical Dimension on Information Quality as the estimate values are 0.142, 0.275, and 0.537 respectively and the P-values are less than 0.05. Moreover, the R^2 is 0.383, which means that the model explains 38.3% of the variation of the Information Quality. Finally, there is a significant positive effect of Mechanical Dimension, and Technical Dimension on Postponement as the estimate values are 0.482, and 0.914 respectively and the P-values are less than 0.05, while, there is an insignificant effect of Human Dimension on Postponement as the P-value is more than 0.05. However, the R^2 is 0.388, which means that the model explains 38.8% of the variation of the Postponement.

The model fit indices are presented in Table 10 which could be observed that are all within their acceptable levels. The SEM model conducted for the effect of Customer Value on Dependent Variables is illustrated in Fig. 3.

Table 11 shows the SEM analysis of the effect of E-Commerce on Supply Chain Management Practices. It could be observed that there is a significant positive effect of E-Commerce on Strategic Supplier Partnership as the estimated value is 0.187 and the P-value is less than 0.05. Further, the R^2 is 0.053, which means that the E-Commerce explains 5.3% of the variation of the Strategic Supplier Partnership. Moreover, there is a significant positive effect of E-Commerce on Customer Relationship as the estimated value is 0.261 and the P-value is less than 0.05. Further, the R^2 is 0.074, which means that the E-Commerce explains 7.4% of the variation of the Customer Relationship. Also, there is a significant positive effect of E-Commerce on Information Sharing as the estimated value is 0.207 and the P-value is more than 0.05. Further, the R^2 is 0.091, which means that the E-Commerce explains 9.1% of the variation of the Information Sharing. Furthermore, there is a significant positive effect of E-Commerce on Information Quality as the estimated value is 0.384 and the P-value is less than 0.05. Further, the R^2 is 0.333, which means that the E-Commerce explains 33.3% of the variation of the Information Quality. Finally, there is a significant positive effect of E-Commerce on Postponement as the estimated value is 0.451 and the P-value is less than 0.05. Further, the R^2 is 0.198, which means that the E-Commerce explains 19.8% of the variation of the Postponement.

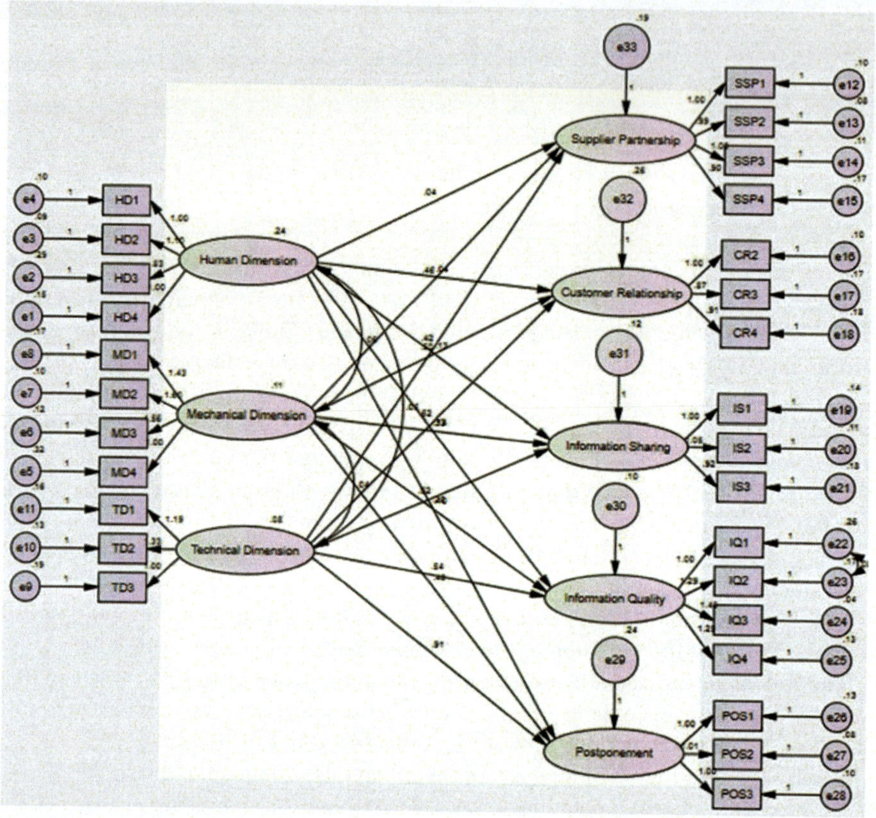

Fig. 3 SEM analysis of the effect of customer experience management dimensions on supply chain management practices

Table 11 SEM analysis of the effect of E-commerce on supply chain management practices

			Estimate	P	R square
Strategic Supplier Partnership	<---	E-Commerce	.187	***	.053
Customer Relationship	<---	E-Commerce	.261	***	.074
Information Sharing	<---	E-Commerce	.207	***	.091
Information Quality	<---	E-Commerce	.384	***	.333
Postponement	<---	E-Commerce	.451	***	.198

Table 12 The model fit indices

CMIN/DF	GFI	AGFI	CFI	RMSEA
3.603	0.834	0.786	0.893	0.091

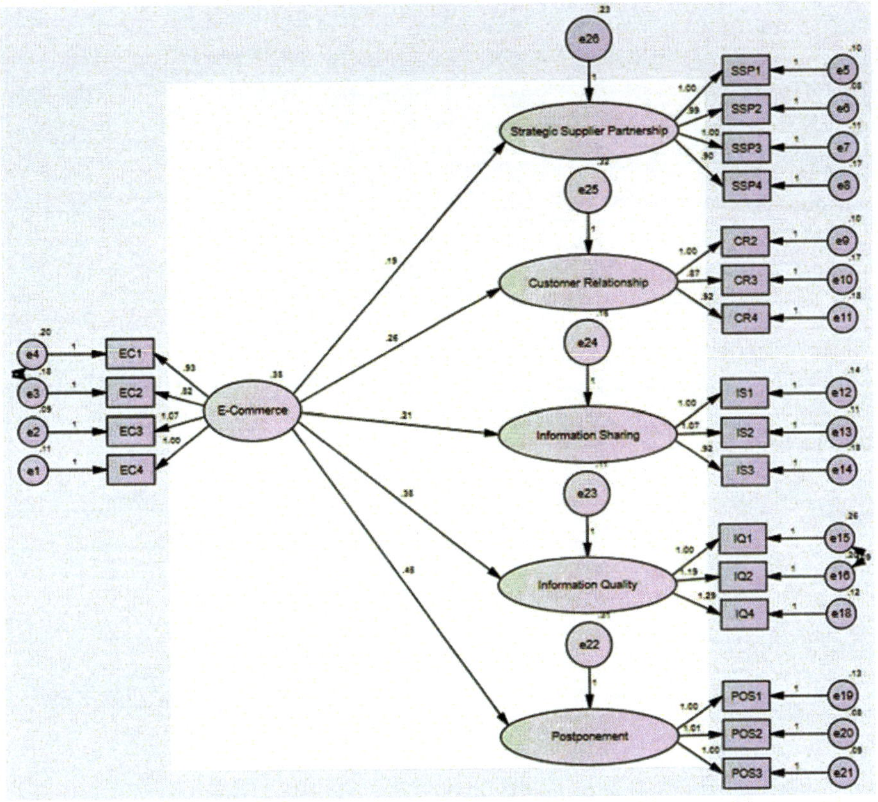

Fig. 4 SEM analysis of the effect of e-commerce on supply chain management practices

The model fit indices are presented in Table 12 which could be observed that are all within their acceptable levels. The SEM model conducted for the effect of Firm Value on Dependent Variables is illustrated in Fig. 4.

Table 13 shows the SEM analysis of the mediation role of Human Dimension. It could be observed that there is a significant effect of Human Dimension on Supply Chain Management Practices as the *P*-value is less than 0.05, which means that there is a direct relation between Human Dimension and Supply Chain Management Practices. Moreover, it could be noted from Table 11 that there is a significant effect of E-Commerce on Supply Chain Management Practices, also, it could be observed from Table 13 that there is a significant effect of E-Commerce on Supply Chain Management Practices, which means that Human Dimension partially mediates the relation between E-CommerceandSupply Chain Management Practices.

The model fit indices are presented in Table 14 which could be observed that are all within their acceptable levels. The SEM model conducted for the effect of Mediation Role of Human Dimension is illustrated in Fig. 5.

Table 13 The mediation role of human dimension

			Estimate	P
Human Dimension	<---	E-Commerce	.101	.046
Strategic Supplier Partnership	<---	E-Commerce	.150	.002
Customer Relationship	<---	E-Commerce	.221	***
Information Sharing	<---	E-Commerce	.169	***
Information Quality	<---	E-Commerce	.333	***
Postponement	<---	E-Commerce	.405	***
Strategic Supplier Partnership	<---	Human Dimension	.263	***
Customer Relationship	<---	Human Dimension	.315	***
Information Sharing	<---	Human Dimension	.320	***
Information Quality	<---	Human Dimension	.354	***
Postponement	<---	Human Dimension	.401	***

Table 14 The model fit indices

CMIN/DF	GFI	AGFI	CFI	RMSEA
2.612	0.859	0.824	0.918	0.071

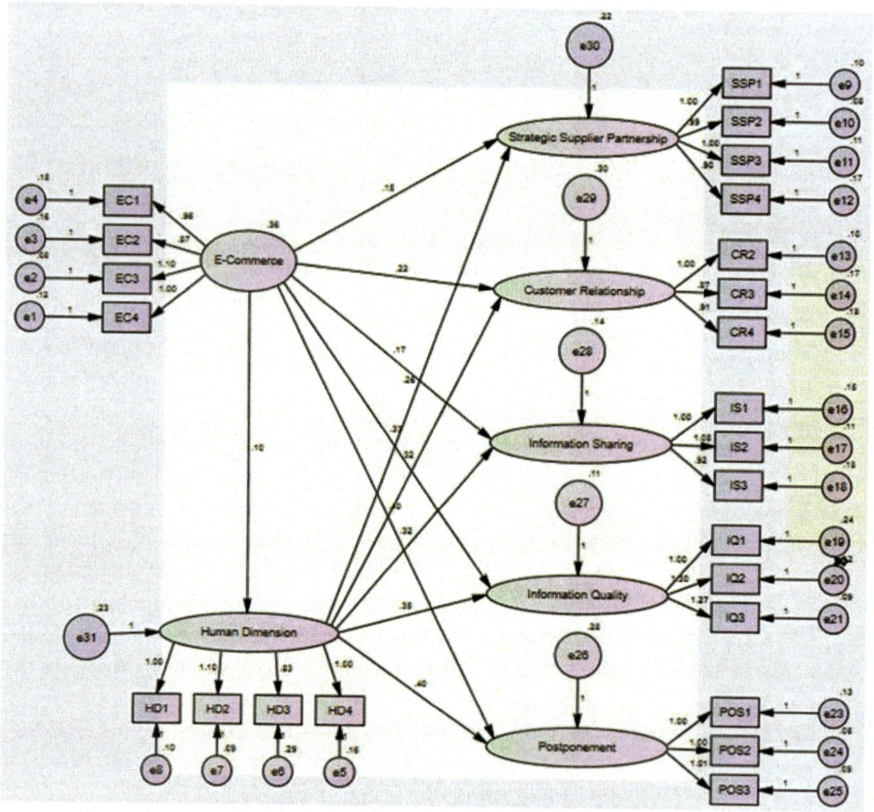

Fig. 5 SEM analysis of the mediation role of human dimension

Table 15 The mediation role of mechanical dimension

			Estimate	P
Mechanical Dimension	<---	E-Commerce	.208	***
Strategic Supplier Partnership	<---	E-Commerce	.020	.695
Customer Relationship	<---	E-Commerce	.070	.257
Information Sharing	<---	E-Commerce	.067	.144
Information Quality	<---	E-Commerce	.249	***
Postponement	<---	E-Commerce	.267	***
Strategic Supplier Partnership	<---	Mechanical Dimension	.690	***
Customer Relationship	<---	Mechanical Dimension	.794	***
Information Sharing	<---	Mechanical Dimension	.608	***
Information Quality	<---	Mechanical Dimension	.504	***
Postponement	<---	Mechanical Dimension	.795	***

Table 15 shows the SEM analysis of the mediation role of Mechanical Dimension. It could be observed that there is a significant effect of Mechanical Dimension on Supply Chain Management Practices as the *P*-value is less than 0.05, which means that there is a direct relation between Mechanical Dimension and Supply Chain Management Practices. Moreover, it could be noted from Table 11 that there is a significant effect of E-Commerce on Supply Chain Management Practices, also, it could be observed from Table 15 that there is a significant effect of E-Commerce on Information Quality, and Postponement, while there is an insignificant effect of E-Commerce on Strategic Supplier Partnership, Customer Relationship, and Information Sharing, which means that Mechanical Dimension partially mediates the relation between E-CommerceandInformation Quality, and Postponement, while, fully mediate the relation between E-Commerce and Strategic Supplier Partnership, Customer Relationship, and Information Sharing.

The model fit indices are presented in Table 16 which could be observed that are all within their acceptable levels. The SEM model conducted for the effect of Mediation Role of Mechanical Dimension is illustrated in Fig. 6.

Table 17 shows the SEM analysis of the mediation role of Technical Dimension. It could be observed that there is a significant effect of Technical Dimension on Supply Chain Management Practices as the *P*-value is less than 0.05, which means that there is a direct relation between Technical Dimension and Supply Chain Management Practices. Moreover, it could be noted from Table 11 that there is a significant effect of E-Commerce on Supply Chain Management Practices, also, it could be observed from Table 17 that there is a significant effect of E-Commerce on Information Quality, and Postponement, while there is an insignificant effect of E-Commerce on Strategic Supplier Partnership, Customer Relationship, and

Table 16 The model fit indices

CMIN/DF	GFI	AGFI	CFI	RMSEA
2.688	0.859	0.824	0.915	0.073

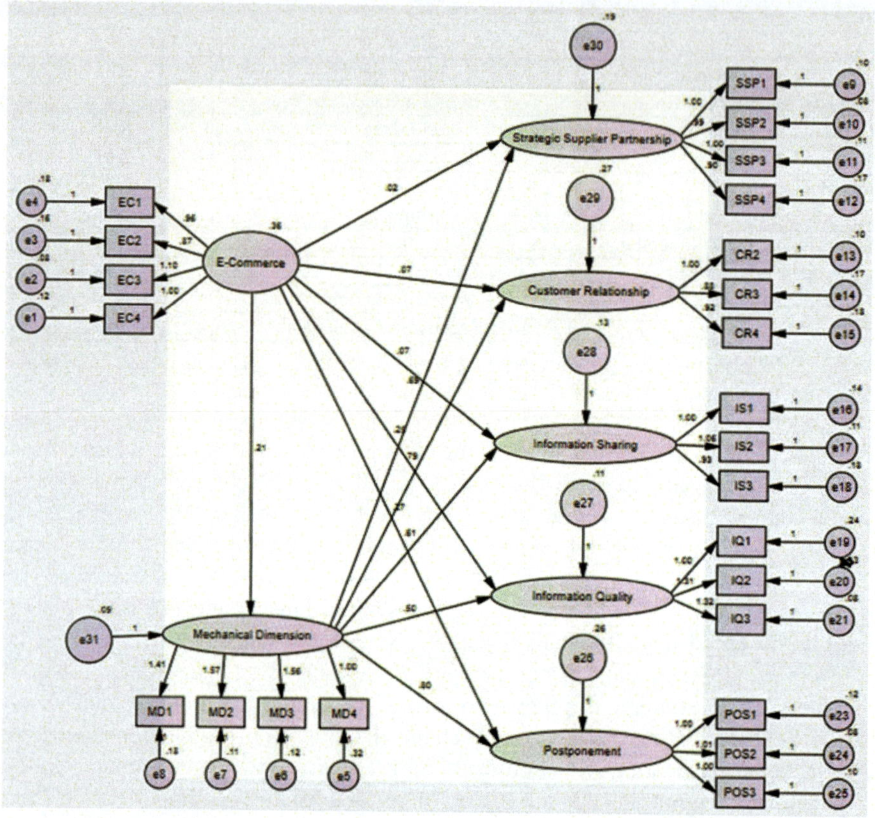

Fig. 6 SEM analysis of the mediation role of mechanical dimension

Table 17 The mediation role of technical dimension

			Estimate	P
Technical Dimension	<---	E-Commerce	.156	***
Strategic Supplier Partnership	<---	E-Commerce	.027	.607
Customer Relationship	<---	E-Commerce	.061	.336
Information Sharing	<---	E-Commerce	.056	.238
Information Quality	<---	E-Commerce	.220	***
Postponement	<---	E-Commerce	.228	***
Strategic Supplier Partnership	<---	Technical Dimension	.846	***
Customer Relationship	<---	Technical Dimension	1.094	***
Information Sharing	<---	Technical Dimension	.864	***
Information Quality	<---	Technical Dimension	.836	***
Postponement	<---	Technical Dimension	1.273	***

Table 18 The model fit indices

CMIN/DF	GFI	AGFI	CFI	RMSEA
2.613	0.871	0.836	0.920	0.071

Information Sharing, which means that Technical Dimension partially mediate the relation between E-Commerce and Information Quality, and Postponement, while, fully mediate the relation between E-Commerce and Strategic Supplier Partnership, Customer Relationship, and Information Sharing.

The model fit indices are presented in Table 18 which could be observed that are all within their acceptable levels. The SEM model conducted for the effect of Mediation Role of Technical Dimension is illustrated in Fig. 7.

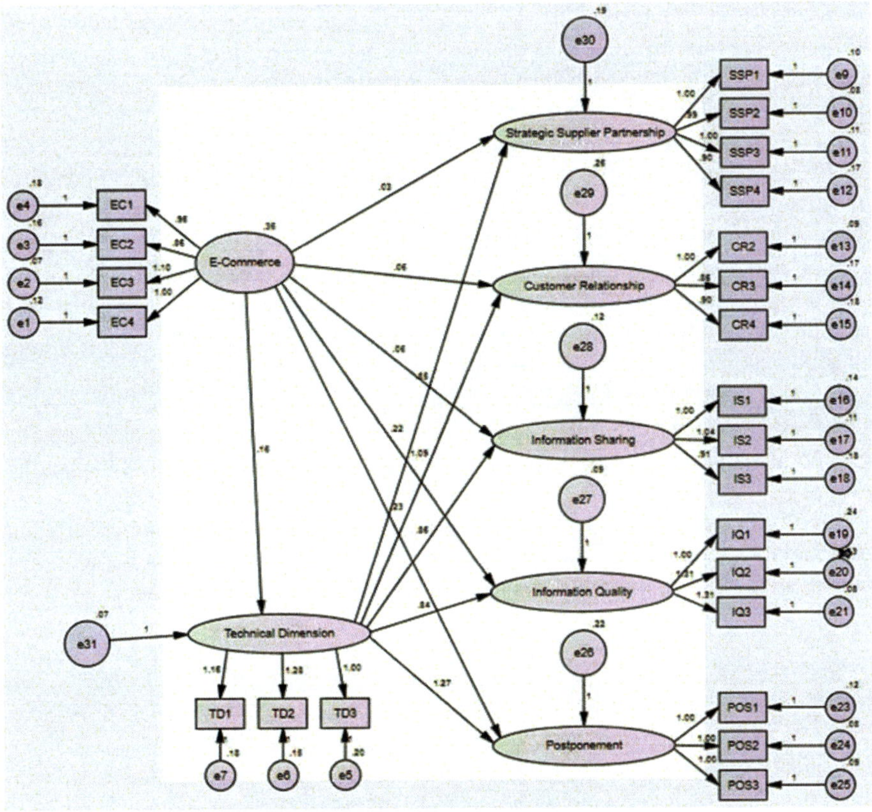

Fig. 7 SEM analysis of the mediation role of technical dimension

6 Conclusion

The main aim of this paper was to give a contribution to increase the body of knowledge in the field of e-commerce and supply chain management practices while mediating customer experience management. E-commerce and supply chain became a vital part of today's world economy and a crucial element of success. Lacking a clear supply chain structure will expose companies to lots of delay and delivery errors. However, with the adoption of e-commerce, enterprises can simply track the shipment flow, delivery, and expand to new markets and reach new customers, which created a new horizon of opportunities and jobs. With the increasing use of e-commerce, some companies started to provide solutions that help all parties of the chain to provide better tasks, cheaper and faster. As web technologies develop, e-commerce became a powerful enabler of supply chain integration and coordination across a wide range of industries. Speed and connectivity of Internet technology have reformed how business are conducted. As e-commerce impact SCM practices, information visibility is reached through connectivity among trading partners, thus, supply chain can be better managed.

The optimization of supply chain management practices can provide lots of benefits for companies. The drawbacks of late delivery, uncontrollable inventory, and transportation high cost could be mitigated through supply chain management. Through effective management, the usage of modern technology can access precisely and timely to make use of the rapid communication between manufacturers, suppliers, and customers showing the entire supply chain advantages and enhancing the competitive advantage of enterprises. Hence, there is no doubt that the supply chain management based on e-commerce platform is the expected development track in the era of information technology.

The efficiency of the supply chain can be assessed from understanding the customer perceptions towards physical, information, and financial flows happening in the supply chain. This is mostly because ultimately customer perception of such logistical support is what determines success and growth of e-commerce. It's not about supply chain but it is about the customer and any change that is needed to be made in supply chain should be the result of customers' needs and wants.

References

1. Johnston R, Clark G (2008) Service operations management: improving service delivery. Pearson Publication, Harlow, England
2. Nambisan P, Watt JH (2011) Managing customer experiences in online product communities. J Bus Res 64(8):889–895
3. Smart A (2008) eBusiness and supply chain integration. J Enter Infor 21(3):227–246
4. Pires SRI (2009) Supply chain management. São Paulo, Atlas
5. Slack N, Chambers S, Johnston R (2010) Operations management. Prentice Hall, Upper Saddler River, NJ

6. Laudon KC, Laudon JP (2011) Management information system. Prentice Hall, Upper Saddler River, NJ
7. Turban E, King D, Lee J, Liang T-P, Turban D (2010) Electronic commerce 2010 a managerial perspective. Prentice Hall, Upper Saddle River, NJ
8. Chaffey D (2009) E-business and e-commerce management, strategy, implementation & practice. Prentice Hall, Bukupedia
9. Christopher M (2011) Logistics and supply chain management. FT Press, Upper Saddler River, NJ
10. Ninlawan C, Seksan P, Tossapol K, Pilada W (2010) The implementation of green supply chain management practices in electronics industry. In Proceedings of the international multiconference of engineers and computer scientists, vol 3, pp 17–19
11. Corrêa HL (2010) Supply network management. São Paulo, Atlas
12. Handfield RB, Nichols EL (2002) Supply chain redesign: transforming supply chains into integrated value systems. FT Press, Upper Saddler River, NJ
13. Li S, Ragu-Nathan B, Ragu-Nathan TS, Rao SS (2006) The impact of supply chain management practices on competitive advantage and organizational performance. Omega 34(2):107–124
14. Donlon JP (1996) Maximizing value in the supply chain. Chief Executive
15. Tan KC, Kannan VR, Handfield RB (1998) Supply chain management: supplier performance and firm performance. J Supply Chain Manage 34(3):2
16. Tan KC (2002) Supply chain management: practices, concerns, and performance issues. J Supply Chain Manage 38(4):42–53
17. Lee HL (2004) The triple-a supply chain. Harvard Bus Rev 82(10):2–11. [Google Scholar, Infotrieve]
18. Chen IJ, Paulraj A (2004) Towards a theory of supply chain management: the constructs and measurements. J Operat Manage
19. Min S, Mentzer JT (2004) Developing and measuring supply chain management concepts. J Bus Logistics 25(1):63–99
20. Tian FY, Li JY, Li FD, Han YZ (2005) The application of the DS12C887 in the SCM system. Electri Measure Instrument 3
21. Power DJ, Sohal A, Rahman SU (2001) Critical Success factors in agile supply chain managment an empirical study. Int J Phys Distribut Logistics Manage 31(4):247–265
22. Hongyoun Hahn K, Kim J (2009) The effect of offline brand trust and perceived internet confidence on online shopping intention in the integrated multi-channel context. Int J Retail Distrib Manage 37(2):126–141
23. Johnson ME, Whang S (2002) E-business and supply chain management: an overview and framework. Prod Operat Manage 11(4):413–423
24. Turban E, King D, Lee J, Warkentin M, Chung H, M (2020) Electronic commerce: a managerial perspective, 2nd edn, Prentice Hall, pp 121–129
25. Sharifi H, Ismail HS, Reid I (2006) Achieving agility in supply chain through simultaneous design of and design for supply chain. J Manuf Technol Manage 17(8):1078–1098
26. Pine B, Gilmore J (1998) Welcome to the experience economy. Harvard Bus Rev 1998:97–105
27. Puccinelli NaM, Goodstein RC, Grewal D, Price R, Raghubir P, Stewart D (2009) Customer experience management in retailing: understanding the buying process. J Retail 85(1):15–30
28. Ding DX, Hu PJ, Verma R, Wardell DG (2010) The impact of service system design and flow experience on customer satisfaction in online financial services. J Serv Res 13(1):96–110
29. Vargo SL, Lusch RF (2004) Evolving to a new dominant logic of marketing. J Mark 68(1):1–17
30. Stuart FI, Tax S (2004) Toward and integrative approach to designing service experiences Lessons learned from the theatre. J Operat Manage 22:609–627
31. Bate P, Robert G (2007) Toward More user-centric OD—lessons from the field of experience-based design and a case study. J Appl Behavioral Sci 41(1):41–66
32. Pickles J, Hide E, Maher L (2008) Experience based design: a practical method of working with patients to redesign services. Clinical Govern Int J 13(1):51–58
33. Schmitt, B.H. 2003. Customer experience management: a revolutionary approach to connecting with your customers. Wiley

34. Gentile C, Spiller N, Noci G (2007) How to sustain the customer experience: an overview of experience components that co-create value with the customer. European Manage J 25(5):395–410
35. Meyer C, Schwager A (2007) Customer experience. Harvard Bus Rev 1–11
36. Shaw C, Dibeehi Q, Walden S (2010) Customer experience: future trends and insights. Palgrave Macmillan New York, New York
37. Rich R (2012) Customer experience: assuring service quality for every customer. TM Forum Insights Research. TM Forum, New Jersey, USA
38. Bridges E, Florsheim R. (2008) Hedonic and utilitarian shopping goals: the online experience. J Bus Res 61(4):309–314
39. Badgett M, Boyce MS, Kleinberger H (2007) Turning shoppers into advocates. IBM Instit Bus Value 35(3):352–369
40. Helkkula A, Kelleher C, Pihlström M (2012) Characterizing value as an experience: implications for service researchers and managers. J Serv Res 15(1):59–75
41. Heskett JL, Jones TO, Loveman GW, Sasser WE, Schlesinger LA (1994) Putting the service-profit-chain to work. Harvard Buss Rev 1994:164–170
42. Manning H, Bodine K (2012) Outside in: the power of putting customers at the center of your business. Forrester Research, Boston NY
43. Meyer C, Schwager A (2007) Understanding customer experience. Harvard Bus Rev 85(2):117–126
44. Rae J (2006) The importance of great customer experiences. Bloomberg Bus Week Mag. Posted on November 26, 2006,
45. Verhoef PC, Lemon KN, Parasuraman A, Roggeveen A, Tsiros M, Schlesinger LA (2009)
46. Coelho MF, Santos PAA (2013) The smartphone experience marketing: cocreating Value through mobile apps. PMKT Brazilian J Market Opinion Media Res 13:46–66. São Paulo, Brasil (ISSN 1983–9456 Print and ISSN 2317-0123 Online)
47. Sharma A (2011) Take-off of online marketing: casting the next generation strategies. Bus Strat Ser 12(4):202–208

Agritech Startup Ecosystem in Ukraine: Ideas and Realization

Vitalina Babenko, Larysa Zomchak, Maryna Nehrey, Abdel-Badeeh M. Salem, and Oleksandr Nakisko

Abstract The digitalization of agriculture is much slower than in other industries, due to the high cost of digital solutions and their complex functionality. Recently, the situation is changing, and now agriculture can get real benefits from technology with simple software solutions. Technologies used in agriculture include sensors, mobile devices, communication networks, drones, robotics, artificial intelligence, and others. These technologies provide an opportunity to improve the business processes of agriculture and increase its efficiency. In 2020, the agritech startup ecosystem of Ukraine has more than 70 startups that have been used for farm management, drones and remote sensing, precision farming, and urban agriculture. SWOT analysis indicate that the agritech startup ecosystem is characterized by more weaknesses and threats than strengths and opportunities. The Ukrainian agricultural sector has significant potential for increasing efficiency and development. Agriculture today needs startups that use information technology to address the balance of economic, environmental, and social efficiency to increase agricultural productivity.

Keywords Agriculture · Agritech · Technological innovation · Startup

V. Babenko (✉)
V. N. Karazin, Kharkiv National University, 4, Svobody Sq., Kharkiv 61022, Ukraine
e-mail: vitalinababenko@karazin.ua

L. Zomchak
Ivan Franko National University of Lviv, Universytetska Str., 1, Lviv 71001, Ukraine

M. Nehrey
National University of Life and Environmental Sciences of Ukraine, Heroyiv Oborony st., 15, Kyiv 03041, Ukraine

A.-B. M. Salem
Ain Shams University, El-Khalyfa El-Mamoun Str., Cairo, Egypt

O. Nakisko
Kharkiv Petro Vasylenko National Technical University of Agriculture, Alchevskikh str., 44, Kharkiv 61002, Ukraine

D. A. Magdi et al. (eds.), *Digital Transformation Technology*, Lecture Notes in Networks and Systems 224, https://doi.org/10.1007/978-981-16-2275-5_19

1 Introduction

In recent decades, the digitalization of agriculture has been much slower than in other industries. The reasons for this were the high cost of software solutions, the long production cycle, and investors' fears of the possibility of complete crop loss due to negative natural phenomena. According to Basso and Antle [1], digital technologies are most actively implemented in large farms and very rarely in small farms. This is due to both the high cost of digital solutions and their complex functionality.

Gradually the situation is changing, and now agriculture can get real benefits from technology using simple software solutions. The number of those who make the most of modern technologies in the form of innovations in agriculture is still small. Often, the introduction of digital technologies is chaotic, which leads to additional costs and losses. Therefore, the agritech startup ecosystem has significant prospects for development.

2 Theoretical Background

By agritech startup ecosystems, we mean a set of startups that develop for agriculture digital and geospatial information technologies for monitoring, assessing, and managing climate, soil, and genetic resources to increase agricultural productivity.

Technologies used in agriculture include sensors, mobile devices, communication networks, drones, robotics, artificial intelligence, and others. These technologies provide an opportunity to improve the business processes of agriculture and increase its efficiency. Classification of main digital agriculture technology is presented in Fig. 1.

3 Materials and Methods

Digitalization has become a sign of the times. In the world and in Ukraine, the issues of implementation of digital transformations and development of digital economy and management come to the fore. The agricultural sector provides up to 30% of Ukrainian exports and about 15–17% of GDP, and in the future, this figure, according to experts, may increase to 25%. But this requires increasing the efficiency of agricultural production, including through the introduction of innovative technologies.

Conducted within the framework of the CGIAR Research Program, the global advancement was made to ensure that IT technologies will play a leading role in the development of agricultural business until 2030 [3, 4]. Recall that the program of sustainable development was adopted by the United Nations Organization, as early as in 2015 there were 17 main goals [5].

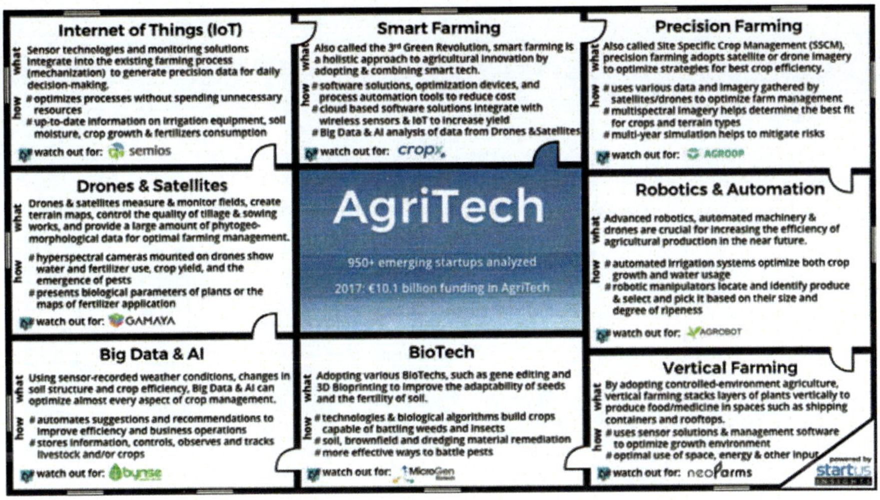

Fig. 1 Classification of main digital agriculture technology [2]

IT solution is responsible for the improvement of productivity and income of agrarian enterprises in a direct line of optimization of business processes and the failure of precision farming systems.

According to the assessment of the World Economic Forum, if 20–25% of agricultural business changes to precision farming, then by 2030, the global yield can increase by 10–15%, and the wikis of greenhouse gases can decrease by 10% [6].

AgFunder, 2020, reports that $17 billion was invested in AgTech technologies in 2018, which is 43% more than in 2017 [7]. There is a growing market for technology companies that collect and analyze big data in agriculture. Agri-FoodTech investment continues to break records, reaching a staggering $20 billion in 2019, up more than 6× from 2012. MarketsAndMarkets expects the AgTech analytics market to grow from $ 585 million in 2018 to $1236 million by 2023 [8].

The researchers [9] argue that the economic impact of the widespread shift to technologies based on big data (for example, precision farming) is estimated at $2.3 trillion. Of these, $250 billion can be obtained only through the introduction of artificial intelligence technologies and data analysis.

Agritech startup funding is growing since 2012 rapidly (Fig. 2).

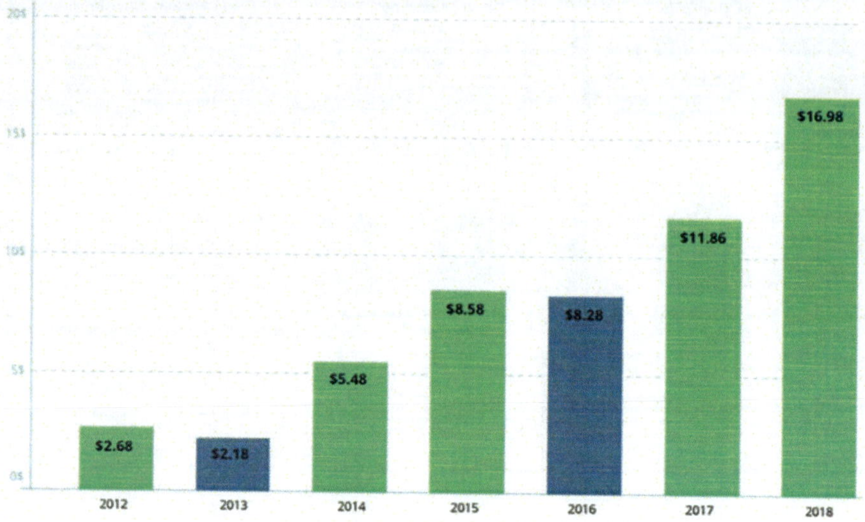

Fig. 2 Agritech startup financing [10]

4 Research Results and Discussion

4.1 Role of the Agricultural Sector in the Economic Development of Ukraine

The production function is an economic and statistical model of the production process in an economic system that expresses a stable regular quantitative relationship between the volume of resources and output. The macro-economic production function is a traditional key element of modern macro-economics. The production level in the general economy or corporate environment is described by the production function. The economists try to find a functional connection between economic "inputs" and "output," i.e., between independent and dependent variables.

For comparison, at the level of the country and agriculture as a separate sector of the economy, models of macroeconomic production function were built, which reflect the dependence of GDP on the working population and capital investment for 2002–2019 and production function for agriculture as the dependence of agricultural output on quantity. Workers involved in agricultural work and capital investment in agriculture. All data are collected from the State Statistics Service of Ukraine.

To construct the production function, the Cobb–Douglas function was chosen as the most popular among researchers

$$Y = a_0 L^{a1} K^{a2},$$

where Y is the volume of GDP, UAH million; L—the number of the working population aged 15–70 years, thousand people; K—capital investments, UAH million.

The result is

$$Y = e^{27.32} L^{-2.6} K^{0.99}.$$

The parameter a_2 that characterizes the quantitative effect of volume capital investment in the economy of Ukraine on its GDP is 0.99. The parameter a_1 that characterizes the quantitative impact of the number of people employed in the economy of Ukraine is -2.6, that is a negative number. That is, there is a straight dependence of the volume of GDP of Ukraine from the dynamics of capital investment, and negative dependence—on the number of employees in the national economy. This means that the increase in capital investment by 1% provides an increase in the nominal national GDP of Ukraine by about 0.99%, and increasing the number of employees can reduce it by 2.6%. Since the sum of the degrees of the model is less than one, Ukraine is characterized by an extensive type of economic growth.

For comparison of the economic development of Ukraine and the agricultural sector of economics development, let us build the same model for the agricultural sector:

$$Y = a_0 L^{a1} K^{a2},$$

where Y is the volume of agricultural production, UAH million; L—the number of the working population in agriculture, thousand people; K—capital investments in agriculture, UAH million.

The result is:

$$Y = e^{-8.78} L^{2.38} K^{0.17}.$$

And it is quite different for the agricultural sector than for the economics of Ukraine. Both parameters are positive, which means that there is a strait dependence between the volume of agricultural production, the number of the working population in agriculture, and capital investments in agriculture. What is interesting, in agriculture, the shortage of skilled workers is particularly noticeable.

By 2020, we have more than 70 agritech startups in Ukraine. Let us get known with the ideas and solutions of the most popular of them (Fig. 3).

Let us look at the ideas and realization of the most successful Ukrainian agritech startups.

Fig. 3 Agritech in Ukraine: market map [11]

4.2 Kray Technologies

Kray Technologies create drones to apply pesticides and fertilizers to the soil [12]. The main product of the company—Kray Protection Unmanned Aerial System—is a multicopter (its size is 2 × 2 m), equipped with a spraying system. The multicopter is controlled by artificial intelligence. The Kray protection unmanned aerial vehicle system allows you to quickly fertilize fields on an industrial scale. This technology can replace ground sprayers and agricultural aircraft.

The basis of the multicopter is a system of auto-navigation and computer vision, as well as an electrostatic spraying system with the use of rotary atomizers. Computer vision helps the drone recognize obstacles and avoid them. Rotary atomizers allow the introduction of pesticides into the soil in ultra-small amounts. This means that the system covers a large area of soil with small drops. The drone can develop a speed of up to 110 km/h.

Kray Technologies presented the world's first automatic drones to apply plant protection products and fertilizers to field crops with industrial productivity (up to 470 ha/day), which reduces the cost of application by 90% to $2/ha (US market). They allow the farmer to avoid crop losses associated with the application and apply new frequent feeding technologies that promise a yield increase of 20–40%.

The drone allows saving in three strategically important areas for farms. First, it is radically lower technological losses of the crop, for example, due to tracks from ground sprayers or in poorly or incompletely cultivated edges of the field, which is typical for the use of agricultural aircraft. Such losses can amount to 4–5% of the crop.

Secondly, it saves pesticides by using ultra-low-volume application technology in its pure form—high-quality monodisperse application. The essence is as follows: to increase efficiency, you need to ensure uniform coverage of plants with drops of drugs. Quality requirements increase with decreasing application rate, such application has the greatest activity at a very small droplet size. But for maximum efficiency and to prevent unproductive losses of chemicals, the droplet size should be in a very narrow range.

Third, it is 5–10 times lower operating and capital costs. Our drone costs an order of magnitude less than agricultural aircraft, the operator is one person, the drone does not require additional infrastructure, such as the runway.

4.3 AGRIEYE

AGRIEYE is a smart landfarming solution for farms. It is a platform for precision farming recommendations based on artificial intelligence and the analysis of large amounts of data from satellites and drones [13]. The service helps farmers get higher yields at lower costs by giving clear indications of the number of seeds, fertilizers, and water needed for optimal plant growth at each point in the field. The main task is to solve the problem of monitoring and management in general, to make them convenient, as well as to give farmers the free time they will spend on education, additional innovations, and development of their business. How it works:

- the farmer registers the account on the service, selects the boundaries of his fields on the map, fills in the necessary information, and pays for the services of analytics in accordance with the size of the fields;
- the farmer gets a fully automated drone and a camera, which are already included in the cost of the service. The drone is fully automated and does not require special skills. The farmer simply throws it into the air, and the drone flies along a pre-programmed route;
- after scanning, the drone is necessary to connect to the laptop and to unload the received data on service where the further analysis with a conclusion of recommendations will take place;
- uploading satellite imagery data for the last 5 years, the system assesses the trends in which a crop has germinated. Based on trends, the program does not go through the stage of determining the soil, but comes immediately to what is needed in order to surpass the result or repeat it. Nominally calculates the performance of each square, both in culture and in general.

Results: the farmer will receive a complete analysis: future crop forecasting, soil moisture, field temperature, automatic analysis by soil type, estimated time of crop germination, and much more. That is, the farmer will see the full history of his field, but without any recommendations. AGRIEYE is a part of ITC R&D.

4.4 PreAgri.com

PreAgri.com is an online service for collecting and analyzing all spatial data of agricultural enterprises, as well as for planning and monitoring the implementation of fieldwork [14]. Using hardware solutions for precision farming in combination with a geoinformation service, farmers will have a complete picture of the processes occurring in the field. Based on the collected data, he will be able to make effective decisions that will save on the number of imported products, execution time, and resources involved.

The solution allows to:

- Collect fieldwork data.
- Plan and process agrochemical analysis of fields.
- Analyze all types of spatial data.
- Get NDVI data.
- Collect telematic data.
- Calculate agricultural reports.

The PreAgri system is integrated with field monitors, weather stations, and telematics solutions, allowing automatic data acquisition and analysis.

PreAgri helps manage risk, increase yield potential, and make more informed decisions while minimizing the use of natural resources.

4.5 Tradomatic

Tradomatic is a platform for agricultural products trading [15]. It is a professional tool for trade in agricultural products. On the platform, you can sell and buy sunflower oil, sunflower, wheat, corn, soybeans, canola, and barley—in one click and at the best price.

The platform has three main functionalities that unite all market players.

Auctions are a tool that allows you to get the maximum price for a product and get the most market coverage. The system automatically notifies the responsible purchase managers of all first-class buyers of the selected product.

Analytics is a section where the user can create an analytics panel which allows you to monitor the selected product on the market. From the Tradomatic analytics panel, you can go from viewing prices to directly concluding a deal, which distinguishes Tradomatic from any bulletin board.

Indicators—functionality for large buyers who are actively purchasing on CPT/DAP bases. Placement of indications allows users to notify all market players of their prices or changes in less than 1 min.

The main large and medium market players (both sellers and buyers) are registered on the platform: ADM, Bunge, Cofco, Cargill, Louis Dreyfus Company, Glencore,

Kernel, Nibulon, Spike Trade, Olam, Agroprosperis, MHP, DPKZU, BayWa Group, Ambar Export, Soufflet Group, HermesTrading, and others.

4.6 BIOsens

BIOsens is a mobile laboratory for rapid diagnostics of product safety, which allows you to conduct the necessary examination in just 15 min [16]. It is a mobile device for the rapid diagnosis of the presence of mycotoxins in agricultural products. Its sensors are able to detect infections at an early stage in corn, wheat, peanuts, feed, etc., in the field. Prompt diagnosis will allow farmers to prevent the spread of the fungus in time, take the necessary measures, and save the crop. In addition, farmers will not spend resources, energy, and time on cultivating crops that are not suitable for further consumption. In this way, the efficiency of the economy increases, and its carbon footprint decreases. The device connects to a smartphone or tablet on which the appropriate program is installed. In the BIOsens device, there are biosensitive plates on which in the form of liquid the product for testing is put, and in 10 min, the ready result appears on the mobile phone which is connected via Bluetooth to the device. Today the biosensor is designed for consumers of the agricultural sector, and in the future, the team plans to develop in the mass user market. The BIOsens device will be applicable in agricultural business: for agronomic researches in the field, for express diagnostics on elevators, and for control of the safety of forages on farms.

4.7 New Ideas and Smart Technologies

Nowadays in Ukraine, there are several ideas and technologies that are just beginning to be implemented in the agricultural sector or are under development:

AERO: a system capable of detecting pests. This is done by digital shooting from the air in the ultraviolet range with a drone. Thus, it is possible to determine the main places of accumulation of pests and spot them.

CropCare: a huge database of various pest control products. The farmer needs to enter his crops and GPS-field data in it, and after that, the selection of optimal drugs will be carried out.

AgroGuard: a system of security poles equipped with infrared sensors. In case of violation of the boundaries of the site or in the event of any other event, the owner promptly receives a message on the phone.

DrT-Tech: Allows you to systematize all data collected from sensors and fields in one convenient structure. You can use the appropriate program on your smartphone to view them.

Fractal: designer of intelligent processes. With its help, it is possible to automate as much as possible working processes of the agrarian and industrial complex, for example, the account of working hours or regulation of certain mechanisms. The program is able to combine all devices into a single network (or multiple networks as needed).

AgromaxEffect: is a program that simulates the future harvest, based on the culture and characteristics of the site.

Koleso: an online platform where farmers can rent idle units. And other farmers who need such equipment—to rent it.

4.8 SWOT Analysis on Agritech Startup Ecosystem of Ukraine

The key task of the agricultural sector in modern conditions is to change, adapt to new conditions, and develop. The analog world is becoming increasingly fragile, and the digital world is becoming antifragile. Studies [17–20] show that the situation of COVID-19 not only posed threats to agriculture but also opened up new opportunities, in particular in digitalization and the introduction of innovative technologies. In our opinion, the agricultural sector of Ukraine needs the introduction of a significant number of agritech, which will accelerate the development of the agricultural sector and increase its efficiency. SWOT analysis of the agritech startup ecosystem is presented in Table 1.

5 Conclusion

The agritech startup ecosystem is characterized by more weaknesses and threats than strengths and opportunities. The Ukrainian agricultural sector has significant potential for increasing efficiency and development. To ensure the realization of this potential, it is necessary to do the following:

- increase government spending on agriculture, in particular on the development of innovative technologies;
- ensure access of agritech to financing;
- accelerate the process of digitalization of agriculture, in particular, to promote the spread of affordable ICT and introduce e-government;
- increase the interest of non-governmental organizations in the introduction of innovative technologies in agriculture;
- create favorable conditions for the development of agritech ecosystems.

Table 1 SWOT analysis on agritech startup ecosystem

Internal factors	
Strengths • The government starts to implement digital technology in the agriculture sector • 27 agricultural universities and colleges which offer extension-related qualifications • Effective solutions that promise an increase in productivity	Weakness • The high cost of innovative technologies • Problems with finding financing • The delayed effect of digital technologies implementation can decrease the effect of its realization • Unreadiness for change. The established habits of farmers and the lack of new skills and abilities • Bureaucratic hurdles for starting a business • Underdeveloped IT infrastructure • Lack of information on the effectiveness of agritech
External factors	
Opportunities • The agriculture sector is a big part of Ukrainian economics, which can demand many agritech projects • The application of agritech produces a lot of data that can be used for agriculture development • Agritech can significantly reduce the need for workers	Threats • If the agritech is not reliable enough and accessible to attackers, the danger may arise for agriculture, even to lead to large-scale catastrophes • The probability of different agritech results in different conditions • Access to data can increase inequalities, impede competition, and create economic barriers

Further development of the agritech startup ecosystem can be a key solution for the sustainable development of agriculture not only in Ukraine but also around the world. In our opinion, agriculture today needs startups, which use information technologies that address the issue of balancing economic, environmental, and social efficiency to increase agricultural productivity.

References

1. Basso B, Antle J (2020) Digital agriculture to design sustainable agricultural systems. Nat Sustain 3(4):254–256
2. Startus-insights. AgriTech Innovation Map Reveals Rising Technologies & Startups. https://www.startus-insights.com/innovators-guide/agritech-innovation-map-reveals-rising-technologies-startups/. Last accessed 2020/11/01
3. CGIAR Research Centers. Agri-Food Systems CGIAR Research Programs. https://www.cgiar.org/research/research-portfolio/. Last accessed 2020/11/01
4. Thornton P, Dinesh D, Cramer L, Loboguerrero AM, Campbell B (2018) Agriculture in a changing climate: keeping our cool in the face of the hothouse. Outlook Agric 47(4):283–290
5. United Nations. Promote Sustainable Development. https://www.un.org/en/sections/what-we-do/promote-sustainable-development/. Last accessed 2020/11/01

6. World Economic Forum. Realizing a New Vision for Agriculture: a roadmap for stakeholders. Switzerland: World Economic Forum.Agfunder. Why FoodTech & AgTech? (2010). https://agfunder.com/. Last accessed 2020/11/01
7. Agfunder. Why FoodTech & AgTech? https://agfunder.com. Last accessed 2020/11/01
8. Marketsandmarkets. https://www.marketsandmarkets.com. Last accessed 2020/11/01
9. Chakravorti B, Chaturvedi RS (2017) Digital planet 2017: how competitiveness and trust in digital economies vary across the world. The Fletcher School, Tufts University, 70
10. Agfundernews. AgriFood Tech in 2018. https://agfundernews.com/agrifood-tech-in-2018_17bn_funding_breakout-year.html. Last accessed 2020/11/01
11. Agritech.unit.city. Ukrainian agritech industry guide. http://agritech.unit.city/guide/f98ewf9fewfw/AgriTech_Industry_guide_en.pdf. Last accessed 2020/11/01
12. Kray.technology. Drone built to replace ag planes. https://kray.technology/. Last accessed 2020/11/01
13. Agrieye. http://agrieye.io/. Last accessed 2020/11/01
14. Preagri. https://preagri.com/index.htm. Last accessed 2020/11/01
15. Tradomatic. https://www.tradomatic.io/. Last accessed 2020/11/01
16. Sens.bio. https://sens.bio/. Last accessed 2020/11/01
17. Elleby C, Domínguez IP, Adenauer M, Genovese G (2020) Impacts of the COVID-19 pandemic on the global agricultural markets. Environ Resource Econ 76(4):1067–1079
18. Kaminskyi A, Nehrey M, Rizun N (2020) The impact of COVID-induced shock on the risk-return correspondence of agricultural ETFs. In: Kiv A (ed) Machine learning for prediction of emergent economy dynamics, Proceedings of the selected papers of the special edition of international conference on monitoring, modeling and management of emergent economy (M3E2-MLPEED 2020), Odessa, Ukraine, 13–18 July 2020, CEUR-WS.org, pp 204–218. http://ceur-ws.org/Vol-2713/paper17.pdf
19. Mokhnenko A, Babenko V, Naumov O, Perevozova I, Fedorchuk O (2020) Mathematical-logistic model of integrated production structure of food production. In: CEUR workshop proceedings, vol 2732, pp 446–454. http://ceur-ws.org/Vol-2732/20200446.pdf
20. Gray RS (2020) Agriculture, transportation, and the COVID-19 crisis. Can J Agric Econ/Revue canadienne d'agroeconomie

The Role of Digital Transformation in Enhancing Business Resilience with Pandemic of COVID-19

Yasmin Elgazzar, Rehab El-Shahawy, and Youssef Senousy

Abstract Many businesses have had trouble keeping pace with technological transition until the COVID-19 crisis exploded. Since the pandemic started, the challenge has intensified and has placed a growing emphasis on the future of work and life more than people thought before. Digital technology will be tomorrow an increasingly crucial aspect of business resilience with virtually every company having to rely on data analytics, digital tools and automation. In recent times, the reports focused on business resilience, but ignore its duality. It is not raised from a crisis, as it is widely known, but it also implies a new reality. The aim of this paper is to present a literature review the effect of digital transformation to the business resilience. Moreover, the basic and main phases of business resilience forced the companies to implement it through the pandemic. Also, the dimensions of business resilience that the businesses should focus on. Finally, the methodology of digital transformation causes a long-term business resilience and profitability.

Keywords Digital transformation · Business resilience · Covid-19 crisis · Business resilience phases · Business resilience dimensions

1 Introduction

The ability to cope with changes generates pressure to react proactively and improve flexibility and systemic agility [1]. The economic disruptions in the 2020s are driving market change and generating a mandate to be agile and resilient. Resilience has become the highest priority for executives today, and it will serve to be a crucial factor in winning the 2020s. As the business environment becomes more unpredictable and volatile, businesses with the ability to handle uncertainty and unpredictability deliberately—in a word, Resilience—are likely to succeed [2]. On the other hand, IBM's recent report indicated in times of uncertainty, consumer demands are increasing,

Y. Elgazzar (✉) · R. El-Shahawy
Canadian International College, Cairo, Egypt

Y. Senousy
National Organization for Social Insurance, Cairo, Egypt

D. A. Magdi et al. (eds.), *Digital Transformation Technology*, Lecture Notes in Networks and Systems 224, https://doi.org/10.1007/978-981-16-2275-5_20

with increased requirements for relevance, speed, continuity of service, and even entertainment in the event of a large house population. They also want to connect brands more explicitly and more empathically. Focus and drive virtual communication strategies for digital major networks is a strong goal for every company [3]. On the report of the latest studies conducted by BCG, approximately 60% of companies expect to report a decline of at least 10% in revenue and earnings in 2020. The suggested explanation is that customer preferences and contact points are increasingly changing, and a substantial change to remote work poses a significant risk to business IT infrastructure [2].

Since the digital economy has increased people's understanding of digital goods and services, causing businesses to look for new competitive advantages in the digital space, digital companies have grown into more than just online sales. Digital companies are using technology to generate new business value styles, consumer service, and organizational capacities supporting their core businesses. The concept encompasses both digital brands and traditional players that turn companies through digital technologies [4]. What's more, the global emergency is placing intense pressure on companies and their digital platforms. Companies in every sector face both opportunities and risks [3].

Accordingly, digital transformation has now become more important than ever before. In a recent BCG report, 75% of managers agreed that in the face of the COVID-19 crisis, digital transformation is increasingly urgent; 65% reported that their investment in digital transformation is anticipating to increase. Meanwhile, most organizations have addressed the immediate priorities of the resilience response process. Vulnerabilities in any field may affect the ability of the company to survive and succeed. Since accelerating the adoption of data and digital platforms is one of the critical dimensions for enabling resilience matters [2], this paper is going to present a literature review the effect of digital transformation to the business resilience. Moreover, the basic and main phases of business resilience forced the companies to implement it through the pandemic. Also, the dimensions of business resilience that the businesses should focus on. Finally, the methodology of digital transformation that causes a long-term business resilience and profitability.

This paper is composed of several sections: Sect. 2 presents the literature review and Sect. 3 the main phases of business resilience; Sect. 4 presents dimensions of business resilience. Section 5 presents the methodology of digital transformation that causes long-term business resilience and, finally, in Sect. 6 the conclusion and the future work.

2 Literature Review

There are important researches related to digital transformation and how it can support business resilience during the pandemic of COVID-19. Recently, digital technologies are seen as an essential determinant of the technological-induced shift in organizations and business models, known as digital transformation. Schmitz et al.

[5] introduced an example of a particular technology-stimulated transition as part of a digital transformation. The study concluded that agile design and implementation are crucial to a healthy digital transformation, taken into account, from the outset, technical, and organizational interdependencies [5]. The use and application of digital technologies also affect large parts of businesses and go further than their borders by influencing goods, business processes, distribution channels, and supply chains [6].

According to Liu et al. [7], digital transformations are "the integration of digital technologies into business processes" [7]. Also, Henriette et al. [8] defined digital transformation as "a business model is driven by the changes associated with the application of digital technology in all aspects of human society." [8]. Hess et al. [9] "a business model is driven by the changes associated the application of digital technology in all aspects of human society." [9]. Although digital transformation requires the use of digital platforms and technology to affect various aspects of the organization in order to generate value, it is also important to consider precisely how various types of digital technology combined with platforms have an effect on particular aspects of the organization. In addition, it is necessary to understand the characteristics of the value created by this transformation [10]. Since the tragic COVID-19 pandemic has affected the world dramatically and homework acts dominate the world, it has become clear how important the digital transformation of many business activities is to ensure the continuity of its operations and thus maintain the corporate competitive advantage.

Most businesses after the pandemic build strategies and plan to make the connection between them easier using digital transformation. Candello et al. [11], presented a paper is to open a discussion with symposium participants, to share perspectives on coping with adversities. Due to the ongoing COVID-19 pandemic and new developments sales professionals' social work practices [11]. Moreover, the efforts to innovate ways to obtain resilient business architecture is presented by Řepa and Svatoš [12]. The authors explained a method for designing a resilient and adaptive minimal business architecture such that business requirements can be clearly described, supported by the functionality of the applications assessed, and the lack of support resolved. They also provided a meta-model specifying how these principles work together in the areas under discussion [12].

As the continuity insurance and risk (CIR) magazine published "WHEN YOU'RE RESILIENT, YOU'RE IN BUSINESS". They described a great organization as one capable of overcoming and continuing hardship, sometimes stronger than ever. This is resiliency. They claimed that the faster corporate customers can shake off a disruption, the faster they are in a position to achieve their goals. The ability of a system to maintain its identity and to adapt its essential structure and function in the face of disturbances is normally considered resilience. In the literature of disaster management, the word organizational resilience is a key term. It points to organizations' ability to react to disruption and to exploit opportunities coming from the changing environment [13]. Centered on a comprehensive analysis of literature, McManus et al. [14] have described business resilience in a complex, dynamic and interconnected setting as the role of an organization being aware of its general situation, managing vulnerabilities and adaptive ability" [14]. According to Zohuri [15], "a system's ability to withstand disruptions and preserve its core role and structure" is best described [15]. A modern risk management strategy can be developed in collaboration with the organization to provide the top decision makers of the organization with constant information on the vulnerability of the earnings drivers [16]. Winnard and Julie [17] also argued that a conventional mix of strategic analysis and technical thinking approach to tackle together resilience and sustainability and provide decision makers with better integrated knowledge is highly recommended [17].

The literature review section presented researches related to four main points; the implementation of digital transformation, and the integration of it into the business processes by using the different technological platforms. Also, showed how the Covid-19 pandemic forced the companies to start in digital transformation. Moreover, mentioned the relation between business and the efforts to innovate ways to resilient business architecture. Finally, the business resilience continuity during the risks that may occur to the organizations.

In the following section we will discuss the main phases of business resilience during the COVID-19 pandemic.

3 Phases of Business Resilience

Resilience and smart systems, along with their engineering, are becoming a new setting for complex system efficiency and maintenance, decision making, and corporate survival. Business resilience, which means that market, industry, or financial change can respond rapidly and inexpensively, without any disturbance, is a major concern for businesses. The phases of business resilience are changed during the COVID-19 pandemic. There are five main phases for business resilience; Recovery, Response, Rethink, Survive, and Thrive [3, 4] (see Fig. 1).

Fig. 1 Phases of business resilience

3.1 *Response*

To prepare for an incident, most organizations have business continuity plans (BCPs) and incident management protocols in place. During the recovery period, the company will have to deal with fluctuations in demand, production, labor markets and the availability of credit. It is unlikely that conventional forecasts and organizational processes will be successful, so organizations must develop data-driven sense-and-response strategies to overcome uncertainty and must learn to respond quickly to rapidly evolving scenarios [18].

3.2 *Recovery*

During the COVID-19 crisis, companies would need to maintain frequent contact with suppliers so as to get a better understanding of their recovery strategies and, if possible, reach out to alternative suppliers. Remaining in regular touch will ensure that you can respond rapidly in order not to interfere with the experience of your customers.

3.3 *Rethink*

Rethink phase in the resilience building process, which includes planning the organization for the future. Permanent changes in customer and employee behavior are likely to arise from the current situation, and certain markets and business models may be irreversibly disrupted. However, as the essence of competition changes, resilient firms will have a clear chance to emerge stronger. For this purpose, by creating resilience now, businesses should begin laying the foundations of sustainable advantage [19].

3.4 *Survive*

The businesses that were hardest hit by the economic downturn do whatever they can to keep the lights up and just stay alive. These businesses are facing a substantial decline in sales with significant personnel and cost reductions, or even a complete loss of business that is threatening their life. Therefore, it must be a plan to survive economic crises like this.

Table 1 Businesses Phases Before and After COVID-19

	Business response	Business rethink	Business recovery	Business survive	Business thrive
Before COVID-19	Uncoordinated and sluggish response	Limited processes for performance and improvement	Inadequate model of recovery contributes to high costs	Poor potential for real-time monitoring of threats and vulnerabilities	In the future, vaccinations and therapies differ, yet instability and uncertainty are persisting
After COVID-19	Creation of pre-determined partnership and action plans	performance metrics are reviewed and used to make improvements	Company recovery as expected quickly	The ability to track and alert early warning 24/7	Sustainable competitive advantage investment Transform to create resilience and thrive throughout the future

3.5 Thrive

When we reach the post-COVID 19 periods, businesses that are quickly speeding up the adoption of AI technology will move from survival to a competitive economy in the future.

After the discussion of the phases, Table 1 illustrates the change of each phase in businesses before and after COVID-19 pandemic.

4 Business Resilience Dimensions

A business needs to integrate sustainability from a market-by-market approach to the most essential infrastructure in every area of the organization. In any region, vulnerabilities may influence the survival and growth of business. The emphasis of business leaders on six dimensions (see Fig. 2) is to (1) protect and increase customers; (2) build agile operations; (3) empower individuals; (4) accelerate data adoption and digital platforms; (5) increase cyber security and (6) improve finance [3]. Resilience derives from any level, but business leaders often need to look at it together so that their relationships can gain resilience.

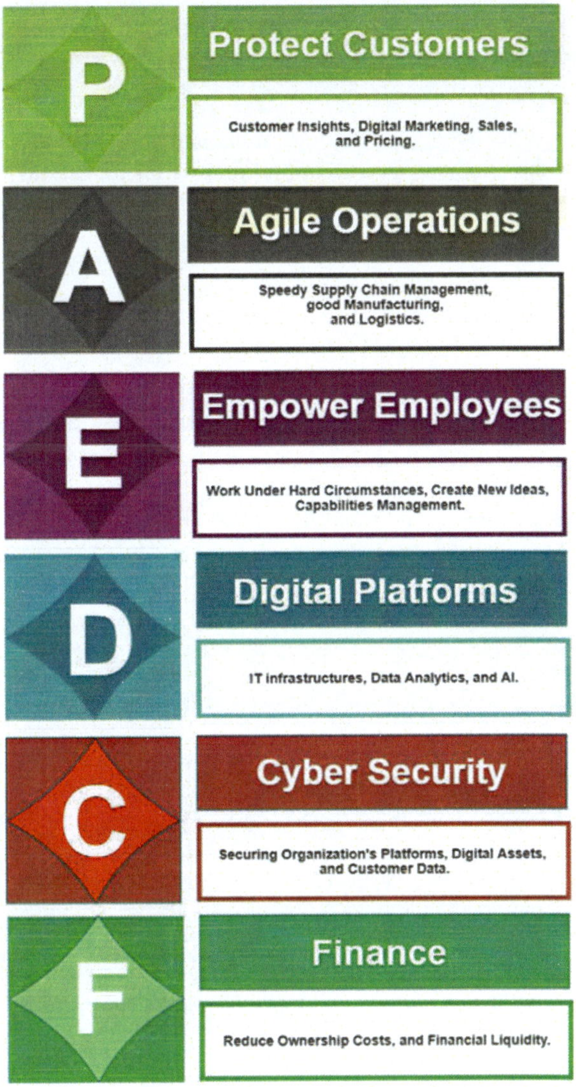

Fig. 2 The six dimensions of business resilience

In building resilience in all six dimensions, digital transformation is important. Resilient businesses may adopt an operational model with an integrated view of the human-technological relation this leads to superior financial performance, almost doubling profit growth and rise in the business value growth rate.

5 Digital Transformation and Improving Business Resilience

New technologies generate tremendous opportunities, but businesses that do not understand and pursue multidimensional transitions may be lacking. Here are the four digital transformation forms to boost long-term business resilience (see Fig. 3):

5.1 Robotic Process Automation (RPA)

A major emphasis of corporate operation has been on business processes. Data, analytics, APIs, machine learning and other innovations give businesses valuable new ways to reinvent processes around the corporation—with the goal of lowering costs, reducing cycle times, or increasing efficiency. Companies introducing technology such as robotic process automation (RPA) to streamline back office operations such as accounting and legal.

Fig. 3 Long-term digital business resilience

5.2 *Modern Business Models*

In several businesses, emerging technologies are used to change conventional business models during COVID-19. While the transition process focuses on the restricted fields of business, the transformation of the business model focuses on the basic elements of how industry value is generated. The dynamic and strategic nature of these opportunities involves input and leadership from strategies and/or companies, and is frequently initiated as separate projects while the conventional organization continues to function. Changing the critical value building blocks would generate substantial new prospects for growth for businesses that achieve transformed business models. This way should be pursued by more businesses.

5.3 *Ecosystems*

The basic technologies that enhance the resilience of the company as mentioned above such as artificial intelligence, machine learning, the Internet of Things (IOT), and enhanced reality can now be accessed not only from the conventional IT supply base, but also from the increasing start-up ecosystem, where most innovation occurs. Corporations that know how to efficiently reach and exploit this innovation, particularly from new sources, are benefiting from new development.

5.4 *Cultural Change*

For long-term digital change, organizational behaviors, processes and talent and abilities in the digital world need to be redefined. Digital recognition by top-class organizations includes agile workflows, a preference against testation and learning, decentralized decision making, and increased dependency on market ecosystems. They are also taking aggressive steps to make their business resilience improve.

6 Conclusion

Digital technology will be tomorrow an increasingly crucial aspect of business resilience with virtually every company having to rely on data analytics, digital tools and automation. In recent times, the reports focused on business resilience, but ignore its duality. It is not raised from a crisis, as it is widely known, but it also implies a new reality. The paper presented a literature review of the effect of digital transformation to the business resilience which consists of four main points; the implementation of digital transformation, and the integration of it into the business processes by using the different technological platforms. Also, showed how the Covid-19 pandemic forced the companies to start in digital transformation. Moreover, mentioned the relation between business and the efforts to innovate ways to resilient business architecture. Finally, the business resilience continuity during the risks that may occur to the organizations. Moreover, the basic and main phases of business resilience forced the companies to implement it through the pandemic such as Recovery, Response, Rethink, Survive, and Thrive. Also, the six dimensions of business resilience like protect and increase customers; build agile operations; empower individuals; accelerate data adoption and digital platforms; increase cyber security and improve finance that the businesses should focus. Finally, the methodology of digital transformation causes a long-term business resilience and profitability.

References

1. Paul S, Seetharaman P, Samarah I, Mykytyn P (2004) Impact of heterogeneity and collaborative conflict management style on the performance of synchronous global virtual teams. In: Information and management, pp 303–321
2. Close K, Grebe M, Andersen P, Khurana V, Franke M, Kalthof R (2020) The digital path to business resilience, BGC Report
3. IBM Homepage, Customers Need More Meaningful Digital Engagement in Times of Crisis. https://www.ibm.com/blogs/services/2020/04/01/customers-need-more-meaningful-digital-engagement-in-times-of-crisis/. Last accessed 1/04/2020
4. Casalino N, Żuchowski I, Labrinos N, Nieto A, Jiménez J (2020) Digital strategies and organizational performances of SMEs in the age of coronavirus: balancing digital transformation with an effective business resilience. SSRN Electro J
5. Schmitz M, Dietze C, Czarnecki C (2019) Enabling digital transformation through robotic process automation at Deutsche Telekom, pp 15–33
6. Matt C, Hess T, Benlian A (2015) Digital transformation strategies. In: Business and information systems engineering. Springer Fachmedien Wiesbaden, pp 339–343
7. Liu DY, Chen SW, Chou TC (2011) Resource fit in digital transformation: lessons learned from the CBC Bank global e-banking project. Manag Decis 49(10):1728–1742
8. Henriette E, Feki M, Boughzala I (2015) The shape of digital transformation: a systematic literature review. In: Conference: 9th mediterranean conference on information systems
9. Hess T, Benlian A, Matt C, Wiesböck F (2016) Options for formulating a digital transformation strategy. MIS Q Exec 15(2):123–139
10. Morakanyane R, Grace A, O'Reilly P (2017) Conceptualizing digital transformation in business organizations: a systematic review of literature. In: 30th Bled eConference: digital transformation—from connecting things to transforming our lives, BLED 2017, pp 427–444
11. Candello F, Ferreira J, Vasconcelos M (2020) Slow down, redesign, and accelerate: adaptations and resilience on business sales transformation amid COVID-19 crisis. New Future of Work 2020
12. Repa V, Svatoš O (2019) Model consistency as a tool for digital business architecture verification. Procedia Comput Sci 159:2144–2153
13. Orchiston C, Prayag G, Brown C (2016) Organizational resilience in the tourism sector. Ann Tourism Res 56:145–148
14. McManus S (2008) Organizational resilience in New Zealand. Univ. of Canterbury, Christchurch
15. Zohuri B (2018) A general approach to business resilience system (BRS), p 2

16. Starr R, Newfrock J, Delurey M (2003) Enterprise resilience: managing risk in the networked economy. Strategy and Business
17. Winnard J, Adcroft A, Lee J, Skipp D (2014) Surviving or flourishing? Integrating business resilience and sustainability. J Strategy Manage
18. Shek H, Leung K (2015) Business resilience & incident response—are you ready? HK Financial Services Business Continuity Management Forum. KPMG
19. Aberdeen Homepage, https://www.aberdeen.com/featured/blog-ecm-3-phases-business-resilience/. Last accessed 6/11/2020

Bankruptcy Prediction Using Artificial Intelligence Techniques: A Survey

Samar Aly, Marco Alfonse, and Abdel-Badeeh M. Salem

Abstract One of the most fundamental topics in finance and investment decision making is the early prediction of credit default events, which could lead to bankruptcy and hence severe consequences on both micro and macroeconomics. Precluding the financial institutions from being bankrupt requires an efficient method for the prediction of bankruptcy. Many researchers have used traditional statistical techniques to predict bankruptcy, while other researchers have been attempting to apply Artificial intelligence (AI) approaches for the same purpose. Most studies make a comparison between traditional statistical techniques and AI techniques. This survey presents several studies in bankruptcy prediction with the aim of comparing the results of the different methods. This comparison focuses on the performance metrices and the accuracy. Our survey will also show that a single methodology cannot be the general superior tool for every bankruptcy situation. Rather, each methodology will exhibit either strong or weak performance base on the situation and the used datasets.

Keywords Machine learning · Bankruptcy prediction · Classifiers · Banks failure · Financial crisis

1 Introduction

Banks and financial institutions have an essential effect on both micro and macroeconomics. So, credit default events could upset the stability of micro and macroeconomics expansion [1]. The indicator for both the development and strength of the global economy is the accurate prediction of the financial institutions credit default

S. Aly (✉) · M. Alfonse · A.-B. M. Salem
Computer Science Department, Faculty of Computer and Information Science, Ain Shams University, Cairo, Egypt
e-mail: samar.aly@cis.asu.edu.eg

M. Alfonse
e-mail: marco_alfonse@cis.asu.edu.eg

A.-B. M. Salem
e-mail: absalem@cis.asu.edu.eg

D. A. Magdi et al. (eds.), *Digital Transformation Technology*, Lecture Notes in Networks and Systems 224, https://doi.org/10.1007/978-981-16-2275-5_21

events. Managing credit risk becomes a top priority for financial institutions, lenders, fund manager, financial regulators, governments, and economic stakeholders [2]. As a matter of fact, financial institutions do not go bankrupt out of a sudden. Rather, there is an early detection for bankrupt events. Systemic risk analysis is like an alarm for other financial institutions when one institution goes bankrupt [3]. Consequently, the early indicators for credit default events are an important step to avoid business institutions from being bankruptcy [4]. Bankruptcy prediction is involved in both computer science and economics [5]. Many researchers used traditional statistical techniques to predict bankruptcy, and currently, they attempt to apply AI models [6]. AI techniques outperform traditional statistical techniques in bankruptcy prediction [7]. Concerning designing an efficient bankruptcy prediction model, the most essential data preprocessing phase is feature selection. Feature selection may influence forecasting performance [8]. However, the superior mixing of feature selection and classification mechanisms was recognized in very few researches [9]. Research in bankruptcy prediction has not reached an end yet, it is still an active and evolving point of interest, despite the already existing several models of bankruptcy prediction. These models were developed under specific conditions and a set of parameters [10], which makes them not suitable for all applications. This paper is organized as follows: Sect. 2 presents the used methodology in our survey and smart tools for bankruptcy prediction; Sect. 3 presents the experimental datasets feature; Sect. 4 presents the analysis of the performance metrics of machine leaning techniques for predicting bankruptcy; Sect. 5 presents the machine learning techniques that are applied for bankruptcy prediction, and Sect. 6 presents conclusions and future work for bankruptcy prediction models.

2 Intelligent Methodologies and Smart Tools for Bankruptcy Prediction

Regarding to the important impact of bankruptcy on micro and macroeconomics, many researches are trying to apply various methodologies in predicting bankruptcy. This review paper serves as a guide and good knowledge for selecting an appreciate model of bankruptcy perdition. It represents varied tools with different situations and datasets. Initially, researchers use some traditional methodologies [11–13] to predict bankruptcy such as Multiple Discriminant Analysis (MDA) and Logistic Regression (LR). After that many studies prove that AI technique especially machine learning methodologies such as Support Vector Machine (SVM), Decision Tree (DT), Random Forest (RF), Naïve Base (NB), Artificial Neural Network (ANN), Genetic Algorithm (GA), etc., have a high performance than traditional methodologies [14–16]. This section will represent varied methodologies in bankruptcy prediction.

Our research totally depends on Web of Science (WoS) database and ScienceDirect Web site to collect our review data. We tried to search in high indexed journals for precious scientific papers to have a good appreciate knowledge. We will review

19 scientific papers with different techniques for credit default events prediction between 2016 and 2020[17].

This section will represent varied smart tools in bankruptcy prediction. We focus on the most commonly used tools in our selected 19 studies. We will present two statistical tools (MDA and LR) and seven AI tools (SVM, DT, RF, NB, ANN, Ensemble method, and GA) especially machine learning techniques. Many studies prove that machine learning models are more efficient and accurate than these traditional methods [2, 6, 14, 15, 18] in bankruptcy prediction.

2.1 Multiple Discriminant Analysis (MDA)

MDA is considered an advanced discovery in the estimation of credit risk because Altman [12] makes a successful study to predict bankruptcy between industrial institutions. It depends on minimizing the variance between members of the same category and maximizing distance between members of the distinct groups [18]. It utilizes a group of financial ratios to differentiate between bankrupt and non-bankrupt institutions [14].

2.2 Logistic Regression (LR)

LR is a traditional statistical technique [19], proposed by Ohlson [13]. It is a very simple tool and has neutral distribution of error, so it is widely used in credit default prediction. It is used for binary classification problems (active and inactive) [20] In order to constitute the posterior probability of each class, LR applies linear methods of the independent ratios. To know which class the institution belongs to (active or inactive) by applying the ratio of log-likelihood, log-likelihood is a linear method. Also, LR can use dependent ratios to apply logistic method for evaluating the failure probabilities of institution by nonlinear most likelihood [10]. Figure 1 represents the LR diagram.

2.3 Support Vector Machine (SVM)

SVM model is the most supervised machine learning technique for classification and regression problems. It is commonly used in classification challenges. It is linearly separable classifier [21]. Its main idea is to find a hyperplane that separates the input features nonlinearly into various classes in high-dimension space by SVM kernel function [20]. Nowadays, SVM is widely used with small and high dimensional data size due to its high performance with them [1]. The SVM main objective is to detect hyperplane with the highest margin (maximum space among input data of each

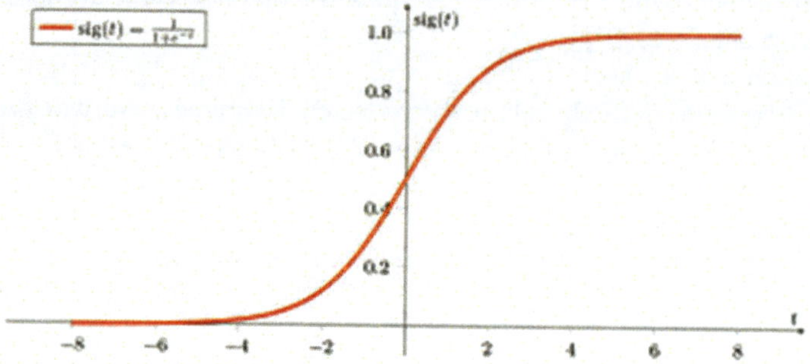

Fig. 1 Binary classification of Logistic regression model by log-likelihood method

Fig. 2 Binary separation of two classes by SVM model

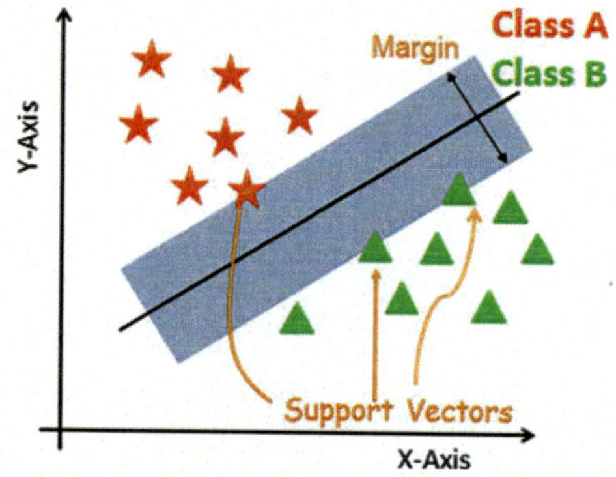

class) [18]. Support vectors are variables near to the hyperplane and are used to state the binary output (active and inactive) of the evaluated institutions. The remaining samples are disregarded and are not included in the decision making [14]. SVM has a high accuracy with less calculations. Many researchers apply SVM to predict banks failure for decision making. Figure 2 shows the binary separation between two classes using SVM model.

2.4 *Decision Tree (DT)*

DT is a very simple and popular supervised classifier model. It uses a tree diagram to classify problems, starting from the root node then followed by branches until leaves [22]. It can be used for classification and regression problems. It has adaptable efficient features, so it can classify varied complicated problems. The most repeatedly used DT models are Iterative Dichotomiser3 (ID3) and Classifier 4.5 (C4.5) [14], and each of them are presented by Quinlan, et al. [23]. Many researchers apply DT in credit default prediction because of its ability of decision making. They used DT

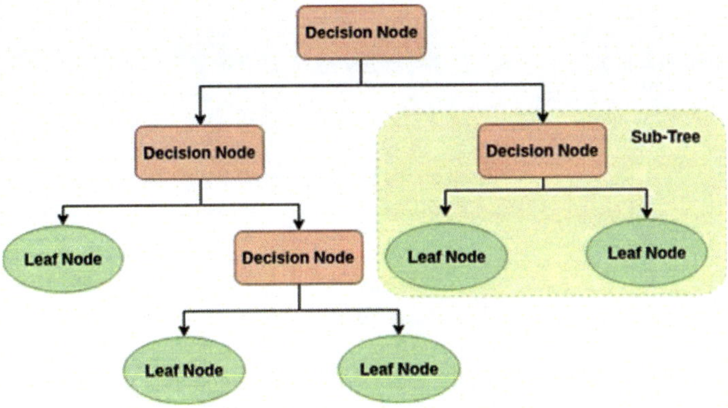

Fig. 3 Decision Tree flowchart diagram to determine the final output

technique to differentiate between two classes (bankrupt and non-bankrupt) [21]. The main parts inside DT are inner node indicates the feature, and each branch indicates a decision rule and leaves indicate the outputs. Figure 3 shows the flowchart diagram of DT that will be used in decision making problems.

2.5 Random Forest (RF)

The RF model is an ensemble tree-based machine learning technique, used for classification and regression problems. It is recognized as Classification and Regression Trees (CART) [18]. It applies Bagging technique on the training dataset to generate random varied sub-trees in the ensemble Decision Tree [24]. It creates a very large number of sub-trees to ensure a good performance by selection random variables [21]. So, the RF technique will generate different DTs every time using the same dataset [20]. Each tree within model makes a prediction about which class will collect more votes to be final class. RF has a high accuracy and robustness as it has varied DTs [18]. Figure 4 shows how ensemble tree of RF make a classification between trees.

2.6 Naïve Base (NB)

NB is an advanced machine learning technique, and it consists of many statistical techniques that depends on Bayes' theory [19]. It is used for classification issues. The NB computes the probability of all classes and the maximum likelihood that a specified feature belongs to ang class by using training data to compute occurrence of every feature in all classes. Bayes' theory suppose that every feature evaluation is not

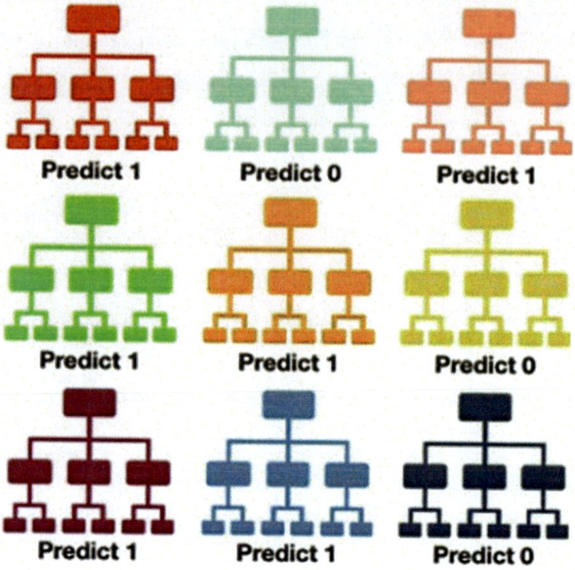

Fig. 4 Six 1 s and three 0 s, model's prediction is 1 (sub-trees) (A sample example for RF technique)

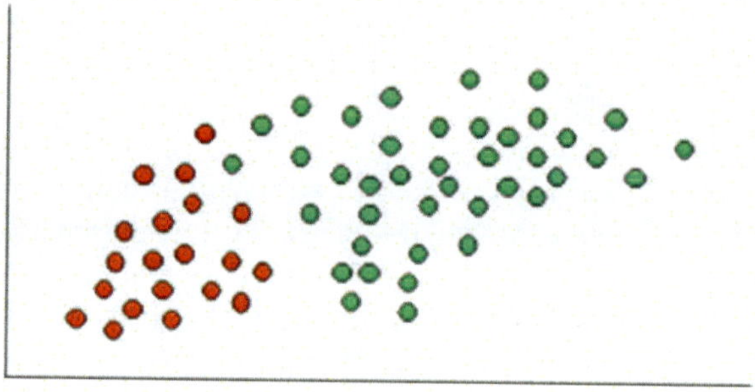

Fig. 5 Naïve Bayes classifier to differentiate between two classes

dependent on other features' properties. NB classifier is very beneficial in very big data sets. Figure 5 represents NB classification model to make a separation between features.

2.7 *Artificial Neural Network (ANN)*

The main idea of ANN is to simulate issues like neural system of the human brain. ANN are multilayer connected network. It contains an input layer, hidden layers, and an output layer which is responsible for decision making [20]. Each node is in any layer is linked to each node in the following layer. ANN is widely used in many fields such as banking, computer vision, speech recognition, and financial. Odom and Sharda [25] are the first researches who applied ANN on banks failure issues

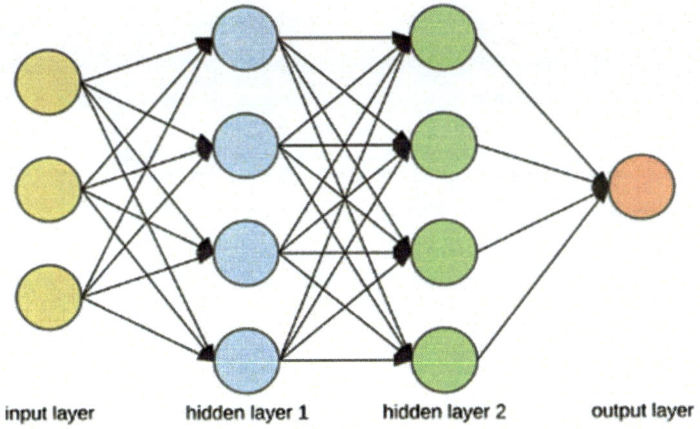

Fig. 6 Artificial neural networks layers using multilayer perceptron to detect the final decision

in 1990. ANN model shows a good performance in some credit default prediction issues. Its ability to manipulate nonlinear data gives it a good chance with complicated financial data to detect effective patterns [1]. It is a machine learning model that can simulate varied relationships. It is depending on the idea of learning by applying many examples by different algorithms as humans do with new cases. Figure 6 shows the multilayers connected NN to decide the final output.

2.8 *Ensemble Classifier*

Ensemble technique merges varied methods with different attitudes for a better performance [20]. As taking different opinions in decision-making problems is better than a single perspective [26]. Many researchers apply ensemble techniques to improve prediction quality [27], but it does not mean that ensemble classifiers are always better than the base single classifiers in all cases. They attempt to use ensemble techniques to overcome vulnerabilities in some cases when applying single classifiers. Bagging (BA) and Boosting (BO) are the most popular ensemble models. BA makes a randomly sampling using training set to produce new datasets. BO is a technique that unites varied weak classifiers to produce a powerful learner. Recently, some studies attempt to use a single-based AI model instead of varied AI models to construct a good ensemble model [21]. Hsieh and Hung make a study of improving ensemble methods by applying preprocessing phase using classification of class-wise concept [28]. Figure 7 shows how parallel ensemble classifier integrate different algorithms to work together.

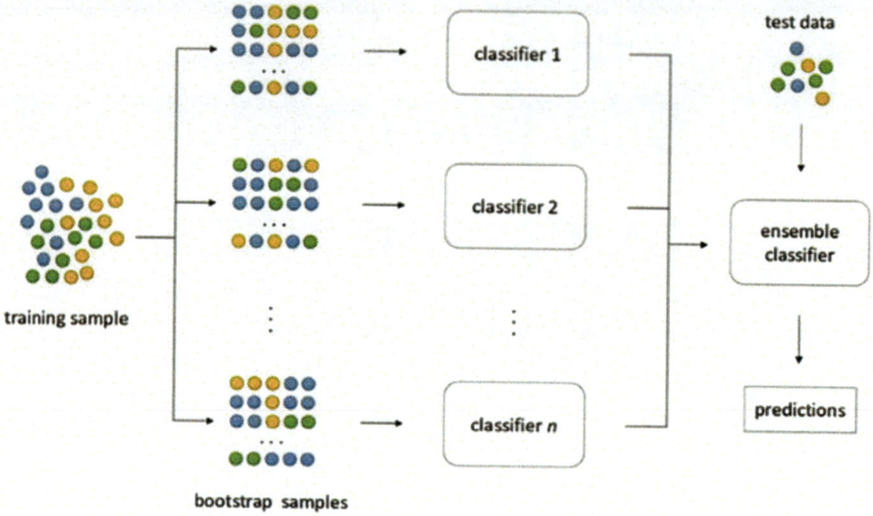

Fig. 7 A parallel ensemble classifier diagram extracted from different classifiers

2.9 *Genetic Algorithm (GA)*

GA is a very effective technique because of its ability to handle large and complicated search spaces and optimization issues. It is commonly used in classification problems [29]. Studies apply GA to predict bankruptcy in qualitative and quantitative problems [30]. GA has ability to produce decision rules by using data for classifying institutions (active or non-active). It simulates Darwin's selection theory to deal with nonlinear and non-convex issues. It can effectively determine the global minimum in a huge space [14].

3 The Experimental Datasets Feature

There are various datasets represented in this review study. This section presents the datasets description. We will explain the most popular datasets in bankruptcy prediction and their characteristics in Table 1. It will easily help you to select your suitable datasets regarding your requirements.

Table 1 Represents the used datasets in bankruptcy prediction and their features

Dataset	Features
Taiwan bankruptcy datasets [31]	Collected from Taiwan Economic journal [32] between 1999 and 2009. It contains 239 inactive and 239 active financial institutions and 95 features
Polish enterprises dataset [33]	Has 43,405 number of cases (2091 inactive firms and 41,314 active firms). It has 64 attributes and 10,503 instances. There are more than 10 thousand active companies during 2000 and 2012, and about 700 inactive companies during 2007 and 2013
Russian enterprises dataset [34]	Covered 912 perceptions. It has a balanced data (456 inactive institutions and 456 active institutions) and 55 attributes (macro/micro business environment factors and financial ratios)
The qualitative_bankruptcy dataset [35]	Has 7 features and 250 records. It has 107 instances are enrolled as bankruptcy and 143 instances are enrolled as non-bankruptcy. It includes 7 qualitative parameters from specialists to predict bankruptcy. Qualitative analysis dataset is obtained from various financial establishments and University of California, Irvine (UCI) [36] machine learning repository
Australian credit dataset [37]	Has 690 instances (400 for active and 290 inactive instances). It has 14 attributes. It is a public dataset as there is a good mix between small and big values
Quantitative ANALCAT bankruptcy dataset [38]	Has 50 instances. It contains 25 instances are enrolled as bankruptcy and 25 instances are enrolled as non-bankruptcy
DIANE [39]	Has 30 features. It contains 540,000 records from active and inactive institutions in France, during the period 2002 till 2006
North American institutions dataset [40]	Collected between 1985 and 2013 by Compustat. Barboza, et al. [18] used Salomon Center database [40] to collect information of bankrupt companies
Bankscope [41]	Has a balanced data. It covers about 30,000 financial institutions that is represented by Bureau van Dijk [39]. The financial records of the dataset have 31 financial attributes
Weislaw [42]	Includes 240 cases. It contains 112 instances which are enrolled as bankruptcy and 128 instances which are enrolled as non-bankruptcy. It has 30 attributes

(continued)

Table 1 (continued)

Dataset	Features
Korean Bankruptcy Dataset [43]	Has 120.048 inactive institutions and 307 active institutions; its balancing rate is small (about 0.0026). Every institution in Korean Bankruptcy Dataset [43] has a lot of financial ratios but only 19 attributes are used
German credit dataset [44]	Has 20 features, 700 and 300 for active and inactive cases, respectively. It is a widely used data in finance and offered by Prof. Hogmann [45]
Federal Deposit Insurance Corporation database [46]	Established in 1993 and provided by Banking Act. It is an independent agency contains deposits in U.S banks
Japanese stock market dataset [47]	Contains 102 bankruptcy institutions and 2062 non-bankruptcy institutions through four financial durations. It was collected from 2002 to 2016

4 Analysis of the Performance Metrics of Machine Leaning Techniques for Predicting Bankruptcy

This section will summary the outcomes and discuss our review analysis from the 19 selected scientific papers in predicting bankruptcy. We try to present the weakness and power of each tool according to the used datasets and their criteria to build a powerful bankruptcy prediction model. This survey shows a performance comparison between the discussed studies in tables and charts. Each researcher measures his performance of tool according to some criteria to show power of the applied tool. We will discuss six criteria that are presented in varied studies. Researchers determine their measure criteria according to the client requirements. We will discuss the six criteria which are accuracy (ACC), area under the curve (AUC), precision, recall, precision/recall plot (PRC), and type I error.

Accuracy: Accuracy considers one of the most fundamental performance measures that distinguish tools from one another. Many researchers focus on the accuracy measurement to build a good prediction model for bankruptcy prediction. We will discuss average of the accuracy from the 19 selected scientific papers that informed an accuracy measure in Fig. 8. The accuracy value was not reported for all papers, we will calculate the average accuracy from the papers that informed an accuracy value. Table 2 represents the number of studies that reported the accuracy as a performance measure. Figure 8 focuses mainly on the most used tools in building a bankruptcy prediction model. It helps the researchers to have a good knowledge in tools accuracy that helps them in their proposed prediction model for bankruptcy. It shows that Ant Colony Optimization (ACO) and GA techniques have the highest accuracy, and SVM shows an average accuracy; however, DT shows the lowest accuracy value.

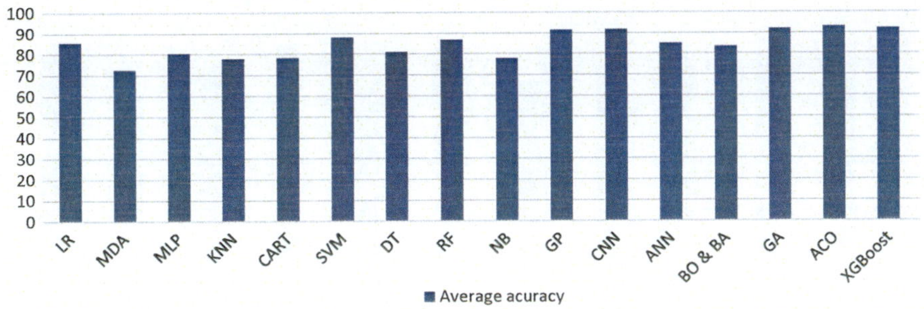

Fig. 8 Results of the average accuracy calculated from the studies that informed the accuracy

Table 2 Summary number of the presented studies for bankruptcy prediction in terms of AUC, ACC, Type I error, precision, recall, and PRC

Technique	Number of authors that applied technique	Number of authors who informed AUC	Number of authors who gave ACC	Number of authors who informed a Type I error	Number of authors who informed precision and recall	Number of authors who informed PRC
LR	6	4	3	-	1	1
MDA	1	–	1	–	–	–
MLP	1	–	1	1	–	–
LDA	2	2	–	–	–	1
KNN	2	1	1	1	–	1
CART	1	–	1	1	–	–
SVM	7	3	5	2	1	1
DT	2	1	1	–	–	–
RF	5	4	3	2	-	–
NB	2	1	1	1	–	–
GP	1	-	1	-	1	-
CNN	2	1	1	–	–	–
ANN	4	3	1	–	–	1
BA & BO	4	2	2	3	–	–
GA	2	–	2	–	1	–
ACO	2	–	2	1	-	–
XGBoost	3	3	1	–	–	–

Gaussian Processes (GP); Linear Discriminate Analysis (LDA); k-Nearest Neighbors (k-NNs)

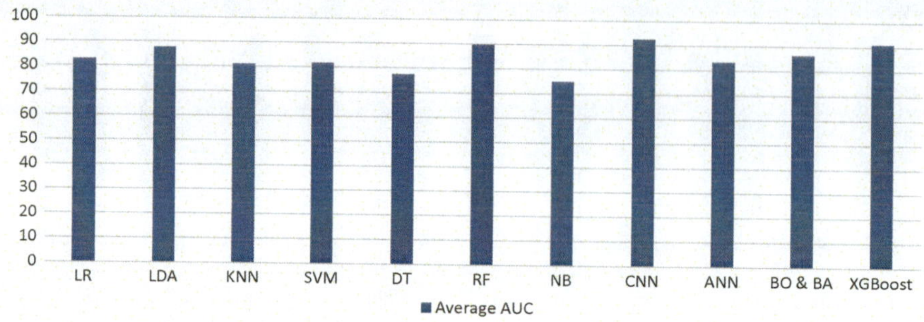

Fig. 9 Results of the average AUC calculated from the studies that informed the AUC

Area under the curve: AUC is the second main performance criterion. It is used in problems of classification with machine learning techniques. Most researchers are interested in the AUC measurement in predicting the performance of bankruptcy model such as [15, 16, 18, 24, 48–50]. This performance measurement informs us how well the model can differentiate between active and inactive firms. Figure 9 presents the average AUC value from the 19 selected studies. It helps researchers to select a good tool with high AUC ratio to distinguish between bankrupt and non-bankrupt institutions. We use Table 2 to collect information about studies that reported AUC value. Figure 9 shows that Convolutional Neural Network (CNN) and Extreme Gradient Boosting (XGBoost) techniques have a high AUC ratio in predicting bankruptcy problem. However, DT and NB techniques have a low AUC ratio in bankruptcy prediction.

Type I error: Error cost criterion is an important measure for the error rate of accuracy. It includes type I error measurement. Type I error are used in the problems of misclassification when bankrupt institution considered a healthy institution. Table 2 shows that most studies do not report type I error measurement. Researchers that report type I error measurement such as [8, 18, 51–54]. We encourage researchers to focus on error cost criterion because it is an important criterion. Figure 10 shows that ACO technique has the lowest misclassification rate in predicting bankruptcy problem, while CART technique has a high type I error rate.

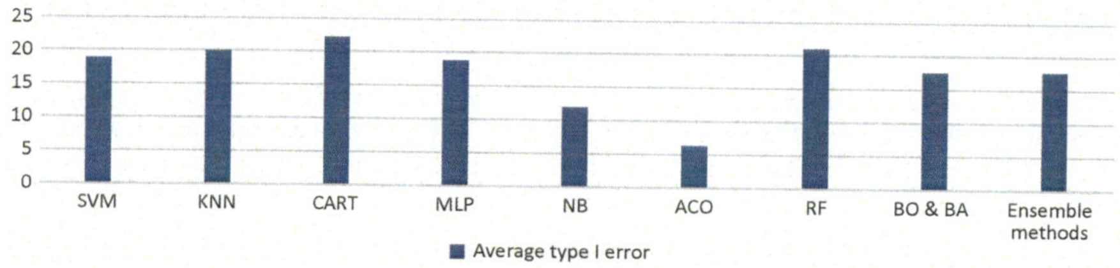

Fig. 10 Results of the average type I error calculated from the studies that informed the type I error

Precision: Precision is a performance matrix. Its meaning is close to the accuracy meaning. Accuracy means how the result is close to the truth value, and precision means how close the number of valid measurements is to each other by the prediction model used [55]. Its ratio is determined by the number of valid measurements according to the total valid values. Its value is between 0 and 1. It has the potential not to classify a health institution as bankrupt institution. Table 2 shows that there are few studies reporting a precision measurement. Only two studies from our reviewed studies report precision value [17, 55].

Recall: Recall is also a performance matrix. It refers to the ability of the predicting model to find a valid measurement. Its meaning is close to the precision. Precision determines how many valid measurements from the returned results. Recall determines the total number of valid measurements that should be returned by the applied prediction model [55]. Table 2 shows that there are only two studies among our selected 19 studies reporting a recall measurement [17, 55].

Precision/Recall plot: PRC is a performance criterion. It is used less often than other criteria. It determines the ratio of valid positive prediction values between all valid values by the prediction model. It can be a good indicator for further more classifications [15] problems. Table 2 shows that there are only one study representing PRC measurement in 2018.

We summary the selected 19 scientific papers in Table 2. Table 2 presents the number of the researchers that used predicating models in bankruptcy problem in details. It presents all applied models in predicting bankruptcy problem. It also determines our selected six criteria (ACC, AUC, precision, recall, Precision, PRC, and type I error). It shows the number of researchers that reporting every criterion.

5 A Study of Machine Learning Models for Bankruptcy Prediction

The early popular predictive bankruptcy models that help organizations to make a good decision in investigations are by William H. Beaver [11] in 1966, Altman [12] in 1968, and Ohlson 1980 [13]. This paper will represent different mechanisms from several studies during (2016–2020) with different techniques for credit default events prediction [17].

Liang, et al. [51] evaluate the bankruptcy prediction performance by adding five varied Financial Ratios (FRs) and five varied Corporate Governance Indicators (CGIs) as input features. They collected data from Taiwan [31], every institution has 95 CGIs and 95 FRs. The experimental results in this study prove that CGIs and FRs are very significant attributes to predict credit default events. But, the efficiency of CGIs and FRs has not been mentioned in many studies.

Zieba, et al. [48] present a new study to predict bankruptcy. This study learns an ensemble of DTs by applying XGBoost technique. The authors introduce artificial features for the representation of data. The artificial features are a group of econometric measures with arithmetic operations. The authors collected data from Polish institutions [33] using UCI repository [36]. The results show that XGBoost outperforms other techniques and has a slightly better performance with artificial features. The performance is increased when the proposed learner used artificial features in the ensemble framework.

Zelenkov et al. [17] apply a two-step classification method (TSCM) with GA to predict bankruptcy. Their technique has two main functions: determination of the related factors and adjusting the designed model to the concerned application. The first step of the designed model is to train classifiers of different models, while the second step is to combine the trained classifiers into the selected ensembles. The GAs are used in both steps; features selection then the determination of weights in the selected ensemble. A balanced dataset from Russian corporations [34] is used for testing the presented technique. The test results obtained from the examined model have shown that the best prediction accuracy is 0.934. Furthermore, the utilized method has the most precise precision–recall rate among some evaluated models. Its recall is 0.953 for bankrupts, and precision is 0.910 for non-bankrupts.

Uthayakumar et al. [52] propose an ACO based on anti-miner technique for quantitative and qualitative bankruptcy examination. Many researchers used quantitative methods to predict bankruptcy and few of them apply qualitative methods for a better performance. Qualitative method attempts to use unstructured data. It depends on the knowledge of the experts, comments, surveys, etc., so it can answer questions like what, how, and when. Quantitative method typically uses only numbers. Three diverse datasets (qualitative_bankruptcy [35], quantitative Australian credits bankruptcy [37], and quantitative ANALCAT bankruptcy [38]) have been utilized to show a reliable result. This research uses UCI machine learning repository [36] to collect dataset. The proposed model has shown that the anti-miner algorithm has a better performance than other traditional techniques such as RF, LR, radial basis function (RBF), and multilayer perceptron (MLP).

Antunes et al. [55] make a comparison in bankruptcy prediction between a probabilistic technique GP, LR, and SVM. The GP technique can efficiently enhance the financial disaster prediction with high performance than other techniques by utilizing actual bankruptcy information of French enterprises. In addition to obtain more trustworthy and accepted conclusion, researchers extend their research with varied datasets of credit default from UCI [36]. The main dataset utilized in this research is DIANE [39]. The GP technique presents a better performance than SVM and LR.

Barboza et al. [18] apply machine learning techniques (RF, SVMs, BO, and BA) to predict credit default events and compare their accuracy with LR, Neural Networks (NNs), and LDA. They collected data from North American institutions and Salomon Center database [40]. The main objective of this research is the enhancement in accuracy utilizing machine learning mechanisms, by adding variables of Altman's Z-score [12]. They utilize novel predictive variables like the change in price-to-book, operating margin, growth measures, and change in return-on-equity by Carton and Hofer [56]. Machine learning mechanisms have about 10% better accuracy than traditional techniques. The RF, BO, and BA techniques show a better performance, and their accuracy are upgraded by adding predictive variables. The authors observe that SVM technique does not have a better accuracy than other machine learning techniques.

Le and Viviani [15] make a comparative study between two traditional statistical techniques (LR and MDA) and three machine learning techniques (SVMs, ANNs, and k-NNs) to predict banks failure. They utilized sample of 3000 United States (U.S.) banks (1562 non-bankrupts and 1438 bankrupts). For every bank throughout a 6-years duration (2008–2014), data were gathered randomly from database of Bankscope [41] (five years before banks being inactive and one year when be inactive). The obtained performance revealed that the most accurate approaches are K-NNs and ANN.

Uthayakumar et al. [57] offer a cluster-based classification paradigm including upgraded K-means clustering and a Fitness-Scaling Chaotic Genetic Ant Colony Algorithm (FSCGACA) based on classification technique. The upgraded K-means technique is used to remove the incorrectly clustered data. After that, a rule-based paradigm is chosen to be suitable for the presented dataset. Eventually, FSCGACA is used to obtain the best parameters of the rule-based paradigm. This research includes three different datasets; qualitative_bankruptcy dataset [35], dataset of Polish [33], and dataset of Weislaw [42]. The outcomes of detailed analysis ensured that the proposed prediction model is the most reliable to different classification paradigm with various measures and with varied dataset.

Jardin [53] attempts to use a technique that depends on a wide variety of situations for credit default events. He presents a technique based on temporal patterns for Kohonen maps to describe the financial situation of some financial institutions. He uses a database administered by Bureau van Dijk [39] to collect samples. Over 9 years, from 2006 to 2014, data were collected. The outcomes evidence that presented models have a better performance than traditional techniques.

Le et al. [49] present a Cluster-based Boosting (CBoost) algorithm for an efficient bankruptcy prediction. They use a robust framework that apply CBoost algorithm and Instance Hardness Threshold (IHT). The presented framework used with imbalance datasets, whereas the number of stable institutions is greater than the number of bankrupts. It restructures the imbalanced dataset applying IHT based on understructuring technique to eliminate the noise samples in the main class that have a high IHT value. The main class will be grouped into clusters, then the weight of each instance will be initialized by calculating the distance between each instance and its nearest centroid. The proposed algorithm tries to find a powerful classifier by combining weak classifiers from different iterations. The presented framework uses Korean Bankruptcy Dataset [43]. The proposed model outperforms various techniques to predict bankruptcy issue with the unbalanced data like (oversampling-based approach using SMOTEENN and GMBoost algorithm and the clustering-based under-sampling approach).

Gogas et al. [58] propose a predicting model for credit default events relies on machine learning techniques. The presented methodology distinguishes between the active banks and bankrupts by a linear decision margin. The data of inactive U.S.

banks were collected between 2007 and 2013. The applied sample of 1433 contains 481 bankrupts. They applied two-step feature selection method with different configurations with SVM to the training and testing phases. The proposed model achieves 99.2% accuracy to predict credit default events.

Alaka et al. [14] write a review including the performance of eight credit default prediction models. They use the Engineering Village (EV), WoS, and Business Source Complete (BSC) databases to collect information. These models contain two statistical techniques: LR and MDA; and six AI techniques: DTs, SVMs, GAs, ANNs, Case-Based Reasoning (CBR), and Rough Sets (RSs). The presented models rely on 13 key criteria. The criteria are deterministic output, variable selection process, data size, transparency of result, accuracy, variable kinds, and extra. It was observed that there is no single model has a better performance than other models based on 13 criteria. Providing a graphical and tabular framework is the best way for selecting the suitable model with varied situations. This review provides a deep comprehension of the characteristic of the models to predict bankruptcy.

Liang et al. [54] present a novel approach of ensemble classifier to minimize the cost of misclassification through model structuring of credit risk analysis. They use Unanimous Voting (UV) technique with combination of varied classifier to produce prediction output. The used datasets are German [44], Australian [37], Chinese [59], and Taiwanese datasets [31] The German and Australian datasets are popular datasets utilized in the literature. The Chinese and Taiwanese datasets have been accumulated from the Economic journal of Taiwan [31]. The UV ensemble technique proves a better performance than single and ensemble classifiers. It has a better accuracy, reduces type I/II errors, and gives the smallest cost of misclassification.

Carmona et al. [24] apply the XGBoost to predict financial institution failure in the U.S. banking section. Key variables have been recognized to expect and prevent financial institution defaults. The authors utilized data during the period 2001–2015 made up of a yearly collection of 30 financial ratios for 156 U.S. national merchant banks (78 inactive banks and 78 active banks). They collected data from the Federal Deposit Insurance Corporation database [46]. XGBoost shows a high predictive performance than RF and LR techniques.

Hosaka [60] applies the CNNs to predict bankruptcy of financial institutions that is often considered as a two-class classification issue. In different fields, CNNs are widely used for supervised machine learning (classification). CNNs are more convenient for images data than numerical input data. He utilizes Japanese stock market [47] to collect data. The presented technique extracts from the input statements a collection of financial attributes and appeared as a grayscale picture. These images are used for training and testing the CNN. The author uses 7520 pictures for bankrupt and non-bankrupt companies in addition to GoogLeNet dataset for training the CNNs. The trained CNN for bankruptcy prediction obtained a better performance than techniques like SVMs, DTs, MLP, Altman's Z-score [61] LDA, or Adaptive Boost (AdaBoost).

Lin et al. [8] propose two types of feature selection techniques comprise of wrapper-based and filter-based techniques, and two types of classification techniques comprise of statistical and machine learning mechanisms to improve the prediction models. They also use two ensemble classifiers: BA and BO for comparison. The results based on three varied datasets (German credit [44], Australian credit [37], and Taiwan bankruptcy datasets [31]) with various input ratios show that feature selection of wrapper-based with GA is superior to filter-based. The German [44] and Australian credit [37] datasets are the most extensively investigated datasets in credit default prediction. Moreover, the results also show that integrating SVM and NB classifiers with GA without BA and BO using provided datasets has the smallest error rates for prediction.

Uthayakumar et al. [62] apply an ACO based on the prediction of the financial problems technique to predict financial crisis. They rely on feature selection phase to upgrade the performance of classification. The proposed ACO consists of ACO based Feature Selection (ACO-FS) technique and ACO based Data Classification (ACO-DC) technique. ACO-FS technique uses five well-known datasets like qualitative [35] and quantitative bankruptcy datasets. The authors show that ACO-FS model is better than Particle Swarm Optimization technique, GA, and Gray Wolf Optimization technique-based feature selection with 20 number of iterations. They show that ACO-DC is better than state-of-the-art techniques to predict bankruptcy. They prove that ensemble ACO model for financial crisis is very robust and efficient than some AI techniques.

Gregova et al. [16] make a comparison between three various models (LR, RF, and NN models) to predict credit financial crisis. Their research is focused on industrial institutions especially in Slovak republic. They collect data from [63]. They analyze financial statements during (2016–2018). Their experimental results show that the three proposed techniques have a convergent performance with high accuracy in bankruptcy prediction. NN model produces a better output with respect to all characteristics of performance. In Slovak, the development of financial crisis prediction methodologies for industrial institutions was not evolved until now. So, this research will participate in the development of prediction techniques in Slovak republic.

Dastile et al. [50] Make a survey to review the strength and guidance for applying statistical and machine learning paradigms in the financial crisis problem. Their survey focuses on 74 studies from different sources like journal articles or conference papers from 2010 to 2018. This survey selects the most used datasets (German [44] and Australian [37] datasets) in credit default prediction. The results show the average performance of different algorithms (BO, ANN, BA, DT, CNN, LDA, LR, NB, RF, SVM, and XGBoost) to predict credit default problems. The experimental results of authors observe that ensemble models have a better performance than single models in the prediction.

Table 3 presents a review of the presented studies in terms of the preprocessing phase, the proposed technique, the used dataset, and the achieved performance.

The proposed studies show that SVM, LR, BO, BA, and NNs techniques are widely used in credit default events prediction. These techniques have different performance corresponding to the used dataset, the surrounding effects, and if they are followed by preprocessing step. Researchers apply preprocessing phase to improve performance. Some researchers apply feature selection process as a preprocessing step. Studies reveal that ensemble-based model for bankruptcy predication has a

Table 3 Analysis of the presented models for bankruptcy prediction

Authors	Preprocessing	Techniques	Datasets	Performance (%)					
				ACC	AUC	Precision	Recall	PRC	Type I error
Liang et al. [51]	Feature selection	SVM KNN CART MLP NB	Taiwan dataset	81.3 77.9 78.2 80.5 77.9	–	–	–	–	16.7 20.1 22.2 18.8 12
Zieba et al. [48]	–	XGBoost using artificial features	Polish institutions	–	μ (10 cross validation folds) = 95.5 δ (10 cross validation folds) = 19	–	–	–	–
Zelenkov et al. [17]	Feature selection	GA	Russian companies	95.1	–	0.932	0.965	–	–
Uthayakumar et al. [52]	–	ACO-based Ant-miner	qualitative_bankruptcy quantitative Australian credits bankruptcy quantitative ANALCAT bankruptcy	100 86.37 100	–	–	–	–	0 6.37 0
Antunes et al. [55]	D-dimensional Random Sample	GP SVM LR	DIANE Database	97.71 96.31 96.81	–	98.03 95.95 95.90	90.38 85.15 87.80	–	–

(continued)

Table 3 (continued)

Authors	Preprocessing	Techniques	Datasets	Performance (%)					
				ACC	AUC	Precision	Recall	PRC	Type I error
Barboza et al. [18]	–	SVM RF BO BA	Salomon Center database and Compustat	79.77 87.06 86.65 85.67	85.17 92.92 92.97 92.48	–	–	–	21.05 16.54 18.80 17.29
Le et al. [15]	–	LDA LR ANN SVM K-NNs	Bankscope	–	77.6 79.6 81.9 71.5 81.14	–	–	75.877.3 80.3 65.5 78.3	–
Uthayakumar et al. [57]	–	K-means FSCGACA	Qualitative bankruptcy Weislaw Polish bankruptcy	97.93 94.72 97.55	–	–	–	–	–
Jardin [53]	–	BA BO RF HM FM	Bureau Van Dijk, Diane	80.92 80.21 80.95 80.55 83.22	–	–	–	–	25.48 26.12 25.29 25.87 23.08
Le et al. [49]	–	CBoost	KBD	–	86.8	–	–	–	–

(continued)

Table 3 (continued)

Authors	Preprocessing	Techniques	Datasets	Performance (%)					
				ACC	AUC	Precision	Recall	PRC	Type I error
Gogas et al. [58]	Feature selection	SVM	U.S. banks	99.22	–	–	–	–	–
Alaka et al. [14]	–	MDA LR NN SVM DT GA	WOS, BSC, and EV databases	72.5 76 85 84 81 89	–	–	–	–	–
Liang et al. [54]	–	UV ensemble model	German Australian Chinese Taiwanese	75.50 86.37 91.27 83.18	–	–	–	–	36 9.61 4.66 7.27
Carmona et al. [24]	–	XGBoost LR RF	Federal Deposit Insurance Corporation	94.74 84.21 92.11	97.8 84 93	–	–	–	–
Hosaka et al. [60]	–	CNNs	Japanese stock exchange	91.6	–	–	–	–	–

(continued)

Table 3 (continued)

Authors	Preprocessing	Techniques	Datasets	Performance (%)					
				ACC	AUC	Precision	Recall	PRC	Type I error
Lin et al. [8]	Feature selection	BA and BO	Australian credit German credit Taiwan	–	–	–	–	–	5.24 8.01 0
Uthayakumar et al. [62]	ACO-FS	ACO-FS and ACO-DC	Qualitative bankruptcy ANALCAT bankruptcy Australian credit German credit Polish bankruptcy	100 94 89.28 76.6 97.69	–	–	–	–	–
Gregova et al. [16]	–	LR RF NN	Register of Financial Statements	–	87.7 88.2 88.6	–	–	–	–
Dastile et al. [50]	–	BO ANN BA DT CNN LDA LR NB RF SVM XGBoost	German credit	–	80 70–80 80 70 90 70–80 60–70 80–90 70–80 70–80	–	–	–	–

(continued)

Table 3 (continued)

Authors	Preprocessing	Techniques	Datasets	Performance (%)					
				ACC	AUC	Precision	Recall	PRC	Type I error
Dastile, et al., 2020 [50]		BO	Australian credit	–	80–90	–	–	–	–
		ANN			80–90				
		BA			80–90				
		DT			80–85				
		CNN			95–100				
		LDA			80				
		LR			85–90				
		NB			80				
		RF			80–90				
		SVM			90				
		XGBoost							

FM failure pattern-based models; *HM* hybrid ensemble-based models; μ mean; δ standard deviation better performance than utilization a single model of the base ensemble model. However, it does not mean that ensemble model is always better than any selected classifier. Many researchers used the UCI machine learning repository to collect their data. The UCI [36] repository contains different financial datasets such as qualitative [35], Australian [37], Polish institutions [33], and German [44].

6 Conclusion and Future Work

Constructing an efficient model for predicting credit default events is a critical issue for financial institutions to take right loan choices. This survey shows most recent studies on credit default events. Some authors compare between two approaches: traditional statistical approach and AI approach in their studies. Most of authors proved that AI models outperform traditional statistical models. Our study attempts to present various techniques with diverse datasets to overcome the drawbacks of the existing techniques for bankruptcy prediction. Even though the many effective studies on credit default detection, rarely probabilistic techniques were executed. Moreover, there is no single methodology which is generally superior than any other tools regarding to all recognized criteria. Each methodology has its strengths and weakness that make it not suitable to every situation. The improvement of the prediction accuracy is the major research issue. It was found that the most important step to improve performance is the preprocessing phase. Within the last 20 years, ensemble models that have been used for bankruptcy prediction are widely studied. The efficiency of an ensemble model for credit default events of an institution is related to the ability of the model to take different credit default events. In future, it is aimed to present a powerful methodology that is suitable for varied situation with different datasets to predict bankruptcy problem in an efficient way.

References

1. Chen N, Ribeiro B, Chen A (2016) Financial credit risk assessment: a recent review. Artif Intell Rev 45(1):1–23. https://doi.org/10.1007/s10462-015-9434-x
2. Le T et al (2019) A fast and accurate approach for bankruptcy forecasting using squared logistics loss with GPU-based extreme gradient boosting. Inf Sci 494:294–310. https://doi.org/10.1016/j.ins.2019.04.060
3. Shi Y, Li X (2019) An overview of bankruptcy prediction models for corporate firms: a systematic literature review. Intang Capit 15(2):114–127. https://doi.org/10.3926/ic.1354
4. Climent F, Momparler A, Carmona P (2019) Anticipating bank distress in the Eurozone: an extreme gradient boosting approach. J Bus Res 101:885–896. https://doi.org/10.1016/j.jbusres.2018.11.015
5. Chou C-H, Hsieh S-C, Qiu C-J (2017) Hybrid genetic algorithm and fuzzy clustering for bankruptcy prediction. Appl Soft Comput 56:298–316. https://doi.org/10.1016/j.asoc.2017.03.014

6. Pal R et al (2016) Business health characterization: a hybrid regression and support vector machine analysis. Expert Syst Appl 49:48–59. https://doi.org/10.1016/j.eswa.2015.11.027
7. Lahmiri S, Bekiros S (2019) Can machine learning approaches predict corporate bankruptcy? Evidence from a qualitative experimental design. Journal 19(9):1569–1577. https://doi.org/10.1080/14697688.2019.1588468
8. Lin WC, Lu YH, Tsai CF (2019) Feature selection in single and ensemble learning-based bankruptcy prediction models. WILEY Expert Syst 36:1–8. https://doi.org/10.1111/exsy.12335
9. Lahmiri S (2016) Features selection, data mining and financial risk classification: a comparative study. WILEY 23:265–275. https://doi.org/10.1002/isaf.1395
10. Veganzones D, Séverin E (2018) An investigation of bankruptcy prediction in imbalanced datasets. Decis Support Syst 112:111–124. https://doi.org/10.1016/j.dss.2018.06.011
11. Beaver WH (1966) Financial ratios as predictors of failure. J Acc Res 4:71–111. https://www.jstor.org/stable/2490171
12. Altman EI (1968) Financial ratios, discriminant analysis and the prediction of corporate bankruptcy. J Financ 23(4):589–609. https://doi.org/10.1111/j.1540-6261.1968.tb00843.x
13. Ohlson JA (1980) Financial ratios and the probabilistic prediction of bankruptcy. J Acc Res 18(1):109–131. https://hdl.handle.net/10.2307/2490395
14. Alaka HA et al (2018) Systematic review of bankruptcy prediction models: towards a framework for tool selection. Expert Syst Appl 94:164–184. https://doi.org/10.1016/j.eswa.2017.10.040
15. Le HH, Viviani J-L (2018) Predicting bank failure: An improvement by implementing a machine-learning approach to classical financial ratios. Res Int Bus Financ 44:16–25. https://doi.org/10.1016/j.ribaf.2017.07.104
16. Gregova E et al (2020) Predicting financial distress of Slovak enterprises: comparison of selected traditional and learning algorithms methods. Sustainability 12(10):39–54. https://doi.org/10.3390/su12103954
17. Zelenkov Y, Fedorova E, Chekrizov D (2017) Two-step classification method based on genetic algorithm for bankruptcy forecasting. Expert Syst Appl 88:393–401. https://doi.org/10.1016/j.eswa.2017.07.025
18. Barboza F, Kimura H, Altman E (2017) Machine learning models and bankruptcy prediction. Expert Syst Appl 83:405–417. https://doi.org/10.1016/j.eswa.2017.04.006
19. Choi H, Son H, Kim C (2018) Predicting financial distress of contractors in the construction industry using ensemble learning. Expert Syst Appl 110:1–10. https://doi.org/10.1016/j.eswa.2018.05.026
20. Xia Y et al (2017) A boosted decision tree approach using Bayesian hyper-parameter optimization for credit scoring. Expert Syst Appl 78:225–241. https://doi.org/10.1016/j.eswa.2017.02.017
21. Ala'raj M, Abbod MF (2016) A new hybrid ensemble credit scoring model based on classifiers consensus system approach. Expert Syst Appl 64:36–55. https://doi.org/10.1016/j.eswa.2016.07.017
22. Zhu Y et al (2019) Forecasting SMEs' credit risk in supply chain finance with an enhanced hybrid ensemble machine learning approach. Int J Prod Econ 211:22–33. https://doi.org/10.1016/j.ijpe.2019.01.032
23. Quinlan R, Francisco S (1993) Obtaining the C4.5 Code. C4. 5: programs for machine learning. Morgan Kaufmann Publishers, pp 10–11. https://doi.org/10.1007/BF00993309
24. Carmona P, Climent F, Momparler A (2019) Predicting bank failure in the U.S. banking sector: an extreme gradient boosting approach. Int Rev Econ Financ 61:304–323. https://doi.org/10.1016/j.iref.2018.03.008
25. Odom MD, Sharda R (1990) A neural network model for bankruptcy prediction. In: IJCNN international joint conference on neural networks. IEEE, San Diego, CA, USA, USA. https://doi.org/10.1109/IJCNN.1990.137710
26. Abellán J, Castellano JG (2017) A comparative study on base classifiers in ensemble methods for credit scoring. Expert Syst Appl 73:1–10. https://doi.org/10.1016/j.eswa.2016.12.020
27. Sun J et al (2018) Imbalanced enterprise credit evaluation with DTE-SBD: decision tree ensemble based on SMOTE and bagging with differentiated sampling rates. Inf Sci 425:76–91. https://doi.org/10.1016/j.ins.2017.10.017

28. Hsieh N-C, Hung L-P (2010) A data driven ensemble classifier for credit scoring analysis. Expert Syst Appl 37(1):534–545. https://doi.org/10.1016/j.eswa.2009.05.059
29. Robu R, Stefan H (2011) A genetic algorithm for classification. In: Recent researches in computers and computing—international conference on computers and computing, ICCC'11
30. Lakshmi TM, Martin A, Venkatesan VP (2016) A genetic bankrupt ratio analysis tool using a genetic algorithm to identify influencing financial ratios. IEEE Trans Evol Comput 20(1):38–51. https://doi.org/10.1109/TEVC.2015.2424313
31. Taiwan Database. FINASIA A Division of Taiwan Journal 1990 AccessDate: November/7/2019; Available from: http://www.finasia.biz/ensite/Default.aspx?TabId=121
32. Taiwan Economic Journal (TEJ). FINASIA 1990, AccessDate: May/8/2020; Available from: http://www.finasia.biz/ensite/
33. Asuncion A, Newman D (2007) Polish companies bankruptcy data Data Set. UCI, AccessDate: March/4/2020; Available from: https://archive.ics.uci.edu/ml/datasets/Polish+companies+bankruptcy+data
34. Russia Bankruptcies. TRADING ECONOMICS 2007, AccessDate: January/2020; Available from: https://tradingeconomics.com/russia/bankruptcies
35. Asuncion A, Newman D (2007) Qualitative_Bankruptcy Data Set. UCI 2007, AccessDate: June/1/2020; Available from: https://archive.ics.uci.edu/ml/datasets/qualitative_bankruptcy
36. Asuncion A, Newman D (2007) UC Irvine Machine Learning Repository. UCI 2007, AccessDate: June/22/2020; Available from: https://archive.ics.uci.edu/ml/index.php
37. Asuncion A, Newman D (2007) Statlog (Australian Credit Approval) Data Set. UCI 2007, AccessDate: December/30/2019; Available from: http://archive.ics.uci.edu/ml/datasets/statlog+(australian+credit+approval)
38. Simonoff JS (2003) analcatdata_bankruptcy. OpenML 2003, AccessDate: June/11/2020; Available from: https://www.openml.org/d/476
39. Diane. Bureau van Dijk 1998, AccessDate: October/30/2019; Available from: https://www.bvdinfo.com/en-gb/our-products/data/national/diane
40. DATABASES. NYU 1987, AccessDate: July/1/2020; Available from: http://pages.stern.nyu.edu/~ealtman/Credit%20&%20Debt%20Markets%20Databases.htm
41. BankFocus. BVDInfo 1990, AccessDate: February/18/2020; Available from: https://banks.bvdinfo.com/version-2020217/home.serv?product=orbisbanks
42. Wieslaw Pietruszkiewicz. dblp 1993, AccessDate: February/18/2020]; Available from: https://dblp.org/pers/hd/p/Pietruszkiewicz:Wieslaw
43. South Korea Bankruptcies. TRADING 1990, AccessDate: February/18/2020; Available from: https://tradingeconomics.com/south-korea/bankruptcies
44. Asuncion A, Newman D (2007) Statlog (German Credit Data) Data Set. UCI 2007, AccessDate: December/13/2019; Available from: https://archive.ics.uci.edu/ml/datasets/statlog+(german+credit+data
45. Prof. Dr. Thomas F. Hofmann. TUM 1868, AccessDate: December/13/2019; Available from: https://www.professoren.tum.de/en/hofmann-thomas/
46. DATA CATALOG. Data.gov 2009, AccessDate: July/22/2020; Available from: https://catalog.data.gov/dataset
47. Japan Stock Market. Siblis Research 2015, AccessDate: April/16/2020; Available from: http://siblisresearch.com/data/japan-shiller-pe-cape/
48. Zieba M, Tomczak SK, Tomczak JM (2016) Ensemble boosted trees with synthetic features generation in application to bankruptcy prediction. Expert Syst Appl 58:93–101. https://doi.org/10.1016/j.eswa.2016.04.001
49. Le T et al (2018) A cluster-based boosting algorithm for bankruptcy prediction in a highly imbalanced dataset. Symmetry 10:1–12. https://doi.org/10.3390/sym10070250
50. Dastile X, Celik T, Potsane M (2020) Statistical and machine learning models in credit scoring: a systematic literature survey. Appl Soft Comput J 91:106–263. https://doi.org/10.1016/j.asoc.2020.106263
51. Liang D et al (2016) Financial ratios and corporate governance indicators in bankruptcy prediction: a comprehensive study. Eur J Oper Res 252:561–572. https://doi.org/10.1016/j.ejor.2016.01.012

52. Uthayakumar J, Vengattaraman T, Dhavachelvan P (2017) Swarm intelligence based classification rule induction (CRI) framework for qualitative and quantitative approach: an application of bankruptcy prediction and credit risk analysis. J King Saud Univ-Comput Inf Sci. https://doi.org/10.1016/j.jksuci.2017.10.007
53. Jardin PD (2018) Failure pattern-based ensembles applied to bankruptcy forecasting. Decis Support Syst 107:64–77. https://doi.org/10.1016/j.dss.2018.01.003
54. Liang D et al (2018) A novel classifier ensemble approach for financial distress prediction. Knowl Inf Syst 54:437–462. https://doi.org/10.1007/s10115-017-1061-1
55. Antunes F, Ribeiro B, Pereira F (2017) Probabilistic modeling and visualization for bankruptcy prediction. Appl Soft Comput 60:831–843. https://doi.org/10.1016/j.asoc.2017.06.043
56. Carton RB, Hofer CW (2006) Measuring organizational performance. Edward Elgar Publishing, Greet Britain
57. Uthayakumar J et al (2018) Intelligent hybrid model for financial crisis prediction using machine learning techniques. Inf Syst e-Bus Manage 1–29. https://doi.org/10.1007/s10257-018-0388-9
58. Gogas P, Papadimitriou T, Agrapetidou A (2018) Forecasting bank failures and stress testing: a machine learning approach. Int J Forecast 34:440–455. https://doi.org/10.1016/j.ijforecast.2018.01.009
59. China Database. FINASIA 1990, AccessDate: March/21/2020; Available from: http://www.finasia.biz/ensite/Default.aspx?TabId=126
60. Hosaka T (2019) Bankruptcy prediction using imaged financial ratios and convolutional neural networks. Expert Syst Appl 117:287–299. https://doi.org/10.1016/j.eswa.2018.09.039
61. Altman EI, Hotchkiss E (2006) Corporate financial distress and bankruptcy, 3rd edn. Wiley, Canada
62. Uthayakumar J et al (2020) Financial crisis prediction model using ant colony optimization. Int J Inf Manage 50:538–556. https://doi.org/10.1016/j.ijinfomgt.2018.12.001
63. Ministry of Finance of the Slovak Republic. Register of Financial Statements 2014, AccessDate: August/3/2020; Available from: http://www.registeruz.sk/cruz-public/domain/accountingentity/simplesearch

A Proposed Framework to Apply Secured Mobile Banking in Egypt

Nevien Makram Labib and Nermin Ibrahim Alarabi

Abstract Banking as a leading sector in Egyptian economy has realized that the investment in technology represents a great leap nowadays and an effective way to control cost, attract more customers, and fulfill consumers' needs for convenience and technical innovation. The remarkable progression of mobile sector all over the world has made an exclusive chance for delivering financial as well as social services through mobile network. Mobile banking is a powerful tool for developing banking services, supporting growth, promoting renovation, and enhancing competitiveness. High Internet usage and mobile phone penetration rate were found out to be the most important potential for the development of mobile banking in Egypt; however, lack of timely approval of new products by regulatory body, lack of interoperability system and lack of aggregators between services and retail agents were the major challenges analyzed. The mobile banking security is enhanced using virtual keyboard, smart cards, one-time password, multi-factor authentication.

Keywords M-Banking · Banking security · Multi-factor authentication

1 Introduction

Innovations have changed all sorts of human life and gradually improved the quality of human life. Their impact has extended from affecting personal life to the way the business is being done. Technology has improved the efficiency and decreases the operating costs in business organizations. Because of so, banks have transformed the way of offering services from branch banking to Internet banking to mobile banking to be able to cope with the emergence of self-service technology [1].

Banks offer mobile banking due to the following reasons lower operating costs, greater geographic diversification, improved or sustained competitive position, increased customer demand for services and new revenue opportunities. Benefits to customers are increased convenience, reduced theft, fraud and mismanagement,

N. M. Labib · N. I. Alarabi (✉)
Sadat Academy for Management Sciences, Cairo, Egypt

D. A. Magdi et al. (eds.), *Digital Transformation Technology*, Lecture Notes in Networks and Systems 224, https://doi.org/10.1007/978-981-16-2275-5_22

ability to maintain electronic records, access to other value-added services and time-saving [2].

Mobile banking from bankers' perspective is a way of offering financial services through smartphones. Whereas from customers' perspective mobile banking refers to the process of accessing bank account, performing transactions and using different banking services in smartphone (mobile) either through banking application or through wireless Internet [1].

Mobile banking security will include data transmission which is important to secure the data of the user to prevent the hackers from attacking or stealing these data [3]. A diverse range of mobile malicious code has appeared to code that exploits vulnerabilities in a smartphone operating system to actually disclose the owner's information and cause financial damages so we propose a secure element of mobile devices which is a smart card and is well-trusted and tamper-proof device can be used for security critical application [4].

Mobile banking adoption in Egypt faces many challenges such as technical and functional (related to the technology itself), physiological and behavioral barriers (related to the users), and the complexity because it is believed that the highly secured systems is very complicated for the ordinary users to deal with.

2 Mobile Banking

Mobile banking is considered one of the most value added and important mobile commerce applications currently available. Mobile banking has can be considered as a subset of e-banking or online banking and refers to use of wireless networks to conduct financial transactions instead of wired networks [5] or is any form of banking transaction that is performed using a portable device, such as a mobile phone or a personal digital assistant [6].

The Ministry of Information and Communication Technology (ICT) reports that mobile usage subscription is reached 99.82 million exceeding the actual number of population till Oct 2017. With a prevalence rate reached 110.37%. The number of users that use mobile Internet reached 27.49 million users whereas the total number of Internet users reached 33.7 million users with a prevalence rate reached 41.3% and Online payments is between 2 and 3% of the total local production which is targeted to reach 10% in the next 3 years, while the total number of the smart wallets in Egypt is 11 million for 9.5 million users [7].

3 Previous Studies of Mobile Banking

The term mobile banking is derived from the idea where your bank can be in your mobile device by access to some special applications provider by financial institutions like banks or microfinance institutions. Mobile banking allows their user to access and perform many financial transactions, facilitate their work, save their time and cost, save the transportation cost as well as reduce the risk of thief. Technological innovation is rapidly changing the traditional banking approach, the transition from the costly physical branch system to modern banking ranging from automated teller machines to 24/7 e-banking is definitely an achievement [8]. Mobile banking offers bill payments, micro-payments international and local remittances, loan re-payments, micro-savings, microcredit, insurance services, and remittances, banks also plan to offer: bill payments, financial education, customer complaint services, airtime top-up, and e-vouchers for rural farmers [9]. Moreover, posit that, mobile banking provides loan repayment, cash-in and cash-out services, savings on deposit account, bill payments, salary payment, air ticket purchase, school fees, government payments, transportation fees and donations [10]. Additionally, a study shows that most urban households would want to avail micro-savings, microcredit and micro-insurance [11]. A study [12] stated that mobile banking provides financial services such as balance inquiry, transaction history, money transfers, and bill payment via a mobile phone, a smartphone or a PDA (personal digital assistant). Also today, mobile banking is the main technological innovation being discussed in economic and social development, especially among low-income groups [11].

The current m-banking solutions neither ensure communication security nor application security, these solution claims to achieve security principles from the bank perspective. On the other hand, the user is seeking the balance between protection, ease of use and availability which means that the level of security must allow reasonable access yet protect against threats. So the traditional m-banking solutions have many problems as follows:

- The client must be known to the bank system previously which means that the user cannot sign up for the application by himself.
- The clients' mobile device should also be known to the bank server which means that if the mobile device has changed or be stolen, the device is vulnerable to be used by unauthorized person.
- Validating the SIM card number cannot be held only by the bank server but also the Mobile Server Subscriber should be involved in the clients' transactions which make the bank server as a whole vulnerable to security breaches and attacks.
- This model is using the same security regulations for both informative and transitive phase without considering that the level of risk in both phases is not the same.
- Providing access control credentials (username/password) through the mobile keyboard may cause the credentials to be vulnerable to keyboard hooking or key logger risks.

4 Threat Analysis

According to computer security, the mobile threat is defined as any malware that targets smartphone and PDA. The vulnerability is a weakness which allows an

Table 1 Impact of risks on mobile applications

Threats		Level of risk		
Name	Type	Low 1	Normal 2	High 3
Malware, viruses, spyware	Application-based threats			✓
Ransomware, worms, rootkit		✓		
Privacy threats			✓	
Vulnerable applications		✓		
SQL injections, cross site scripting (XSS)				✓
Phishing scams	Web-based threats			✓
Trojan horses			✓	
Party posting as a legitimate services, drive by download, browse exploits		✓		
Network exploits, Wi-Fi sniffing, mobile network services	Network-based threats			✓
Collision			✓	
Man-in-the-middle-attack (MITM)	Network-based threats		✓	
Replay attack, impersonation attack		✓		
Active eavesdropping				✓
E-fraud, lost or stolen devices, using public Wi-Fi	Physical threats			✓
Computing resources		✓		
Device limitations, Storage limitations, Platform limitations	Smartphone-based threats			✓
Bluetooth				✓
Keyboard hooking				✓

attacker to reduce system's assurance. This can be done through the insertion of three elements: a system susceptibility or flaw, attacker access to the flaw, and attacker capability to exploit the flow. To exploit vulnerability, an attacker must have at least one applicable tool or technique that can connect to the system weakness [10].

Mobile threats and vulnerabilities are classified into five types: application-based threats, web-based threats, network-based threats, physical threats, and smartphone-based threats. In order to deal with these risk and find a suitable integrated solution for them, the impact of these risks were classified and graduated from 1 (low risk) to 3 (high risk) which means the damage that may occur on the system as a result to this threat. Table 1 shows the risk assessment (the impact of the risks on mobile applications).

The Vulnerabilities and risks mentioned above may cause security weakness to the system which needs a countermeasure to control the damage that may occur as a result for them. The security controls aim to:

- Create a secure session for the user.
- Verify user identity.
- Verify device-in-use identity.
- Ensure that the users' financial data are fully secured.
- Ensure that the users' transactions are performed correctly in a secure channel.

5 Proposed Model of Mobile Banking

5.1 Traditional Mobile Banking Model

The traditional mobile banking model is using the traditional method for securing financial data. This method is depending basically on the bank server security protocols which is combined with a trust negotiation method. This method authenticates the client, and the mobile device used in accessing the bank account information and processing transactions as shown in Fig. 1.

The model starts the operation by establishing a secure session between the clients' mobile device and the bank server where the clients' financial data are actually stored.

The model is using a traditional single factor authentication method in which the client must provide the user credentials (username and password) which was previously created by the bank representative which means that the user cannot sign up for this model online from the web application.

Authenticating the device-in-use process is run silently in the background without the user interface. This process requires the exchange of digital credentials related to the device. The International Mobile Equipment Identity (IMEI) and the Subscriber Identification Module (SIM) Serial numbers are the credentials used for this purpose. IMEI is a unique number used for identifying the device and has no permanent or semi-permanent relation to the user. A SIM number is burned on an integrated circuit of the actual SIM card that securely stores to the Mobile Server Subscriber. The combination of IMEI and SIM numbers will lead to the verification of the mobile in use. Upon the successful identification of the client and the mobile in use, an access to the clients' financial data is then generated.

This model has many problems listed below:

- The client should be known to the bank system previously which means that the user cannot sign up for the application by himself.
- The clients' mobile device should also be known to the bank server which means that if the mobile device has changed or be stolen, the device is vulnerable to be used by unauthorized person.

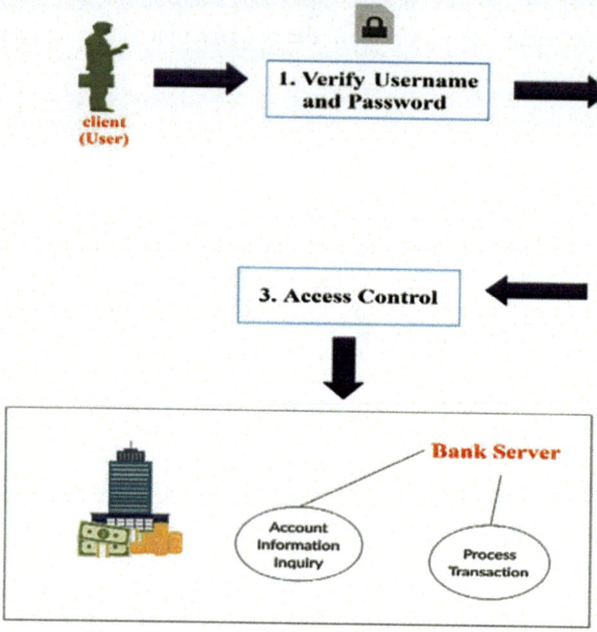

Fig. 1 Traditional mobile banking model

- Validating the SIM card number cannot be held only by the bank server but also the Mobile Server Subscriber should be involved in the person which make the bank server as a whole vulnerable to security breaches and attacks.
- This model is using the same security regulations for both informative and transitive phase without considering that the level of risk in both phases is not the same.
- Providing access control credentials (username/ password) through the mobile keyboard may cause the credentials to be vulnerable to keyboard hooking or key logger risks.

5.2 *Enhanced Mobile Banking Model*

The enhanced mobile banking model shown in Fig. 2 is proposed to solve the problems appeared in the traditional model. This model is on two levels from the risk point of view. The security controls used on each phase is selected based on the accessed data or the processes allowed. This model is designed to achieve the technical goals and eliminate the security risks on web application.

The enhanced model is designed taking into consideration that the highly risk alerts are not allowed to appear in the testing phase. The security controls which is

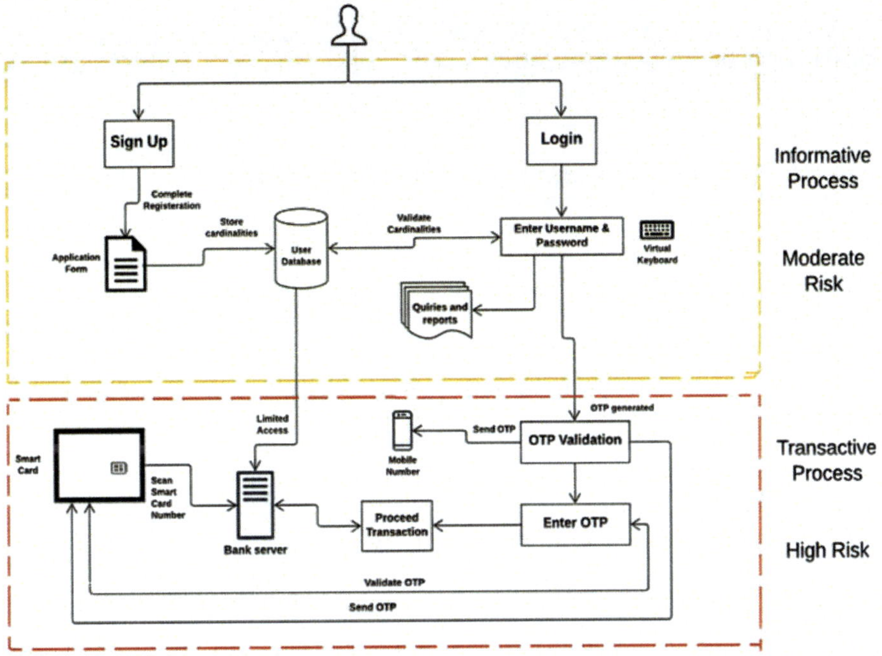

Fig. 2 Enhanced mobile banking model

proposed in the Technical Requirements Phase are considered in the design model and then executed in the implementation phase.

This model is implemented in the form of web application that can be run on any kind of portable devices such as smartphones, tablets, handhelds, and laptops. The web application works on a client–server model. A server is a web server and a client is a web browser. The user sends HTTPS requests through the browser (client) to the web server and the web server accomplishes the requested tasks and sends the required information back to the browser. To allow support by all kinds of browsers, it uses HTML or XHTML as a format to generate documents.

The first Level or (The Moderate Risk Level) is an informative process in which the user can provide data about his account in the case of new user or get information about his account in the case of current user—financial transactions are not allowed on this level—in this phase the user either choose to:

Sign Up: For the users who use the application for the first time, the user is asked to complete the registration by filling an application form which is automatically stored in the user database which have limited access to the bank server.

Login: For the users that already completed their registration, the user is asked to enter username and password on a virtual keyboard designed as an on-screen keyboard. After login, the user is having an access for his account information through queries ex. (Short Bank Statement), provided by user database.

The second level or the high risk level is providing a transitive process to user, and the user can access this level only after login in order to proceed a transaction.

5.3 *Model Implementation and Security Techniques*

The proposed framework is designed to use different security techniques on three graduated phases in order to create a secure session for the user, verify user identity, verify device-in-use identity, and ensure that financial data and transactions for each client are fully secured as shown in (Fig. 3).

Sign Up phase

Application form: When the user creates an account for the first time, the user is asked obligatory to fill a form containing the following data: (first name, family name, valid phone number, valid E-mail address, User ID, and Account Password), the previous data are stored automatically in the User database.

User Database: This database is a simple relational database as shown in Fig. 4 which has a limited access to server. The purpose of creating such database is minimizing the risk of attacking bank server. In addition to storing the user data provided in the application form, it provides user with information for simple queries such as short bank statement and current balance.

Smart Card Recognition: A smart card is a physical card that has an embedded integrated chip that acts as a security token. Smart cards are typically the same size as a driver's license or credit card and can be made out of metal or plastic. They connect to a reader either by direct physical contact (also known as chip and dip) or through a short-range wireless connectivity standard such as radio-frequency identification (RFID) or near-field communication (NFC). The chip on a smart card can be either a microcontroller or an embedded memory chip which is used here as an extra tool to store the login credentials. Smart cards are designed to be tamper-resistant and use encryption to provide protection for in-memory information. Those cards with

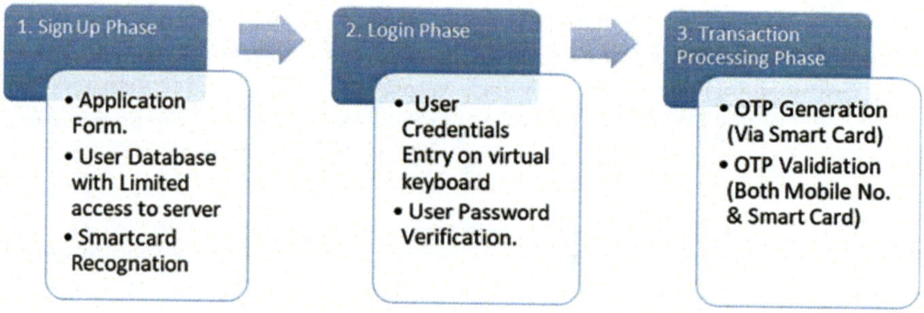

Fig. 3 Model implementation and security techniques

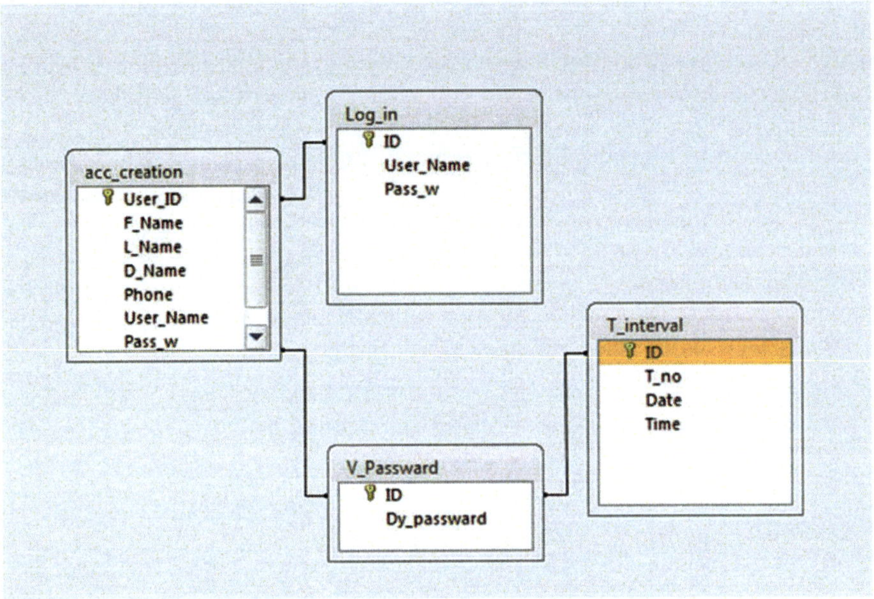

Fig. 4 User database

microcontroller chips can perform on-card processing functions and can manipulate information in the chip's memory.

Login Phase

User Credentials Entry on Virtual Keyboard: When user attempts to login to his account, it is obligatory recommended to provide his login credentials (username/password) using on-screen built-in keyboard as a part of the application design. It's not allowed for the user to use the device keyboard to prevent the Keyboard hooking, key loggers, and Cookies hijacking attacks. The application design doesn't have the ability of storing sign-in credentials for more security.

User Password Verification: According to the traditional advice, the strong password should contain at least 12 characters (combination of Upper and lower case letters, symbols, and numbers), it should not be a dictionary word or combination of them, and does not rely on obvious substitution. Each time the user is attempt to login the system compares the clients. password with the password stored in the user database (the regular way of authentication).

Transaction Processing Phase

OTP Generation Via*Smart Card*: One-time password systems provide a mechanism for logging on to a network or service using a unique password which can only be used once, as the name suggests. This prevents some forms of identity theft by making sure that a captured user name/password pair cannot be used a second time. Typically, the users' login name stays the same, and the one-time password changes with each login. One-time passwords are a form of so-called strong authentication, providing much better protection to online bank accounts, corporate networks and other systems containing sensitive data. So OTP is Used with the regular authentication method to form a multi-factor authentication technique in the case of processing financial Transaction [13]. More advanced hardware tokens use microprocessor-based smart cards to calculate one-time passwords. Smart cards have several advantages for strong authentication including data storage capacity, processing power, portability, and ease of use. They are inherently more secure than other OTP tokens because they generate a unique, non-reusable password for each authentication event, store personal data, and they do not transmit personal or private data over the network.

OTP Validation: The OTP is received as an SMS on the mobile number, entered manually by the user and then be validated via smart card.

5.4 Model Testing and Results Discussion

Testing process is executed for both traditional and enhanced model to reach the minimum accepted risks using OWSAP ZAP testing proxy tool. The Zed Attack Proxy (ZAP) is an easy to use integrated penetration testing tool for finding vulnerabilities in web applications. ZAP scans the web application and output a list of alarms graduated in the level of risk from high to low [14].

After Scanning completes, ZAP shows the result summary in the form of different categories of alerts. Basically, alerts are potential vulnerabilities and have been categorized as high priority, medium priority, low priority, and informational priority, which indicates the degree of associated risks. A high priority alert means that an issue under this category is more serious than other priority alerts. Likewise, medium priority alerts, low priority alerts, and informational priority alerts are consecutively less and less serious. Alerts categories are indicated by different color flags. The risk alerts is classified into Red Flags for high risk, Orange Flags for medium risk, Yellow Flags for low risk, and Blue Flags indicate no risk.

The Firefox browser was used for testing both traditional and proposed banking application and simple steps that ZAP tutorial suggests were followed to start the process. ZAP has been set to protective mode in order to avoid any potential errors in attacking real Web sites which are not allowed to test. The browser was configured to proxy via ZAP, the application was set in a context, and the application was explored manually almost every pages. After that, ZAP normal Spider first and then AJAX Spider was started to crawl the application. And finally, active scanner was run. The results of those scans are shown in Figs. 5 and 6.

Comparing testing results for both traditional and proposed model is showing that the high risk alerts which represent the major security problem threaten the financial systems which have totally disappeared in the enhanced model, and also the number of medium to low risk has been minimized to a very low percentage which is not

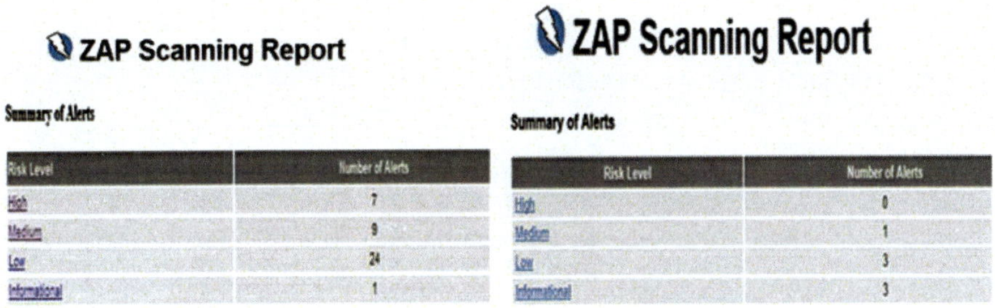

Fig. 5 Traditional model versus proposed model scanning reports

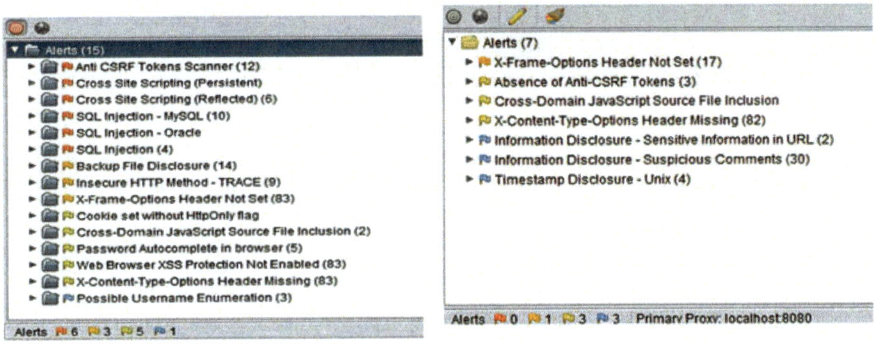

Fig. 6 Traditional model versus proposed model risk alerts

exceeding 3% compared to 88% in the traditional model. Figures 7 and 8 are showing the risk analysis for both models.

6 Conclusion and Future Directions

6.1 Conclusion

This work proposed to provide bank clients with a secure platform that could be used to access their personal and business accounts anywhere and at any time. The proposed solution has achieved competitive advantage rather than the traditional solutions. It has solved the main security problems that affect the efficiency and effectiveness of the web application. Using the enhanced model, the client can sign up for a mobile banking account directly through the application after filling the application form. The authentication of device is achieved through smart card in which the credentials are only accessed by the bank server and the mobile server subscriber is not involved in the process. Moreover, using a virtual keyboard designed

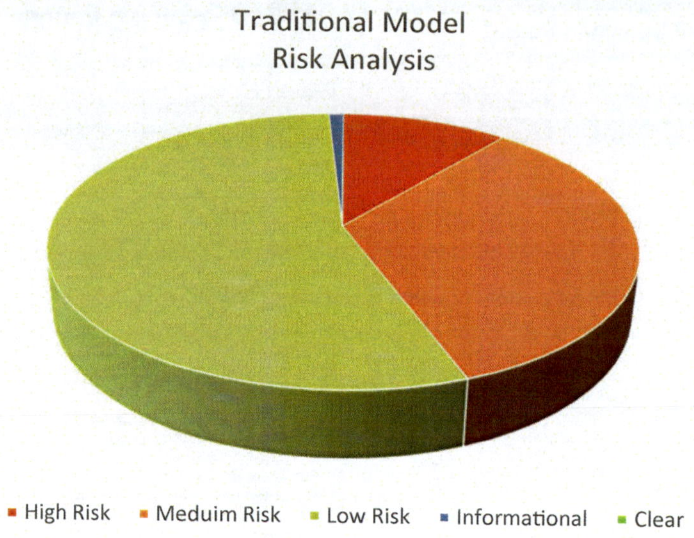

Fig. 7 Traditional model risk analysis

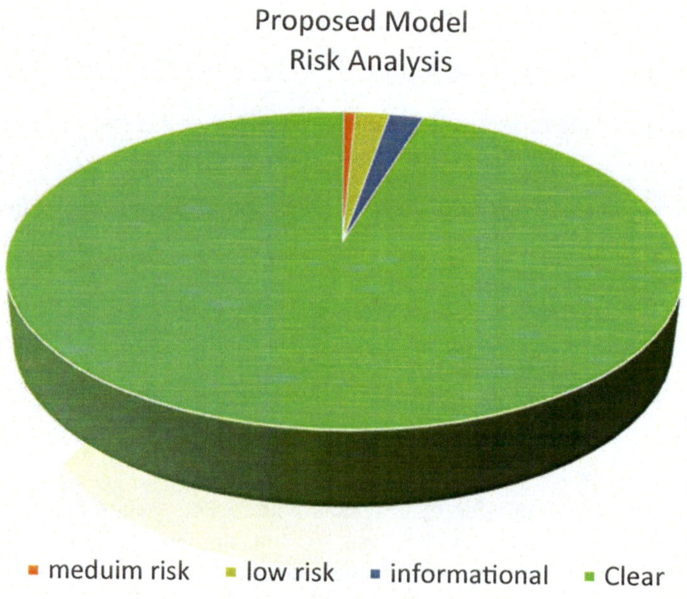

Fig. 8 Enhanced model risk analysis

as an on-screen built-in tool in the web application ensures that keyboard is clear and cannot store the credentials or make it vulnerable to unauthorized access and using multi-factor authentication techniques (sign-in credentials and OTP) is enhancing the overall security of the model. As for Generating OTP through OTP generation algorithm on the smart card is considered an innovative way to make a close circle for the process without involving the mobile server subscriber in. Finally, OTP validation is done through mobile device which ensure security and through smart card which represent an alternative way to keep track of the operations and transactions done on the clients' accounts in addition to the records kept on the bank server. The variation of techniques used in this model is achieving a high level of security and improving the overall performance of the m-banking application.

6.2 Future Directions

Our future work will cover the security issues that were out of the scope of this study which include: the limitations with mobile banking applications such as performance and delays, the service running on the bank side for mobile banking such as servers running back end services, fraud detection, and support processes. The future work may also include the network issues and how to integrate this approach with a location verification method to improve security in the case of mobile stolen or loss.

References

1. Sahu A, Deshmukh GK (2020) Mobile banking adoption: a review. J Crit Rev 7(14)
2. Tam C, Oliveira T (2017) Literature review of mobile banking and individual performance. Int J Bank Market 10(11)
3. Hayikader S, Nurafiqah F (2016) Issues and security measures of mobile banking apps. Int J Sci Res Publ 6(1)
4. Mishra S (2014) E-banking: an effective tool of CRM in banking sector. Int J Res Commer Manage 5(07)
5. Bojan D, Mutu S (2010) Electronic banking advantages for financial services delivery. Int J Bank Market 6(4)
6. Hun Ha K, Canedoli A, Baur AW, Bick M (2012) Mobile banking—insights on its increasing relevance and most common drivers of adoption. Electron Markets
7. http://www.mcit.gov.eg/. Last accessed 2020/9/19
8. Majuad H, Ahmed E, Samikon S (2020) Mobile banking adoption in organization: review of empirical literature. Int J Innov Sci Res Technol 5(9)
9. Alrousan MK, Al-Adwan AS, Al-Madadha A, Al Khasawneh MH (2020) Factors affecting the adoption of E-marketing by decision makers in SMEs. Int J E-Bus Res 16(1):1–27
10. Ammar AA, Ahmed EM (2016) Factors influencing Sudanese microfinance intention to adopt mobile banking. Cogent Bus Manage 3(1)
11. Giotopoulos I, Kontolaimou A, Korra E (2017) What drives ICT adoption by SMEs? Evidence from a large-scale survey in Greece. J Bus Res 81(7)
12. Oliveira T, Thomas M, Espadanal M (2014) Assessing the determinants of cloud computing adoption: an analysis of the manufacturing and services sectors. Inf. Manage 51(5)

13. Thishesaayee A, Sumathly (2014) OTP encryption techniques in mobiles for authentication and transaction security. Int J Innov Res Comput Commun Eng 2(10)
14. Beyer C (2014) Mobile security: literature review. Int J Comput Appl 97(8)
15. Elkhodr M (2012) A proposal to improve the security of mobile banking applications. Presented at 10th international conference on ICT and knowledge engineering
16. https://www.owsap.org/. Last accessed 2019/11/25

Modeling a Business Intelligent System for Managing Orders to Supplier in the Retail Chain with Unified Model Language

Silvia Parusheva and Daniela Pencheva

Abstract The study focuses on some aspects related to the established trends in the design of business intelligent systems (BIS), specialized for use in retail chains for fast-moving consumer goods. Current concepts concerning the design of information systems and their application in commercial systems for managing orders to suppliers are considered. Modern methods have been applied for building a business intelligent module model in order to increase the productivity of the commercial system and to refine the product quantities needed for sales. The proposed model was created with the help of the object-oriented modeling language Unified Model Language (UML), referring to the leading trends in the design of business intelligent systems. Technical diagrams that illustrate the main functionalities of the developed model and represent the interaction between the user and the information system are presented.

Keywords System design · Business intelligent system · Retail chain · Fast-moving consumer goods · UML modeling · Business process model and notation (BPMN) · Decision-making and intelligent information systems

1 Introduction

In recent years, companies are increasingly focusing on the enhancement and modernization of their information systems with the help of innovative information technologies. By automating the processes, the efficiency of the system increases, and so does its productivity [1]. Minimizing manual work is a prerequisite for reducing the number of errors and improving the quality of the processes performed [2]. There is a growing interest from companies in optimizing and improving business processes in order to increase their competitiveness and performance in globalized economic markets [3–5]. On the other hand, the retail sector, does not lag this trend [6, 7]. It also has processes related to striving to increase the efficiency of trade business

S. Parusheva (✉) · D. Pencheva
University of Economics—Varna, 77, Knyaz Boris I Blvd., 9002 Varna, Bulgaria
e-mail: parusheva@ue-varna.bg

D. A. Magdi et al. (eds.), *Digital Transformation Technology*, Lecture Notes in Networks and Systems 224, https://doi.org/10.1007/978-981-16-2275-5_23

processes through automation, development, and improvement of commercial information systems, quality control, attracting new customers, cost optimization, etc. [8–10].

Before proceeding to the development and implementation of new functionalities to the information system in order to improve it, it is necessary to make a detailed description of workflows and to design a visual model of business processes. In the context of these important tasks, a number of authors [11, 12] express their view that business process modeling occupies a key position in Research and development operations. Modeling is considered to be the starting point in creating the requirements that innovations in the information system must meet because it recreates the whole system or parts of it with visual means that are standardized and easy to understand. The authors emphasize that the two most widely used standards for visual presentation of a business model are Unified Model Language and Business Process Model and Notation (BPMN). In essence, the two standards have a number of similarities, but also many differences, which will be presented in the text of the first section of the paper.

The actuality of the researched topic is determined by the tendency of the companies in the field of retail trade in fast-moving goods to invest in improvement and optimization of their current information systems, in order to achieve higher productivity, minimize manual labor, and reduce costs. Technological advances are having a positive impact and enabling business operations to be digitized and improved through business intelligence methods and approaches that provide added value in retail chains.

The purpose of this paper is to develop a conceptual framework for a **model of a business intelligent system for managing orders to suppliers in the Fast-moving consumer goods (FMCG) retail chain.**

The paper is structured in four sections. The first section introduces the relevance of the topic related to the need to develop a business intelligence order management system and the importance of modeling to achieve this goal. The next section presents the specific possibilities of modeling as a tool by which information systems that automate business processes in the commercial sector can be improved. The third section offers a model of a business intelligent vendor order management system, as the model is created using the object-oriented approach and visual methods of the Unified Model Language standard. The model follows a level of abstraction that includes only basic functionalities related to supply chain activities related to managing orders to suppliers. In order to fulfill the set goal and to offer a model for building a business intelligent order management system to suppliers in a retail chain, the means of object-oriented approach and the visual methods of UML are used. Many authors point out that the use of appropriate visual methods to depict business models, such as UML, is part of the application of good research practices [13–15]. The applied set of diagrams that describe the model is as follows: business case diagrams, class diagram, sequence diagram, state diagram, and business process diagram. Finally, the last section draws the conclusion and summarizes the proposal made for conceptual framework for a model of a business intelligent system for managing orders to suppliers in FMCG retail chain. In addition, it provides suggestions for future studies.

2 Modeling as a Method for Improving Information Systems, Automating Commercial Activities

The consistent aspiration of the companies towards technological growth and relevance of the digitalization of the business processes in relation to the latest achievements of IT determine the possibility to achieve higher productivity and quality of the offered products or services. This rule also applies in full force to commercial activity. Through commerce and the processes within it, business connections are made between market participants and the created products reach the end users. New information and communication technologies and digitalization are changing the nature and ways of exchanging goods between market agents and have a direct impact on commercial operations, including systems for ordering products sold. A number of authors study both already established and innovative concepts applied in the management of companies operating in various fields, as one of the most frequently considered areas is trade.

In the retail sector, a number of companies are implementing projects related to the reorganization of the internal business processes, incl. reducing the degree of vertical integration [16, 17], with the aim of increasing sales and achieving greater business efficiency. Authors point out that concepts such as virtual companies, extended enterprises, strategic alliances, and corporate networks are increasing their distribution, forcing supply chains to be flexible and process-supporting information systems to have multiple interfaces, real-time data transfer, and embedded intelligence technologies [18–20].

The actions of commercial companies aimed at improving their existing information system, incl. the addition of new modules to it or aimed at the implementation of a new commercial information system, requires a detailed study of the current state of the system. At the next stage, an action plan is created, which aims to develop an advanced future system or expand the existing one with new modules. In addition, it is fundamental to develop a conceptual model that represents innovation **through graphical representation**. It can be presented as diagrams or charts, and also in tabular form. Some authors, such as Unhelkar [21], compare the modeling of information systems with a bridge that connects business analysis and actual development. In essence, the models provide research and development teams with opportunities to detect gaps and errors, as well as a comprehensive view of the future design of the commercial information system.

The study by Bouzidi et al. [13] states that among the first steps taken in developing a conceptual model is the choice of an appropriate language for business modeling, which will represent the requirements of the system in the most correct way. There are different approaches and visual tools for modeling information systems. According to some authors [3, 22, 23], the two most used methods for visual representation of a business system model are Business Process Model and Notation (BPMN) and Unified Model Language (UML). It is also stated that diagrams using the UML standard are easy to perceive by managers in organizations that do not always have the technical background [24, 25]. Due to its wide application possibilities and variety of visual objects, UML is one of the preferred languages for modeling information systems among software developers. It presents an illustration of the defined scenarios in the information system, which is the subject of consideration [26]. UML is believed to be the result of the collective efforts of prominent practitioners, researchers, and authors who unify the methodology for its application. Its distinctive advantages are visual presentation, possibility for specialization, constructiveness, documentation, and possibility for maintenance.

Visual modeling is within the research interest of a number of authors [2, 3, 27, 28]. In his book "Enterprise process management systems. Engineering Process-Centric Enterprise system using BPMN 2.0." Vivek [2] presents an in-depth study of the key benefits of visual modeling. He points out that, in essence, BPMN is a relatively younger standard. BPMN was introduced a few years after UML and brings together concepts for graphically depicting the implementation of a business process. The starting point for its creation is already existing visual programming languages such as UML Activity Diagram, UML Business Process, Activity-Decision Flow Diagram, Rosetta New, and Integrated Definition (IDEF). The main goal of BPMN is to create a kind of bridge between designers, developers, and end users of the company's system. The standard is fully focused on recreating business processes in a simple way with a well-defined notation.

There are a number of similarities between the two standards, as well as many differences. The similarities between them are related to the fact that both methods graphically represent different perspectives of the information system. The notation of visual modeling languages has been simplified to be easily understood and accessible to non-developers. Both standards are considered to be flexible, as they can recreate a variety of business activities and can be transformed into program code using specialized software. According to studies by authors [26, 29–32] the differences that distinguish UML from BPMN are significant. One of them is related to the presentation of the seven key aspects that characterize information systems as a whole object. They are systematized in the following groups:

1. Requirements review—covers the requirements of the process and the stakeholders involved.
2. Information view—captures the artifacts (products) that are produced and consumed by the process and shows the connections between the artifacts.
3. Stakeholder view—captures the stakeholders involved in the process.
4. View of the structure of the process—captures the structure and terminology of the process; forms the basis for any kind of mapping between different processes and standards, which is important when performing audits and evaluations.
5. Process content view—defines the process content in terms of the artifacts and activities that make up this process.

Table 1 Comparative characteristics between the standards for modeling information systems UML and BPMN

Indicators	UML	BPMN
Graphical representation	• Contains a greater variety of visual components	• Contains a small number of visual components
Semantics	• Focused on designing complete models of information systems	• Specialized in business process design
Purpose	• Language for visual modeling, presentation, construction, documentation, and specification of the artifacts involved in the information system	• Language for graphical presentation of business processes, in order to create an accessible visual interpretation suitable for people without technical experience
Modeling approach	• An object-oriented approach is applied	• A business process-oriented approach is applied
Language representation	• Presents all seven structural aspects of modeling	• Presents only two structural aspects of modeling

6. View of the behavior of the process—defines the behavior of the process: how the activities are arranged, the artifacts entering and leaving the activities, and the stakeholders involved in the process.
7. Process instance view—captures a sequence of processes and defines a scenario that can be used to verify process requirements.

The main characteristics of the two visual standards UML and BPMN are systematized in Table 1 [29–33].

From the point of view of visual interpretation of processes and their decomposition into successive tasks, UML has wider possibilities for graphical representation of the states and structure of objects from other categories participating in the system.

On the other hand, the standard is oriented towards the design of complete models of information systems. It presents all seven structural aspects of modeling, which are essential in the design of an information system. Based on the presented in paper, model of a business intelligent system for managing orders to suppliers through the visual methods of UML more fully illustrates its structure, adjacent objects, their states, and existing relationships. BPMN has a complementary function in the current study.

3 Creating a Model of a Business Intelligent System for Managing Orders to Suppliers Using the UML Standard

Each industrial sector has specific characteristics that are recommended to be taken into account when developing an information system for a company [34]. In their book, the authors Berman et al. [35] present a comprehensive study of the key characteristics of business processes and their impact on the use of digital technologies such as web applications, mobile technologies, social media, and others. Through the large number of communication channels and daily network connectivity, users receive real-time information on promotional campaigns or guidance on how to make an online purchase. The authors emphasize the high levels of competition in retail sales, dictated by the ability of consumers to choose the way, time and place of interaction with retailers. Commercial companies in this sector are expected to be active and at the same time adaptable to anticipating consumer needs and at the same time to provide a universal approach to seamless shopping, regardless of how the purchase is made.

The paper-presented model of a business intelligent order management system in retail chains covers the functionalities related to the forecasting of the amount of consumer demand and the automated sending of orders to suppliers. An important distinguishing feature of business intelligent systems is their main functionality, aimed at supporting decision-making processes. According to Burgin [36], three types of models can be distinguished, describing information systems related to decision-making processes:

1. **Operational model**—describes in detail the decision-making processes in the analytical module.
2. **Descriptive model**—presents the properties of the objects involved in the system and the relationships between them.
3. **Impact model**—reflects the influence of the objects during the implementation of the decision-making processes.

Regarding the types of systems related to managing orders to suppliers, a number of authors [37–39] accept the thesis that when choosing an ordering system have the greatest importance the moment of placing the order, the size of the order, and the frequency of the orders. According to these factors, they separate the systems for managing orders to suppliers in the following groups, presented in Table 2 [35, 37, 38]. The authors of the paper summarize the presented classification of commercial systems intended for managing orders to suppliers in tabular form.

The object-oriented standard for visual interpretation UML is composed of visual elements that can describe each of these models to build the overall vision of the proposed model of business intelligent order management system to suppliers in the retail chain. Research by authors [21, 40] summarizes the stages in the modeling, development, and maintenance of information systems that use UML tools. They are presented as a connected process that can be iterative.

In the present study, the stages of development of a supplier order management information system are presented in the form of a diagram (see Fig. 1) using the UML standard [21, 40].

Each of the mentioned stages is distinguished by the following characteristic features:

1. **Vizualization**—it is the main goal of the UML standard, as its notations and diagrams provide a standard mechanism for representing requirements,

Table 2 Classification of systems for preparation of orders to suppliers

Name of system	Features
Rhythmic ordering system	Orders are shipped at regular intervals, and the quantity of goods ordered may vary. Inventory is controlled only on the date of the order, and the quantity is calculated as the difference between stock and desired stock
Order point system	The level of inventory is determined. Upon reaching it, an order is prepared. It is necessary to build an information system that constantly monitors inventory level
Optimal system	It combines the "Rhythmic orders" and "Order point" systems. It is necessary to determine the exact time for placing an order
System with two levels of inventory	The lower and upper limits of the stock are determined. The order is placed when the lower limit of the stock is reached, and the requested quantity is the difference between the maximum and the actual level of the stock

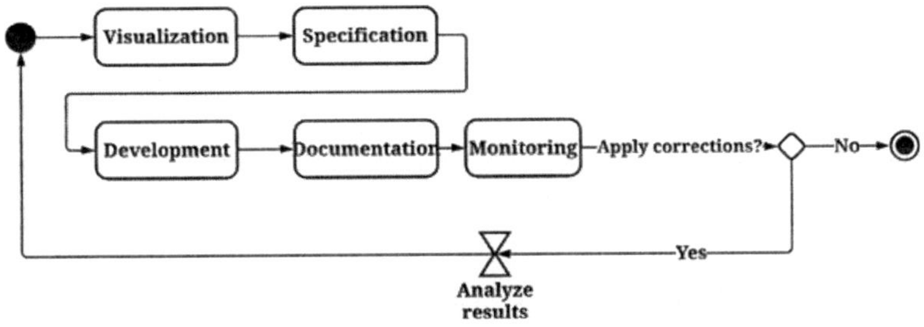

Fig. 1 Stages of development of an order information system for managing orders to suppliers with UML

processes, solution design, and architecture. Visualizations are created using development environments that allow to share visualizations within a team of developers.

2. **Specification**—UML helps facilitate the modeling of artifacts. In the context of computer information systems, the term artifact is used to denote the products used in the development of computer information systems. An example of this is the specifications of business cases, classes, attributes, and operations for which special notations are provided. These specifications help to eliminate inaccuracies during the development of the system, as each of them emphasizes a specific aspect of it.
3. **Development**—UML is used to develop software because it allows the generation of program code when imported depending on the applied software.
4. **Documentation**—UML provides additional and detailed documentation of requirements, architecture, design, project plans, tests, and prototypes to improve specifications and visual images.

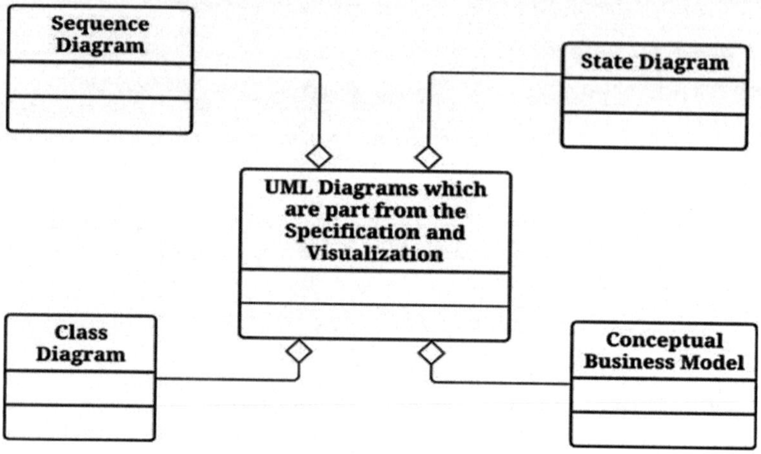

Fig. 2 Diagrams involved in the graphical representation of the commercial intelligence system model

5. **Support**—well-documented UML models provide significant assistance in the ongoing maintenance of software systems. The models allow easy visualization of an existing system, its architecture, and technological design. This allows developers to identify the correct places in the system that are subject to change and to assess the effect of the implemented change on the rest of the system.

The diagrams developed using the UML standard, graphically representing the models presented by Burgman, include conceptual business model, business case diagram, class diagram, sequence diagram, and state diagram. The diagrams are positioned in the first two stages of the full development cycle of the information system—visualization and specification The specified set of diagrams, which will be discussed in the next presentation, are presented in Fig. 2 and will be presented in the next statement as well.

3.1 Conceptual Business Model

The conceptual business model is an integral part of the business modeling process. It shows the specific purpose of the considered system in its entirety. The developed conceptual model of the process is the basis for the object-oriented presentation of elementary processes [41]. These are processes that cannot be decomposed into simpler ones. They define the roles and activities performed, as well as the artifacts that are obtained in the development of information systems [42].

As mentioned, the main purpose of the study is to propose a conceptual model of a business intelligent order management system in a retail chain. According to the authors' opinions presented above, the evolution of technologies related to supply management provides new opportunities for updating and upgrading business processes, which makes the choice of subject area relevant. In their study, Martinez-Rolan, and Otero-Pineiro [43] define business intelligence as the ability to transform a set of data into information, and information into knowledge that can be applied to business decision-making processes of solutions. From the perspective of commercial systems and supply management, business intelligence can be applied by integrating a module through which the process of managing orders to suppliers is automated and the required quantities of products for future periods are calculated based on forecast values using statistical methods [44]. According to the concept presented by Knolmayer et al. [17], the forecasting process implemented in a trading system can be performed by forecasting the required quantities based on historical data containing the amount of sales made.

In line with the increased competition between retail chains in the market, it is particularly important that the projected quantities of goods meet the level of consumer demand [45]. This fact gives a serious reason to optimize the processes of supply planning with the development of a model of business intelligent order management system to suppliers.

On the other hand, in the trade of fast-moving consumer good,s there are specifics in drawing up orders to suppliers. One of the most important requirements is that they meet consumer demand. When forecasting the required quantities for sale, it relies mainly on the forecast of customer demand, and it is necessary to set a framework that includes a set of influencing factors. These include: Point of Sales (POS) data, loyalty card data, and historical sales data [38, 46], as well as existing data in some retail chains digital discount card implemented through free mobile applications. In order to achieve a more accurate result in the preparation of the processes for planning the required quantities, additional values may be included, which will help to specify the final values, such as calendar data, demographic data, and information about promotional campaigns of competitors, etc.

1. **Data from loyalty cards**—this type of data also belongs to the group of demographic data but provides more accurate information about consumers in the specific area where the store is located.
2. **Transaction data from Point of Sales system**—using POS data from the in-house system it is possible to determine the best-selling products for a specific period, average customer account size, and in combination with data from loyalty cards, it is possible to define a group of preferred products for a specific customer group.
3. **Meteorological conditions**—the climatic season can be attributed to the determining factors for human behavior. Detailed weather forecast for a specific period can be accessed on certain Internet sites, such as. https://www.accuweather.com/.
4. **Calendar**—calendar information related to the distribution of the days of the week, national holidays, rites, and customs is also relevant to consumer attitudes related to the consumption of goods.

5. **Population data**—demographic and social statistics are available on the website of the national statistical institution, which represents the average purchasing power, consumption, purchased quantities of goods, etc.
6. **Promotional campaigns of competitors**—another factor determining the demand for products is the supply of promotional goods by competitors located in the immediate vicinity. Geographic data for specific sites can be used by sites with electronic maps such as https://www.google.bg/maps, and competing promotional products from the company page of the chain or from aggregator sites, i.e. these are sites that make comparisons of the prices of the products participating in the promotional campaigns.

In their book, the authors Knolmayer et al. [17] share the opinion that the participating parameters and their quality are essential for building the forecast model. In this publication, it is proposed that the data extracted from Internet sources go through a module to improve their quality, as the processes set out in the module include several stages. These actions are as follows: cleansing, enrichment, and validation of the data flow, after which the processed data should be directed to the business intelligence module to create a forecast model. The graphical representation of the module is presented in Fig. 3, by means of the BPMN standard.

The discussed module for improving data quality is an essential component in the conceptual model of business intelligent system for managing orders to suppliers. The graphical representation of the proposed conceptual model is visualized in Fig. 4. The diagram shows the entities that make up the order management system. It is focused on presenting the relations between its constituent objects, using the UML standard to create it. Each object that makes up the system is presented independently. Because the diagram illustrates objects without their associated properties, emphasis is placed on the relationships between them and their organization.

It does not define fields, attributes, or methods. The "Network Controller", "Local System", "File System", "Data Purification", "Forecasting", "External Sources", "Order", "Comment", "Note", "Secure Order", and "Encryption objects" are described in Sect. 3.3. Class Diagram and presented in Fig. 5.

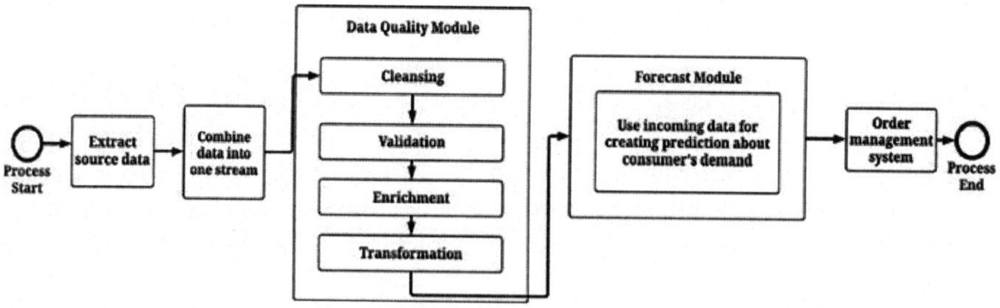

Fig. 3 Model of a data cleansing module in a business intelligent supplier order management system

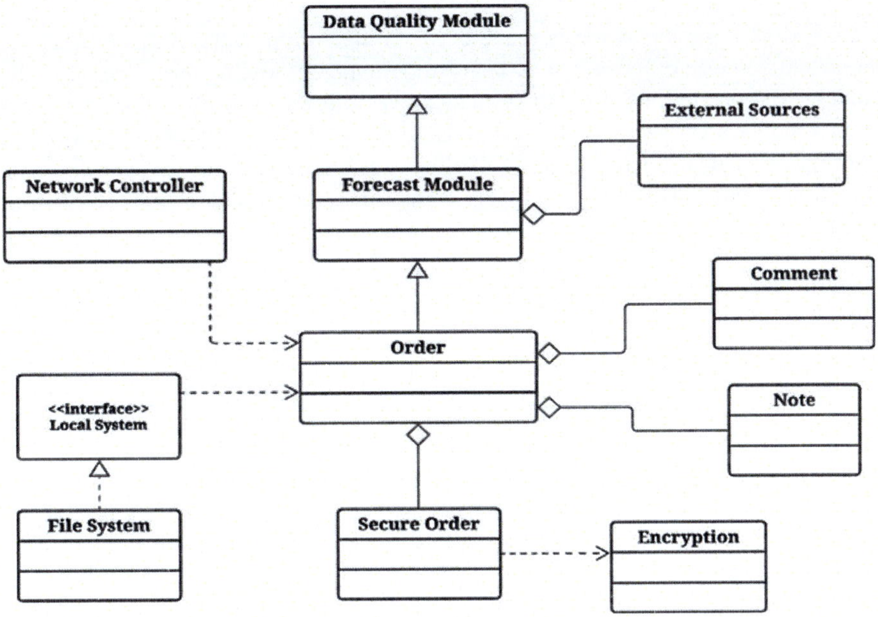

Fig. 4 Conceptual model of a supplier order management system

3.2 *Business Case Diagrams*

The graphical representation of the system is based on the object-oriented approach laid down in UML methods. Some authors [25, 47] point out that the application of UML is methodologically assured and supported by a number of development environments. For this reason, its use is constantly expanding its scope and several IT companies are initiating its application. This determines the prospects in the application of UML for the purposes of business modeling in the context of the present development. According to the requirements for creating a UML model, before starting the actual development of diagrams, it is necessary to gather detailed information from future users of the system. This is followed by a study and summary of the information obtained, which sets out the basic needs of stakeholders. Usually, the needs are identified with the need for additional functionalities that the new system needs to have. After defining them, each user defines the most frequently performed operations related to its position in the organization. The process described so far is referred to as the collection of requirements for building a prototype system [48]. According to another study [49], the capabilities of the system can be categorized as externally accessible services that it provides to customers or users who work directly with it. The author recommends, along with the possibilities, to clearly define the limitations of the functionalities of the system.

In addition, a comparison is made between the current state of the system and the defined criteria based on the collected requirements. Among the common shortcomings of outdated information systems are as follows: slow execution of current operations, lack of process automation, lack of a module for automatic sending of information. The presented prototype includes several basic operations that could be part of an automated business intelligent system. They are structured in the following groups:

1. Automatic creation, modification, and sending of orders to suppliers by e-mail or Electronic Data Interchange.
2. Protection orders to suppliers that contain sensitive information.
3. Automatic storage of prepared orders to suppliers in the cloud space (using cloud services such as Google Cloud, MS Azure, etc.).

The listed functionalities can be differentiated in the following directions—creation and change of an object (order), security, and synchronization with the help of cloud technologies. Based on the already defined requirements, the possible business scenarios for interaction with the system can be highlighted. Each of them can be represented graphically by means of UML, which can be the subject of a detailed graphical presentation in a future publication.

3.3 Class Diagram

Some authors [50] point out that in the object-oriented approach, the most used diagrams for representing systems are class diagrams. They represent each of the entities that make up the system, with their associated attributes. From the point of view of the order management system, the entities that will build the model described so far are the following (see Fig. 5):

1. **Data cleansing**—the entity is characterized by methods for enriching, cleansing, and validating the entire incoming data stream. After successful implementation of the methods, the data passes to the object "Forecasting".
2. **Forecasting**—this entity has an adjacent "External Sources" attribute. As this attribute sends additional data directly related to the preparation of a forecast model, it is depicted as a separate entity. The "Forecasting" entity is determined by a forecasting method, the result of which will be reflected in the "Order" entity.
3. **Order**—with associated attributes "Content", "Comment" and "Note". Because the last two attributes, "Note" and "Comment", can contain additional content, they are also shown as separate entities. The "Order" is characterized by methods for setting and adding additional content.
4. **Secure order**—with a related attribute for storing an encrypted password. This entity is directly related to the "Encryption entity", which has a method for setting and encrypting a password.

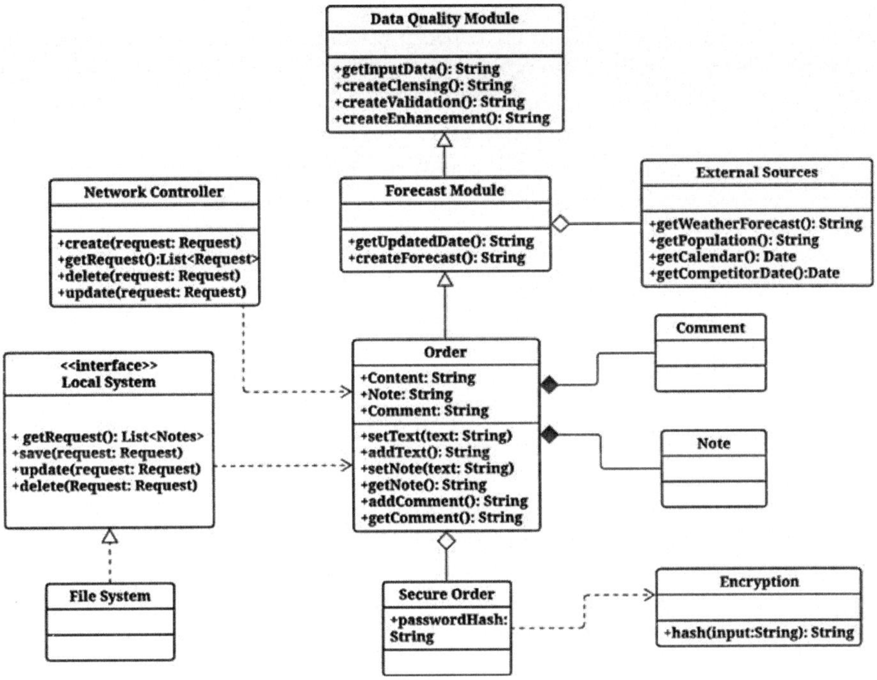

Fig. 5 Class diagram of a business intelligent supplier order management system

5. **Local provisioning**—an entity-type interface that takes function of the core of the system. The operations it contains are related to the main commands used when working with orders, namely—create, save, modify, and delete. The system's "Local Security" interface is invariably linked to the "File System", which is shown as a separate entity.
6. **Network controller**—an entity with methods for managing the creation, modification, deletion, and access of an order. The "Network controller" manages the requests to the system made by its users.

The "Prediction" class includes a method for creating a forecast ("+ Create-Forecast (): String"). Due to the complexity and variability of the forecasting process, the use of a business intelligence algorithm is recommended for the software implementation of the method.

3.4 Sequence Diagram

The graphical representation of the system model is related to the reproduction of the actions when executing commands. In his book, Lano [49] shares the view that developing a "sequence diagram" helps to reveal the roles in the process that coordinate its implementation. The interactions between the objects are presented graphically and give a visual example of the behavior of the system in terms of communication between its constituent objects. Each of them is presented as an independent entity that can send and receive signals from other entities (see Fig. 6).

The naming of objects follows a standard in which the first character of the name is a reference to the sequence of the object within its family. The sequence of objects is marked with a letter symbol, as "a" is an indicator of the first position. If there is a second object of the corresponding type, it will be named with the first letter "b". The objects forming the diagram are the following:

- visual presentation of the actions "aPreview";
- a controller that sends and receives responses from "aController" commands;
- "Order" order;
- file manager, part of the operating system "aFileManager";
- online manager for cloud storage management "aOnlineManager";
- e-mail server for managing the sending of orders to "E-mail server" providers.

The sequence of actions related to the presented objects is visualized graphically in Fig. 6 and includes the following technical steps: when creating a new request, a signal is sent to the controller, which is forwarded to the object "aOrder". After the completion of the work on the respective order, it is saved, and a signal is sent to "aController" to report the event. The newly created order is ready to be attached to the cloud space. A signal is sent to "offline Manager", and after the action is completed, a response is returned to the controller. After receiving a response for completed action, the order can be sent to a supplier by email. A signal is sent to "e-Mail server" to send an order. Upon completion of the action, a confirmation is sent to "aController" for the end of the activity.

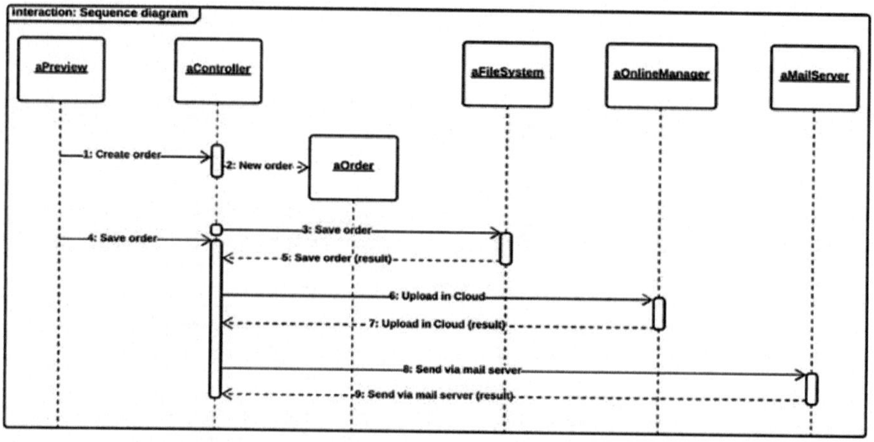

Fig. 6 Business sequence diagram of a supplier order management system

3.5 State Diagram

The state diagrams designed with UML means show in general the basic concept of the system to be considered. They semantically combine business scenarios, consistency, and possible states in the system [50–52].

The developed diagram shows the activities that are part of the prototype of the automatic order management system. Each activity is presented as a separate object, and the connections show the possible actions and consequent outcomes. Variants for successful and unsuccessful execution of an occurred event are defined (see Fig. 7). The automation of the process is determined by the active connection between each of the objects and the sequence in performing the actions that make up the presented model. The start of the process is determined by the action to create an "order". The data extraction module is activated and through data transmission infrastructure they are directed to the cleansing processes. After improving the quality of the data, the data flow moves to the next module in the chain which is called "Forecasting". It is

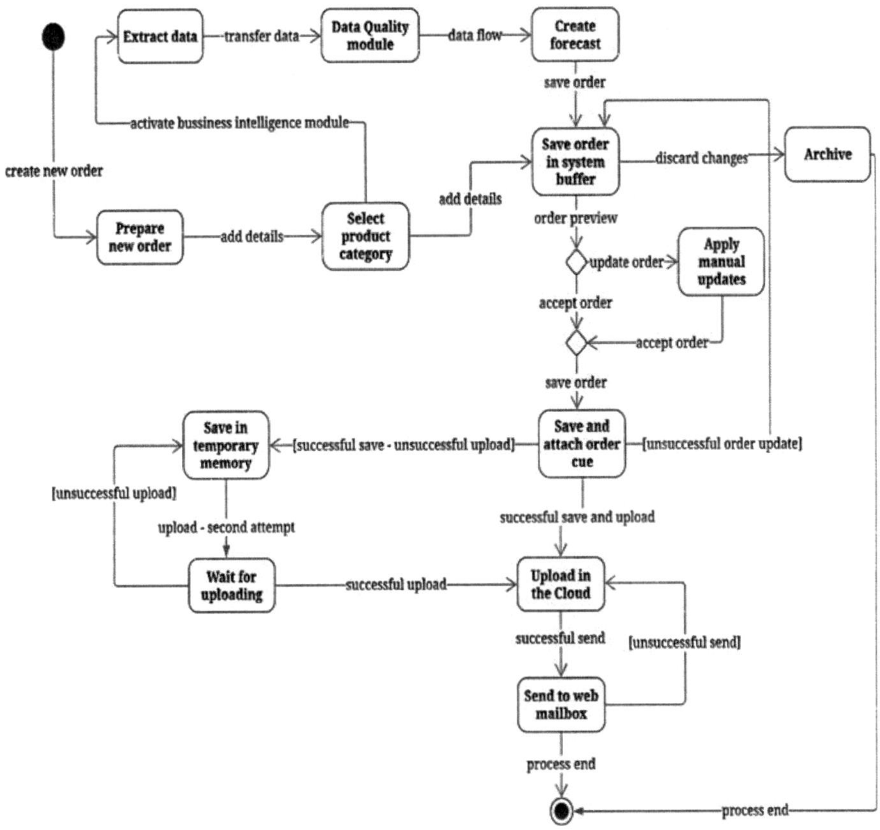

Fig. 7 Business state diagram of the of an automatic order management system to suppliers

possible to apply additional details to the "order" or, if necessary, manually adjust the values obtained relating to the quantities of the product. The system's user has the functionality to make adjustments if necessary, as it is possible that the results obtained after calculating the estimated quantities have minimal deviations.

In the situation of an unexpected interruption of the program, the "order" will be archived in order to minimize the loss of information. Otherwise, it will be stored in the "System buffer". If the user decides to completely reject the creation of a "New order", the process will end. In case the user decides to continue working on the "Order", it will be saved in the "Local system" and will be prepared for storage in the cloud. If the process is performed without related errors, the "Order" will be sent to the appropriate "Supplier". The business status diagram also describes events that occur in the presence of system disturbances. The "save and attach an order" action may cause a network problem, so it will be rejected, and execution will return to "Save to system buffer".

During the "Cloud storage process", an Internet connection may be disrupted, then the request will be stored in temporary memory and will be in standby mode. The same information retention measures will apply if there is a problem sending an order by email to a supplier.

The business state diagram is the last visual element that shows aspects of the developed model of the business intelligent order management system to suppliers in retail chains, presented in the paper. This type of diagram completes the system specification stage according to the methodology developed by the authors Weilkiens et al. and Unhelkar [21, 40] for graphical representation of information systems using UML.

4 Conclusion and Future Work

This paper discusses the methods used to model a business intelligent system. The paper proposes a conceptual model of a business intelligent order management system for suppliers in a retail chain. The presentation of subprocesses and related activities is performed by means of UML diagrams. A diagram describes the conceptual business model, which illustrates the objects that are part of the system, as well as the relationships between them. The developed model of a business intelligent order management system to suppliers in a retail chain of fast-moving consumer goods presents the entities involved in the model with a high level of abstraction.

Next, in the paper, a proposal is made for a graphical representation of the decomposed processes that make up the system, as each process or sequence is presented with an appropriate diagram. The developed diagrams belong to the following types: class diagram, sequence diagram, and state diagram. Each diagram is accompanied by a description of the implementation related to the entities, objects and/or activities involved. Due to one of the great advantages of the UML standard, namely its structure, the diagrams based on this method are easy to understand and read.

The UML standard is one of the fundamental visual object-oriented languages for modeling business information systems. Based on it, the foundation concept of the model of business intelligent order management system to suppliers in the retail chain of fast-moving consumer goods framework has been made. The main distinguishing features of the model are compliance with the stages in the development of a model with UML standard, focusing on the process of predicting customer demand, taking into account local factors and adding a module for data cleansing to improve quality of the information managed by the system.

Part of the future research intentions are related to an in-depth study of the factors directly or indirectly related to the processes of making purchases in retail chains, as well as the structure of incoming data flows and methods related to their extraction and processing. Future research may focus on testing and validating various artificial intelligence methods to be used to determine the volume and frequency of deliveries.

References

1. Todoranova L, Nacheva R, Sulov V, Penchev B (2020) A model for mobile learning integration in higher education based on students' expectations. Int J Interact Mob Technol (iJIM) 14(11):171–182. https://doi.org/10.3991/ijim.v14i11.13711
2. Vivek K (2019) Enterprise process management systems. In: Engineering process-centric enterprise systems using BPMN 2.0. CRC Press
3. Geambaşu CV (2012) BPMN vs. UML activity diagram for business process modeling. Account Manag Inf Syst 11(4):637–651
4. Melo PN, Machado C (2020) Business intelligence and analytics in small and medium enterprises. CRC Press
5. Nakamo M (2020) Supply chain management: strategy and organization. Springer, Kyoto
6. Deloitte (2017) Fast moving consumer goods analytics framework. Deloitte. https://www2.deloitte.com/content/dam/Deloitte/nl/Documents/consumer-business/deloitte-nl-cip-fmcg-analytics-framework.pdf. Last accessed 2020/09/11
7. Wu Y (2019) Achieving supply chain agility: information system integration in the Chinese automotive industry. Palgrave MacMillan
8. Vasilev J, Stoyanova M (2019) Information sharing with upstream partners of supply chains. In: 19th international multidisciplinary scientific GEO conference SGEM 2019, geoinformatics and remote sensing, vol 19(2.1). STEF92 Technology Ltd., Sofia, pp 329–336. https://doi.org/10.5593/sgem2019/2.1
9. Walker WT (2005) Supply chain architecture: a blueprint for networking the flow of material, information, and cash. CRC Press
10. Wati FA et al (2019) Enterprise architecture for designing human resources application standard reference. Int J Innov Technol Explor Eng 8(12):5173–5178. https://doi.org/10.35940/ijitee.L2772.1081219
11. Hitpass B, Astudillo H (2019) Editorial: industry 4.0 challenges for business process management and electronic-commerce. J Theor Appl Electron Commer Res 14:1–3. https://doi.org/10.4067/S0718-18762019000100101
12. Santos J, Graciano Neto V, Nakagawa E (2020) Business process modeling in systems of systems. In: Anais do II Workshop em Modelagem e Simulação de Sistemas Intensivos em Software, pp 26–35
13. Bouzidi A, Haddar N, Kais H (2019) Traceability and synchronization between BPMN and UML use case models. Ing Syst d'Inf 24(2):215–228. https://doi.org/10.18280/isi.240214

14. Munteanu FI, Purcarea AA (2019) A theoretical comparison upon the use of UML or BPMN in the modeling of logistic processes and the management of logistical operations. Sci Bull Naval Acad 22:53–60. https://doi.org/10.21279/1454-864X-19-I2-006
15. Suchenia A et al (2019) Towards knowledge interoperability between the UML, DMN, BPMN and CMMN models. Krakow. https://doi.org/10.1051/matecconf/201925202011
16. Agrawal N, Smith S (2015) Retail supply chain management quantitative models and empirical studies, 2nd edn. Springer
17. Knolmayer GF, Mertens P, Zeier A, Dickersbach JT (2009) Supply chain management based on SAP systems. Architecture and planning process. Springer
18. Hamid K, Ahmed H (2018) Integrating business process concepts into UML activity model. SUST J Eng Comput Sci (JECS) 19:57–63
19. Sachs JD et al (2019) Six transformations to achieve the sustainable development goals. Nat Sustain 2:805–814. https://doi.org/10.1038/s41893-019-0352-9
20. Yun JJ, Won D, Park K, Jeong E, Zhao X (2019) The role of a business model in market growth: the difference between the converted industry and the emerging industry. Technol Forecast Soc Change 146(C):534–562. https://doi.org/10.1016/j.techfore.2019.04.024
21. Unhelkar B (2018) Software engineering with UML. CRC Press
22. Achouri A, Ayed JbL (2016) A formal semantic for UML 2.0 activity diagram based on institution theory. Int J Soft Comput Softw Eng [JSCSE] 3(3.30). https://doi.org/10.7321/jscse.v3.n3.30
23. Peixoto D et al (2008) A comparison of BPMN and UML 2.0 activity diagrams. In: VII Simpósio Brasileiro de Qualidade de Software
24. Suchenia A, Łopata P, Wisniewski P, Stachura-Terlecka B (2019) Towards UML representation for BPMN and DMN models. https://doi.org/10.1051/matecconf/201925202007
25. Wautelet Y, Poelmans S (2017) An integrated enterprise modeling framework using the RUP/UML business use-case model and BPMN. In: Poels G, Gailly F, Serral Asensio E, Snoeck M (eds) The practice of enterprise modeling. PoEM 2017. Lecture notes in business information processing, vol 305. Springer. https://doi.org/10.1007/978-3-319-70241-4_20
26. Rajagopal D, Thilakavalli K (2017) A study: UML for OOA and OOD. Int J Knowl Content Dev Technol 7(2):5–20. https://doi.org/10.5865/IJKCT.2017.7.2.005
27. Bridgeland DM, Zahavi R (2009) Business modeling: a practical guide to realizing business value. Morgan Kaufland Publishers
28. Cvetković D (2019) Modeling and computer simulation. IntechOpen, London
29. Akkiyat I, Souissi N (2019) Building a process meta model extended for cycles. In: Proceedings of the Arab WIC 6th annual international conference research track (ArabWIC 2019). Association for Computing Machinery, New York, NY, pp 1–7, Article 17. https://doi.org/10.1145/3333165.3333182
30. Ciccozzi F, Malavolta I, Selic B (2019) Execution of UML models: a systematic review of research and practice. Softw Syst Model 18:2313–2360. https://doi.org/10.1007/s10270-018-0675-4
31. El Hayat SA, Toufik F, Bahaj M (2020) UML/OCL based design and the transition towards temporal object relational database with bitemporal data. J King Saud Univ Comput Inf Sci 32(4):398–407. https://doi.org/10.1016/j.jksuci.2019.08.012
32. Sabatucci L, Cossention M (2019) Supporting dynamic workflows with automatic extraction of goals from BPMN. ACM Trans Auton Adapt Syst (TAAS) 14(2):1–38. https://doi.org/10.1145/3355488
33. Planas E, Cabot J (2020) How are UML class diagrams built in practice? A usability study of two UML tools: Magicdraw and Papyrus. Comput Stand Interfaces 67. https://doi.org/10.1016/j.csi.2019.103363
34. Zhaoxia G (2016) Intelligent decision-making models for production and retail operations. Springer Verlag, Berlin, Heidelberg
35. Berman B, Evans JR, Chatterjee P (2018) Retail management. A strategic approach, 13th edn. Pearson

36. Burgin M (2019) Evaluation of information in the context of decision-making. In: Bosse E, Rogova GL (eds) Information quality in information fusion and decision making. Springer, pp 279–294
37. Andarwati M, Amrullah F, Thamrin E, Muslikh AR (2020) An analysis of point of sales (POS) information systems in SMEs with the black box testing and PIECES method. IOSR J Bus Manag 22(9):20–25. https://doi.org/10.9790/487X-2209052025
38. Dolgui A, Ivanov D, Sethi S, Skolov B (2018) Control theory applications to operations systems, supply chain management and industry 4.0 networks. Elsevier, Bergamo
39. Hajiabbas MP, Mohammadi-Ivatloo B (2020) Optimization of power systems problems: methods, algorithms and MATLAB codes. Springer
40. Weilkiens T, Lamm JG, Roth S, Waler M (2015) Model-based system architecture. Wiley
41. Petter S, DeLone W, McLean E (2008) Measuring information systems success: models, dimensions, measures, and interrelations. Eur J Inf Syst 17:236–263
42. Fjeldstad OD, Snow CC (2017) Business model and organization design. Long Range Plan 51(1):32–39. https://doi.org/10.1016/j.lrp.2017.07.008
43. Martinez-Rolan X, Otero-Pineiro T (2020) Marketing analytics: why measuring web and social media matters. In: Business intelligence and analytics in small and medium enterprises. CRC Press, pp 75–88
44. Salles M (2015) Decision-making and the information system, 3rd edn. Wiley
45. Dimitrova V (2017) Gross margin and buyer power in Bulgarian food retailing. Economic studies. Bulgar Acad Sci 26(6):172–189
46. Khade AA (2016) Performing customer behavior analysis using big data analytics. Procedia Comput Sci 79:986–992. https://doi.org/10.1016/j.procs.2016.03.125
47. Miles R, Hamilton K (2006) Learning UML 2.0. O'Reilly Media
48. El-Adaileh AN, Foster S (2019) Successful business intelligence implementation: a systematic literature review. Successful BI implementation. J Work-Appl Manag 11(2):121–132. https://doi.org/10.1108/JWAM-09-2019-0027
49. Lano K (2017) Agile model-based development using UML-RSDS. CRC Press
50. Nyisztor K (2018) UML and object oriented design foundations. Leakka
51. Booch G, Rumbaugh J, Jacobson I (2005) The unified modeling language user guide. Addison-Wesley Professional
52. Vasile E, Croitoru I, Simon DO (2016) Using information systems in decisions for businesses. Intern Audit Risk Manag 11:91–106

Forecasting in Enterprise Resource Planning (ERP) Systems: A Survey

Hassan ElMadany, Marco Alfonse, and Mostafa Aref

Abstract Using machine learning in ERP enables the organization to store, manage, and analyze their data to get the right decisions and gain valuable insights that were previously unmanageable. There are many industries and lines of business that contain large volumes of data such as manufacturing, finance, and healthcare … etc. One of the most usages of machine learning in ERP is forecasting which refers to the prediction of the future based on past data. This paper surveys different approaches for forecasting in ERP to gain profit for the business process.

Keywords Machine learning · ERP · Forecasting · Sales forecasting · Production forecasting · Demand forecasting

1 Introduction

Enterprise Resource Planning (ERP) is a business management software that is used to facilitate the business process cycle by integrating all business modules and functions like Human Resource (HR), Purchasing, Sales, Inventory Control (IC), Finance, Production, Customer Relationship Management (CRM), Accounting, … etc., into a single system, so your business operates as a seamless one. It covers the business needs and challenges in an effective, efficient, and productive way [1, 2]. ERP holds a large amount of data called big data. In the past, mainstream research in this area is focused on the implementation of critical success factors, ERP upgrade, project management practices, etc. Future research is needed to explore the potentials of ERP systems to be linked to new technology like big data and machine learning to

H. ElMadany (✉) · M. Alfonse · M. Aref
Computer Science Department, Faculty of Computer and Information Sciences, Ain Shams University, Cairo, Egypt
e-mail: hassanelmadany@cis.asu.edu.eg

M. Alfonse
e-mail: marco_alfonse@cis.asu.edu.eg

M. Aref
e-mail: mostafa.aref@cis.asu.edu.eg

D. A. Magdi et al. (eds.), *Digital Transformation Technology*, Lecture Notes in Networks and Systems 224, https://doi.org/10.1007/978-981-16-2275-5_24

get accurate decisions. There are many benefits of using machine learning in business such as faster and accurate decision making, data adaptability in real-time, improved efficiency, achieve a high level of automation, and forecasting.

In this survey, the authors focused on using forecasting in ERP systems. Forecasting can be defined as the process to predict future based on past data. It is used to gain profit for the business process [1]. It can be classified into three main categories as below [2]:

- Economic forecasting: predicting financial indicators such as cash supply, inflation rates, and interest rates.
- Technological forecasting: predicting technological advancement and innovation.
- Demand forecast: used to forecast future demand for the goods or services of a company.

There are two main types of forecasting methods in these three categories as follows:

- Qualitative methods: these kinds of forecasting methods are based on judgments, views, intuition, feelings, or personal experiences. They do not depend on strict computations of mathematics. They could be one of the following:
  - Executive opinion: it is a strategy in which a group of executives meets and jointly create a forecast.
  - Market survey: it is a strategy using interviews and surveys to evaluate client preferences and demand assessment.
  - Sales force composite: it is an approach in which each seller estimates sales in his or her area.
  - Delphi method: it is an approach in which consensus agreement is reached among a group of experts.
- Quantitative techniques: mathematical methods are based on these kinds of prediction techniques. They could be one of the following:
  - Time-series models: they look at previous information trends and try to predict the future based on the underlying trends in that information.
  - Associative models: models suppose that the predicted variable is linked to other environmental factors.

This paper is organized as follows: Sect. 2 presents forecasting methods in ERP, Sect. 3 presents a comparison between the approaches discussed in the survey, and finally, the conclusion and future work are reported in Sect. 4.

2 Forecasting in ERP

Forecasting in ERP can be categorized into three main categories as follows:

- Production forecasting.
- Sales forecasting.
- Demand forecasting.

Production forecasting is a critical step in ERP, including determining the quantity of inventory to be held on hand, how much raw material should be bought, or how much a product should be produced. The authors in [3] created a model to improve production forecasting in the automotive industry. The model is a combination between the gray model (GM) system and support vector machine and least squares Support vector machine models (SVM/LSSVM). The model has three connection methods as follows:

- Series connection: it is used to construct a multidimensional LSSVM model using the distinct single-dimensional GM. Because the GM targets the single-dimensional series, it can perform even in multiple samples and LSSVM is good at handling multiple-dimensional and large samples (as shown in Fig. 1).
- Parallel connection: it is simultaneously applying GM and LSSVM (as shown in Fig. 2).
- Residual connection: first apply the GM, then use the output to be the next input of the LSSVM model (as shown in Fig. 3).

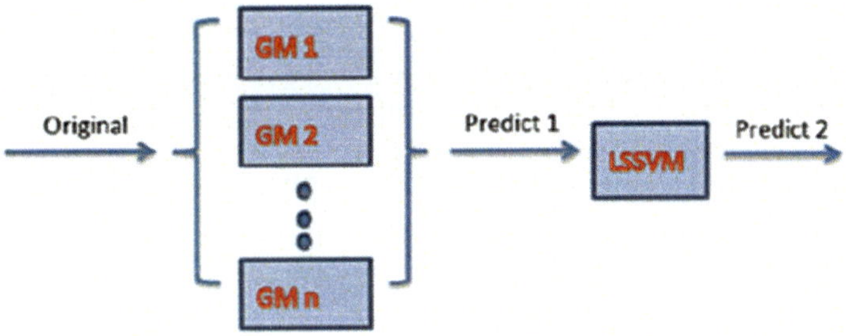

Fig. 1 Series connection [3]

Fig. 2 Parallel connection [3]

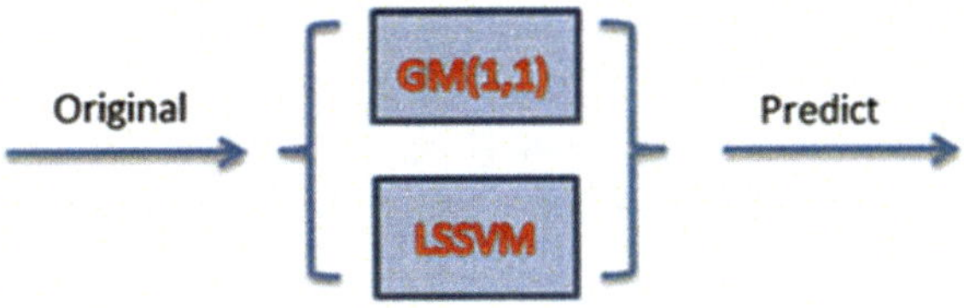

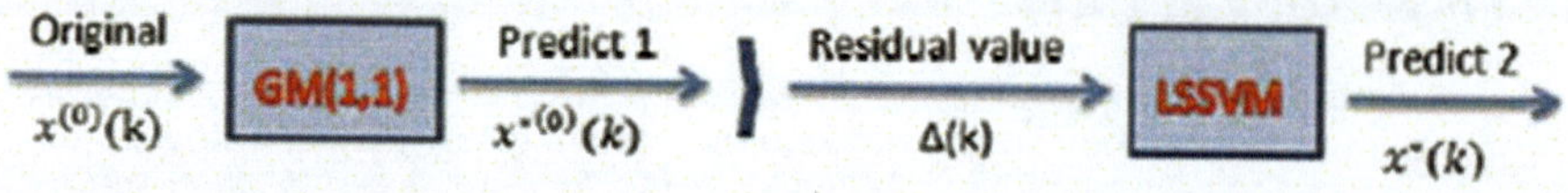

Fig. 3 Residual connection [3]

The combination model in [3] using the single GM and LSSVM model proves that it has a more powerful forecasting ability for the small-scale samples than single SVM or LSSVM. The advantages of the combination model are as follows:

- High accuracy.
- Best performance.
- The forecasting model will be more reliable.

In the business strategy, sales forecasting plays a very important role. Efficient sales forecasts can help decision makers measure production and material costs, selling prices, strategic planning, etc.

The authors in [4] describe a new hybrid sales forecasting model based on back-propagation neural networks (BPNN) and adaptive learning networks called FCBPN which stands for fuzzy clustering and back-propagation neural networks. They used data that were obtained from the Taiwan sales case study known as a printed circuit board (PCB). The approach consists of three stages as shown in Fig. 4

- Data preprocessing stage using Winter's Exponential Smoothing method.
- Extracting clusters memberships fuzzy system.
- Adaptive networks evaluating stage.

Using a fuzzy clustering enables the clusters to be larger and therefore improves the reliability of the forecast process results to a certain degree, allowing that information to be captured by a cluster.

In [5], the authors develop a hybrid model using the self-organization map (SOM) neural network, genetic algorithms (GA), and fuzzy rule base (FRB) to forecast the future sales of a printed circuit board factory. Monthly sales data from a real-world PCB Company in Taiwan from January 1999 to December 2003 have been collected for model training and testing. The model consists of four stages as shown in Fig. 5:

- Data collecting stage.
- Data clustering stage.
- Evolving fuzzy rule stage.
- Output forecasting stage.

In [6], the authors define a smart ERP-FCBPN hybrid sales forecasting scheme based on ERP architecture through Delphi, fuzzy clustering, and BPNN with adaptive learning rate. The model is experimented using data come from an industrial company that manufactures packaging in Tangier from 2001 to 2009. The approach consists of three phases as shown in Fig. 6

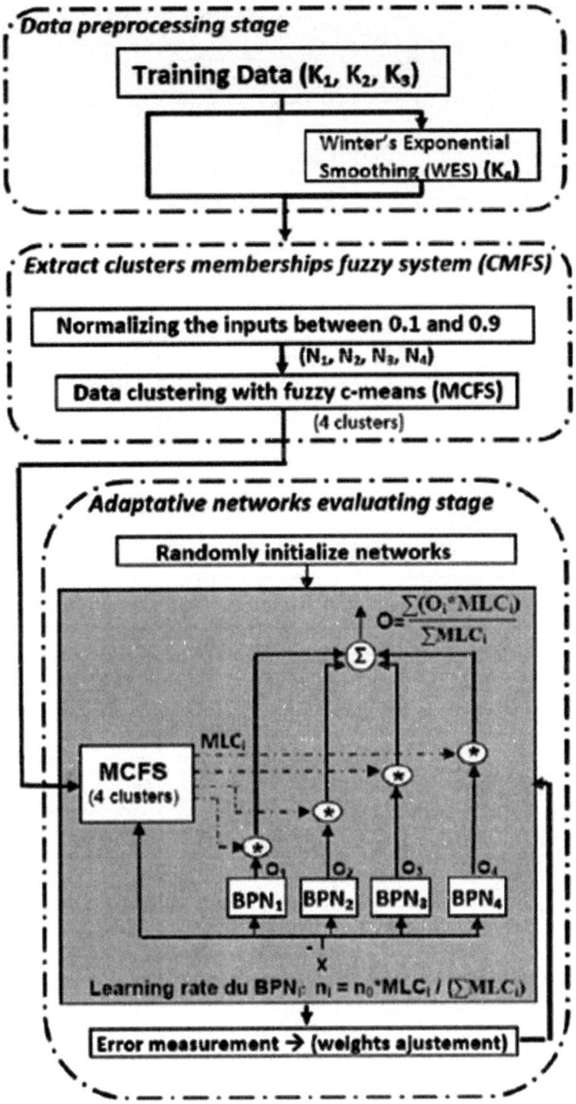

Fig. 4 Architecture of FCBPN model [4]

- **Data collection stage**: using Delphi method to determine the key factors that affect the sales volume of all the possible factors collected.
- **Stage of data pre-processing**: using Winter's Exponential Smoothing method by preprocessing all the historical data and use them to predict the production demand, so it can be used to get a preliminarily forecast the quantity of sales.
- **Stage of learning by FCBPN**: using Delphi-based hybrid sales forecasting scheme, fuzzy clustering, and BPNN with adaptive learning rates.

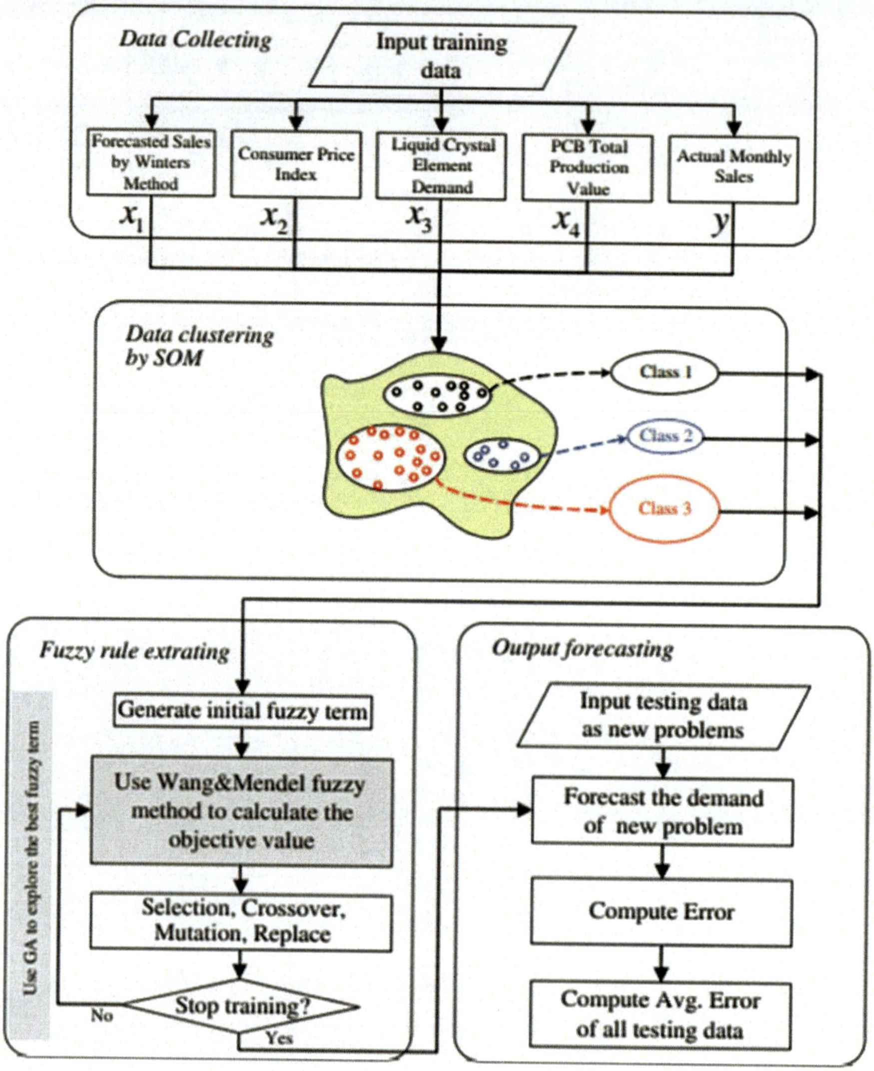

Fig. 5 Framework of the hybrid model by clustering and evolving fuzzy rules [5]

In [7], the authors defined forecasting as the fundamental step of demand management that optimizes customer satisfaction through supply chain forecasting capabilities as a prediction or estimate of the actual value in the future or another situation. Forecasting has an impact on customer requirements fulfillment, risk reduction, and improvement in the supply chain measurement. Researchers have been attracted by the role of forecasting in the rational decision-making process for future events. Figure 7 illustrates the general flow of forecasting. Team materials management (MM) is the forecast custodian. MMs role is to process, verify, and manage the

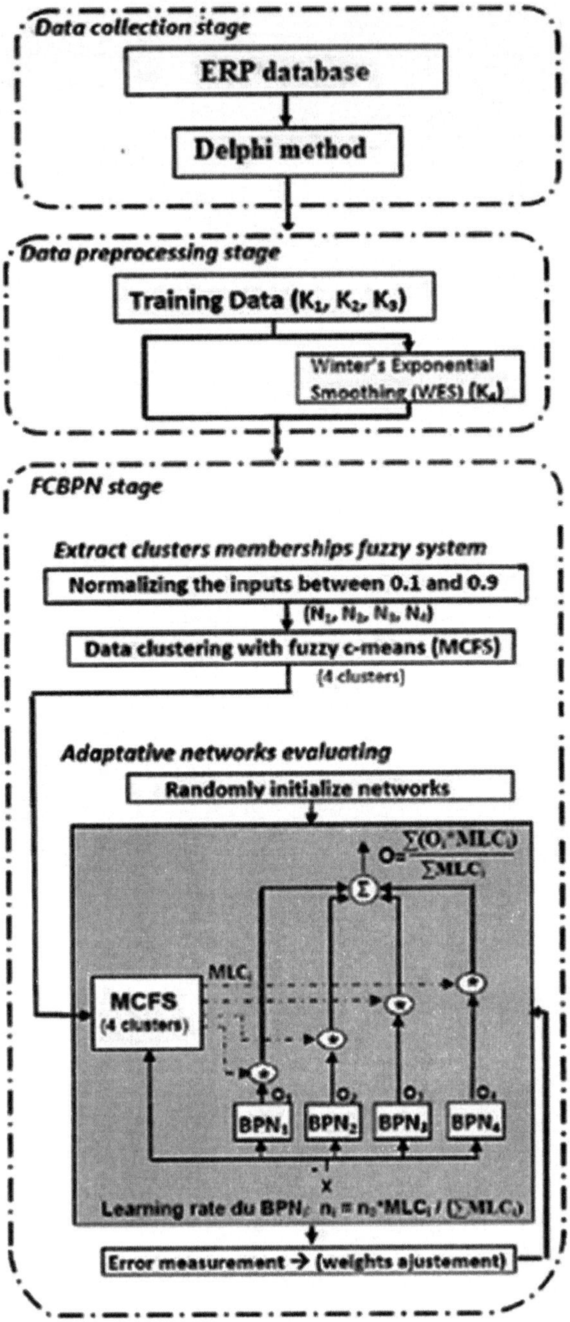

Fig. 6 Architecture of ERP-FCBPN [6]

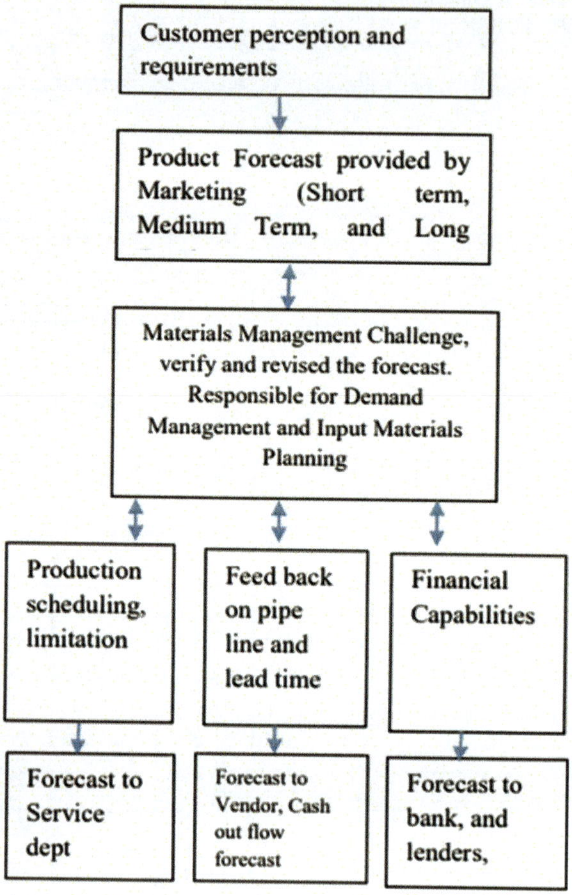

Fig. 7 General flow of forecasting [7]

product forecast to accelerate the organization's production of raw materials and finished goods and inventory planning.

MM managers check the forecast value with historical data after receiving the forecast and review it after considering the purchase delay, late shipment, production limitation, etc. User departments (production, marketing, finance, and supply chain management) share the revised forecast for comment. The final forecasting then converted to requirements of input materials through ERP or similar type software for sharing. MM team sometimes receives a new product forecast and an urgent forecast to meet any special requirements. Once the forecast figure has been agreed and finalized, all users will be demanded. MM Manager is the demand management anchor point. In these sectors, managers coordinate and share the information regarding all activities as mentioned in the general forecast flow. Even in case of any deviation, the managers make an alternative plan for ensuring maximum customer satisfaction. The managers continuously follow up all activities of implementation of forecasting in the organization. In some organizations, there is a weekly formal follow-up system regarding the materials management job. The forecasting management model is shown in Fig. 8.

In [8], the authors demonstrate the link between model predictive control (MPC) and the forecasting system and will investigate its effectiveness in managing the supply chain. Figure 9 illustrates the block diagram MPC scheme for inventory control. They use the following forecasting methods:

- Naive forecasting due to comparative purposes.
- Linear auto-regression (Linear AR).
- Holt-Winters.
- Radial basis function (RBF) neural network.

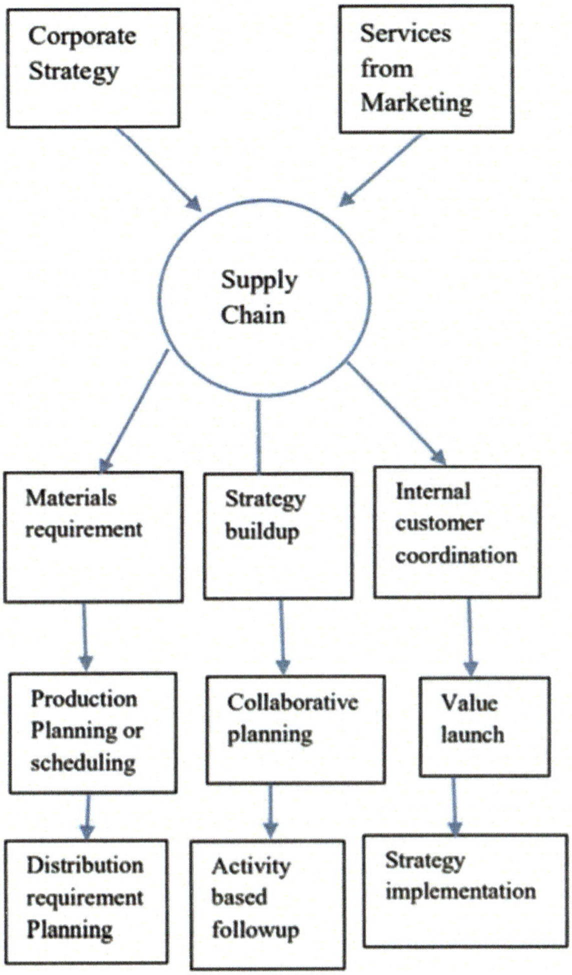

Fig. 8 Forecasting management model [7]

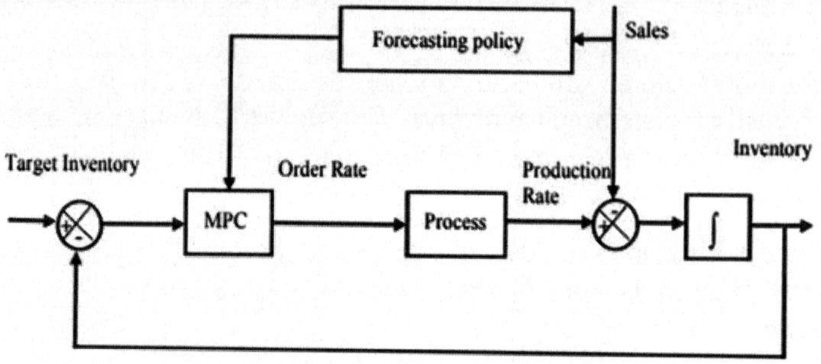

Fig. 9 Block diagram of MPC schema for inventory control [8]

The forecasting framework was tested in four test cases where, in each case, a different forecasting method as mentioned above was used to generate sales forecasts.

3 Discussion and Comparison

This section presents a discussion and comparison between approaches discussed in the survey as given in Table 1, and the comparison was based on four criteria: objective, methodology, dataset, and performance results. We presented this survey to promote the usage of forecasting in ERP systems in the following modules:

- Sales.
- Production.
- Demand planning.

Three main methods can be used to measure the performance:

- **Mean Absolute Error (MAE)**: it measures the average magnitude of the errors in a set of predictions without considering their direction. It is the average over the test sample of the absolute differences between prediction and actual observation where all individual differences have equal weight (see Eq. 1).
- **Root Mean Squared Error (RMSE)**: it is a quadratic scoring rule that also measures the average magnitude of the error. It is the square root of the average of squared differences between prediction and actual observation (see Eq. 2).
- **The Squared Correlation Coefficient**: it is a measure of the linear correlation (dependence) between two variables X and Y, giving a value between +1 and − 1 inclusive (see Eq. 3).

$$\text{MAE} = \frac{1}{n}\sum_{j=1}^{n}[y_j - \hat{y}_j] \quad (1)$$

Table 1 Comparison between the forecasting methodologies

References	Methodology	Dataset	Results
[3]	Gray model + LSSVM model	• The data set used from the automotive industry collected from the USA, Germany, China, Japan, Korea, and India • It includes 4 types as follows: – Cars – Light trucks – Medium trucks – Heavy trucks • Data were collected from 2004 to 2012	Using LSSVM model $R^2 = 0.758$ RMSE = 0.314 MAE = 0.256 Using Gray model $R^2 = 0.863$ RMSE = 0.264 MAE = 0.226 Using Gray + LSSVM $R^2 = 0.965$ RMSE = 0.151 MAE = 0.104
[4]	• BPNN • Adaptive learning networks	• The test data collected from PCB industry in Taiwan • Training data collected from January 1999 to December 2002 • Testing samples were from January 2003 to December 2003	MAPE = 1.97 RMSE = 18,009
[5]	• SOM neural network • GA • FRB	• Monthly sales data from a real-world PCB Company in Taiwan from 1999 to 2003	MAPE = 0.022 RMSE = 21,346.0
[6]	• Delphi method • Fuzzy clustering • BPNN with adaptive learning rate	• The data test collected from sales forecasting case study of manufactures packaging in Tangier • Data collected from January 2001 to December 2008 while • The total number of testing samples was from January 2009 to December 2009	MAPE = 3.00 RMSE = 201
[8]	• Naive forecasting due to comparative purposes • Linear AR • Holt-Winters • RBF neural network	• The test data was supplied by a leading Greek dairy products manufacturer	Average forecasting error: RBF ANN = 0.0534 Holt = 0.0958 Linear AR = 0.1020 Naï ve model = 0.1988

$$\text{RMSE} = \sqrt{\frac{1}{n}\sum_{j=1}^{n}(y_j - \hat{y}_j)^2} \quad (2)$$

where y_j refers to the observation, $\hat{y}_j$ refers to the predicted values of a variable, and n is the number of observations available for the analysis

$$R^2 = \frac{\left(n\sum x_i y_i - \sum x_i \sum y_i\right)^2}{\left[n\sum x_i^2 - (\sum x_i)^2\right]\left[n\sum y_i^2 - (\sum y_i)^2\right]} \quad (3)$$

where y_i, x_i are the set of observations n is the number of observations.

4 Conclusion and Future work

We presented this survey to promote the use of forecasting in ERP system in the following modules: sales, production, and demand. The works presented here achieved its main aim to effectively increase business profitability by using forecasting in such modules in ERP systems. Further research will concern on developing an approach to enhance using forecasting in ERP systems in different industries such as manufacturing and healthcare.

References

1. Pekša J (2018) Extensible portfolio of forecasting methods for ERP systems: an integration approach. Inf Technol Manag Sci 21
2. Kurzak L (2012) Importance of forecasting in enterprise management. Adv Logist Syst 6(1):173–182
3. Wong SF, Ho WI, Lin B, Huang Q (2014) A new model for production forecasting in ERP. Int J Ind Manuf Eng 8(4):555–560
4. Hicham A, Mohamed B (2012) An improved approach based on fuzzy clustering and back-propagation neural networks with adaptive learning rate for sales forecasting: case study of PCB industry. Int J Comput Sci Issues (IJCSI) 9(3):404
5. Chang PC, Liu CH, Wang YW (2006) A hybrid model by clustering and evolving fuzzy rules for sales decision supports in printed circuit board industry. Decis Support Syst 42(3):1254–1269
6. Hicham A, Mohammed B (2012) Sales forecasting based on ERP system through Delphi, fuzzy clustering and back-propagation neural networks with adaptive learning rate. Int J Comput Sci Issues (IJCSI) 9(6):24
7. Albarune AR, Habib MM (2015) A study of forecasting practices in supply chain management. Int J Supply Chain Manag 4(2):55–61
8. Doganis P, Aggelogiannaki E, Sarimveis H (2006) A model predictive control and time series forecasting framework for supply chain management. World Acad Sci Eng Technol

Real-Time Machine Learning-Based Framework for the Analysis of Banking Financial Data

Ahmed Mahmoud, Ahmed Salem, and Emad Elsamahy

Abstract Due to the rapid increase of banking applications and their impact on the day-by-day life of a huge number of individuals, the volume of created data generated from these applications is growing significantly. Therefore, an ever-increasing number of individuals use these applications to complete their daily transactions should provide their feedback on the afforded services. This feedback can contain important information that will be helpful in improving the afforded services. By analyzing what is common between most of customers, their concerns, visions, and tendencies, their interest in the offered services and products could be decided. However, managing financial data that are generated in a huge amount continuously needs a suitable architecture which does not depend on storing data on storage media but it must be handled as soon as it arrives, aiming at providing decision within a very specific and short period of time. Many studies in the domain of real-time data analytics based on different approaches have been published. This paper will attempt to develop a framework for banking financial data analytics in real time with the ability of in-memory processing using machine learning classification techniques. The steps that were followed in building the proposed real-time framework include: data preprocessing, features selection, identifying the best machine learning model, and model validation. Furthermore, the proposed framework can increase its efficiency through the utilization of continuous learning. The proposed framework was validated using a dataset is related to direct marketing campaigns of a Portuguese banking institution.

Keywords Real-time analysis · Machine learning · Continuous learning

---

A. Mahmoud · A. Salem · E. Elsamahy (✉)
Arab Academy for Science, Technology, and Maritime Transport, El Moshir Ismail St., P.O. Box 2033, Cairo, Egypt
e-mail: e.elsamahy@aast.edu

A. Salem
e-mail: a.salem@aast.edu

D. A. Magdi et al. (eds.), *Digital Transformation Technology*, Lecture Notes in Networks and Systems 224, https://doi.org/10.1007/978-981-16-2275-5_25

1 Introduction

With the enormous measures of information being gathered from different sources, analysis of data became very essential in decision making for organizations, logical investigation, or the improvement of innovation [1].

Right now, everything relies upon countless online transactions. The information is expanding at a fast rate on World Wide Web in addition to mobile applications. As the data is developing persistently from various, heterogeneous sources comprising of structured, unstructured and semi-structured, analytics of such enormous volumes of datasets has revealed a lot of significance for up-and-coming research points in different specialties. New information is deduced based on analysis of the various and heterogeneous mix of huge datasets, which is helpful to grow new measurable models, procedures and techniques to predict the user actions in different uses of diverse fields. Creating new methods and techniques for the fine-tuning of huge datasets are useful in making speedier and more accurate decisions. The working efficiency of data handling and execution of predictive patterns utilizing modern techniques can be upgraded [2].

Using the traditional techniques, the data was initially analyzed after storing inside the storage media which is characterized by its large access time. Managing enormous streaming of data makes the use of storage media unsuitable, and thus, there has been an ongoing change in the process of storing data using memory instead of the normal storage media. Since real-time applications depend upon continuous information flow and quick analysis to land at a decision or activity inside a short and quite certain time, in-memory handling reduces the amount of access time which will have a pivotal job when performing data analysis in real time [1].

Honestly, taking advantage of the gathered data has been constantly considered by academics and experts. However, the high-speed volume and heterogeneity of huge flow of real-time data drive the boundaries of the present storage, handling, and administration abilities [3].

Dealing with real-time data depicts various issues. The most important ones being the handling, managing and expectation of streaming data arrival, which require hardware assets and tedious for analytics purposes [4].

To solve these issues, distributed computing systems are generally utilized including Hadoop MapReduce. One of the primary detriments of MapReduce is incompetence when performing the iterative algorithms, as it is not intended for iterative procedures. The mappers read similar data again which is already existed on storage media. Hence, after every iteration, the outputs must be saved to the storage media for the following iteration, which seriously leads to a very large time consumption. Therefore, MapReduce is not adequate enough as it is not designed to handle streaming data [4].

Decision making and problem solving in the business segment are incredibly relying upon getting to and overseeing information. Today, with enormous amounts of information produced each day, by having many different sources of data collection, it has gotten fundamental for business and financial sector to proficiently oversee and process their insight. Therefore, it will be able to offer best types of assistance, achieve operational greatness and upgrade the basic leadership process, through making the right decisions in real time [5].

Real-time processing applications are intended for managing and analyzing stream data. The real-time data applications speed up the calculation rapidly with the assistance of in-memory registering. Moreover, real-time applications require to handle data in consecutive structure as soon as they arrive then produce instant decisions. In streaming systems, the machine learning algorithms need to analyze a nonstop succession of data continuously [6].

The major goal of the machine learning is to permit a model that is able to learn from a previous or current timeframe and utilize the results to take decisions or predictions concerning upcoming anonymous events. Multiple algorithms of machine learning are used to discover and extract useful information from the data through training and testing datasets. The two main approaches of these algorithms are classified into supervised and unsupervised [6].

Nowadays banks operate in a world of time-sensitive opportunities. Therefore, being able to provide instant notifications will be one of the highest priorities in any financial institution in order to attract more customers. Moreover, with the existence of too much data that are generated and collected about customers more than ever, analyzing such amount of data became very difficult to be achieved. Real-time analytics is representing one of the top research areas that can provide helpful solution in the financial sector, which enables the institution to respond to customer's data upon their arrival. Consequently, the establishment of real-time data analysis systems became essential, which enables any financial organization to boost its financial performance and improves customer experiences.

This paper presents a real-time framework using machine learning model for analyzing real bank marketing dataset downloaded from UCI machine learning repository Web site [7]. This paper is organized as follows: Sect. 2 presents an overview of related work. The proposed architecture is illustrated in Sect. 3. In Sect. 4 a discussion for the obtained results is depicted. Some conclusions were drawn together with some recommendations for future research work are shown in Sect. 5.

2 Related Work

In ongoing decades various kinds of technologies and methodologies for real-time analytics have been offered. A significant number of these approaches were focused on the utilization of data mining platforms for predicting results [4]. Recently huge number of researchers are utilizing machine learning algorithms for analytical tasks such as data mining, predictive analysis, and pattern recognition [4].

Babak Yadranjiaghdam, Seyedfaraz Yasrobi and Nasseh Tabrizi presented a framework that is based on in-memory processing for analyzing structured and unstructured Twitter dataset. The real-time data analytics system utilized Kafka to import tweets, and to categorize them based on topics over consumers' hubs. Furthermore, Apache Spark is also utilized to get to these consumers legitimately and to analyze these data [1].

Van-Dai Ta, Chuan-Ming Liu and Goodwill Wandile Nkabinde have utilized generic architecture for big data healthcare analytics by merging the benefits of batch processing which was carried out based on Hadoop cluster and the generated data is stored in HBase database. Using stream processing empowers real-time processing by integrating Kafka, Storm group and NoSQL Cassandra [8].

M. Mazhar Rathore et al. have proposed a framework which processes a considerable volume of several social networks data for monitoring earth events, accidents, medical diseases, user directions, and opinions to get future real-time decisions. The application was developed utilizing a Hadoop single node cluster for offline data analysis with Apache Spark for streaming data analysis [9].

Ms. D. Jayanthi and Dr. G. Sumathi have introduced a comparison between several real-time systems: Hadoop, Apache Spark, Apache Storm, Apache Samza, Flume, Scribe, S4, HStreaming, All-RiTE, Impala, Amazon Kinesis and Kafka. Furthermore, an architecture for real-time analytics was introduced based on machine learning. The proposed architecture includes: data sources, data integration, real-time analytics, data storage, and data visualization [10].

Sasan Amini, Ilias Gerostathopoulos and Christian Prehofer have introduced a case study of controlling the opening and shutting down of an expressway hard shoulder path in microscopic traffic simulation. The proposed work created overall and flexible architecture that depend on distributed processing framework for real-time traffic control. This architecture utilizes Kafka as the layer that decouples distributers and subscribers from the analytics engine [3].

Gautam Pal, Gangmin Li and Katie Atkinson have shown a case study of catching click-stream data to represent the real-time data recuperation utilizing machine learning techniques for real-time analysis with a natural parameterization. Moreover, a comparison among Kafka and Flume for real-time data ingestion and processing technique is presented [11].

Ramanna Hanamanthrao and S. Thejaswini have built up a complete solution for streaming, processing, and analyzing click-stream data which incorporate numerous apparatuses for big data platforms. The proposed solution utilized Apache Kafka to transfer data, Spark streaming for processing of incoming data. Moreover, Elasticsearch was utilized for index, search, and performing analytics, Kibana utilized for data visualization, while Hadoop was utilized to store and disconnect batch process of enormous datasets as MapReduce job [12].

Abderrahmane Ed-Daoudy and Khalil Maalmi have presented a framework for real-time data analytics of healthcare data. The proposed application includes data processing, combining socket streams and Spark streaming. The data is sent for real-time analytics through the connected devices as socket streams. A machine learning model for the prediction of the health status after receiving healthcare data through spark streaming was identified [4]. Furthermore, Abderrahmane Ed-Daoudy and Khalil Maalmi have proposed another real-time heart disease prediction system. This system utilized Apache Spark as a distributed processing platform for streaming data event. The system consists of two major sub-parts. The first part predicts heart disease using Spark as streaming processing with Spark MLlib as analytics tool. The second part visualizes the analysis results after storing huge amount of generated data in Cassandra database [13].

Chaman Verma, Veronika Stoffova, Zoltan Illes, Sudeep Tanwar and Neeraj Kumar have proposed a machine learning classification model which is identified using optimization techniques to tune its hyper-parameters. This system provides a real-time identification tool, in order to facilitate the identification process for the students' origin. The most prominent features were selected based on the use of principal component analysis (PCA) [14].

Amir Bahador Parsa et al. have used eXtreme Gradient Boosting (XGBoost) to develop a model that can be used for accident detection based on real-time data such as traffic, network, demographic, land use, and weather features. The main objectives of this study are: identifying the influence of individuals' features on accident occurrences and providing real-time prediction for the occurrence of accidents [15].

Based on the above survey, two important conclusions can be drawn. First, most of the research work is based on the combination of traditional data analytics tools to build a complex framework which only focuses on data collection and storage. Moreover, data visualization which represents the output of the analysis process was introduced without mentioning details about the process of building a real-time analysis framework. Second, none of the approaches that were surveyed focuses on real-time analytics in banking services industry although there are too many daily transactions that need to be analyzed in real time. This study suggests a mechanism to cover the gaps that are previously mentioned, through building a real-time machine learning-based framework for data analytics of bank marketing dataset.

3 Proposed Model

The motivation behind this study is to build up a framework for analyzing banking marketing data in real time. This framework is created on same data analytics approaches. The main idea is to analyze a lot of transactions arriving in a short amount of time and provide the results as fast as possible. The proposed framework is mainly based on the use machine learning classification model that will be able to classify data in real time.

This framework will be responsible for managing any processing in memory to analyze streaming data in real time instead of using MapReduce jobs that are inadequate enough to handle streaming data and then store the results into data warehouse for visualizing [4]. Furthermore, the possibility of improving efficiency through the process of continuous training using cluster historical data, taking into consideration previous training processes.

The proposed framework as shown in Fig. 1 consists of three sections: (A) building up a real-time analytics model, (B) apply the model to perform a real-time analysis, and (C) improve the model efficiency.

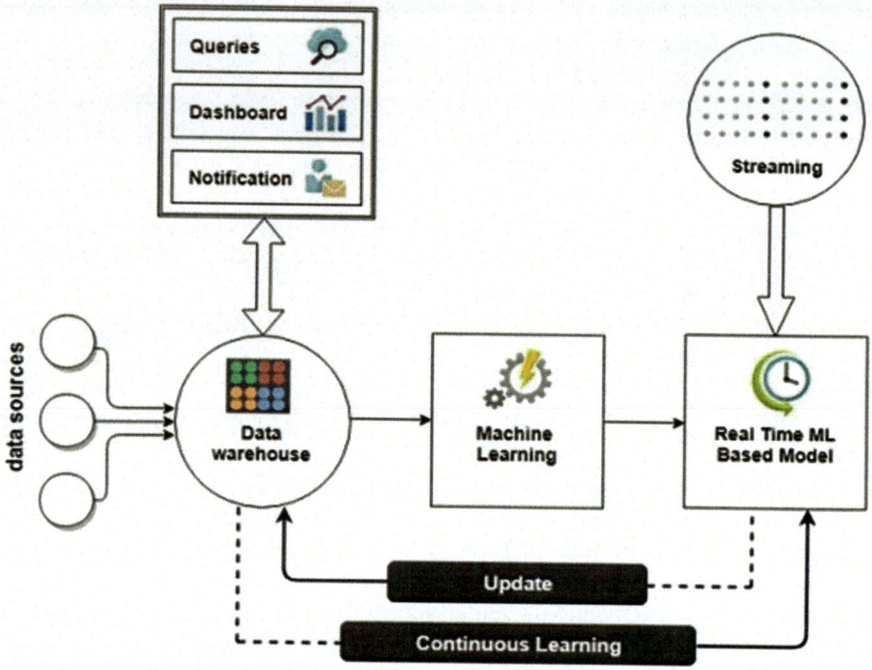

Fig. 1 The proposed real-time machine learning-based framework

Building up a real-time analytics model has been accomplished by using machine learning with many comparisons and validations to select the best algorithm. Moreover, this process was carried out based on using 70% of the entire dataset as training data subset, and 30% of the entire dataset as testing data subset. Complete information about the used dataset can be found in [7].

Real-time analytics model has been applied by utilizing the model to analyze the streaming data with testing results of the prediction process using testing data subset.

Model efficiency has been improved by utilizing the model for additional training using 30% of the testing data subset and update the model in case the efficiency is improved after testing the accuracy using 70% of the testing data subset.

3.1 Building Up a Real-Time Analytics Framework

After analyzing the bank marketing dataset, data preprocessing phase was found to be essential. This phase includes certain steps which are: (1) predicting missing values, (2) categorizing non-numerical data, (3) splitting data into independent variables (X) and dependent variable (Y).

Table 1 Features scoring

No.	Feature	Score	Selected
1	nr.employed	5926.610601	Yes
2	Pdays	4861.173272	Yes
3	euribor3m	4309.479048	Yes
4	emp.var.rate	4023.829925	Yes
5	Previous	2304.257088	Yes
6	Contact	881.7070703	Yes
7	cons.price.idx	778.5897949	Yes
8	Poutcome	705.6781644	Yes
9	Campaign	182.1566729	No
10	Education	135.5191524	No
11	cons.conf.idx	124.4099747	No
12	Marital	86.76112457	No
13	Age	38.09465929	No
14	Job	26.76022941	No
15	day_of_week	10.50309588	No
16	Housing	5.06158775	No
17	Month	1.5149397	No
18	Loan	0.821520785	No
19	Default	0.380881764	No

Furthermore, the identification of the processing was carried out using Python with the utilization of the functions of the machine learning package (sklearn). The most dominant features were selected using SelectKBest, which was found to be (8) features (shaded in gray) which scored the highest scores as shown in Table 1. The (8) features were selected as selecting more features provide no significant improvement in the accuracy of prediction. Finally, the dataset was split into training and testing data subsets using sklearn-train_test_split.

The python automated machine learning tool, which is called tree-based pipeline optimization tool (TPOT), was utilized in order to optimize machine learning pipelines using genetic programming. Using this tool will automate most of the tedious part of machine learning by intelligently exploring thousands of possible pipelines to find the best algorithm that can effectively classify the data with high accuracy, see Fig. 2.

Ten iterations of genetic programming were generated to select the best algorithm for iteration, which achieves the highest accuracy and less log loss, see Table 2. Based on the results shown in Fig. 3, the algorithms that achieved an accuracy above 90% and achieved log loss less than 0.4 (shaded in gray) are GradientBoostingClassifier, XGBClassifier, RandomForestClassifier, and ExtraTreesClassifier. It is worth noting that the selection of 90% accuracy and 0.4 limit for the log loss function is based

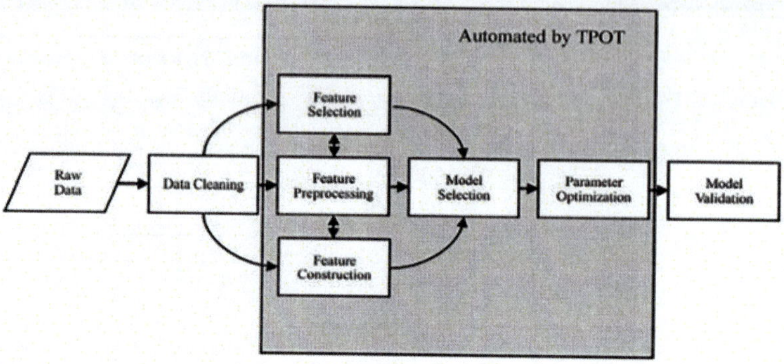

Fig. 2 Steps for the machine learning model identification [16]

Table 2 Results of genetic programming

Iteration No.	Classifier	Log loss	Accuracy	Precision	Recall	F_1-score
1	XGB	0.274	89.981	0.88	0.9	0.88
2	Gradient boosting	0.275	90.038	0.88	0.9	0.88
3	XGB	0.365	90.022	0.88	0.9	0.88
4	XGB	0.366	89.917	0.88	0.9	0.88
5	Random forest	0.278	90.046	0.88	0.9	0.88
6	Extra trees	0.273	90.022	0.88	0.9	0.88
7	Decision tree	0.277	89.828	0.88	0.9	0.88
8	XGB	0.364	89.973	0.88	0.9	0.88
9	XGB	0.271	89.981	0.88	0.9	0.88
10	Random forest	0.271	89.981	0.88	0.9	0.88

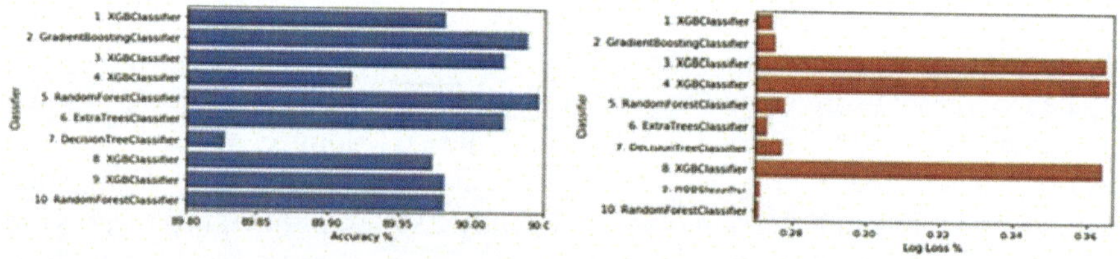

Fig. 3 Accuracy and log loss results of genetic programming

on published researches in this area, which guarantee the high performance of the model.

In order to validate the achieved results, the traditional steps for identifying a machine learning model are carried out. Ten different machine learning algorithms were implemented and trained using the training data subset. These algorithms were tested against the test data subset. The results of the classification results are shown in Table 3. Applying the same rules that were previously mentioned in selecting the best algorithms (shaded in gray), GradientBoostingClassifier and XGBClassifier were found to be the best ones (see Table 3 and Fig. 4).

Based on the results shown in Tables 2 and 3, the algorithms which achieved the best performances using both the automatic and manual identification techniques, are GradientBoostingClassifier and XGBClassifier. Therefore, the log loss as well as the accuracy for both algorithms were compared in order to select the best algorithm. Table 4 shows a checklist that was constructed to select the best algorithm which is the GradientBoostingClassifier.

Table 3 Comparison between different algorithms using the traditional steps

Classifier	Log loss	Accuracy (%)	Precision	Recall	F_1-score
Logistic regression	0.289	89.795	0.88	0.9	0.87
Kneighbors	1.602	89.188	0.87	0.89	0.88
SVC-Kernel	0.313	89.811	0.88	0.9	0.88
GaussianNB	1.382	82.528	0.85	0.83	0.84
Decision tree	2.433	88.929	0.87	0.89	0.88
Random forest	1.045	89.172	0.87	0.89	0.88
Gradient boosting	0.273	90.07	0.88	0.9	0.88
Linear discriminant analysis	0.321	89.374	0.87	0.89	0.88
Quadratic discriminant analysis	3.786	88.541	0.84	0.89	0.84
XGB	0.271	90.054	0.88	0.9	0.88

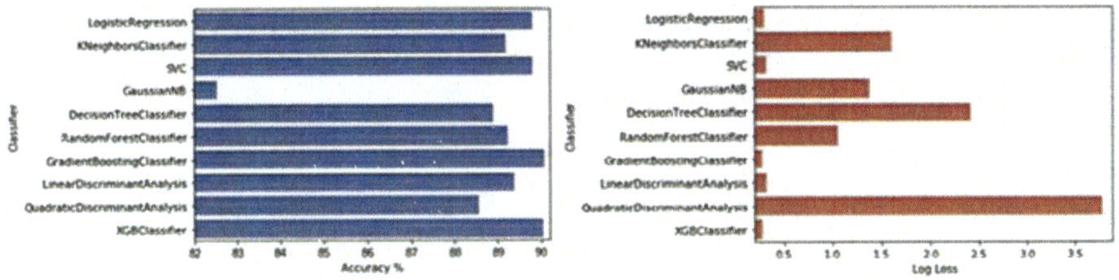

Fig. 4 Accuracy and log loss results of comparison of different algorithms

Table 4 Checklist-based for selecting the best machine learning algorithm

Feature	GB	XGB
Common algorithm that has highest accuracy using genetic programming	✔	
Common algorithm that has highest accuracy using different algorithms	✔	
Common algorithm that has lowest log loss using genetic programming	✔	
Common algorithm that has lowest log loss using different algorithms		✔

Moreover, aiming to achieve more improvement in the classification's accuracy of the machine learning model, a re-evaluation process was applied on the unselected features. The methodology that was carried out is based on testing the classification accuracy of the machine learning model, against the testing subset, after adding one additional feature to the (8) features previously selected (the features that were not selected—the remaining 11 features), see Table 1. Consequently, the accuracies after adding each feature separately were calculated, if the maximum accuracy of the (9) features is greater than the accuracy of the (8) features, this feature is added to the model's inputs otherwise the process is stopped. This feature is marked as "selected" and the process is repeated for the remaining unselected features. Figure 5 shows a flowchart illustrating the steps that were followed in the re-evaluation process for testing the impact of addition on one feature.

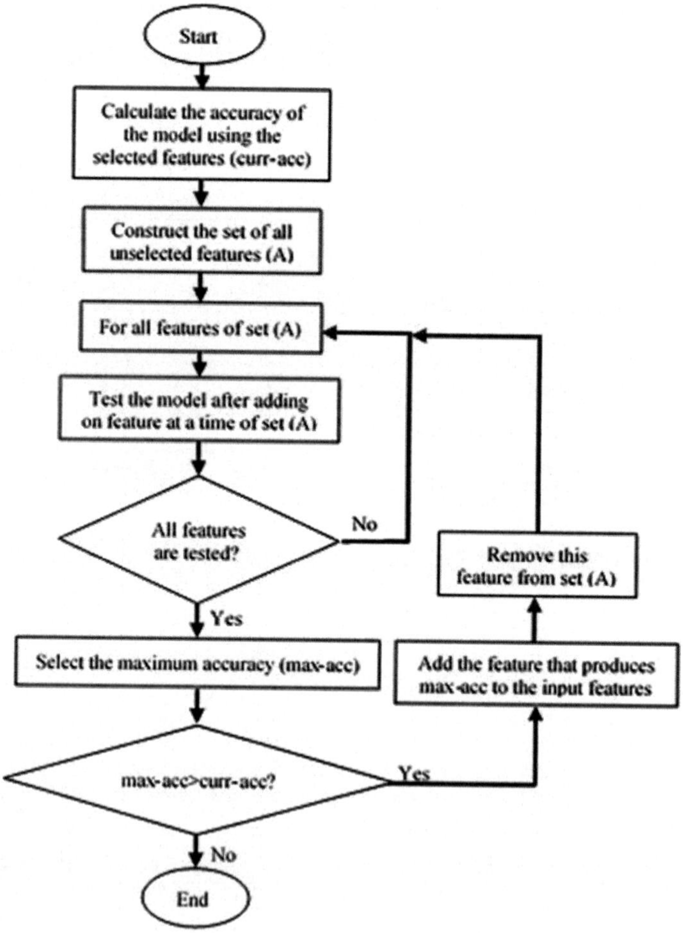

Fig. 5 Flowchart for the features' re-evaluation process

Table 5 Final result of best features

Features	Count	Accuracy
education, contact, month, day_of_week, campaign, pdays, previous, poutcome, emp.var.rate, cons.price.idx, cons.conf.idx, euribor3m, nr.employed	13	90.2323%

Table 6 Final result of best features

Actual/Predicted classes	Predicted class 1	Predicted class 0	Total actual classes
Actual class 1	(TP) 364	(FN) 1024	1388
Actual class 0	(FP) 183	(TN) 10,786	10,969
Total predicted classes	547	11,810	12,357

Table 5 shows the final results of the classification accuracy after applying the re-evaluation process. It can be seen that after adding (13) features, the accuracy is increased to 90.2323% compared to the use of only (8) features that produce an accuracy of 90.07% (see Table 3).

Furthermore, the model validation was carried out using confusion matrix and k-fold cross-validation. The confusion matrix is used to predict the performance of a certain classification model based on test data that have known actual values. It can be used to calculate certain parameters which helps in validating the proposed classification model. A summary of the incorrect and correct predictions is presented along with their count values.

In Table 6, class 1 and class 0 represent positive and negative responses. On the other hand, the observations are symbolized as (P) and (N) to represent the positive and negative classes. Moreover, (TP), (FP), (TN), and (FN) were used to represent the following categories: correctly predicted as positive, falsely predicted as positive, correctly predicted as negative, and falsely predicted as negative. Furthermore, the calculation of the confusion matrix provides insight into the types of errors that are being made. The results in Table 6, show that there exists an imbalance in the testing data subset and therefore, the accuracy will not be able to correctly validate the model. Consequently, the F_1-score, which is a combined parameter of both precision and recall, can successfully validate our model. Therefore, most performance measures are computed from the confusion matrix which include: precision, recall and F_1-score in addition to the accuracy using Eqs. 1–6 (see Table 7). In spite of producing an accuracy of 90.23%, the F_1-score should be checked due to the dataset imbalance. Achieving 88%, the model proves itself as a good classifier.

$$\text{Precision of Class 0} = \text{TP}/(\text{TP} + \text{FP}) \tag{1}$$

Table 7 Final result

Accuracy	Precision	Recall	F_1-score
0.902323	0.89	0.9	0.88

Table 8 Results of K-fold cross-validation

K-Value	Accuracy (%)
1	89.81
2	89.08
3	90.46
4	90.77
5	90.08
6	89.77
7	90.43
8	89.66
9	90.50
10	90.35

$$\text{Precision of Class 1} = \text{TN}/(\text{TN} + \text{FN}) \tag{2}$$

$$\text{Recall of Class 0} = \text{TP}/(\text{TP} + \text{FN}) \tag{3}$$

$$\text{Recall of Class 1} = \text{TN}/(\text{TN} + \text{FP}) \tag{4}$$

$$f_1\text{-score} = (2 * \text{Recall} * \text{Precision})/(\text{Recall} + \text{Precision}) \tag{5}$$

$$\text{Accuracy} = (\text{TP} + \text{TN})/(\text{TP} + \text{TN} + \text{FP} + \text{FN}) \tag{6}$$

To overcome the limitation of limited data samples, the cross-validation technique is utilized. This technique is based on a resampling procedure for the validations of machine learning models. A group splitting is applied to the data which is controlled by a parameter called (k) that represents the required number of groups. This parameter was selected to be (10) as this value was found to be extensively used in most of the literature. Table 8 shows the results of k-fold cross-validation.

The results presented in the above table show that the variance of accuracies for the different folds is 0.234 which reproves the validity of the identified model.

3.2 *Using the Model for Real-Time Analysis*

After identifying the machine learning model based on 70% of the dataset records as a training data subset, the model was validated against the testing data subset. The validation results show that the model succeeded in the prediction phase with high accuracy. Moreover, to complete the proposed framework, the model will be tested against the real-time data.

Table 9 Results of testing the developed model before continuous learning

Classes	Precision	Recall	F_1-score	Support
0	0.91	0.98	0.94	3263
1	0.7	0.27	0.39	444
AVG	0.88	0.9	0.88	3707

Table 10 Results of testing the developed model after continuous learning

Classes	Precision	Recall	F_1-score	Support
0	0.92	0.99	0.95	3263
1	0.86	0.35	0.5	444
AVG	0.91	0.91	0.9	3707

The testing data subset (30% of the original dataset) was divided into 30% for the representation of the real-time data (streaming data) and 70% for the validation purposes. Once the streaming data arrives, the model will be loaded in memory to analyze the streaming data using memory processing then update the data warehouse with the results to be visualized.

3.3 *Improving the Model Efficiency*

In this section, the model is retrained using the 30% dataset of the testing data subset (3707 out of 12,357 records) to simulate the real-time training. If the accuracy is increased, the model is updated otherwise no update is needed. The validation results show that the accuracy (against 70% of the test data subset) is 89.857% before applying real-time training (see Table 9). However, after the application of real-time training, the accuracy reaches 91.557% and the model will be updated (see Table 10).

4 Discussion

Applying the concept of real-time training within the terminology of machine learning has proved itself in enhancing the performance of the machine learning-based identified models. Tables 9 and 10 showed the improvement that appears in the model performance.

Moreover, in constructing the machine learning model, the technique of re-evaluation of features shows good results in the form of improving the model performance which can be used for the enhancement of feature selection phase.

5 Conclusion and Future Work

Nowadays there is a great value for the real-time data analysis especially in the business and financial sectors. In this paper, the real-time approach was utilized to develop a framework which has the ability to analyze streaming data. This framework is based on a machine learning model that will be able to predict what will happen to support decision making in real time with high accuracy that can be continuously enhanced whenever new data is available.

The machine learning classification algorithms were implemented using python. Gradient Boosting algorithm was chosen to develop the proposed model after testing many algorithms on bank marketing datasets.

An enhanced features selection analysis was introduced that successfully improve the machine learning model's performance after its application.

The proposed work can be enhanced using message brokering system to receive and send streaming data to the model for performing high-speed ingestion of streaming data in real time. Furthermore, the prediction accuracy can also be increased by the use of clustering techniques to perform real-time analysis on clusters data instead of the data itself.

References

1. Yadranjiaghdam B, Yasrobi S, Tabrizi N (2017) Developing a real-time data analytics framework for twitter streaming data. In: IEEE International congress on big data (BigData congress). IEEE, pp 329–336
2. Rao TR, Mitra P, Bhatt R, Goswami A (2018) The big data system, components, tools, and technologies: a survey. Knowl Inf Syst 60:1–81
3. Amini S, Gerostathopoulos I, Prehofer C (2017) Big data analytics architecture for real-time traffic control. In: IEEE International conference on models and technologies for intelligent transportation systems (MT-ITS). IEEE, pp 710–715
4. Ed-Daoudy A, Maalmi K (2018) Application of machine learning model on streaming health data event in real-time to predict health status using spark. In: International symposium on advanced electrical and communication technologies (ISAECT). IEEE, pp 1–4
5. Sanyour RA, Abdullah M (2019) Real time data analysis and visualization for the breast cancer disease. Periodicals Eng Nat Sci 7(1):395–407
6. Zheng T, Chen G, Wang X, Chen C, Wang X, Luo S (2019) Real-time intelligent big data processing: technology, platform, and applications. Sci China Inf Sci 62(8):82101
7. Bank Marketing Data Set from, https://archive.ics.uci.edu/ml/datasets/bank+marketing. Last accessed 15 Dec 2020
8. Ta V-D, Liu C-M, Nkabinde GW (2016) Big data stream computing in healthcare real-time analytics. In: IEEE International conference on cloud computing and big data analysis (ICCCBDA). IEEE, pp 37–42
9. Rathore MM, Paul A, Ahmad A, Imran M, Guizani M (2017) Big data analytics of geosocial media for planning and real-time decisions. In: IEEE International conference on communications (ICC). IEEE, pp 1–6
10. Jayanthi MD, Sumathi G, Sriperumbudur S (2016) A framework for real-time streaming analytics using machine learning approach. In: Proceedings of national conference on communication and informatics-2016

11. Pal G, Li G, Atkinson K (2018) Big data real time ingestion and machine learning. In: IEEE Second international conference on data stream mining & processing (DSMP). IEEE, pp 25–31
12. Hanamanthrao R, Thejaswini S (2017) Real-time clickstream data analytics and visualization. In: IEEE International conference on recent trends in electronics, information & communication technology (RTEICT). IEEE, pp 2139–2144
13. Ed-Daoudy A, Maalmi K (2019) Real-time machine learning for early detection of heart disease using big data approach. In: International conference on wireless technologies, embedded and intelligent systems (WITS). IEEE, pp 1–5
14. Verma C, Stoffová V, Illés Z, Tanwar S, Kumar N (2020) Machine learning-based student's native place identification for real-time. IEEE Access 13(8):130840–130854
15. Parsa AB, Movahedi A, Taghipour H, Derrible S, Mohammadian AK (2020) Toward safer highways, application of XGBoost and SHAP for real-time accident detection and feature analysis. Accid Anal Prev 136:105405
16. Olson RS, Bartley N, Urbanowicz RJ, Moore JH (2016) Evaluation of a tree-based pipeline optimization tool for automating data science. In: Proceedings of the genetic and evolutionary computation conference, pp 485–492

E-Commerce in Ukraine: Place, Range of Problems, and Prospects of Development

Vitalina Babenko, Iryna Perevozova, Oleksandr Prystemskyi, Olha Anisimova, Alexandr Fedorchuk, and Iryna Balabanova

Abstract The place of Ukraine in the global electronic business and electronic commerce is considered. The place, tendencies, problems, and ways of solving Ukrainian problems of electronic marketing are investigated on the basis of comparative information for different years and countries. The subject of the article is Ukrainian e-business and e-commerce, and its place in the global e-business. The goal is to study and search for ways to develop and improve the electronic market of Ukraine in the conditions of the domestic Ukrainian market. The task is to find a modern place for Ukrainian e-commerce, problems, and trends, as well as find ways to solve problems. Within the framework of this study, methods of comparative analysis are used on the basis of statistical materials, regulatory documents, and scientific publications for descriptive and structural analysis to assess the state of the Ukrainian electronic market and methods for its formation of trends in their development and prospects. As a result of the study, it was found that at this stage, the main limiting factor is the weak level of informatization of society, in particular the low quality and unevenness of the Internet coverage. It is proposed to use the achievements of technological and informational progress and the legislative platform to solve existing problems. A list of the most popular e-commerce business models used in the country

V. Babenko (✉)
V. N. Karazin Kharkiv National University, 4, Svobody Sq., Kharkiv 61022, Ukraine
e-mail: vitalinababenko@karazin.ua

I. Perevozova
Ivano-Frankivsk National Technical University of Oil and Gas, Karpatska Str., 15, Ivano-Frankivsk 76018, Ukraine
e-mail: perevozova@ukr.net

O. Prystemskyi · I. Balabanova
Kherson State Agricultural University, Str. Stritenskaya, 23, Kherson 73006, Ukraine
e-mail: i.o.balabanova@ukr.net

O. Anisimova
Vasyl' Stus Donetsk National University, Vinnytsia, Ukraine
e-mail: o.anisimova@donnu.edu.ua

A. Fedorchuk
Kherson State University, University Street 27, Kherson 73000, Ukraine
e-mail: 15961980@ukr.net

D. A. Magdi et al. (eds.), *Digital Transformation Technology*, Lecture Notes in Networks and Systems 224, https://doi.org/10.1007/978-981-16-2275-5_26

has been formed, the largest subjects of these relations have been identified as well as opportunities for the further development of the e-commerce market and attracting new users.

Keywords E-commerce · B2B · B2C · G2C · Electronic platform · Marketplace · Competitiveness of electronic business · Electronic brand of country

1 Introduction

In the last years, there is passing to the global use of digital electronic facilities with the aim of exchange information and of realization of transaction operations with the aim of the acquisition of a commodity, that must satisfy consumer needs. In this connection, the division of labour of the international market takes place between those subjects that before did not take participation in the process. The modern development of world trade is impossible without e-commerce. As one of the key drivers of the development of world trade electronic business comes forward in a modern world economy [1]. In the conditions of modern scientific and technical progress and increase of requirements to speed of exchange information between the subjects of trade activity effective organization of electronic business of the country is able to promote her competitiveness. In modern terms electronic business in Ukraine is on the stage of becoming and develops constantly, acquiring new turns and increasing the amount of money, but in the last years, the rates of height began to go down [2]. Besides, a disturbance causes the rapid rates of integration, that in the conditions of the emerging market can meet certain problems, that will need or need a decision that causes the necessity of realization of the electronic market of Ukraine and selection of certain problems and decisions analysis in the conditions of contemporaneity. The development of e-commerce in the daily life of Ukraine is an interesting research problem and can serve as an experience for the integration of e-business in countries with emerging market economies. The relevance of the research in this article is based on the fact that the theoretical and methodological foundations in assessing the effectiveness of e-commerce in Ukraine have not been fully investigated in the Ukrainian and foreign literature, which does not allow studying its place, range of problems and development prospects in full.

2 Literature Review

The problems of the formation of e-commerce in different countries and in the world as a whole are investigated in the scientific papers of various researchers on the topic of e-business. The authors study the impact of various sectors of e-commerce on the country's economy. Some scientists are engaged in the investigation of ecommerce in Ukraine, although most of the publications are devoted to separate problems, and not to the analysis as a whole. A number of Ukrainian and foreign scientists are engaged in the problem of e-commerce in Ukraine. The authors of the papers [3, 4] investigate the problem of modern trends in e-commerce based on the factors of the development of international e-commerce in the context of globalization. The scientific works of Hlinenko L. and Dainovskyi Yu. [5] study the tendencies of Ukrainian e-commerce, its role and tendencies. The most common business models of Ukrainian electronic merchants and their comparison in individual business processes with the world's leading Internet merchants are presented in the publications of A. Lukyanchenko [6]. The author examines the problem of e-commerce by describing the specifics and tools typical of Ukrainian e-commerce. In the article of V. Oliinyk et al. [7, 8] ten tendencies of ecommerce in Ukraine are investigated, of which the main ones are: loyalty and personalization; regionality; change of categories; sector; mobilization; social commerce; work with the community; expansion of China; efficient logistics; multichannel. Chankhi O. and Ozemblovskyi V. [9] describe the B2C vector of e-commerce in Ukraine and identify the European vector as the main direction for the development of e-commerce in Ukraine. Apopiy V. and Shaleva A. [10] highlight the socioeconomic aspects of e-commerce, examining the main problems of Ukrainian retail. The issues related to the country's information infrastructure as a basis for the development of e-commerce are considered too. The processes of informatization and their mechanisms were studied by the following scientists: Kulchik Z. et al. [11], Perevozova I. et al. [12], who studied the tools of information and communication technologies in the country's economy. Podgorna I. et al. [13] consider electronic markets as a tool for improving the Ukrainian economy. Hrabovskyi Y. et al. [14] examines informatization and its implementation in the Ukrainian economy.

Defending the place of Ukraine and world electronic business and commerce, this paper analyzes the modern state of electronic business in Ukraine, to define the basic problems of development of electronic business in Ukraine and to give ways to their decision.

3 Methodology

In the article, the following methodology is used: method of analysis at the research of statistics, method of synthesis at the leading out of new tendencies, method of the scientific abstracting and method of generalization.

4 Research Results and Discussion

Emerging cyberspace is expanding at a very rapid rate. It not only stores and processes electronic information resources but also provides lively exchange of them. Therefore, the main task now is to keep up with the global processes taking place in the modern world in view of the development of information and telecommunication technologies. The world is on the cusp of the era of online e-business (hereinafter referred to as the OEB), which generalizes various opportunities for interaction between suppliers and consumers; in addition, it makes them independent, unattached to fixed devices, enabling them to make purchases, make payments, participate in the auction even with just a cell phone or a handheld device [15].

Web services can always be used, regardless of time or location.

That is, we can talk about the emergence and development of a new market, namely the electronic market on the basis of which more and more transactions and services are carried out.

The global e-commerce market is one of the most dynamic and steadily growing markets. According to the eMarketer report [16], in 2019, the volume of world retail e-commerce reached 12.8% of retail trade in general (Fig. 1).

In Fig. 1, we can see a steady increase in retail sales worldwide through the Internet, which presents to us the importance and extent of e-business development in the world. We can see that in the last five years, sales have doubled from $1.67 billion to $3.578 billion. There is also a steady upward trend that is stable and growing by half a billion US dollars each year. That is to say, the prospect and continued development of e-commerce in the e-commerce segment worldwide, it can also be noted that sales growth is related to the proliferation of Internet and customer trust in marketplaces, and therefore, the next growth will be related to the penetration of the Internet to countries, developing countries, and third world countries through the inclusion of new markets in the world market, which will lead to a constant and stable growth of this industry.

According to Fig. 2, we see the growth rate of online purchase agreements relative to their purchase in real life. Based on this chart, we can conclude that buying goods online is gaining in popularity and growing little by little, picking up retail positions in real life. Over the past five years, the percentage of electronic retail transactions has more than doubled. We can also see an increase in the growth rate, albeit small if the increase from 2014 to 2015 was 1.1%, then in the period from 2018 to 2019, this increase was 1.4%, and the upward trend is visible throughout the chart. Based on the trends of already known growth, as well as the improvement of the technological and electronic sphere, we can speak about ever-increasing growth in the future.

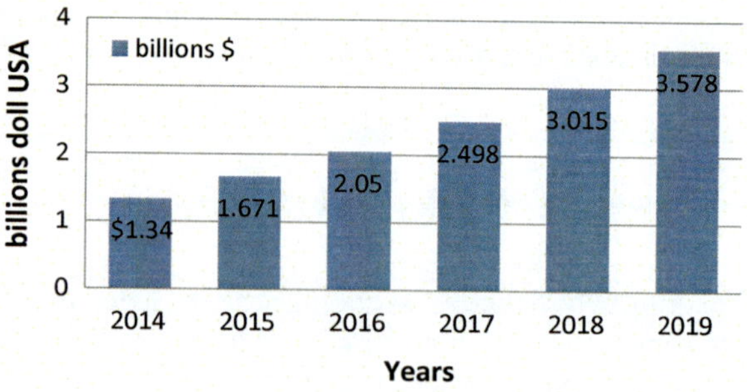

Fig. 1 Retail in the e-commerce market in the world [16]

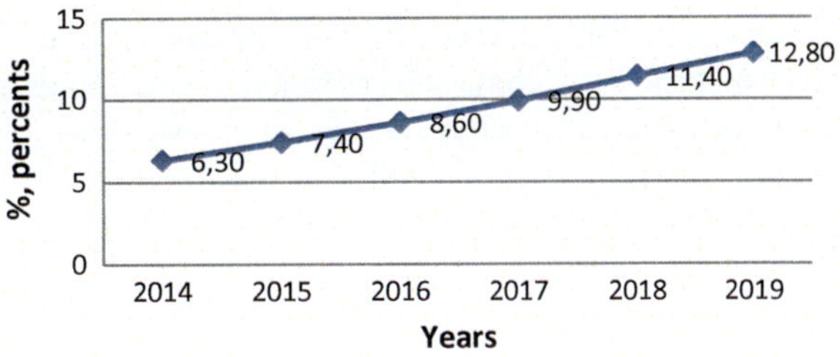

Fig. 2 Part of e-commerce from the global retail level [16]

Speaking of e-business, it is necessary to note its segmentation. Figure 3 shows that the most popular in the world is the B2B segment, followed by the B2C segment. That is, we can say that e-business is more promising for providing services to another business than e-business for providing services and/or goods to the buyer. It should also be noted that this trend persists in many individual countries. This trend is also present in Ukraine. However, the e-commerce market in Ukraine is one of the most successful Ukrainian markets in general. This is the only industry whose growth rate has been the highest in Europe in the last two years, providing highly optimistic forecasts and anticipating more and more new entrants. However, compiling and delivering a business development strategy based on only a linear extrapolation of the previous years' statistics is very dangerous, which makes the study of the real state and trends of the e-commerce market highly relevant [17, 18].

Through analyzing e-commerce in Ukraine, it is possible to speak about its ambiguity. In spite of its obvious advantages (openness, democracy, globality, etc.), e-business has an aggressive range of points for doing business in it, the success of which depends on the degree of the thoughtfulness of the business strategy. Indeed, the fundamental foundations of e-business are the same as the foundations of classic business: its purpose is to generate cash flow, generate profits, and create competitive advantages. The main difference between e-commerce and traditional methods of business is that it flows faster, and therefore, its competitive advantages are faster [19].

There are several reasons for this situation: insufficient Internet access, inactive use of plastic cards, underdevelopment of electronic payment systems, lack of preparedness for customers to work online, and low purchasing power of the population. The poor organization of the delivery of goods and the lack of confidence in

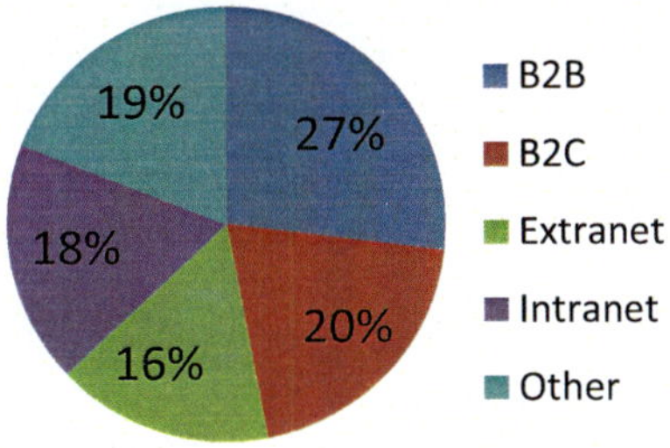

Fig. 3 World structure of marketing segments in e-business (compiled by the authors)

the systems of prepayment are also deterrents. The obvious consequence of this is that most B2C (business-to-consumer) e-commerce businesses are storefronts, not a source of income. Thus, at the current stage of development of e-commerce in Ukraine, the functioning of online stores does not go beyond the experiment of using the opportunities of the network.

However, all this does not deny the future of Ukrainian retail on the Internet. There are markets for goods and services for which e-commerce is far more profitable than traditional commerce, and the general problems (imperfection of hardware and technology bases, delivery and payment systems, low solvency of the population, etc.) are not as significant (Table 1). These are digital goods and information for which prompt delivery is important such as luxury goods (cars, antiques, and jewelry), travel services, and rare goods, for which considerable expenses are there for search and purchase (spare parts, special literature) [20].

As can be seen from Table 1, the dynamics of e-commerce indicators in Ukraine is much better than the dynamics of retail indicators in general. The total Internet penetration rate in Ukraine is now around 62%, for users with above-average income, it reaches almost 100%, similar to the situation in the age group of 15–45 years. From this group, there are three categories of Ukrainians who use the Internet: schoolchildren and students, military personnel, and business owners who provide the lion's share of active online shoppers [21].

Figure 4 presents data on the number of Internet users according to the State Statistics of Ukraine. For 2019, the number of Internet users is 26,066 thousand people, which is 62% of the total population of the country, which speaks of insufficient provision of Internet resources and gives a "ceiling effect" for e-business. However, it should be noted that in the last five years, the number of Internet users has increased more than four times. It is fair to say that this trend is unlikely to occur, but if we do not take into account the periods with jumping graphs, namely 2014–2015 and 2016–2017, we get an average increase of 1.5 million users per year.

Table 1 Dynamics of indicators of development of e-commerce of Ukraine [20]

Indicator	2017	2018	2019	2020
Retail trade in Ukraine, billion UAH	901.9	1031.7	1159.3	1228.9
Retail trade in Ukraine, billion USD	76	47.4	45.4	43.5
Volume of Internet trade in Ukraine, billion UAH	12.3	25.5	38.4	48
Internet trade volume in Ukraine, billion USD	1.04	1.17	1.5	1.7
Annual retail growth index in USD equivalent (%)	1.5	14.4	12.4	6
Annual growth index of retail trade volume in UAH equivalent (%)	−31.5	−37.6	−4.3	−4%
Annual index of growth of Internet trade in UAH equivalent (%)	75.2	107.6	50.4	23
The annual index of growth of Internet trade in USD equivalent (%)	18.3	13.2	28	13
Internet penetration in Ukraine (%)	1.4	2.5	3.3	4

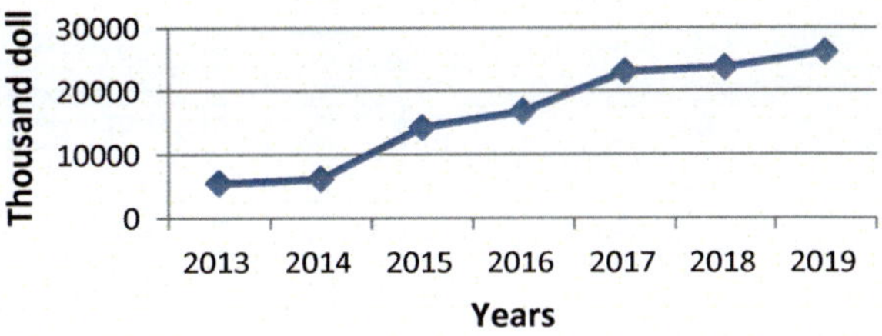

Fig. 4 Number of Internet users in Ukraine (compiled by authors by materials [21])

Figure 5 shows us Ukraine's position in relation to the world and its neighboring countries. Although Ukraine exceeds the average number of Internet users in the world, it lags behind its neighbors by 17% of the population coverage and 37% behind the world leader—Norway. In terms of growth trends, we can see the similarity of growth rates and positions of the most graph of Ukraine to the world average: in the last year, it is shown the world has grown by 5% while Ukraine has grown by 5.9%; from 2015 to 2016, Ukraine has grown by 4.1% and the world by 3%; from 2014 to 2015, Ukraine has grown by 2.7%, the world by 1.8% from 2013 to 2014, Ukraine has grown by 5.3%, and the world by 3.1%. We can conclude that although the growth rate is similar to the world average, but still Ukraine exceeds them by an average of 1% if you do not take into account the growth of Ukraine from 2013 to 2014. That is, the average growth of Ukraine is 4.5% per year while the average growth of Belarus is 7.5%, and the average Poland's growth rate is 3.3%. The growth rate of Ukrainian Internet users exceeds that of Poland and the world, but 1.5 times lower than the growth rate of Belarus.

In addition, it should be noted that as of 2019, the average Internet coverage in the regions of Europe is: Northern Europe—94%, Western Europe—90%, Southern Europe—77%, Eastern Europe—74%. At that time for Ukraine in 2019, this figure was 56%, which is 18% less than the average result in our region (see Fig. 4).

If we are talking about the level of development of e-business and e-commerce in the world and Ukraine, we need to sink such an important indicator as the index of development of information and communication technologies (hereinafter ICT). Table 2 shows that Ukraine ranks 79th in the world, although in 2015, Ukraine ranked 71st, and in 2016 ranked78th place with an index of 5.31, i.e., a decrease in

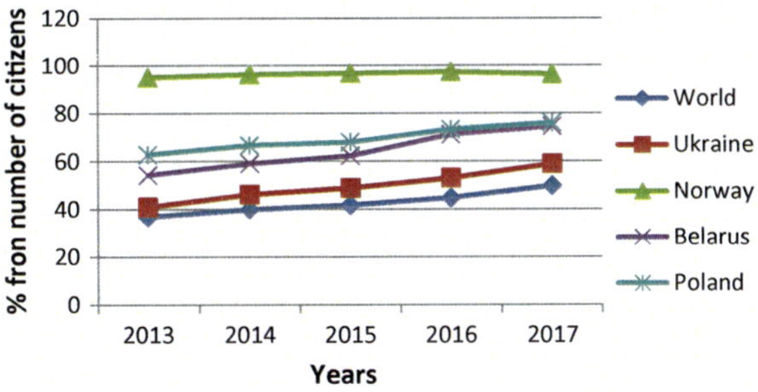

Fig. 5 Number of Internet users in percent of the population (compiled by authors by materials [22])

Table 2 ICT countries ranking [23]

Rate	Country	Index
1	Iceland	8.98
2	South Korea	8.85
3	Switzerland	8.74
4	Denmark	8.71
5	United Kingdom	8.65
6	Hong Kong	8.61
7	Netherlands	8.49
8	Norway	8.47
9	Luxembourg	8.47
10	Japan	8.43
76	Antigua and Barbuda	5.71
77	Dominican Republic	5.69
78	Thailand	5.67
79	Ukraine	5.62
80	China	5.60
81	Iran	5.58
173	Guinea-Bissau	1.48
174	Chad	1.27
175	Central African Republic	1.04
176	Eritrea	0.96

the ranking is not due to a deterioration of ICT conditions but to a slow improvement when compared to the world, but still, such a low position does not improve the situation for the development of e-business in Ukraine, but only slows it down. The biggest problem is not even the low position of Ukraine, but the very slow growth rates because they directly represent the opportunities and rates of e-business development in Ukraine. It is quite telling that in some Atlantic islands, the development of ICT is better than the Ukrainian.

Figure 6 shows the position of Ukraine on the ICT index from 2013 to 2017. From this chart, we can conclude that there is no continuous improvement and introduction of new communication and information technologies in the territory of Ukraine. The biggest improvement was in 2015, in the future, the pace of ICT improvement in Ukraine slowed down. That is, it is possible to conclude that there is no stable technology for improving ICT in Ukraine and its spontaneous nature or its inefficiency in Ukrainian realities.

Ukraine ranks 51st in the e-commerce rankings in the world, which is twelve positions higher than in 2017. These data show that the development of e-commerce in Ukraine has recently intensified due to the quality improvement of postal services and Internet services. This can predict healthy positive trends for its further development, but the situation is still in a difficult situation because the main incentives for growth

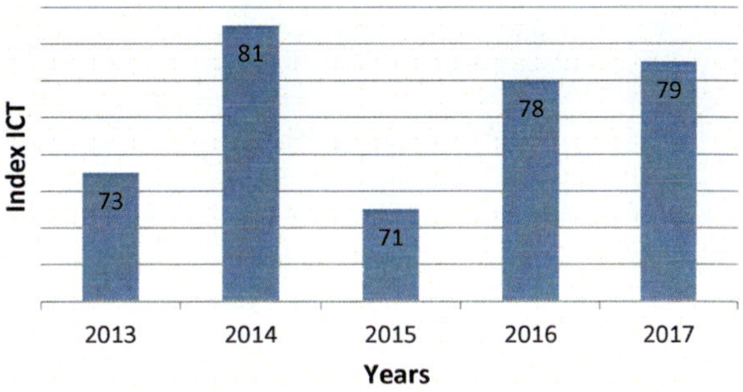

Fig. 6 Ukraine's position on the ICT index (compiled by authors based on materials [20, 24])

are the number of Internet users and the number of active Internet users [25]. Only 53% of the population of Ukraine use the Internet, so the B2C can reach a far from complete market, which will slow its development and create a ceiling effect [21]. Therefore, it is possible to point out the urgent need for a policy of informatization of regions in Ukraine.

It can be said that Ukraine is a mediocre place in e-business, and although Ukraine's indicators by many factors exceed the world average, they remain low relative to not only developed countries, but also among neighboring countries and countries of the Eastern European region. According to the ICT and e-commerce index, Ukraine ranks 79th and 63rd, respectively, which is the world average result, and the 79th place in the ICT index is generally below the average. Although Ukraine has positive trends in the development of e-business, especially in e-commerce and the B2B sector, it is comparable to their real counterparts, but despite the fact that these results are above average in the world, their development rates are small in the context of the countries of the same group with Ukraine. That is, Ukraine's position in e-business is insufficient to speak of its significant contribution to the global e-business sector.

5 Conclusions

Taking into account the analysis of the main trends in the development of the sphere of e-commerce in Ukraine, it is possible to form the following long-term prospects for e-commerce entities:

1. The constant increase in Internet users significantly expands the possibilities for attracting consumers of goods and services.
2. The consciousness of consumers is changing more and more every day toward the inclination to implement information technologies in all spheres of life, including the implementation of transactions via the Internet.
3. The e-commerce subject always has the potential to expand by covering new, different from the usual, market segments, as well as attracting other segments of the population.

4. The management of the service provider enterprise is not limited either by time or space.

If we talk about the short-term prospects for the development of e-commerce in Ukraine, then this, of course, is the improvement of the regulatory framework, and the transfer of the experience of foreign-trading platforms and their adaptation to the conditions of e-commerce within the country. It is this revision of the terms of e-commerce with Ukrainian platforms that will attract the audience that is an active user of foreign, especially Chinese, sites.

Thus, on the basis of the study, it was determined that electronic commerce in the world market is identified with the concept of electronic commerce. According to the Law of Ukraine "On Electronic Commerce," electronic commerce is defined as economic activity in the field of electronic purchase and sale, the sale of goods remotely to a buyer by making electronic transactions using information and telecommunication systems. A statistical analysis of the development of electronic commerce in Ukraine has been carried out, the main trends that have formed in this area have been investigated. It was determined that Ukrainians are increasingly beginning to use the information space to carry out various operations. A list of the most popular e-commerce business models used in the country has been formed, the largest subjects of these relations have been identified, as well as opportunities for the further development of the e-commerce market and attracting new users.

References

1. Sustained sales success in the B2B sector. Tack International. Retrieved from: http://www.tackinternational.com/ua/
2. Marketing. Promotion channels in the B2B market. My business. Retrieved from: https://msb.aval.ua/ru/news/?id=25062
3. Babenko V, Syniavska O (2018) Analysis of the current state of development of electronic commerce market in Ukraine. Technol Audit Prod Reserves 5(4):43. https://doi.org/10.15587/2312-8372.2018.146341
4. Babenko V, Kulczyk Z, Perevozova I, Syniavska O, Davydova O (2019) Factors of development of international e-commerce in the context of globalization. In: CEUR workshop proceedings, vol 2422, pp 345–356. http://ceur-ws.org/Vol-2422/paper28.pdf
5. Hlinenko LK, Dainovskyi YA (2018) Status and prospects of e-commerce development in Ukraine. Market Innov Manag 1. Retrieved from: http://mmi.fem.sumdu.edu.ua/sites/default/files/mmi2018_1_83_102.pdf
6. Lukianchenko AY (2017) Problems of e-commerce system development in Ukraine. Int Sci J "Internauka", 35–37
7. Oliinyk V (2017) Optimal management of the enterprise's financial flows. J Adv Res Law Econ VIII, Fall 6(28):1875–1883. https://doi.org/10.14505/jarle.v8.6(28).22
8. Caporale GM, Plastun A, Oliinyk V (2019) Bitcoin fluctuations and the frequency of price overreactions. Fin Markets Portfolio Mgmt 33(2):109–131. https://doi.org/10.1007/s11408-019-00332-5
9. Chankhi OS, Ozemblovskyi VO (2017) B2C in e-commerce: European vector of Ukraine. Int Sci J "Internauka" Ser Econ Sci 7(7). Retrieved from: https://elibrary.ru/

10. Apopyi VV, Shaleva AY (2017) Retail of Ukraine in the electronic business system: socio-economic aspects. Problems and prospects of electronic business. Belarusian Trade and Economic University of Consumer Cooperation, Gomel, Belarus
11. Kulczyk Z, Babenko V, Perevosova I, Syniavska O, Davydova O (2019) Factors of the development of international e-commerce under the conditions of globalization. SHS Web Conf 65:10–16. https://doi.org/10.1051/shsconf/20196504016
12. Perevozova I, Daliak N, Babenko V (2019) Modeling of financial support for the competitiveness of employees in the mining industry. In: CEUR workshop proceedings, vol 2422, pp 444–454. http://ceur-ws.org/Vol-2422/paper36.pdf
13. Podgorna I, Babenko V, Honcharenko N, Sáez-Fernández FJ, Fernández JAS, Yakubovskiy S (2020) Modelling and analysis of socio-economic development of the European Union countries through DP2 method. WSEAS Trans Bus Econ 17:454–466 (art 44). https://doi.org/10.37394/23207.2020.17.44
14. Hrabovskyi Y, Babenko V, Al'boschiy O, Gerasimenko V (2020) Development of a technology for automation of work with sources of information on the internet. WSEAS Trans Bus Econ 17(25):231–240. https://doi.org/10.37394/23207.2020.17.25
15. Ukraine Economy card. ICT Development Index. Retrieved from: https://www.itu.int/net4/ITU-D/idi/2017/
16. The World Bank. Individuals using the Internet (% of population). Retrieved from: https://data.worldbank.org/indicator/
17. Everybody to the Web: the development of e-commerce in Ukraine and its prospects. Delo.ua. Retrieved from: https://delo.ua/business/
18. Rating ratings of Ukraine by network readiness index. Economic Discussion Club. Retrieved from: http://edclub.com.ua/analityka/
19. Results and plans: e-commerce of Ukraine 2018/2019. Evo.business. Retrieved from: https://evo.business/
20. ITU ICT development index 2017. Retrieved from: https://www.itu.int/net4/
21. E-commerce of Ukraine: results of 2018 and forecasts for 2019 (infographic). Ukrainian Retail Association. Retrieved from: https://rau.ua/novyni/
22. State Statistics Service of Ukraine. Retrieved from: http://www.ukrstat.gov.ua/
23. Worldwide retail eCommerce sales: Emarketer's updated estimates and forecast through 2019. Emarketer. Retrieved from: http://www.emarketer.com/public_media/docs/eMarketer_eTailWest2019
24. UNCTAD B2C E-commerce Index 2018: focus on Africa, 12 (2018). Retrieved from: https://unctad.org/en/PublicationsLibrary/
25. Rating of popular sites for September 2019. Kantar. Retrieved from: https://tns-ua.com/news/

AI and IoT in Healthcare

Hybrid Rough-Genetic Classification Model for IoT Heart Disease Monitoring System

Mohammed M. Eisa and Mona H. Alnaggar

Abstract One of the most reasons for mortality all around the world is heart disease. It is preventable in the event that it can be analyzed more early. In this manner, it is significant to support doctors to analyze the disease and its related dangers. Recently medical classification systems depend on Machine learning algorithms and using IoT sensors makes good monitoring and follow-up for patients. The gathered Medical datasets using Internet of Things IoT sensors are prone to having redundant features, these features tend to decrease the accuracy of the classification. Thus, removing the redundant features is a must to gain better accuracy. This paper presents a model that works based on the sensors' data and provides better classification accuracy depending on a hybrid model for heart disease classification using genetic algorithm (GA) and support vector machine (SVM) techniques. The genetic algorithm is used in feature selection to optimize the dataset by eliminating the features that have no impact or decrease the accuracy. The SVM is used for classification on the optimized dataset. The data used in this paper is Cleveland heart disease dataset, taken from the UCI machine learning repository. It consists of 303 cases with 13 features. The paper also presents a comparative analysis between the proposed model and another heart disease classification models. The experimental results show that the proposed hybrid GA-SVM model has better accuracy than the other models.

Keywords Knowledge discovery · IoT · Heart disease · SVM · Genetic algorithm · Rough sets · Data pre-processing · Classification · Cardiology · Artificial intelligence · Diagnosis · Diagnostic medicine · Data mining

M. M. Eisa (✉)
Ahram Canadian University, Cairo, Egypt
e-mail: Mohammed.mamdouh@acu.edu.eg

M. H. Alnaggar
Kafrelsheikh University, Kafr El-Shaikh, Egypt

D. A. Magdi et al. (eds.), *Digital Transformation Technology*, Lecture Notes in Networks and Systems 224, https://doi.org/10.1007/978-981-16-2275-5_27

1 Introduction

Heart diseases are the main reason for a huge number of death re-phrase in the world over the last few decades and have emerged as the most life-threatening disease. It is the hardening of the blood vessels by fatty deposits called plaque [1]. The heart must get oxygen and supplements to working well. Blood conveys the oxygen and supplements to the heart through the blood vessels called arteries. As the plaque builds up, blood flow to the heart muscle is diminished. At the point when the blood flow is diminished, it can cause chest pain, shortness of breath, or a heart assault to happen [2].

Today, many applications depend on applying data mining techniques, such as cost prediction of corporate claims, financial analysis, manufacturing process control, insurance, risk management, health care and others [3].

1.1 Problem Statement

Applying knowledge discovery in medical informatics databases founded in the hospitals known by the hospital information systems (HIS) containing massive amounts of information which include patient information, data from laboratories that keep on growing year after year. One of the most important medical data is heart disease datasets. With the help of knowledge discovery in database methods, useful patterns of information can be found within the heart disease data, which will be utilized for further research and evaluation of reports. The other question that emerges is the manner by which to classify or assemble this enormous amount of data. The automatic classification technique is done dependent on similitudes present in the information. The Classification process is possibly demonstrated productive if the end that is drawn by the automatic classifier is acceptable to the cardiologist [4].

This paper presents a model based on using IoT sensors to monitor patients and machine learning algorithms such as algorithms to analyze the gathered data from these sensors. This model based on supervised learning algorithms such as genetic algorithm (GA) and support vector machine (SVM).

1.2 Internet of Things

Recently, IoT becomes one of the most used sensors, and these sensors facilitate our life by using it connected to our smartphones or other devices. IoT devices are used in medicine for remote health monitoring and emergency notification system. Building new systems depending on IoT and storing data or signals from the used sensors to help humans in different issues has become a recent request. The obtained data from sensors are uploaded and shared with others on the cloud. Data obtained through sensors are processed and accessed through a smartphone. We can detect or track the heart attacks or health diseases and save the lives of many people using IoT on heart attack detection and heart rate monitoring.

1.3 Rough Set Theory

Rough set theory is one of the important tools for different data analysis and conceptualization. Rough set attracted Machine Learning Engineers' attention [5]. It is a mathematically powerful tool for rule induction because of its ability for describing attributes' dependencies and handling incomplete and not useful data (uncertainty and vagueness) [6]. As Rough set interact with uncertainty, the used dataset can be easily reduced. Pawlak discussed the main concept of Rough set [7]. For any arbitrary subset of the universe, he constructed the lower and upper approximation for an indiscernibility relation.

A dataset is represented in a table called information system in which each row in the table represents a case, a patient record or object. Each column represents an attribute (a variable or property) it can be formulated as:

$$\text{IS} = \{U, A, V\} \tag{1}$$

where U is non-empty set of objects called the universe, A is non-empty finite set of attributes and V is called value set of A.

Rough set theory base on philosophy of classifications so information system should be expressed by dividing non-empty finite set of attributes A into two subsets condition attribute C and decision attribute D (known as supervised learning) and in this case information system called decision table.

This table is formed as:

$$\text{DT} = \{U, C \cup D, V\} \tag{2}$$

where C is a list of condition attributes and $D \notin C$ is decision attribute.

The indiscernibility issue is the foundation of the rough set theory in the decision table, two objects in a decision table are indiscernible if one cannot differentiate between them according to a set of given attributes. The indiscernibility function of the attributes through consideration. The relation of indiscernibility for a set of attributes is a mathematically equivalence relation.

The equivalence relation is a binary relation $R \subseteq X \times X$ which is

- reflexive (xR_x for any object x),
- symmetric (if xR_y then yR_x),
- and transitive (if xR_y and yR_z then xR_z).

The equivalence class of an element consists $[x]_R$ of all objects $x \in X$ such that xR_y.

Let IS = (U, A) be an information system, then with any $B \subseteq A$ there is an associated equivalence relation:

$$\mathrm{IND}_{\mathrm{IS}}(B) = \{(x, x') \in U^2 | \forall a \in B, a(x) = a(x')\} \tag{3}$$

where $\mathrm{IND}_{\mathrm{IS}}$ (B) is called the B-indiscernibility relation.

If $(x, x') \in \mathrm{IND}_{\mathrm{IS}}(B)$, then objects x and x' are indiscernible from each other by attributes from B.

$[x]_B$ denotes the B-indiscernibility relation for the equivalence classes.

1.4 Support Vector Machine

A support vector machine (SVM) is AI calculation that examines information for arrangement and relapse examination. SVM is a regulated learning technique that takes a gander at information and sorts it into one of two classifications. The SVM yields a guide of the arranged information with the edges between the two as far separated as conceivable it makes a level limit called a hyperplane, which isolates the space to make genuinely homogeneous parcels on either side. SVMs are utilized in content order, picture characterization, penmanship acknowledgment and in the science [8].

The target of the support vector machine algorithm is to discover a hyperplane in N-dimensional space (N—the number of features) that have the best classification process to classify data points. To isolate the two classes of data points, there are numerous conceivable hyperplanes that could be chosen. Finding a plane that has the maximum margin and maximum distance between data points of classes is our objective. The confidence of classifying the future data points can be increased if the margin distance maximization provides set of reinforcement [9].

1.5 Genetic Algorithm

Genetic algorithm (GA) is one of the optimization techniques which are based on applying natural selecting principles. It's used to find the optimal or near-optimal solutions to tough problems that otherwise would take a very long time to solve. It's often used to solve optimization issues in machine learning. The genetic algorithm consists of five phases as shown in Fig. 1 [10].

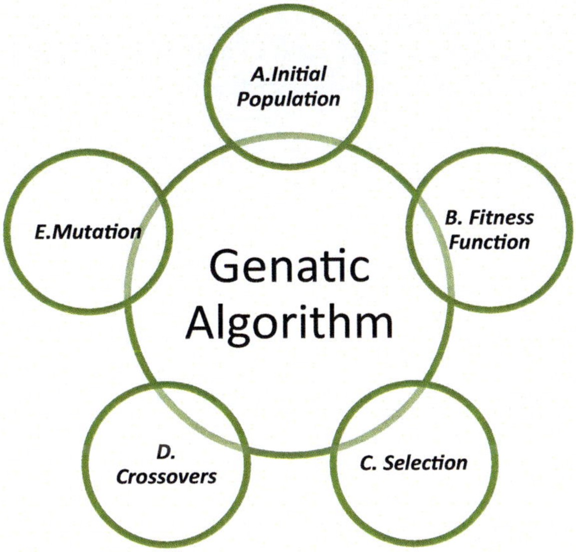

Fig. 1 Genetic algorithm five phases

1.5.1 Initial Population

Initial population is the first phase in the genetic algorithm process, a population is a set of chromosomes in the current generation. The initial population is the very first set of chromosomes which is usually random.

1.5.2 Fitness Function

The fitness function evaluates the fitness of an individual. A fitness score is given for each individual, the probability of being chosen in the selection phase is based on the fitness score.

1.5.3 Selection

Selection is the process of selecting pairs of individuals randomly to produce a new offspring. The higher the fitness score of an individual the higher their chance of being chosen.

1.5.4 Crossover

Crossover is the process of producing an offspring, it is done by selecting a random point for each pair and exchanging genes between the two individuals.

1.5.5 Mutation

Like biological mutation, a mutation is random a change that occurs in one or more genes of an individual. The genes value change from its initial state to another state, but the chance of mutation is very small. The purpose of mutation is producing new solutions to the problem [10, 11].

2 Previous Work

Amid the most recent decade, an enormous measure of issues identifying with coronary heart disease was examined. The treatment of heart disease is a standout among the most significant issues. Numerous analysts have handled medical researchers in the area of treatment of heart disease and new information appears frequently.

In 2011, B. K. Tripathi and et al., a framework for intelligent medical diagnosis using rough set with formal concept analysis. In this study, the rule generation algorithm of rough set theory generates 91 rules. With intelligence domain support, it was reduced to 72 candidate rules and 65 rules by the value of threshold and validation process.

Further exploring suitable rules is to identify the affected characteristics on the relationship between heart disease and the used attributes by the formal concept analysis. This helps the decision maker a priori detection of the heart disease. Accuracy reached to 88% [12].

In 2012, S. Muthukaruppan—A hybrid particle swarm optimization-based fuzzy expert system for the diagnosis of coronary artery disease have offered. Mamdani fuzzy deduction framework and have utilized the triangular membership functions, their proposed classification accuracy was 93.27% [13].

In 2015, V. Krishnagar and his colleagues in their study entitled "Diagnosis of heart disease using fuzzy classification techniques" used a combination of fuzzy techniques and *k*-nearest neighbor, the proposed algorithm accuracy was 83.7% [14].

In 2015, T. Santhanam his colleagues proposed a heart disease prediction model depending on hybrid genetic algorithm and fuzzy technique for building a system to support physicians with more accurate prediction of heart disease which deployed by computing techniques. Like genetic algorithms and fuzzy logic and achieved an accuracy of 86% [15].

In 2018, Kathleen H. Miaoa and colleagues developed an enhanced deep neural network (DNN) learning to aid patients and healthcare professionals and to increase the accuracy and reliability of heart disease diagnosis and prognosis in patients and an accuracy of 83.67% was achieved [16].

In 2018, Navdeep Singh and et al., heart disease prediction system using Hybrid Technique of Data Mining Algorithms. In this study, our aim was to design a heart disease prediction system using various data mining techniques and to perform the analysis of the results obtained for all implemented techniques. So, for the completion of the heart disease prediction model survey, we have evaluated the popular and effective heart disease prediction methods from the literature survey and finally choose the most effective technique of genetic algorithm and naïve Bayes according to their performance analysis on predicting the heart disease. The performances of the models were evaluated using the genetic algorithm and naïve Bayes and achieved an accuracy of 97.14% [17].

In 2020, Jian et al. proposed diagnosis system for heart disease. Their system was implemented to achieve good accuracy based on using set of different classification algorithms. They solved the feature selection problem and increased the accuracy to be 92.37% [18].

Simanta proposed Deep Learning Modified Neural Network (DLMNN) to assist in medication specially in heart diseases diagnosis by using fixed wearable Internet of Things sensor device then storing these normal and abnormal signals on cloud. His proposed methodology enhanced the accuracy if it compared with previous [19].

In addition to the proposed work by Mohammad using IoT framework to increase accuracy of heart disease evaluation by Modified Deep Convolutional Neural Network (MDCNN). His proposed system shows better performance for hug number of records [20].

3 Proposed Model

This paper presents a new model depending on IoT sensors to collect physiological data from the body surface and transmitting these data to the IoT cloud through a wireless channel. Wearable ECG sensors are usually adopted in this system, in this system, the ECG sensors are usually wearable and adopted with the system. This adaptation affected user's daily life is very simple which enables the ECG data to be easily recorded over long hours. Through this means, ECG data can be recorded over long hours. The IoT-based ECG monitoring system is consisting of three parts, the network of ECG sensor, IoT cloud, GUI IoT cloud. ECG monitoring system contains four functional modules, data cleaning, data storage, data analysis and disease warning [21] (Fig. 2).

ECG Sensing Network
The foundation of monitoring system is the network of ECG sensing. This network is responsible for physical data collection from the surface of the body and data transmission to the IoT cloud using the wireless channel. Then, the ECG signals are processed through a series of procedures, such as amplification and filtering, to improve the signal quality and to meet the requirements of wireless transmission (Fig. 3).

The ECG data gathered from sensors are transmitted to the IoT cloud via a specific wireless protocol, e.g., Wi-Fi, Bluetooth, etc. All the three protocols can provide enough data rates for transmitting ECG signals with satisfying energy consumption. However, due to the limited communications ranges of Bluetooth, a smart terminal (such as a mobile phone or computers) is usually needed to receive the ECG data

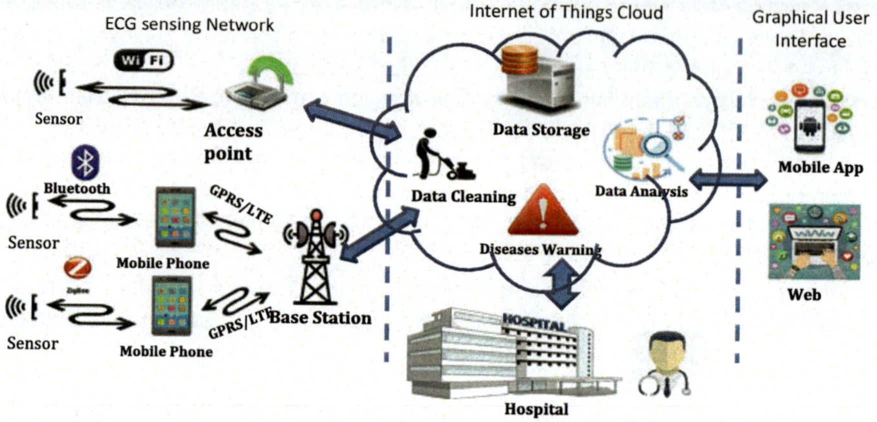

Fig. 2 The architecture of the IoT-based ECG monitoring system

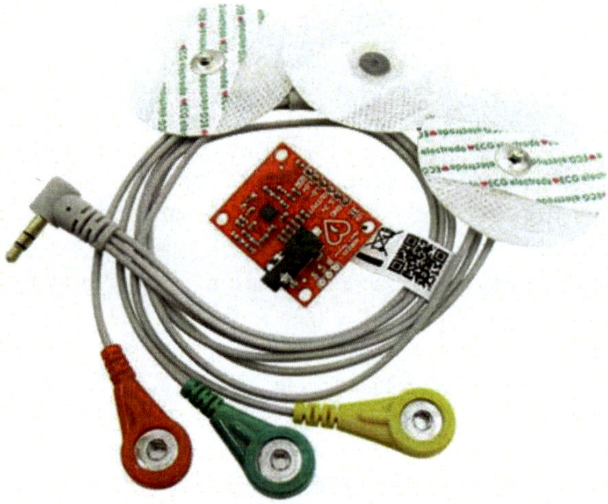

Fig. 3 ECG sensor. https://www.indiamart.com/proddetail/ad8232-ecg-sensor-module-15222184555.html

and then send the data to the IoT cloud through the wireless protocols of the General Packet Radio Service (GPRS), or Long-Term Evolution (LTE).

In this paper, Cleveland heart disease dataset is used obtained from the UCI repository. The dataset contains large and continues values, so we used RSBR data pre-processing to convert continuous values to discrete and convert large values to small values. Feature selection with genetic algorithm is then used to remove irrelevant features to reach the optimal result. Now the dataset is ready for classification, since our dataset is nonlinear, we used support vector machine with kernel trick for classification, SVM works by maximizing the hyperplane between multiple classes. All of these steps shown in Fig. 4 [22].

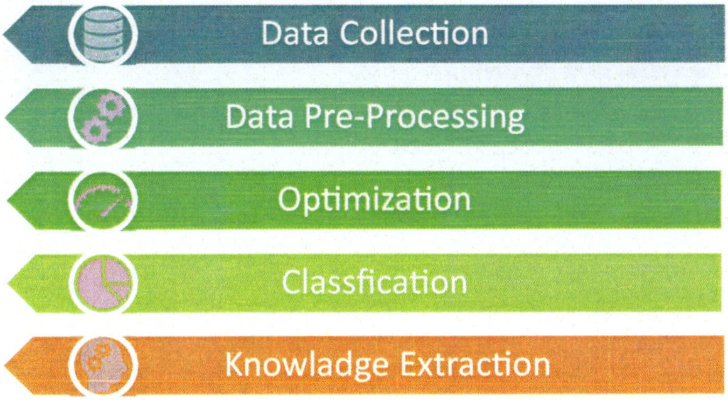

Fig. 4 Proposed model

3.1 Data Collection

The Coronary heart disease dataset is obtained from the UCI repository. The Coronary heart disease dataset contains 13 attributes and 1 decision class. Most of attributes are numerical and there are only three attributes are binary attributes and all attributes have no missing value.

3.2 Data Pre-processing

Data pre-processing is needed because the dataset contains continuous values. One method of pre-processing is RSBR discretization algorithm. Discretization is the process of transforming continuous values to discrete values. The goal of discretization is finding a set of cut points to shrink the ranges of values to the smallest number of intervals possible. These intervals are then used to find an equation and use Boolean algebra to simplify the equation as much as possible, from this equation the cut points can be found. RSBR discretization algorithm is shown in Table 1 [22].

3.3 Data Optimization

Optimization is needed to gain better accuracy, there are many techniques of optimization one of the most advanced techniques is feature selection using genetic algorithm (GA). Feature selection is the way toward finding the most relevant variables for a predictive model. These techniques can be utilized to recognize and remove unneeded, irrelevant and redundant features that do not contribute or decrease the

Table 1 RSBR discretization algorithm

RSBR Discretization algorithm

Input

Information system table (S) with real valued attributes A_{ij}

And n is the number of inter values for each Attribute.

Output

Information table (ST) with discretized real Valued attribute.

Begin procedure

(1) For $A_{ij} \in S$ do

(2) Define a set of Boolean variables as follows:

$$\mathrm{BV}(U) = \left\{ \sum_{i=1}^{n} C_{ai}, \sum_{i=1}^{n} C_{bi}, \sum_{i=1}^{n} C_{ci}, \ldots, \sum_{i=1}^{n} C_{ni} \right\}$$

Where $\sum_{i=1}^{n} C_{ai}$ corresponds to a set of intervals Defined on the variables of attributes a.

(3) End for

(4) Create new information table T^P by using set of intervals

(5) Find minimal subset of C_{ai} that discerns all the objects in decision class D using the following formula:

$$\Phi^U = \wedge \{\psi\ (i,j)\ : d(x_i) \neq d(x_j)\}$$

where $\psi\ (i,j)$ is the number of minimal cuts which must be used to discern two different instances x_i and x_j in the information table.

End procedure

accuracy of the prescient model. Genetic algorithm optimization steps as shown in Table 2 [23].

3.4 *Data Classification*

There are many data mining techniques to classify data, but the support vector machine is one of the best techniques to classify nonlinear problems. A support vector machine (SVM) is one of the machine learning algorithms used in data analysis in classification and regression analysis. SVM essentially functions as the direct

Table 2 GA optimization steps

1. Generate of n chromosomes population randomly (problem suitable solutions)
2. Evaluate $f(x)$ as the fitness function for all chromosome x in the population
3. Repeating the following steps to create a new population until completeness of the new population
4. According to the chromosomes fitness, choose two parent chromosomes from the population (the better fitness, the bigger chance to be selected)
5. Cross over the parents for forming a new child with a probability crossover probability. If there is no performed crossover, children (offspring) is a copy of parents
6. Mutate a new offspring for all locus with the mutation probability (chromosome position)
7. Add the new offspring in a new population
8. for a further run of algorithm Use new generated population
9. Stop if the satisfaction occurred at the end of the condition, and return the best solution in the current population
10. Go to step **2**

separator between two data points to distinguish two distinct classes in the multidimensional condition. SVM utilizes an extremely enormous set of nonlinear features that is task independent. They have a clever approach to prevent over-fitting. They have a very clever way to utilize number of features without requiring nearly as much computation as seems to be necessary. The prime target of this methodology is to maximize the margin between the classes and to minimize the separation between the hyperplane points classification method shown in Table 3.

Table 3 SVM classification map

1. Data pre-processing involves Dividing the data into attributes and labels and dividing the data into training and testing sets
2. all the columns of the bank data frame are being stored in the X variable except the "Class" column, which is the label column. The drop () method drops this column
3. The class column is only being stored in the y variable. y variable has corresponding labels while X variable has attributes
4. Once the data is divided into attributes and labels, the final pre-processing step is to divide data into training and test sets. Luckily, the model selection library of the Scikit-Learn library contains the train, test, split method that allows us to seamlessly divide data into training and test sets
5. divided the data into training and testing sets
6. Scikit-Learn contains the SVM library, which contains built-in classes for different SVM algorithms. To perform the task of classification we will use support vector classifier, which is included in Scikit-Learn's SVM library

3.5 Knowledge Extraction

Coronary heart diseases are most generally because of the atherosclerotic impediment of the coronary arteries. Atherosclerosis is a procedure that can include huge numbers of the body's veins with an assortment of presentations. Half of all deaths in the developed world and quarter of deaths in the developing world are because of heart disease which is included hypertension and the diseases caused heart disease [11].

By this paper, now we can extract knowledge from the dataset and can decide each patient may have or have not coronary heart disease before doing Cardiac catheterization that will help The cardiologist to stands on the patient's condition because Cardiac catheterization has many risks on the patient health.

4 Experiment Design

In this experiment, the medical data identified with heart diseases considered. This dataset was acquired from Cleveland database. This is openly accessible dataset on the Internet. Cleveland dataset concerns the classification of a person into normal and abnormal person regarding heart diseases.

4.1 Dataset Description

Used dataset to test our model, we used the Cleveland Heart Database taken from UCI learning dataset repository. The dataset contains 13 numerical attributes and 1 decision attribute described in Table 4 [10].

4.2 Results and Comparative Study

In this experiment, a total of 303 records and 13 attributes (features) gathered from the UCI repository Cleveland Coronary heart disease dataset the features were first optimized, the redundant and irrelevant features were removed which was done randomly, thus leaving us with an optimized dataset that is ready for classification.

In this paper, a hybrid genetic algorithm-SVM model for to classify heart disease and predict with heart disease treatment through the GA algorithm optimizing and doing feature selection by passing dataset into five phases of GA algorithm then SVM algorithm classify the features that been selected by GA algorithm to predict heart disease treatment. A comparative analysis between the proposed model and other heart disease classification models then to determine proposed model efficiency. The

Table 4 Dataset description

Attribute	Description	Range
Age	Age in years	Continuous
Sex	(1 = male; 0 = female)	0, 1
Cp	– Value 1: typical angina – Value 2: atypical anginal – Value 3: non-anginal pain – Value 4: asymptotic	1, 2, 3, 4
trestbps	Resting blood pressure (in mm Hg)	Continuous
chol	Serum cholesterol in mg/dl	Continuous
fbs	(Fasting blood sugar 0.120 mg/dl) (1 = true; 0 = false)	0, 1
restecg	Electrocardiography results – Value 0: normal – Value 1: having ST-T wave abnormality (T wave inversions and/or ST Elevation or depression of > 0.05 mV) – Value 2: showing probable or definite left ventricular hypertrophy by Estes' criteria	0, 1, 2
Thalach	Maximum heart rate achieved	Continuous
Exang	Exercise induced angina (1 = yes; 0 = no)	0, 1
Old peak	ST depression induced by exercise Relative to rest	Continuous
Slope	The slope of the peak exercise ST segment – Value 1: up sloping – Value 2: flat – Value 3: down sloping	1, 2, 3
Ca	Number of major vessels (0–3) Colored by fluoroscopy	Continuous
Thal	Normal, fixed defect, reversible defect	3, 6, 7
Target	0: Normal person 1: Person with heart disease	0, 1

records were split 80% training dataset and 20% testing dataset the training accuracy was 92% and the testing accuracy was 97.7% and the average is 95%.

"A Hybrid Particle Swarm Optimization Based Fuzzy Expert System for The Diagnosis of Heart Disease" was published, it used particle swarm and fuzzy logic and an accuracy of 93.27% was achieved [13]. Then in this paper "Diagnosis of heart disease using fuzzy classification techniques" was published it used fuzzy logic and an accuracy of 83.7% was achieved [14].

The fuzzy logic and genetic algorithms are used to achieve accuracy 86% when building a prediction model of heart disease [17].

Then in the paper "Coronary Heart Disease Diagnosis using Deep Neural Networks" was published it used deep neural network and an accuracy of 83.67% was achieved [15].

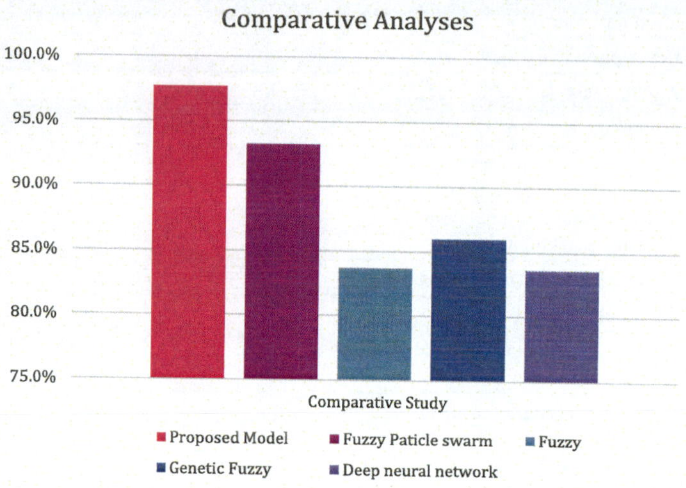

Fig. 5 Comparative analysis

Table 5 Comparative analysis between the proposed model and previous work

Model	PM	FPS	Fuzzy	GA-F	DNN
Accuracy (%)	97.7	93.27	83.70	86	83.67

We made a comparative study between set of different proposed models. The comparative study shows that the proposed model classification accuracy is higher than the other models. Comparative results shown in Fig. 5 and Table 5.

5 Conclusion and Future Work

Data Mining is a set of methodology that applies to giant and complicated databases. this can be to eliminate the randomness and see the hidden pattern. As these data processing ways square measure nearly always computationally intensive. We tend to use data processing tools, methodologies, and theories for revealing patterns in knowledge. Data mining can be used by medical experts to identify heart disease and help them make decisions.

In this paper, genetic algorithm and support vector machine hybrid model used for classification and the model achieved a high accuracy compared to different heart disease classification models. In the future, we are looking forward to using a larger dataset which may increase our accuracy.

References

1. Ramalingam V, Dandapath A, Raja MK (2018) Heart disease prediction using machine learning techniques: a survey. Int J Eng Technol 7(2.8):684–687
2. Betancourt JR et al (2005) Cultural competence and health care disparities: key perspectives and trends. Health Aff 24(2):499–505
3. Dangare CS, Apte SS (2012) Improved study of heart disease prediction system using data mining classification techniques. Int J Comput Appl 47(10):44–48
4. Palaniappan S, Awang R (2008) Intelligent heart disease prediction system using data mining techniques. In: 2008 IEEE/ACS International conference on computer systems and applications. IEEE
5. Dai J et al (2017) Uncertainty measurement for incomplete interval-valued information systems based on α-weak similarity. Knowl-Based Syst 136:159–171
6. Tan A et al (2018) Intuitionistic fuzzy rough set-based granular structures and attribute subset selection. IEEE Trans Fuzzy Syst 27(3):527–539
7. Pawlak Z (1982) Rough sets. Int J Comput Inform Sci 11(5):341–356
8. De Brabanter J et al (2002) Least squares support vector machines. World Scientific
9. Babaoglu İ, Findik O, Ülker E (2010) A comparison of feature selection models utilizing binary particle swarm optimization and genetic algorithm in determining coronary artery disease using support vector machine. Expert Syst Appl 37(4):3177–3183
10. Goldenberg DE (1989) Genetic algorithms in search, optimization and machine learning. Addison Wesley, Reading, MA
11. Eissa MM et al (2014) Hybrid rough genetic algorithm model for making treatment decisions of hepatitis C. In: 2014 International conference on engineering and technology (ICET). IEEE
12. Tripathy B, Acharjya D, Cynthya V (2013) A framework for intelligent medical diagnosis using rough set with formal concept analysis. arXiv preprint arXiv:1301.6011
13. Muthukaruppan S, Er MJ (2012) A hybrid particle swarm optimization based fuzzy expert system for the diagnosis of coronary artery disease. Expert Syst Appl 39(14):11657–11665
14. Krishnaiah V et al (2014) Diagnosis of heart disease patients using fuzzy classification technique. In: International conference on computing and communication technologies. IEEE
15. Santhanam T, Ephzibah E (2015) Heart disease prediction using hybrid genetic fuzzy model. Indian J Sci Technol 8(9):797
16. Miao KH, Miao JH (2018) Coronary heart disease diagnosis using deep neural networks. Int J Adv Comput Sci Appl 9(10):1–8
17. Dewan A, Sharma M (2015) Prediction of heart disease using a hybrid technique in data mining classification. In: 2015 2nd International conference on computing for sustainable global development (INDIACom). IEEE
18. Li JP et al (2020) Heart disease identification method using machine learning classification in E-healthcare. IEEE Access 8:107562–107582
19. Sarmah SS (2020) An efficient IoT-based patient monitoring and heart disease prediction system using deep learning modified neural network. IEEE Access 8:135784–135797
20. Khan MA (2020) An IoT framework for heart disease prediction based on MDCNN classifier. IEEE Access 8:34717–34727
21. Hou L et al (2016) Internet of things cloud: architecture and implementation. IEEE Commun Mag 54(12):32–39
22. Chao S, Li Y (2005) Multivariate interdependent discretization for continuous attribute. In: Third international conference on information technology and applications (ICITA'05). IEEE
23. Rhodewalt F, Smith TW (1991) Current issues in type A behavior, coronary proneness, and coronary heart disease

Diagnosis and Detection of Liver Cirrhosis Based on Image Analysis

Ahmed Gaber, Alaa Hamdy, Hammam Abdelaal, and Hassan Youness

Abstract A non-invasive technique for diagnosis and identification of liver cirrhosis based on ultrasound images and texture features will be presented in this paper. The proposed system first accepts the acquired US liver image, extracts features, and thereafter, classifies the liver tissue of an input image. The proposed method is based on selecting a variable number of dynamic-sized Region of Interests that are distributed within the liver tissue. The multiple Region of Interests (ROIs) are selected to reduce the required computations. From each ROI, efficient six features (homogeneity, variance, correlation, entropy, contrast, and energy) are selected and extracted. Finally, applying a voting-based classifier using multiple Region of Interests. The proposed method optimizes all the parameter thresholds automatically using the genetic algorithm. The recognition accuracy of this investigation is of 90%.

Keywords Classifier · Liver cirrhosis · Feature extraction · Genetic algorithm

1 Introduction

Liver diseases are among the most life-threatening diseases global. Among the different categories of liver lesions, liver cancer has an excessive incidence rate and abnormal mortality rate in Melanesia, East Asia, the countries of Southeast Asia, and sub-Saharan Africa [1]. Liver cirrhosis (LC) represents the end phase of many forms of chronic hepatitis (CH) of different etiologies and with quite variable courses [2]. In chronic liver disease (CLD), the wound-healing process progresses toward

A. Gaber (✉) · H. Youness
Department of Computers and Systems Engineering, Faculty of Engineering, Minia University, Minia, Egypt

A. Hamdy
Department of Communications, Electronics and Computer, Faculty of Engineering, Helwan University, Cairo, Egypt

H. Abdelaal
Department of Information Technology, Faculty of Computers and Information, luxor University, luxor, Egypt

D. A. Magdi et al. (eds.), *Digital Transformation Technology*, Lecture Notes in Networks and Systems 224, https://doi.org/10.1007/978-981-16-2275-5_28

fibrosis due to the persistence of a noxa patogena [3]. LC can be either micro- and macro-nodular. It is a chronic advanced clinical condition, leading to portal hypertension and liver cell failure, which in turn creates a micro-environment that favors the onset of hepatocellular carcinoma (HCC). This new condition, therefore, modifies the natural history of Cirrhosis CLD [2].

Clinically, LC may be distributed into two stages: compensated; and decompensated [4]. Decompensated (DLC) is certainly diagnosed as it presents with a series of clinical and laboratory signs (ascites, esophageal varices, encephalopathy, thrombocytopenia, and gastrointestinal bleeding), while compensated (CLC) may be challenging to discriminate from chronic hepatitis. Once more, LC may be divided into two shapes according to the absence/presence of portal hypertension [4]. As a result, perfect evaluation of the fibrosis stage or of the appearance of overt cirrhosis is essential not only to reach a right diagnosis but also to begin correct treatment to allow an initial diagnosis of the frequent and fatal complications of LC, such as HCC and esophageal varices [5]. There are many diagnosing imaging techniques, such as computerized tomography (CT) scanning, magnetic resonance imaging (MRI) scanning, and ultrasound (US) scanning [6], which are used for liver disease detection. MRI and liver biopsies are both expensive and uncommon procedures [7] nevertheless, an image-acquisition US device is simple to operate, radiation-free, inexpensive, and obtainable in a wide range of radiology centers [6]. One of the major drawbacks related to using the ultrasound is the low quality of the obtained images compared to those of CT and MRI [8, 9].

In [10], a method for characterizing normal liver, cirrhotic liver, and hepatocellular carcinoma (HCC) formed on the cirrhotic liver was suggested. The dataset contains 56 liver ultrasound images taken from 56 subjects. An experienced participating radiologist extracts 60 regions of interest (ROIs) from each object class. In [11], a technique to categorize fatty, normal, and LC images was developed. Five based texture features were extracted and used in a five-layered feed-forward neural network, and overall accuracy of 82.67% was obtained. In [12], a texture analysis method known as texture feature co-occurrence matrix was suggested to categorize hepatitis, healthy liver, and LC. The overall accuracy of 86.7% was obtained using the maximum likelihood classifier. In [13], a mixture of discrete wavelet transformation and contrast enhancement for speckle noise elimination from a dataset of 30 LC and 25 healthy liver images was used. The *K*-means clustering with a minimum Euclidian distance followed by artificial neural networks (ANNs) was utilized.

The research work by Wu et al. [14] suggested a technique to obtaining an automated LC diagnosis method using evolutionary feature construction, and the accuracy achieved was 80%. Four datasets were proposed in [15], which were used instead of three by adding a set of hepatoma images. By using a mixture of gray-level run length statistics, spatial gray-level dependency matrix, and texture analysis of fractal dimension with the *k*-nearest neighbor classifier, the researchers' accuracy was 80%.

The paper will be organized as follows. In Sect. 2, description of the image-acquisition technique and methodology are presented. Then, in Sect. 3, the experimental results are given. Finally, in Sect. 4, the conclusion and limitations will be presented.

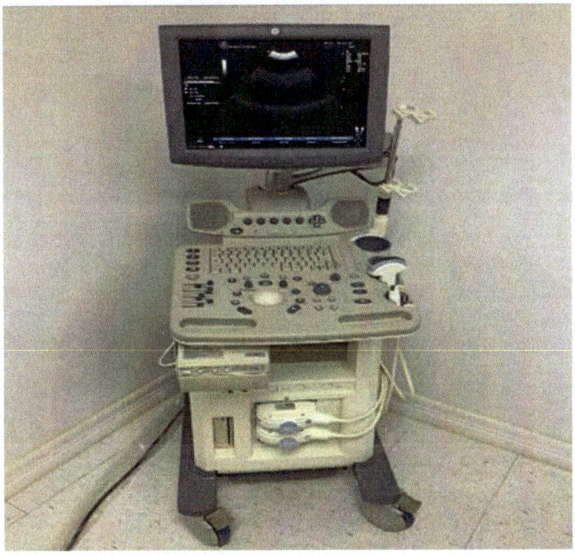

Fig. 1 GE logiq p5 ultrasound machine

2 Materials and Methods

2.1 Image Acquisition

In this study, the ultrasound liver images were obtained at El-Rouad Radiography Center (Menouf, Egypt) via a *GE Logiq P5* equipment ultrasonic machine with a 3 MHz convex probe (see Fig. 1) [16].

In this study, the size of all ultrasound images has 684 × 552 pixels. There are 70 liver images obtained from 70 patients; structured as follows: 30 normal, and 40 cirrhosis. An example of acquired liver images is demonstrated in Fig. 2.

2.2 Methodology

The proposed system for developing and designing an automatic digital system processes and classifies digital ultrasound images for the human liver tissues that involve several parts; datasets that are divided into training and testing sets, the second part is the features extraction, and the final stage is the classifier, as shown in Fig. 3.

One weakness of using a single Region of Interest (ROI) for liver disease diagnosis is that the system is unsuccessful when the classification end result uses insufficient features. Consequently, multiple ROIs are used to additional enhance the performance of the human liver disease diagnosis. Moreover, the accuracy achieved by using multiple ROIs voting was greater than that of only one (single) ROI. In this paper,

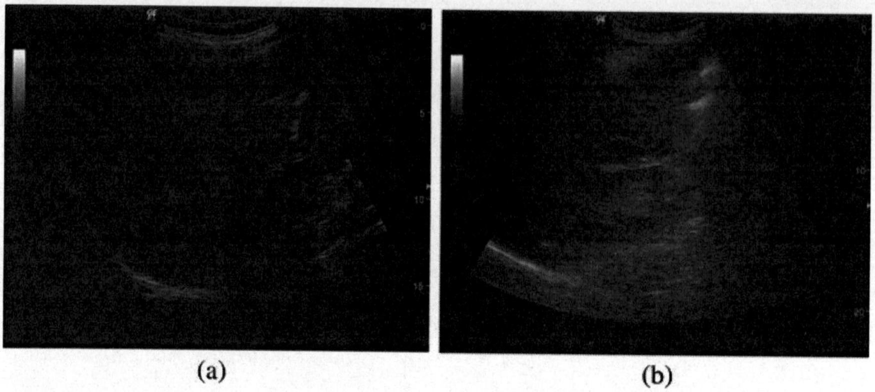

Fig. 2 Examples of an acquired ultrasound image: **a** normal liver, **b** liver cirrhosis

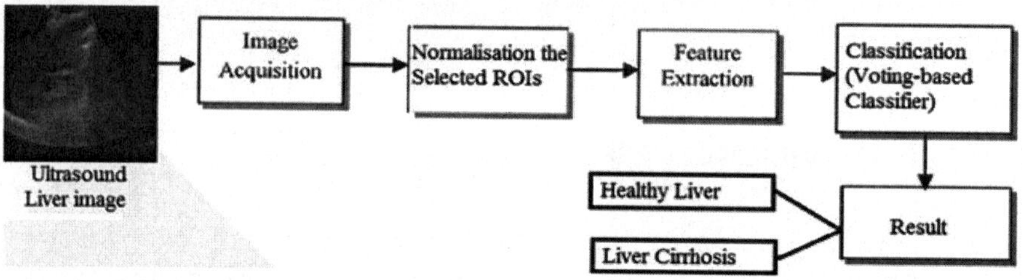

Fig. 3 Block diagram of our proposed system

separately selected ROI is transformed from RGB (color) to gray-level pixels. Then, only six features are obtained; variance, contrast, homogeneity, entropy, correlation, and energy. These features are described in Table 1. All six features are computed for each ROI. Based on the values of these features individually, the type of ROI can be recognized as cirrhosis or healthy liver. It is observed that the normal human liver has a homogenous gray level, low entropy, high energy, and high correlation and is characterized by low contrast as well as low variance. Else, it is considered or counted as LC.

2.3 *Classification of Tissues as Normal/Cirrhosis*

In this section, each selected ROI is classified as normal tissue or cirrhosis tissue. It is observed that the normal liver has a homogeneous grey level, and is characterized by low contrast as well as low variance. Otherwise, it is considered to be a cirrhosis liver, as shown in Fig. 4. In this figure, some tissues falsely appear as cirrhosis, an ROI is classified as normal tissue, and a normal ROI is classified as cirrhosis tissue.

Table 1 Description of the features

Features	Equations
Correlation (Cor)	$\text{Cor} = \frac{\sum_{i=0}^{G-1}\sum_{j=0}^{G-1}(i\times j)\times P(i,j)-\{u_x\times u_y\}}{\sigma_x\times\sigma_y}$
Homogeneity (Homo)	$\text{Homo} = \sum_{j=0}^{2G-2}\frac{1}{1+j^2}\text{Hd}(j\|\Delta x, \Delta y)$
Entropy (Ent)	$\text{Ent} = -\sum_{i=0}^{G-1}\sum_{j=0}^{G-1} p(i, j) * \log(p(i, j))$
Variance (Var)	$\text{Var} = \sum_{i=0}^{G-1}\sum_{j=0}^{G-1}(i-\mu)^2 P(i, j)$
Energy (Ene)	$\text{Ene} = \sum_{i=0}\sum_{j=0} N_d^2(i, j)$
Contrast (Con)	$\text{Con} = \sum_{n=0}^{G-1} n^2\left\{\sum_{i=1}^{G}\sum_{j=1}^{G} P(i, j)\right\}, \|i-j\| = n$

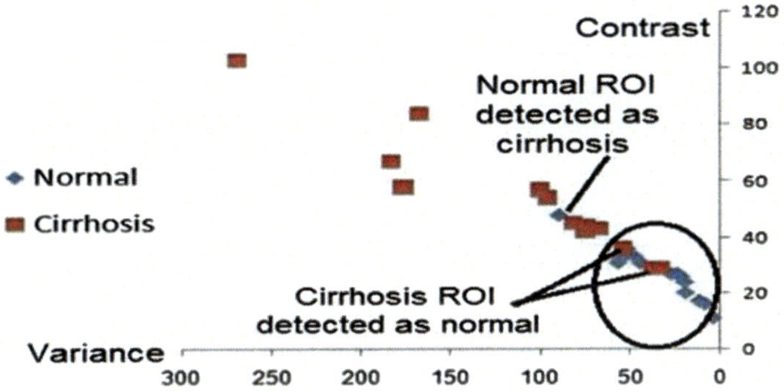

Fig. 4 Feature space of normal/cirrhosis classification indicating false negative and false positive

Contrast can be estimated from the histogram, by comparing the histogram with a threshold known as the histogram threshold. In Fig. 5a, the obtained histogram of a certain ROI within the normal tissue has a narrower range than the cirrhosis tissue in Fig. 5b. Therefore, the contrast should be compared to another threshold referred to as the contrast threshold. Similarly, the variance is compared to a threshold known as the variance threshold. The histogram, contrast, and variance thresholds can be determined manually.

However, it is better to be able to select these thresholds automatically. In this study, a GA is used to optimize the selection of the thresholds. A voting function as a classifier is used to categorize the type of human liver tissue. In this paper, a voting threshold is used to establish and calculate the least possible number of

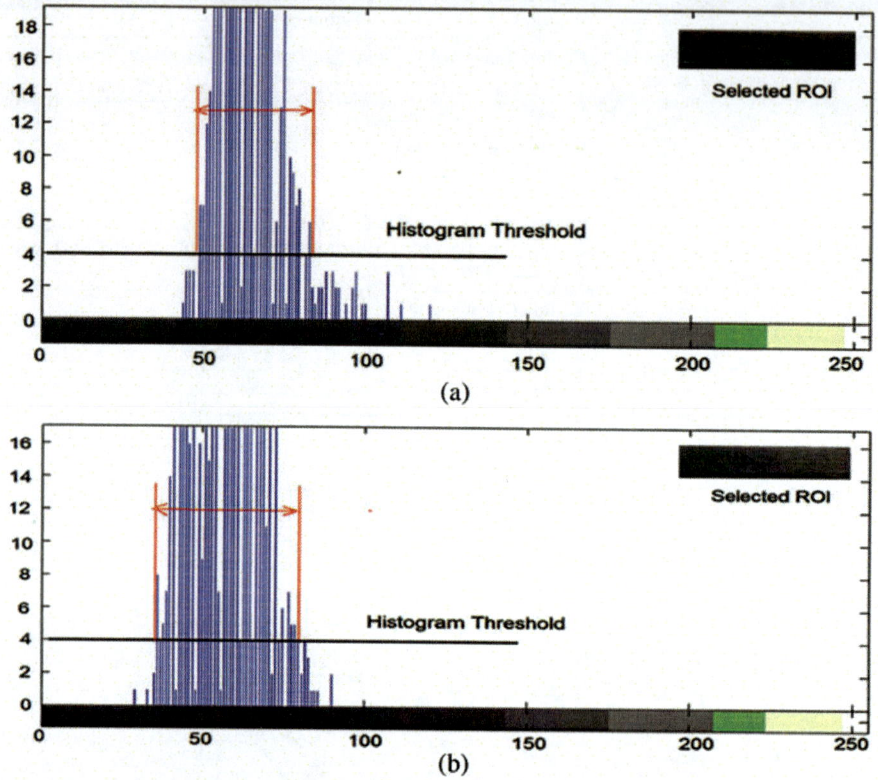

Fig. 5 Histograms of a selected ROI: **a** healthy liver and **b** cirrhosis liver

ROIs to categorize a given liver as a healthy liver. For avoiding the manual selection of all the thresholds and other parameters declared previously, a GA is used. This process of histogram, variance, entropy, contrast, homogeneity, energy, and correlation calculations and comparisons will be repeated for wholly other ROIs. To emulate the specialist's and expert's vision, a different Number of ROIs (No. of ROIs) is carefully chosen for each image under test to investigate the category of the liver tissue in different or changed positions. Also, the size of each chosen ROI (ROI rows × ROI columns) is variable to each image under test to aid accurate diagnosis and detection of diffuse liver diseases. To automatically adjust the parameters, the genetic algorithm (GA) is used with the next chromosome. In this investigation, 17 parameters are shown in Fig. 6.

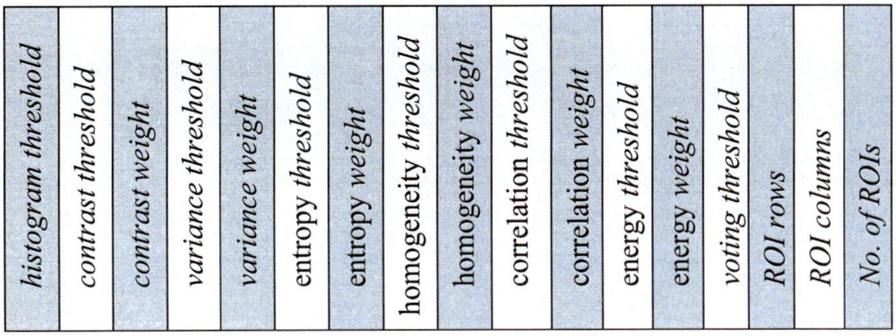

Fig. 6 Chromosome structure congaing the 17 parameters

3 Experimental Results and Discussion

In this study, several experiments have been operated. Multiple ROIs of a maximum of 9 ROIs (determined by GA) per each ultrasound liver image are used. The total number of ROIs from the healthy liver and LC liver are 270 (= 30 $\times$ 9) and 360 (= 40 $\times$ 9), respectively. The original acquired image by GE Logiq P5 ultrasonic machine. The datasets were collected from 70 persons; 30 normal and 40 cirrhosis; of known pathology. All ultrasound images are then transformed to grayscale images (normalization process). Selecting multiple ROIs from ultrasonic images was under the assistance of liver disease specialists. The dataset is allocated into two sets, testing and training sets. The training set of 20 normal liver images and 30 cirrhosis liver images is used for training the system, whereas the testing set of 10 normal liver images and 10 LC images is used. Finally, all ultrasound images are categorized by two medical doctors. The fitness function is the number of wrongly classified ultrasound images. The generation increases as the number of wrongly classified images or false negatives decrease, as shown in Fig. 7.

In the testing phase, it is observed from Table 2 that the obtained result in the case of using the six features; homogeneity, variance, contrast, entropy, correlation, and energy, is 90%. The recognition rate decreased by 9.5% by using two features only. According to the obtained experimental results, it is recommended to use the six features instead of using fewer features.

It is observed that where two features are combined (e.g., contrast and variance), the recognition ratio can be improved, as given in Table 3.

Compared to the case where these features are used independently (see Table 4), it should be noticed that the highest recognition rate in Table 4 is 69.7%, whereas, in Table 3, it is 80.5%.

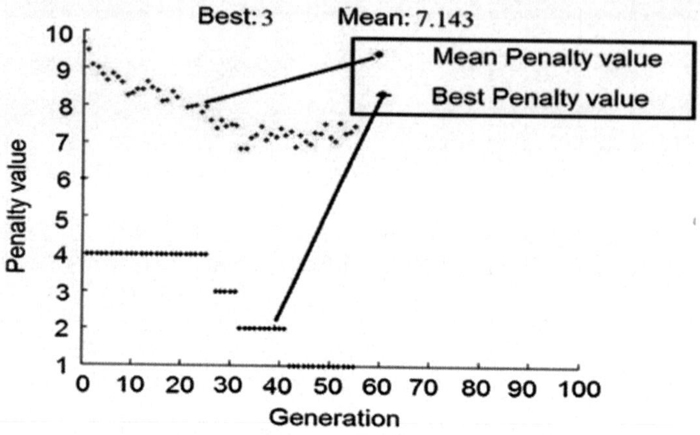

Fig. 7 Seventeen parameters are optimized to increase the perfectly classified ultrasound images

Table 2 Recognition rate in the testing phase with/without feature selection

	The recognition rate (%)
Six features (without feature selection)	90
Two features (with feature selection)	80.5

Table 3 Recognition rate in percentage in case of selecting two features

	Con	Var	Ent	Hom	Cor	Ene
Con		80.5	66.4	65.2	67.5	64
Var	80.5		67.5	66.4	68.7	65.2
Ent	66.4	67.5		62.8	65.2	61.6
Hom	65.2	66.4	62.8		64	61.6
Cor	67.5	68.7	65.2	64		62.8
Ene	64	65.2	61.6	61.6	62.8	

Contrast (Con), Variance (Var), Entropy (Ent), Homogeneity (Hom), Correlation (Cor), Energy (Ene)

Table 4 Recognition rate in case of extracting each feature separately

Extracted feature	Recognition rate (%)
Contrast	67.4
Variance	69.7
Entropy	62.6
Homogeneity	63.8
Correlation	65
Energy	63.8

4 Conclusion

In this paper, a diagnostic method for liver tissues is proposed. The unenhanced ultrasound images of liver tissues are categorized as normal or cirrhosis. The ultrasound images (obtained via a non-invasive process) are used for the identification and diagnosis of abnormal and normal liver tissue. The number of normal ultrasound images dataset used is 30 and cirrhosis images is 40. The proposed work is established on choosing dynamic multiple ROIs to decrease computations. Six features, homogeneity, variance, correlation, entropy, contrast, and energy, are extracted from each selected ROI. The normal liver is characterized by low contrast as well as low variance, low homogeneity, low correlation, low entropy, and low energy. Then, the voting function is applied to categorize tissues as cirrhosis or normal. The values of the parameters are automatically optimized via the evolutionary algorithm (GA). The final decision is based on the voting-based classifier is 90%.

A limitation of this paper is that there is no visibly obtainable database that can be employed to further validate the method. Moreover, the poor quality of the ultrasound images is a big challenge.

References

1. Xu SSD, Chang CC, Su CT, Phu PQ (2019) Classification of liver diseases based on ultrasound image texture features. Appl Sci 9(2):342
2. Blachier M, Leleu H, Peck-Radosavljevic M, Valla DC, Roudot-Thoraval F (2013) The burden of liver disease in Europe: a review of available epidemiological data. J Hepatol 58(3):593–608
3. Hytiroglou P, Snover DC, Alves V, Balabaud C, Bhathal PS, Bioulac-Sage P, Nakanuma Y (2012) Beyond "cirrhosis" a proposal from the international liver pathology study group. Am J Clin Pathol 137(1):5–9
4. Garcia-Tsao G, Friedman S, Iredale J, Pinzani M (2010) Now there are many (stages) where before there was one: in search of a pathophysiological classification of cirrhosis. Hepatology (Baltimore, Md.) 51(4):1445
5. Ginés P, Angeli P, Lenz K, Møller S, Moore K, Moreau R (2010) European association for the study of the liver. EASL clinical practice guidelines on the management of ascites, spontaneous bacterial peritonitis, and hepatorenal syndrome in cirrhosis. J Hepatol 53(3):397–417
6. Gaber A, Attiya G, Hamdy A, Elsayed T (2016) Recognition of diffuse liver cirrhosis based on image analysis. Imaging Sci J 64(3):152–159
7. Larrey D, Meunier L, Ursic-Bedoya J (2017) Liver biopsy in chronic liver diseases: is there a favorable benefit: risk balance? Ann Hepatol 16(4):487–489
8. Hou F, Wang R, Han B, Bai Z, Li Y, Qi X (2020) Liver cirrhosis. In: Liver diseases. Springer, Cham, pp 239–245
9. Boraschi P, Tarantini G, Pacciardi F, Donati F (2020) Dynamic and multi-phase contrast-enhanced CT scan. In: Liver diseases. Springer, Cham, pp 479–491
10. Virmani J, Kumar V, Kalra N, Khandelwal N (2013) SVM-based characterization of liver ultrasound images using wavelet packet texture descriptors. J Digit Imaging 26(3):530–543
11. Pavlopoulos S, Kyriacou E, Koutsouris D, Blekas K, Stafylopatis A, Zoumpoulis P (2000) Fuzzy neural network-based texture analysis of ultrasonic images. IEEE Eng Med Biol Mag 19(1):39–47
12. Horng MH, Sun YN, Lin XZ (2002) Texture feature coding method for classification of liver sonography. Comput Med Imaging Graph 26(1):33–42

13. Hambire VV, Ganorkar SR (2015) Classification of liver disease based on US images. Int Res J Eng Technol 4(4):452–456
14. Wu H, Huang JY, Cheng SC, Yang CK, Lin CL (2011) Evolutionary feature construction for ultrasound image processing and its application to automatic liver disease diagnosis. In: 2011 International conference on complex, intelligent, and software intensive systems. IEEE, Seoul, South Korea, pp 565–570
15. Kyriacou E, Pavlopoulos S, Koutsouris D, Zoumpoulis P, Theotokas I (1997) Computer assisted characterization of liver tissue using image texture analysis techniques on B-scan images. In: Proceedings of the 19th annual international conference of the IEEE engineering in medicine and biology society. Magnificent milestones and emerging opportunities in medical engineering. IEEE, Chicago, USA, pp 806–809
16. https://www.dotmed.com/listings/search/equiment.html?key=GE%20Logiq%20P5&description=259. Last accessed 28 Oct 2020

Breast Cancer Classification in IoT Framework

Mohammed M. Eissa, Magdy Zakaria, and Abeer Hekal

Abstract In the last decade, Breast Cancer (BC) is one of the widespreading diseases for women and causing of the death in the world wide. Early detection of BC can enhance the changes of survival. Now there are many techniques and methodologies for detecting BC in an appropriate time. Machine learning (ML) technique can be used to observe and classify the abnormalities that happened inside the breast such as micro-calcifications, masses, and asymmetry. With the help of ML techniques, it has become easier to distinguish tumor from the affected breast and diagnose BC. This paper aims to classify breast tumor (benign or malignant) in Internet of Things (IoT) health environment framework. Both Support Vector Machine (SVM) and back-Propagation classifiers are used for classification of breast tumor. In this paper, three performance evaluation metrics are used to evaluate, which are; classification accuracy, sensitivity, specificity.

Keywords Breast cancer · SVM · Back propagation · Classification · IoT

1 Introduction

Many patients in the world suffer from cancer. There are different kinds of cancer; BC is one of the most common types that affect women [1]. BC considers the second reason leading to death among women in the world and it nearly occurs in one every eight women [2]. Adequate and early detection of BC can save or at least improve the condition of the injured person. Development of Machine Learning techniques can improve detecting of BC and survivability of cancers, whereas many researchers have been proved that algorithms are the best key solution in detecting BC diagnosis [3].

M. M. Eissa (✉)
Ahram Canadian University, Cairo, Egypt
e-mail: Mohammed.mamdouh@acu.edu.eg

M. Zakaria · A. Hekal
Faculty of Computers and Information, Mansoura University, Cairo, Egypt

D. A. Magdi et al. (eds.), *Digital Transformation Technology*, Lecture Notes in Networks and Systems 224, https://doi.org/10.1007/978-981-16-2275-5_29

1.1 Breast Cancer (BC)

BC is the most common noncause of female mortality [4]. It occurs when cells in breast being nearby tissues or spread throughout the body [5]. BC typically has no symptoms when the malignant tumor is small and most easily treated [6]. Hence, early detection of BC poses a great challenge to medical experts and researchers [7]. Machine learning plays an important role in diagnosis of many diseases like BC, since medical data has become digitized [8].

BC can be diagnosed by classifying tumors. There are two different types of breast tumors; one of them called "*Benign tumors*", and the other called "*Malignant tumors*" [9]. Benign tumors are non-cancerous growths in the body, which have no danger effect on the breast or the whole body. Unlike cancerous tumors (Malignant tumors) that can be spread to other parts of the body and damage surrounding tissue [10].

Breast cancer has many different reasons that can be summarized as being a woman, getting older, starting menopause, being overweight after menopause or gaining weight as an adult, personal and family history of breast cancer, having high breast density on a mammogram, Inability to conceive, having the first child after the age of 35, having a previous biopsy showing atypical hyperplasia, radiation exposure, frequent X-rays in youth, and postmenopausal hormone use (current or recent use) of estrogen or estrogen plus progestin [11].

1.2 Support Vector Machine (SVM)

Support Vector Machine (SVM) is a supervised machine learning algorithm, which can be employed for both classification problems and regression purposes [4]. The idea of SVMs is to find a hyper-plane that best divides a dataset into two classes.

As shown in Fig. 1, support vectors are the data that points nearest to the line called hyper-plane that linearly separates and classifies a set of data. The points of

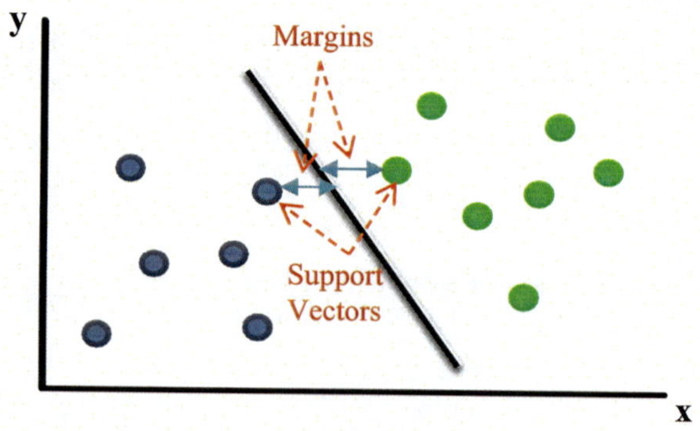

Fig. 1 Idea of support vector machine

data set are considered the critical elements of a data set. So, if they are removed they would alter the position of the dividing hyper-plane [12]. And therefore, the greater the distance between the data points and the hyperplane, the more correct the classification. The distance between the hyper-plane and the nearest data point from either set is known as the margin [13]. The goal of SVM algorithm is to choose a hyper-plane with the greatest possible margin between the hyper-plane and any point within the training set, giving a greater chance of new data being classified correctly [14].

The kernel is a set of mathematical functions that are used by SVM algorithms. Taking data as input and transform it into the required form is the function of the kernel [15]. There are different types of kernel functions in SVM such as; *Gaussian kernel, radial basis function (RBF), sigmoid kernel function, Hyperbolic tangent kernel*, and *Bessel function of the first kind kernel.* Gaussian kernel function is used when there is no prior knowledge about the data. RBF has localized and finite response along the entire *x*-axis so, it is considered the most used type of kernel functions [16]. Sigmoid kernel function can be used as a proxy for neural networks. Hyperbolic tangent kernel used in neural networks, and Removing the cross term in mathematical functions can be used by Bessel function [17].

1.3 Back Propagation Neural Network (BPNN)

In a neural network, the process of updating weights and biases of the neurons based on the errors is called Back Propagation (BP) [18]. It is an algorithm for supervised learning of artificial neural networks. It uses gradient decent, which is considered as an algorithm used to find the local minimum of a function. The gradient of error function is calculated with respect to the neural network weights by giving an artificial neural network and an error function [19].

The activation function gives a decision by calculating the weighted amount and adding more bias with neuron, whether it should be activated or not. The activation function aims to introduce non-linearity into the output of a neuron. Neural network has neurons that work in correspondence of weight, bias, and their respective activation function. The activation functions make back propagation possible as if gradients are provided along with the error to update the weights and biases [20].

As shown in Fig. 2, there are three layers of nodes or neurons in the back-propagation neural networks, which are input, hidden, and output layer [21]. Each node in the input and hidden layers is connected to each of the nodes in the next layer (hidden or output). All connections between nodes are directed, which means that the information flows only one way, and the nodes in a particular layer are not connected with each other. A weighting factor is associated with each connection between nodes. Back-propagation algorithm can modify those weights during the training process to produce learning [22].

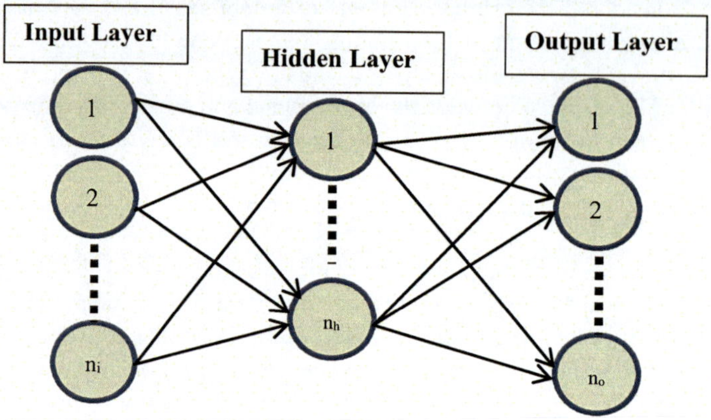

Fig. 2 Idea of back-propagation neural networks

1.4 Internet of Things (IoT)

Internet of Things (IoT) is the integration of objects with the world of internet, by adding hardware or/and software to be smart and to be able to communicate with each other and participate effectively in all an aspects of daily life as shown in Fig. 3 [23]. In other words, it is the ability of connecting anything in the world with the internet to make it smart and easy to use.

Over the past few years, IoT has witnessed a life transformation which provides a way to analyze real-time data and past data through the emerging role of artificial intelligence technologies and data mining in the IoT healthcare environment. The early stage of BC is not effectively diagnosis by the current state-of-the-art methods, and most of the women suffered from this dangerous disease [24].

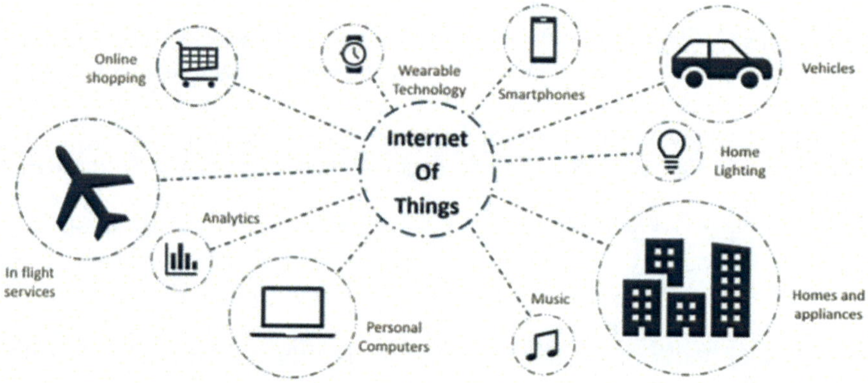

Fig. 3 Examples of internet of things uses

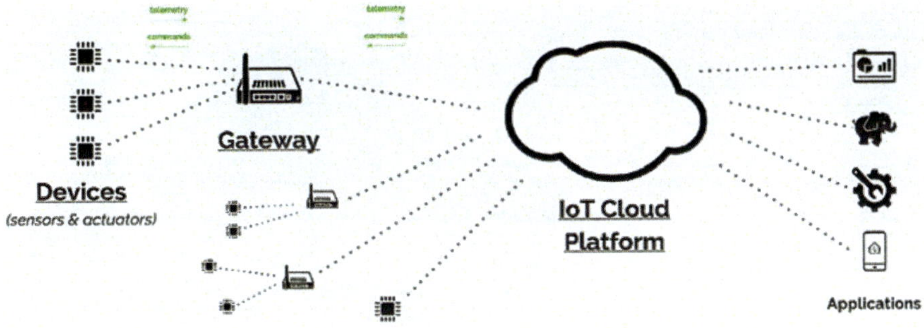

Fig. 4 Architecture of IoT

As shown in Fig. 4, the entire working process of IoT begins with the device itself, such as smartphones, digital watches, electronic devices that communicate securely with the IoT platform. The platforms collect and analyze the data from all multiple devices and platforms and transfer the most valuable data with applications to devices [25].

1.4.1 Of Sensors Used in BC

Researchers have provided many methods and technologies for diagnosing BC, including mammography, magnetic resonance imaging, ultrasound, computed tomography, and biopsy. However, some limitations have appeared on those technologies such as being expensive, time-consuming, and unsuitable for young women. Therefore, developing a high-sensitive and rapid method for diagnosing BC in its early stage is urgent. In recent years, researchers have been paid their attention to developing biosensors to detect breast cancer using various biomarkers [26].

A Nano Wire (NW) Surface-Enhanced Raman Scattering (SERS) has been reported to detect telomerase activity in various cancer cells and tissues. NWs are constructed by depositing nanoparticles on single crystalline NWs and provided highly reproducible SERS spectra [27]. The NW sensors show robust SERS signals only in the presence of excised tumor tissue from 16 tumor-bearing mice, while minimal signals in the presence of hot tumor tissue or normal tissues [28].

There is another type of thermal sensors, that opted for the micro-sensor "CMS type" (shown in Fig. 5), which provides a high sensitivity and the possibility of being integrated in the bra. Furthermore, micro-sensor will be considered as a pixel of the thermal image. Also, it needs a low energy supply of 2.7–3.3 V, with a very low current consumption, of 0.6 mA. Temperature measurements are converted to digital format with high resolution (0.00390625) °C. The accuracy is 0.1 °C, with a thermometric specification relating to ASTM E1112 when soldered to the PCB. Communication takes place via a serial I2C BUS interface. To optimize the consumed energy, in rest time, it is stopped. The operating temperature range is between 0 and + 50 °C [29].

Fig. 5 Package land pattern 8 TDFN 3 × 3 mm^2

2 Previous Work

Liu et al. [30] has been proposed an approach called "*Improved Cost-Sensitive Support Vector Machine*" (ICS-SVM) classifier for breast cancer diagnosis. Not only was it fully considered for the cost of misclassification for the diagnosis of breast cancer, but it also used the IG algorithm for feature selection and a meta-heuristic method was used to improve the classifier. Two data sets of Wisconsin Diagnostic Breast Cancer (WDBC) and Wisconsin Breast Cancer (WBC) were used in this proposed approach, which was is obtained from the UCI Machine Learning Repository. The accuracy of the suggested method according to WDBC is 98.83% and according to WBC is 98.04%.

Islam et al. [31] has proposed a breast cancer prediction model based on SVM and K-Nearest Neighbors (K-NN), which are supervised machine learning techniques for detecting breast cancer through training features. The proposed model was also used for tenfold cross-validation to obtain an accurate result. A WBC dataset has been used, which obtained from UCI Machine Learning Repository. SVM was achieved with an accuracy of 98.57 and 97.14% according to K-NN.

Aličković and Subasi [32] has been proposed a model based on Rotation Forest (RF) with Genetic Algorithm (GA)—based 14 features. This model has two stages, in the first stage GA has been used to extract informative and significant features, which reduced the computational complexity and speed up the data mining process. In the second stage, various data-mining techniques were used to decide on two subjects with or without breast cancer. SVM and multiple classifiers systems have been used in the second stage in order to construct accurate system for breast cancer classification. WDBC dataset has been used. The accuracy of this model reached 97.23 by SVM classifier.

Memon et al. [24] has been proposed a system using SVM classifier in order to classify the malignant and the benign tumors. A recursive feature selection algorithm has also been used to select more suitable features from BC dataset, and Recursive

Feature Elimination (REF) algorithm for suitable and related feature selection for correct target classification. WDBC data set has been used in this proposed system.

Kaur et al. [8] has been proposed a healthcare system based on random forest classifier and Internet of Things (IoT). The machine learning techniques that have been used are *K*-NN, SVM, Decision tree, RF, and MLP. WBC dataset has been used. The accuracy of SVM reached to 96.3%.

3 Proposed IoT Framework for Breast Cancer Classification Model

As shown in Fig. 6, the proposed model consists of two phases. In the first phase of the proposed model (IoT), At least eight thermal sensors are built in each side of patient's suit. Those thermal sensors will measure the temperature of breast tissues as cancerous tissues have higher temperature than benign tissues. The data obtained by sensors will be enhanced so the type of tissue will be classified as normal or benign or malignant tumor. Then the final classification will be transformed via internet to mobile application that installed in patient, doctor, and hospital's devices. This paper

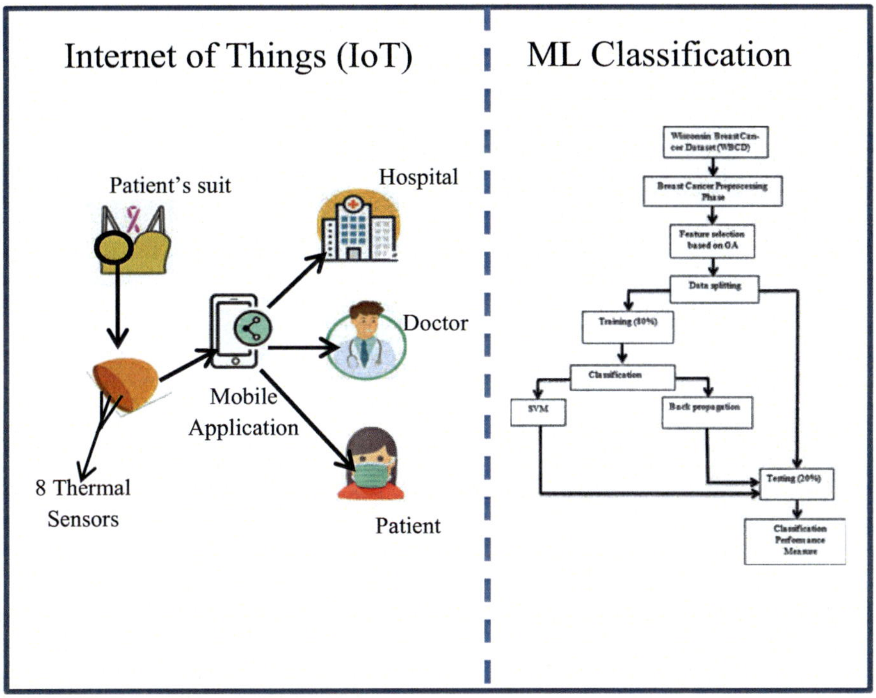

Fig. 6 The proposed framework model

will be focused on the second phase of the proposed model (ML Classification). This part classifies the type of tumor using two classifiers SVM and back-propagation classifiers. The dataset that is used in this model is taken from Wisconsin breast cancer (WBC) from UCI machine learning repository. After preprocessing WBC dataset by normalizing its data and ignoring their missing values, we split the data of the transformed WBC dataset by training 80% and testing 20%. As shown in Fig. 7.

As shown in Fig. 7, the second phase of the proposed model. Breast cancer data set must be preprocessed by removing missing values or calculating their mean, by normalizing data, and removing independent attributes. After transforming BC dataset, the dependent features or attributes will be selected by using Genetic Algorithm (GA). Then the data of the selected features will be splitted (80%) for training

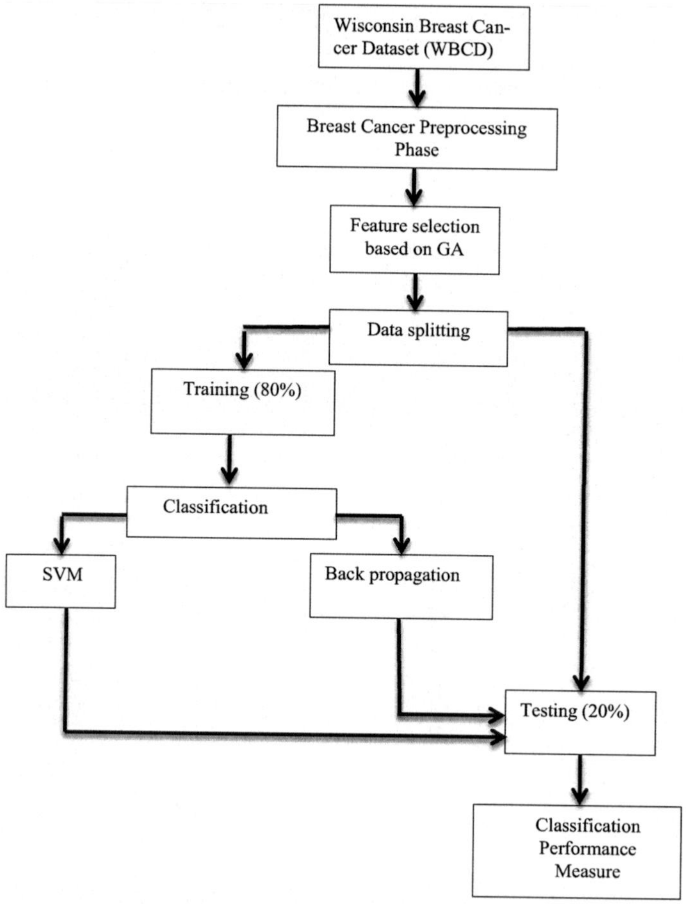

Fig. 7 Proposed breast cancer classification model in IoT framework

and (20%) for testing. The training data will be classified using both SVM and ANN-BP classifiers. Then the results of both classifiers will be also tested, and finally the classification performance measure of both classifiers will be measured.

- **Performance Evaluation Matrices**
  Evaluation metrics used to evaluate the performance of the classifiers. In this paper, three performance evaluation metrics were used, which are; classification accuracy, sensitivity, specificity. This paper computes the following metrics and mathematically expressed in Eqs. (1)–(3), respectively.

  (1) TP (true positive) if the subject is classified as BC
  (2) TN (true negative) if a healthy subject is classified as healthy
  (3) FP (false positive) if a healthy subject is classified as BC
  (4) FN (false negative) if a BC is classified as healthy.

- **Classification Accuracy**
  The accuracy shows the overall performance of the classification system. Accuracy is the diagnostic test probability that is correctly performed.

$$\text{Accuracy} = \frac{\text{TN} + \text{TP}}{\text{TP} + \text{TN} + \text{FP} + \text{FN}} * 100 \tag{1}$$

- **Sensitivity**
  It is the ratio of correctly classified heart patient subjects to all number of heart patient subjects.

$$\text{Sensitivity} = \frac{\text{TP}}{\text{TP} + \text{FN}} * 100 \tag{2}$$

- **Specificity**
  Specificity shows that a diagnostic test is negative, and the person is healthy.

$$\text{Specificity} = \frac{\text{TN}}{\text{TN} + \text{FP}} * 100 \tag{3}$$

- ***F_1*-Score**
  The traditional *F*-measure or balanced *F*-score is the harmonic mean of precision and recall

$$F_1\text{-score} = 2 * \frac{\text{specificty} * \text{sensitivity}}{\text{specificty} + \text{sensitivty}} \tag{4}$$

4 Experimental Results Analysis and Discussion

4.1 Breast Cancer Dataset

The experiments consisted of three main steps, which were data gathering, data preprocessing, and performance evaluating. The dataset used in this experiment was Breast Cancer Wisconsin Dataset obtained from UCI machine learning repository. The data has 699 instances with 10 attributes plus the class attributes. The class distribution are 65.5% (458 instances) for benign and 34.5% (241 instances) for malignant. The attribute information can be seen in Table 1. The Attributes Domain 1 Sample Code Number id number 2 Clump Thickness 1—10 3 Uniformity of Cell Size 1—10 4 Uniformity of Cell Shape 1—10 5 Marginal Adhesion 1—10 6 Single Epithelial Cell Size 1—10 7 Bare Nuclei 1—10 8 Bland Chromatin 1—10 9 Normal Nucleoli 1—10 10 Mitoses 1—10 11 Class 2: benign; 4: malignant.

Before starting with a code part, we should be realizing the importance of splitting data into two subsets (training and testing BC dataset). Training data is a complete set of attributes or independent attributes as shown in Table 1 and target attribute or the dependent attribute, so that the proposed model is able to learn the value of target attribute (benign or malignant). When encountered with a large set of data we use the major portion of data as a training set (80%). In testing part, the minor portion of data (20%) has to be tested in order to know how much the proposed model has learned from that data.

Data set is preprocessed, which is a technique that is used to convert the raw data into a clean data set. In other words, whenever the data is gathered from different sources it is collected in raw format which is not feasible for the analysis. The raw

Table 1 Attribute information of WBC dataset

Number of attributes	Attributes	
1	Sample code number	Id number
2	Clump thickness	1–10
3	Uniformity of cell size	1–10
4	Uniformity of cell shape	1–10
5	Marginal adhesion	1–10
6	Single epithelial size	1–10
7	Bare nuclei	1–10
8	Bland chromatin	1–10
9	Normal Nucleoli	1–10
10	Mitoses	1–10
11	Class	2: benign, 4: malignant

Table 2 Transformed WBC dataset

Number of attributes	Attributes	
2	Clump thickness	1–10
3	Uniformity of cell size	1–10
4	Uniformity of cell shape	1–10
5	Marginal adhesion	1–10
6	Single epithelial size	1–10
7	Bare nuclei	1–10
8	Bland chromatin	1–10
9	Normal Nucleoli	1–10
10	Mitoses	1–10
11	Class	0: benign 1: malignant

dataset was preprocessed to produce well-from data that suitable for training and testing process. The first attribute, sample code number, was removed because it was not relevant to the diagnosis. The last attribute was transformed to 0 (benign) and 1 (malignant) such that it can be properly fitted to the standard BP as shown in Table 2.

After transformation of data set, the features of the attributes of data are selected using Genetic Algorithm (GA) for appropriate feature selection. Machine learning provides two different classifiers; SVM and back-propagation, which are used to classify the type of tumor in the breast (benign or malignant). The dataset "Wisconsin Diagnostic Breast Cancer (WDBC)" is available at the UCI machine learning repository [33]. The dataset is split into 80% for training and 20% for testing purposes in these experiments. In order to check the predictive model performance, various evaluation measures of performances are used such as classification accuracy, specificity, and sensitivity. All the performance metrics are computed automatically before applying feature selection algorithm and predictive model on data, preprocessing techniques are deployed on a dataset for the betterment of the dataset. Furthermore, all these experimental results are reported in tables and for better understanding, some graphics are also designed.

4.2 SVM Results

The SVM has been used and the performance measures of the proposed model have been checked on both different selected attributes, which are selected by GA algorithm and all attributes of BC dataset. The SVM obtained 93% accuracy, 94% specificity, 93% sensitivity, 94 F_1-score. It was noticed that by increasing number of selected attributes the accuracy of SVM increases also.

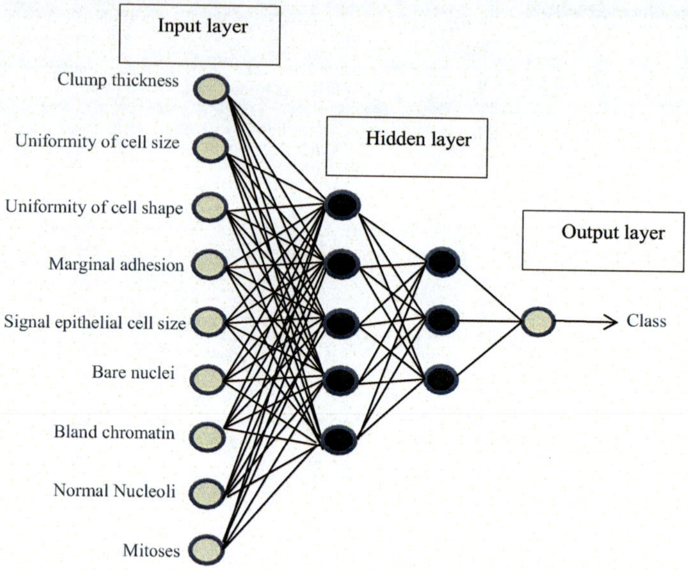

Fig. 8 Back propagation design

4.3 *Artificial Neural Network Backpropagation (ANN-BP) Results*

Backpropagation has been used as a classifier to classify breast tumor. As shown in Fig. 8, nine attributes in BC dataset are used as inputs in the input layer, which are Clump Thickness, Uniformity of Cell Size, Uniformity of Cell Shape, Marginal Adhesion, Single Epithelial Cell Size, Bare Nuclei, Bland Chromatin, Normal Nucleoli, and Mitoses attribute. The output layer is only one attribute, which is the class attribute. There are three hidden layers the first one has five preceptors and three preceptors for the second one. BP has a higher accuracy compared to SVM. Its accuracy reached to 98.98%.

4.4 *Comparative Study*

A comparison study is illustrated in Table 3 for breast cancer classification. The accuracy achieved by the "*Improved Cost-Sensitive Support Vector Machine*" (ICS-SVM) classifier for diagnosis of breast cancer [30] is 98.83%. The authors in [31] have been proposed a model to predict breast cancer using SVM and K-NN, where SVM has been achieved an accuracy of 98.57 and 97.14% according to *K*-NN. The accuracy of the model that is proposed in [32] reached to 97.23%. The accuracy obtained by the proposed a healthcare system based on random forest classifier and Internet of

Table 3 Comparative study

Previous work	Techniques used	Dataset used	Accuracy	Sensitivity	Specificity	F_1-Score
[30]	SVM	WDBC	98.83%	99.01%	95.71%	97.33
		WBC	98.04%	97.88%	98.38%	98.12
[31]	SVM	WBC	98.57%	100%	95.65%	97.77
	K-NN		97.14%	100%	92.31%	96.01
[32]	SVM	WDBC	97.23%	–	–	–
[8]	SVM	WBC	96.3%	–	–	–
The proposed model	SVM	WBC	93%	93%	94%	94
	ANN-BP		98.98%	98.6%	98.2%	98.3

Things (IoT) [8] reached to 96.3%. However, the classification accuracy achieved by SVM and backpropagation in this paper reached to 93% and 98% respectively. According to that comparative study, ANN-BP has the highest accuracy compared to other techniques.

5 Conclusion

This paper proposed framework model for diagnosis of BC. The proposed model is based on two machine learning classifier techniques are used in the IoT framework healthcare. This paper is focused on machine learning techniques in the whole model. The data set is obtained from Wisconsin Breast Cancer (WBC), which is available at UCI machine learning repository. Both SVM and ANN-BP techniques are for the diagnosis of BC. The accuracy obtained by ANN-BP is higher than that of SVM, where 93% accuracy for SVM while 98.98% for ANN-BP. Other techniques are compared to Both SVM and ANN-BP, and it was noticed ANN-BP has the highest accuracy among them.

References

1. Nahid A-A, Mikaelian A, Kong Y (2018) Histopathological breast-image classification with restricted Boltzmann machine along with backpropagation
2. Azmi MSBM, Cob ZC (2010) Breast cancer prediction based on backpropagation algorithm. In: 2010 IEEE Student Conference on Research and Development (SCOReD). IEEE
3. Meenalochini G, Ranjani R, Vijayalakshmi V. Breast cancer diagnosis using AI technologies—a survey
4. Rejani Y, Thamarai Selvi S (2009) Early detection of breast cancer using SVM classifier technique. arXiv preprint arXiv:0912.2314
5. Ng EYK et al (2007) Detection and differentiation of breast cancer using neural classifiers with first warning thermal sensors. Inf Sci 177(20):4526–4538

6. Breast cancer basic facts (search in Google Scholar)
7. Breast cancer detecting in the IoT health environment using modified recursive feature selection (search on Google Scholar)
8. Kaur P, Kumar R, Kumar M (2019) A healthcare monitoring system using random forest and internet of things (IoT). Multimedia Tools Appl 78(14):19905–19916
9. Gayathri BM, Sumathi CP, Santhanam T (2013) Breast cancer diagnosis using machine learning algorithms-a survey. Int J Distrib Parallel Syst 4(3):105
10. Sharma GN et al (2010) Various types and management of breast cancer: an overview. J Adv Pharm Technol Res 1(2):109
11. Florence AO et al (2016) An assessment of the knowledge and practice of self breast examination (BSE) amongst university students. Health 8(5):409–415
12. Hussain M et al (2011) A comparison of SVM kernel functions for breast cancer detection. In: 2011 Eighth international conference computer graphics, imaging and visualization. IEEE
13. Gholami R, Fakhari N (2017) Support vector machine: principles, parameters, and applications. In: Handbook of neural computation. Academic Press, pp 515–535
14. Widodo A, Yang B-S (2007) Support vector machine in machine condition monitoring and fault diagnosis. Mech Syst Signal Process 21(6):2560–2574
15. Ozer S, Chen CH, Cirpan HA (2011) A set of new Chebyshev kernel functions for support vector machine pattern classification. Pattern Recogn 44(7):1435–1447
16. Sullivan KM, Luke S (2007) Evolving kernels for support vector machine classification. In: Proceedings of the 9th annual conference on genetic and evolutionary computation
17. Moghaddam VH, Hamidzadeh J (2016) New Hermite orthogonal polynomial kernel and combined kernels in support vector machine classifier. Pattern Recognit 60:921–935
18. Siregar SP, Wanto A (2017) Analysis of artificial neural network accuracy using backpropagation algorithm in predicting process (forecasting). IJISTECH Int J Inf Syst Technol 1(1):34–42
19. Al-Sammarraie NA, Al-Mayali YMH, El-Ebiary YAB (2018) Classification and diagnosis using back propagation artificial neural networks (ANN). In: 2018 International conference on smart computing and electronic enterprise (ICSCEE). IEEE
20. Chen G (2016) A gentle tutorial of recurrent neural network with error backpropagation. arXiv preprint arXiv:1610.02583
21. Hu F et al (2018) A time simulated annealing-back propagation algorithm and its application in disease prediction. Mod Phys Lett B 32(25):1850303
22. Pereira FC, Borysov SS (2019) Machine learning fundamentals. In: Mobility patterns, big data and transport analytics. Elsevier, pp 9–29
23. Altantawy DA, Obbaya M, Kishk S (2014) A fast non-local based stereo matching algorithm using graph cuts. In: 2014 9th International conference on computer engineering & systems (ICCES). IEEE
24. Memon MH et al (2019) Breast cancer detection in the IoT health environment using modified recursive feature selection. Wirel Commun Mobile Comput
25. https://www.javatpoint.com/iot-introduction
26. Loo SW, Pui T-S (2020) Cytokine and cancer biomarkers detection: the dawn of electrochemical paper-based biosensor. Sensors 20(7):1854
27. Wei H, Abtahi SMH, Vikesland PJ (2015) Plasmonic colorimetric and SERS sensors for environmental analysis. Environ Sci Nano 2(2):120–135
28. Eom G et al (2017) Nanogap-rich Au nanowire SERS sensor for ultrasensitive telomerase activity detection: application to gastric and breast cancer tissues diagnosis. Adv Funct Mater 27(37):1701832

29. Elouerghi A et al (2020) A novel approach for early breast cancer detection based on embedded micro-bioheat ultrasensitive sensors: IoT technology. In: 2020 International conference on electrical and information technologies (ICEIT). IEEE
30. Liu N et al (2018) Improved cost-sensitive support vector machine classifier for breast cancer diagnosis. Math Probl Eng
31. Islam MM et al (2017) Prediction of breast cancer using support vector machine and K-nearest neighbors. In: 2017 IEEE Region 10 humanitarian technology conference (R10-HTC). IEEE
32. Aličković E, Subasi A (2017) Breast cancer diagnosis using GA feature selection and rotation forest. Neural Comput Appl 28(4):753–763
33. Wolberg WH (1995) Wisconsin Diagnostic Breast Cancer (WDBC), University of Wisconsin School of Computer Science, UCI Machine learning Repository, Madison, WI, USA

Comparative Study of Machine Learning Models for Onset Sepsis Prediction

Zeina Rayan, Marco Alfonse, and Abdel-Badeeh M. Salem

Abstract Sepsis is one of life-threatening diseases that is caused by unbalanced body response to some chemicals that the body release into its blood stream when fighting an infection. Early prediction of sepsis can decrease mortality rates. Machine learning techniques can improve the accuracy of early sepsis prediction. This paper presents a comparative study for recent machine learning models that can be used for sepsis prediction using datasets provided by international Challenge, named PhysioNet/Computing in Cardiology 2019.

Keywords Sepsis prediction · Machine learning · Artificial intelligence · Intensive care unit · Medical informatics · Intelligent algorithms

1 Introduction

Sepsis is one of the leading causes of mortality in hospitalized patients and is responsible for placing an enormous cost burden on the health care system [1]. Sepsis is a potentially life-threatening condition caused by the body's response to an infection [2]. Despite this fact, a reliable means of predicting sepsis onset remains elusive. Early sepsis onset predictions could allow more aggressive and targeted therapy while maintaining antimicrobial stewardship. Existing detection and prediction methodologies suffer from low performance and often require time-consuming laboratory test results [3].

Clinical decision support tools, using the clinical values of many tests in ICU, can help predict those at risk of having future sepsis. Since there are many clinical values

Z. Rayan (✉) · M. Alfonse · A.-B. M. Salem
Department of Computer Science, Faculty of Computer and Information Sciences, Ain Shams University, Cairo, Egypt
e-mail: zeinarayan@cis.asu.edu.eg

M. Alfonse
e-mail: marco_alfonse@cis.asu.edu.eg

A.-B. M. Salem
e-mail: absalem@cis.asu.edu.eg

D. A. Magdi et al. (eds.), *Digital Transformation Technology*, Lecture Notes in Networks and Systems 224, https://doi.org/10.1007/978-981-16-2275-5_30

collected from critically ill patients in the ICU, it made these data a good reference for a lot of research topics [3, 4]. This huge amount of data can be used by machine learning techniques to help in accurate prediction of Sepsis [4].

In order to predict sepsis onset, the organizers of PhysioNet/Computing in Cardiology (CinC) Challenge 2019 [5] set up a challenge to develop automated model for the early detection and prediction of sepsis using physiological data collected from ICU patients.

This paper is constructed as follows: Sect. 2 contains a brief description of the dataset and a comparative study between recent machine learning models that are applied for sepsis prediction, Sect. 3 contains conclusion and future work.

2 Machine Learning Models for Sepsis Prediction

This section presents a comparative study between many different machine learning techniques that are applied to ICU data to predict sepsis onset. All of these works utilize the same datasets which is provided by PhysioNet/CinC challenge 2019 [5].

The PhysioNet/CinC challenge 2019, provided three datasets from ICU patients in three separate hospitals. Datasets from two hospitals (A and B) were publicly available to the public and used for the algorithm training. Dataset A contains 20,336 patients, and dataset B contains 20,000 patients. The third dataset is not publicaly available. Each file in the datasets contains the records of one patient during the stay in ICU where samples were collected hourly, generating several time-series features. These features (total 40 features) consist of three groups which are: vital signs (8 features), laboratory values (26 features), and demographics (6 features) [5]. Figure 1 shows histogram plots for each of these features in dataset A.

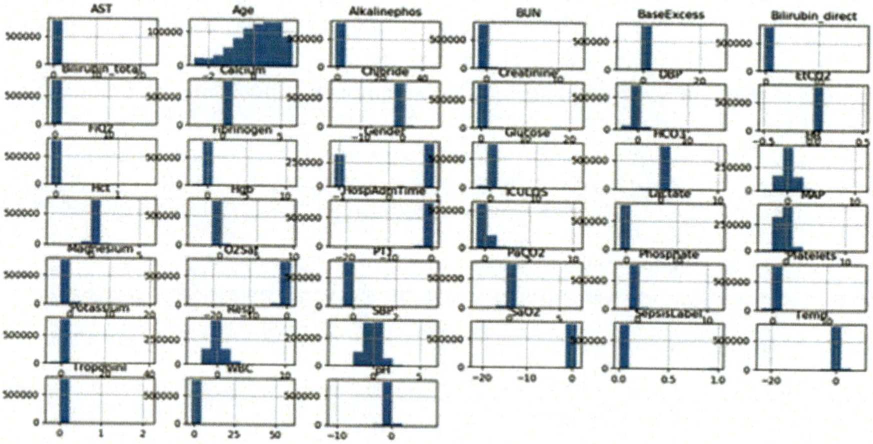

Fig. 1 Histogram plots for each input variable for dataset A

Vollmer et al. [6] proposed a time-specific ensemble learner model based on XGBoost [7], Gradient Boosting Machines (GBM), Distributed Random Forest (DRF), Generalized Learning Model (GLM) and deep learning. Dataset unbalancing was handled by removing outliers, and missing values were handled by a relation between the cumulative sum and the last observation calculated in percent. The cross-validated Area Under the Receiver Operating Characteristic (AUROC) curve for the proposed model was 0.823 and the utility score was 0.394. A utility score is defined by the challenge organizers, rewarding early predictions of sepsis and penalizing too early/late/failing prediction of sepsis or false sepsis prediction in a non-sepsis patient [5].

Fu et al. [8] proposed an ensemble model combining three different predictive models which are LightGBM, XGBoost, and Random Forest (RF). Dataset unbalancing was handled by removing outliers, and missing values were handled using the "pad" method, called carry-forward, where the missing value was replaced by previous non-missing value. The overall feature mean values were calculated to fill the missing values. If a patient feature values are all missing, then they are filled with mean value of that feature for all patients, otherwise they are replaced by the previous record values. Not all features were selected; as 6 variables were excluded: End Tidal Carbon Dioxide (EtCO2), Unit1, Unit2, Gender, BaseExcess, and Hours between Hospital admit and ICU admit (HospAdmTime), which were not related to sepsis or lack of values in hospital system. The proposed model achieved Area Under Curve (AUC) of 0.792 and utility score of 0.087.

Vicar et al. [9] proposed a model based on Long Short-Term Memory (LSTM) network consists of 7 blocks, where each block consists of LSTM layer followed by 3 fully connected layers. Dataset unbalancing was handled by a specialized lost function, which can automatically set the weights appropriately for each of the classes [10], while missing values were handled by replacement of missing values with the average value in most of the cases. The proposed model achieved a normalized utility score of 0.281.

Firoozabadi and Babaeizadh [11] proposed a model based on ensemble of bagged decision trees, which consists of 100 decision trees, maximum number of splits is set to 100, and minimum leaf size of 3. Dataset unbalancing was handled by removing the outliers through applying a plausibility filter to each feature where a range of valid values for each feature was defined. Missing values were imputed by forward filling if a value was available in past. The remaining missing values with no previous values were replaced by the population mean, calculated from the public datasets A and B after applying the plausibility filter. Important features and optimized hyper-parameters were selected in an iterative optimization approach of the utility score. A total of 15 features were selected: Heart Rate (HR), Temperature (Temp), Mean Arterial Pressure (MAP), Respiration Rate (Resp), Measure of Excess Bicarbonate (BaseExcess), Fraction of Inspired Oxygen (FiO_2), Blood Urea Nitrogen (BUN), Calcium, Creatinine, Hematocrit (Hct), Leukocyte Count (WBC), Platelets, Unit1or2, HospAdmTime, and ICU Length of Stay (ICULOS), where the achieved accuracy by the proposed model was 0.871 for set A, 0.912 for set B, and 0.754 for set C.

Liu et al. [12] proposed a model based on bidirectional LSTM, where the final outputs of two directions are summed up and forwarded to a single dense layer with sigmoid activation function. Feature were extracted using Heterogeneous Event Aggregation (HEA). The proposed model achieved utility score of 0.321, and AUC of 0.8224.

Tran et al. [13] proposed a model based on an ensemble of deep neural network (called AEC-Net), Random Forest, and Gradient Boosting Decision Trees (GBDT). The missing values were handled by replacing it with the mean value of that column for the dataset. The dataset imbalance was handled by using all data with sepsis and only 5% of data without sepsis for the training process. The proposed model achieved accuracies of 0.836 for set A, 0.894 for set B, and 0.709 for sec C, and utility scores of 0.378 for set A, 0.347 for set B, and − 0.262 for set C.

Biglarbeigi et al. [14] proposed a model based on ensemble of four K-Nearest Neighbors (KNN) classifiers. The missing values were handled using Markov Chain Monte Carlo (MCMC) [15] simulation; the Not a Number (NaN) values were replaced by the mean of that feature. A total of 8 features were selected: Temp, HR, Systolic Blood Pressure (SBP), Resp, Platelets, Creatinine, Glucose, and WBC, where these variables have the least number of NaNs and are clinically important in diagnosing sepsis. The dataset imbalance was handled through under-sampling. The proposed model achieved a utility score of 0.27.

He et al. [16] proposed a model based on ensemble of XGBoost and GBDT. The missing values were handled by replacement with mean values. The feature extraction is done by using LSTM recurrent neural network. The dataset imbalance was handled by discarding the data with the label 0 with a probability of 1/5 in the data reading phase, and the data was randomly down sampled in the training phase. The proposed model achieved a utility score of 0.313.

Sweely et al. [17] proposed a model based on Random Forest ensemble consisting of 150 learners (decision trees) where Random Under-Sampling Boosting (RUSBoost) taking the information from each learner and using it to help construct the next learner (tree), with max split of 50. As noticed, the size of the non-sepsis records were 54 times larger than the sepsis records; so 54 fully balanced datasets were created to train the model, using padding of the data records with the previous time-stamped records. This was done by looking at 8 previous timestamps (records) and concatenating those to the current record that was being used to train the model (a way similar to over-sampling). The missing values were replaced with the previous nearest value, and if all the values of a feature were missing, then these missing values were replaced by zeros. The proposed model achieved utility score of 0.399, and accuracy of 87.7%.

Morrill et al. [18] proposed signature-based, machine learning approach for sepsis prediction. All 40 features were used, in addition to two new features from the work of [19]: "ShockIndex", which is defined as the heart rate divided by the systolic blood pressure and "BUN/CR" which is the ratio of levels of bilirubin to creatinine. The missing values were handled using forward filling, and the value is left as NaN if no previous value exists for the feature. The proposed model achieved utility score of 0.36.

Yang et al. [20] proposed real-time ensemble technique using multi-feature fusion based on XGBoost learning and Bayesian optimization [21]. All features were selected except Bilirubin Direct (Bilirubin_direct), Troponin I, and Fibrinogen as these features have more than 99% missing values. The missing values were handled using forward filling, and the values will not be processed if there is no previous recorded value for this feature. The dataset imbalance was handled by down-sampling the excessive data records of the class with majority records. The proposed model had utility scores for set A, set B, and set C of 0.43, 0.422, and − 0.048 respectively.

Nejedly et al. [22] proposed a model based on LSTM neural network, and the soft-max activation function transforms the output into probabilities for each classification class. The optimal probability threshold for the model output was evaluated by differential evolution [23] genetic algorithms. The missing values were handled using forward filling. The proposed model had utility scores for set A, set B, and set C of 0.361, 0.327, and − 0.182 respectively.

Pou-Prom et al. [24] proposed a model based on ensemble of two Convolutional Neural Networks (CNN), and a random forest. The Principal Components Analysis (PCA) is used to obtain a lower dimensional feature space. The missing values were handled using forward filling, and if no previous value exists the value is set to − 1. The dataset imbalance was handled using under-sampling. The proposed model had utility scores for set A, set B, and set C of 0.378, 0.315, and − 0.318 respectively.

Deogire [25] proposed a model based on weight assignment for each feature, then calculating the total weight of all selected features, if the total weight is greater than 2 then the record is assigned to class "have sepsis". Six features were selected: Resp, SBP, MAP, Creatinine, Bilirubin, and Platelets. Threshold criteria were applied to each selected feature to differentiate between sepsis and non-sepsis patients. The proposed model achieved accuracies of 0.968 for set A, 0.978 for set B, and 0.984 for sec C, and utility scores of 0.012 for set A, 0.013 for set B, and 0.017 for set C.

Macias et al. [26] proposed a model based on LSTM. The missing values were handled through a window of N hours, if one value is inside the window then the missing value is replaced by it, if more than one value is found inside the window then the missing value is replaced by their mean. Autoencoders were used to extract the most relevant features from the data records, and to reduce dataset dimensionality. The proposed model achieved AUROCs of 0.761 for set A, 0.766 for set B, and 0.762 for sec C, and utility scores of 0.344 for set A, 0.267 for set B, and -0.247 for set C.

Li et al. [27] proposed a model based on ensemble of CNN, and LSTM. A total of 11 features were selected based on their ranking importance which was generated from training a subset using random forest: HR, Pulse Oximetry (O2Sat), Temp, SBP, MAP, Diastolic Blood Pressure (DBP), Resp, EtCO2, Serum glucose (Glucose), Gender, and HospAdmTime. The dataset imbalance was handled using over-sampling. The proposed model achieved accuracy of 0.927, AUROC of 0.964, and utility score of 0.786.

Pawar et al. [28] proposed a model based on logistic regression. A total of 34 features were either selected or derived. The derived features included Pulse Pressure (PP = SBP-DBP), estimated Cardiac Output (CO = PP * HR), Shock Index (SI = HR/SBP), modified Shock Index (mSI = HR/MAP), and Cardiac Output Variation

(COV = PP/MAP). The selected features included MAP, SBP, troponin, O2Sat, $EtCO_2$, lactate, platelets, Partial Thromboplastin Time (PTT), WBC, and Temp. The missing values were handled by forward filling. The proposed model achieved accuracies of 0.795 for set A, 0.889 for set B, and 0.815 for sec C, and utility scores of 0.296 for set A, 0.268 for set B, and 0.007 for set C.

Yu et al. [29] proposed a model based on EasyEnsemble [30] an ensemble of LightGBMs. The dataset imbalance was handled using under-sampling. The features were selected using EasyEnsemble algorithm. The missing values were imputed by forward filling and if no previous values exist, then the missing value was replaced by the mean of that specific feature. The proposed model achieved accuracies of 0.835 for set A, 0.912 for set B, and 0.765 for sec C, and utility scores of 0.401 for set A, 0.278 for set B, and − 0.207 for set C.

Roussel et al. [31] proposed a model inspired from the work of [32]. The missing values were replaced by − 1. An embedded module consisting of three linear layers was used for reducing feature dimensionality. A recurrent neural network model is used to predict sepsis. The network consists of three LSTMs followed by two linear layers. The proposed model achieved utility scores of 0.387 for set A, 0.365 for set B, and − 0.148 for set C.

Scherpf et al. [33] proposed a model consists of an ensemble of stacked RNN and then ensemble of Decision Trees trained with LightGBM. The missing values were handled using forward filling and if no previous value exists the value is set to zero. Features were extracted based on their importance which was calculated using LightGBM. The proposed model achieved utility score of 0.114.

Abromavičius and Serackis [34] proposed a model based on Decision Tree classifier. The missing values were replaced by zero. The dataset imbalance was handled using Adaptive synthetic sampling (ADASYN) algorithm [35]. A total of 10 features were selected: HR, Temp, SBP, DBP, Resp, EtCO2, Calcium, Lactate, and Hct. The proposed model achieved utility scores of 0.036 for set A, 0.013 for set B, and 0.078 for set C.

Li et al. [36] proposed a time-phased model (TASP) for sepsis prediction. It is composed of three parts: early stage (1–9 hours) where a gradient boosting tree is trained on early hours in training set to predict the sepsis risk of each hour, middle stage (10–49 hours) where more time-related features have been used as the input for another tree model to make prediction, and late stage (50 + hours) where a Recurrent Neural Network (RNN) [37] is presented as the predictor. The missing values were handled using forward filling in the early and middle stages, and an extension of GRU-D [38] was used in the late stage. Different features were selected and set of different cut-offs for the different sub-models were used. The proposed model achieved a utility score of 0.420 for set A, 0.419 for set B, and 0.337 for set C.

Li et al. [39] proposed a model adopted from LightGBM [40], which is an effective and efficient gradient boosting tree algorithm. Missing values were handled using last-occurrence-carry-forward strategy; to impute the missing value of a specific variable with its most recent values. The cross-validated AUROC for the proposed model was 0.845 and the utility score was 0.354.

Table 1 Machine learning models in ICU Sepsis prediction

References	Handling dataset imbalance	Handling missing values	Feature extraction	Prediction model	Evaluation
Vollmer et al. [6]	Removing outliers	Relation between the last observation and relative cumulative sums		Time-specific ensemble learner	Normalized utility score = 0:394
					AUROC = 0.823
Fu et al. [8]	Removing outliers	The "pad" method, called carry-forward	6 variables were excluded: $EtCO_2$, Unit1, Unit2, Gender, BaseExcess, and HospAdmTime	Ensemble model which combined three different predictive models	Utility score = 0.087
					AUC = 0.792
Vicar et al. [9]	Specialized lost function	Replacement of missing values with the average value	6 variables were excluded: $EtCO_2$, Unit1, Unit2, Gender, BaseExcess, and HospAdmTime	LSTM network	Normalized utility score = 0.281

(continued)

Table 1 (continued)

References	Handling dataset imbalance	Handling missing values	Feature extraction	Prediction model	Evaluation
Firoozabadi and Babaeizadh [11]	Removing outliers	Either by forward filling or replaced by the population mean	A total of 15 features were selected: HR, Temp, MAP, Resp, BaseExcess, FiO_2, BUN, Calcium, Creatinine, Hct, WBC, Platelets, Unit1or2, HospAdmTime, and ICULOS	Ensemble of bagged decision trees	utility scores of set A, set B, and set C are 0.335, 0.268, and −0.226
					AUROCs of set A, set B, and set C are 0.764, 0.768, and 0.741
					Accuracies of set A, set B, and set C are 0.871, 0.912, and 0.754
Liu et al. [12]			Heterogeneous event aggregation (HEA)	bidirectional LSTM	Utility score = 0.321
					AUC = 0.8224
Tran et al. [13]	Only 5% of data without sepsis were used for training	Replacement with mean value of that column for the entire dataset		Ensemble of AEC-Net, Random Forest, and GBDT	Utility scores of set A, set B, and set C are 0.378, 0.347, and −0.262
					Accuracies of set A, set B, and set C are 0.836, 0.894, and 0.709

(continued)

Table 1 (continued)

References	Handling dataset imbalance	Handling missing values	Feature extraction	Prediction model	Evaluation
Biglarbeigi et al. [14]	Under-sampling	MCMC	A total of 8 features were selected: Temp, HR, SBP, Resp, Platelete, Creatinine, Glucose, and WBC	Ensemble of 4 KNN classifiers	Utility score = 0.27
					Accuracy = 0.97
He et al. [16]	Discarding 1/5 of data with label 0 in the reading phase, and randomly under-sampled in the training phase	Replacement with mean value	LSTM recurrent neural network	Ensemble of XGBoost and GBDT	Normalized utility score = 0.313
Sweely et al. [17]	Timestamp padding	Missing values were replaced with the previous nearest value, and if no previous values then missing values were replaced by 0 s	All 40 features were used	Random forest ensemble	Accuracy = 87.7%
					Utility score = 0.399
Morrill et al. [18]		Forward filling. If no previous value exists the value is left as NaN	All 40 features + Two new features "ShockIndex", and "BUN/CR"	Signature-based, machine learning approach	Utility score = 0.36
					AUROC = 0.868

(continued)

Table 1 (continued)

References	Handling dataset imbalance	Handling missing values	Feature extraction	Prediction model	Evaluation
Yang et al. [20]	Down-sampling	Forward filling, and the values will not be processed if no previous recorded value for the feature	All features except Bilirubin_direct, Troponin I, and Fibrinogen	Ensemble of XGBoost, and Bayesian optimization	Utility scores of set A, set B, and set C are 0.43, 0.422, and − 0.048
Nejedly et al. [22]		Forward filling	All 40 features were used	LSTM neural network	Utility scores of set A, set B, and set C are 0.361, 0.327, and − 0.182
Pou-Prom et al. [24]	Under-sampling	Forward filling, and if no previous value exists, the value is set to − 1	PCA was used to obtain a lower dimensional feature space	Ensemble of 2 CNNs, and random forest	Utility scores of set A, set B, and set C are 0.378, 0.315, and − 0.318
Deogire [25]			A total of 6 features were selected: Resp, SBP, MAP, creatinine, bilirubin, and platelets	Weight Assignment for each feature then calculating the total weight. If total weight > 2 then the record is assigned to class "have sepsis"	Utility scores of set A, set B, and set C are 0.012, 0.013, and 0.017 Accuracies of set A, set B, and set C are 0.968, 0.978, and 0.984
Macias et al. [26]		Window of N h, if one value is inside the window then it is assigned to the missing value, if more, then their mean is assigned to the missing value	All 40 features were used, and it was found possible to reduce 27% of the features through autoencoders	LSTM	Utility scores of set A, set B, and set C are 0.344, 0.267, and − 0.247 AUROCs of set A, set B, and set C are 0.761, 0.766, and 0.762

(continued)

Table 1 (continued)

References	Handling dataset imbalance	Handling missing values	Feature extraction	Prediction model	Evaluation
Li et al. [27]	Over-sampling		A total of 11 features were selected: HR, O2Sat, Temp, SBP, MAP, DBP, Resp, $EtCO_2$, Glucose, Gender, and HospAdmTime	Ensemble of CNN, and LSTM	Utility score = 0.786
					AUROC = 0.964
					Accuracy = 0.927
Pawar et al. [28]		Forward filling	A total of 34 features were either selected or derived	Logistic regression	Utility scores of set A, set B, and set C are 0.296, 0.268, and 0.007
					Accuracies of set A, set B, and set C are 0.795, 0.889, and 0.815
Yu et al. [29]	Under-sampling	Either by forward filling or replaced by the mean	EasyEnsemble algorithm for feature selection	EasyEnsemble: an ensemble of LightGBMs	Utility scores of set A, set B, and set C are 0.401, 0.278, and − 0.207
					Accuracies of set A, set B, and set C are 0.835, 0.912, and 0.765
Roussel et al. [31]		Replaced by − 1	Embedded module consisting of 3 linear layers	RNN	Utility scores of set A, set B, and set C are 0.387, 0.365, and − 0.148

(continued)

Table 1 (continued)

References	Handling dataset imbalance	Handling missing values	Feature extraction	Prediction model	Evaluation
					Accuracy of all sets = 0.854 ± 0.009
Scherpf et al. [33]		Forward filling, and if no previous value exists, the value is set to 0	LightGBM	Ensemble of stacked RNN and then ensemble of decision trees trained with LightGBM	Utility score = 0.114
Abromavičius and Serackis [34]	ADASYN algorithm	Replaced by 0	A total of 10 features were selected: HR, Temp, SBP, DBP, Resp, $EtCO_2$, Calcium, Lactate, and Hct	Decision tree classifiers	Utility scores of set A, set B, and set C are 0.036, 0.013, and –0.078
Li et al. [36]		Forward filling in early, and middle stages, and an extension of GRU-D was used in late stage	Different features and set of different cut-offs for different sub-models used	TASP	Set A score = 0.420
					Set B score = 0.419
					Set C score = 0.337
Li et al. [39]		Last-occurrence-carry-forward strategy		LightGBM	Utility score = 0.354
					AUROC = 0.845

Table 1 shows a comparison between different machine learning approaches used in sepsis prediction, using the PhysioNet/Computing in Cardiology Challenge 2019s datasets.

From Table 1, we can deduct that different ML approaches are used for proposed models that help in sepsis prediction of ICU patients. These models are applied to many patients' data records with different intervals of time. The Ensemble technique, and LSTM is the most commonly used machine learning approaches, and forward filling strategy used for imputing the missing values; proved to get high evaluation performance in most cases.

3 Conclusion and Future Work

Sepsis prediction is a challenging problem and still till now despite many years of research and development efforts because its manifestation is often unclear until later stages. This paper presented a comparative study for sepsis prediction using the PhysioNet/CinC challenges 2019s datasets where all features came from clinical variables that are regularly collected in ICUs. Based on this analysis, we can conclude that to predict sepsis onset using the challenge's dataset you need to balance the dataset as it is highly imbalanced, impute the missing values as there are some features that have high rate of missing values, you may also select features, and finally pick a machine learning algorithm then evaluate the model. In future work, we are going to put more efficient methodology to predict sepsis for the challenge's datasets trying to improve accuracy.

References

1. Angus DC, Linde-Zwirble WT, Lidicker J, Clermont G, Carcillo J, Pinsky MR (2001) Epidemiology of severe sepsis in the United States: analysis of incidence, outcome, and associated costs of care. Crit Care Med 29(7):1303–1310
2. Singer M, Deutschman CS, Seymour CW, Shankar-Hari M, Annane D, Bauer M, Bellomo R, Bernard GR, Chiche J-D, Coopersmith CM, Hotchkiss RS, Levy MM, Marshall JC, Martin GS, Opal SM, Rubenfeld GD, van der Poll T, Vincent J-L, Angus DC (2016) The third international consensus definitions for sepsis and septic shock (Sepsis-3). Jama 315(8):801–810
3. Desautels T, Calvert J, Hoffman J, Jay M, Kerem Y, Shieh L, Shimabukuro D, Chettipally U, Feldman MD, Barton C, Wales DJ, Das R (2016) Prediction of sepsis in the intensive care unit with minimal electronic health record data: a machine learning approach. JMIR Med Inform 4(3):e28
4. Rayan Z, Alfonse M, Salem A-BM (2019) Intensive care unit (ICU) data analytics using machine learning techniques. Int J Inf Theor Appl 26(1):69–82
5. Reyna MA, Josef C, Seyedi S, Jeter R, Shashikumar SP, Westover MB, Sharma A, Nemati S, Clifford GD (2019) Early prediction of sepsis from clinical data: the PhysioNet/computing in cardiology challenge 2019. In: 2019 computing in cardiology (CinC). IEEE, Singapore
6. Vollmer M, Luz CF, Sodmann P, Sinha B, Kuhn S-O (2019) Time-specific metalearners for the early prediction of sepsis. In: 2019 computing in cardiology (CinC). IEEE, Singapore

7. Chen T, Guestrin C (2016) Xgboost: a scalable tree boosting system. In: Proceedings of the 22nd ACM SIGKDD international conference on knowledge discovery and data mining, pp 785–794
8. Fu M, Yuan J, Lu M, Hong P, Zeng M (2019) An ensemble machine learning model for the early detection of sepsis from clinical data. In: 2019 computing in cardiology (CinC). IEEE, Singapore
9. Vicar T, Novotna P, Hejc J, Ronzhina M, Smisek R (2019) Sepsis detection in sparse clinical data using long short-term memory network with dice loss. In: 2019 computing in cardiology (CinC). IEEE, Singapore
10. Crum WR, Camara O, Hill DLG (2006) Generalized overlap measures for evaluation and validation in medical image analysis. IEEE Trans Med Imaging 25(11):1451–1461
11. Firoozabadi R, Babaeizadeh S (2019) An ensemble of bagged decision trees for early prediction of sepsis. In: 2019 computing in cardiology (CinC). IEEE, Singapore
12. Liu L, Wu H, Wang Z, Liu Z, Zhang M (2019) Early prediction of sepsis from clinical data via heterogeneous event aggregation. In: 2019 computing in cardiology (CinC). IEEE, Singapore
13. Tran L, Nguyen M, Shahabi C (2019) Representation learning for early sepsis prediction. In: 2019 computing in cardiology (CinC), Singapore
14. Biglarbeigi P, McLaughlin D, Rjoob K, Abdullah A, McCallan N, Jasinska-Piadlo A, Bond R, Finlay D, Ng K, Kennedy A, McLaughlin J (2019) Early prediction of sepsis considering early warning scoring systems. In: 2019 computing in cardiology (CinC). IEEE, Singapore
15. Ni D, Leonard JD (2005) Markov chain monte carlo multiple imputation using bayesian networks for incomplete intelligent transportation systems data. Transp Res Rec 1935(1):57–67
16. He Z, Chen X, Fang Z, Yi W, Wang C, Jiang L, Tong Z, Bai Z, Li Y, Pan Y (2019) Early sepsis prediction using ensemble learning with features extracted from LSTM recurrent neural network. In: 2019 computing in cardiology (CinC). IEEE, Singapore
17. Sweely B, Park A, Winter L, Liu L, Zhao X (2019) Time-padded random forest ensemble to capture changes in physiology leading to sepsis development. In: 2019 computing in cardiology (CinC). IEEE, Singapore
18. Morrill J, Kormilitzin A, Nevado-Holgado A, Swaminathan S, Howison S, Lyons T (2019) The signature-based model for early detection of sepsis from electronic health records in the intensive care unit. In: 2019 computing in cardiology (CinC). IEEE, Singapore
19. Henry KE, Hager DN, Pronovos PJ, Saria S (2015) A targeted real-time early warning score (TREWScore) for septic shock. Sci Transl Med 7(299):1946–6242
20. Yang M, Wang X, Gao H, Li Y, Liu X, Li J, Liu C (2019) Early prediction of sepsis using multi-feature fusion based XGBoost learning and Bayesian optimization. In: 2019 computing in cardiology (CinC). IEEE, Singapore
21. Shahriari B, Swersky K, Wang Z, Adams RP, Freitas ND (2015) Taking the human out of the loop: a review of Bayesian optimization. Proc IEEE 104(1):148–175
22. Nejedly P, Plesinger F, Viscor I, Halamek J, Jurak P (2019) Prediction of sepsis using LSTM neural network with hyperparameter optimization with a genetic algorithm. In: 2019 computing in cardiology (CinC). IEEE, Singapore
23. Price K, Storn RM, Lampinen JA (2006) Differential evolution: a practical approach to global optimization. Science & Business Media, Springer, Berlin/Heidelberg
24. Pou-Prom C, Yang Z, Sidhaye M, Dai D (2019) Development of an early warning system for sepsis. In: 2019 computing in cardiology (CinC). IEEE, Singapore
25. Deogire A (2019) A low dimensional algorithm for detection of sepsis from electronic medical record data. In: 2019 computing in cardiology (CinC). IEEE, Singapore
26. Macias E, Boquet G, Serrano J, Vicario J, Ibeas J, Morel A (2019) novel imputing method and deep learning techniques for early prediction of sepsis in intensive care units. In: 2019 computing in cardiology (CinC). IEEE, Singapore
27. Li X, Ng GA, Schlindwein FS (2019) Convolutional and recurrent neural networks for early detection of sepsis using hourly physiological data from patients in intensive care unit. In: 2019 computing in cardiology (CinC). IEEE, Singapore

28. Pawar R, Bone J, Ansermino JM, Görges M (2019) An algorithm for early detection of sepsis using traditional statistical regression modeling. In: 2019 computing in cardiology (CinC). IEEE, Singapore
29. Yu Q, Huang X, Li W, Wang C, Chen Y, Ge Y (2019) Using features extracted from vital time series for early prediction of sepsis. In: 2019 computing in cardiology (CinC). IEEE, Singapore
30. Liu X-Y, Wu J, Zhou Z-H (2008) Exploratory undersampling for class-imbalance learning. IEEE Trans Syst Man Cybern Part B (Cybern) 39(2):539–550
31. Roussel B, Behar J, Oster J (2019) A recurrent neural network for the prediction of vital sign evolution and sepsis in ICU. In: 2019 computing in cardiology (CinC). IEEE, Singapore
32. Tomašev N, Glorot X, Rae J, Zielinski M, Askham H, Saraiva A, Mottram A, Meyer C, Ravuri S, Protsyuk I, Connell A (2019) A clinically applicable approach to continuous prediction of future acute kidney injury. Nature 572(7767):116–119
33. Scherpf M, Goldammer M, Malberg H, Gräßer F (2019) Sepsis onset prediction applying a stacked combination of a recurrent neural network and a gradient boosted machine. In: 2019 computing in cardiology (CinC). IEEE, Singapore
34. Abromavičius V, Serackis A (2019) Sepsis prediction model based on vital signs related features. In: 2019 computing in cardiology (CinC). IEEE, Singapore
35. He H, Bai Y, Garcia EA, Li SA (2008) Adaptive synthetic sampling approach for imbalanced learning. In: 2008 IEEE international joint conference on neural networks (IEEE world congress on computational intelligence)
36. Li X, Kang Y, Jia X, Wang J, Xie G (2019) TASP: a time-phased model for sepsis prediction. In 2019 computing in cardiology (CinC). IEEE, Singapore
37. Bengio Y, Courville A, Vincent P (2013) Representation learning: a review and new perspectives. IEEE Trans Pattern Anal Mach Intell 35(8):1798–1828
38. Che Z, Purushotham S, Cho K, Sontag D, Liu Y (2018) Recurrent neural networks for multivariate time series with missing values. Sci Rep 8(1):1–12
39. Li X, Xu X, Xie F, Xu X, Sun Y, Liu X, Jia X, Kang Y, Xie L, Wang F, Xie G (2020) A time-phased machine learning model for real-time prediction of sepsis in critical care. Crit Care Med 48(10):e884–e888
40. Ke G, Meng Q, Finley T, Wang T, Chen W, Ma W, Ye Q, Liu T-Y (2017) LightGBM: a highly efficient gradient boosting decision tree. Adv Neural Inf Process Syst:3146–3154

A Systematic Literature Review of DNA-Based Steganography Techniques: Research Trends, Data Sets, Methods, and Frameworks

Mohammed A. Farahat, A. Abdo, and Samar K. Kassim

Abstract Clinical information can be considered as the most sensitive and confidential details connected to a person. While various effective policies, protocols, and regulatory standards are in place to protect patient information, privacy, and security, vulnerabilities remain key issues for electronic healthcare record (EHR) systems. There are several commonly used techniques such as cryptography and steganography to protect confidential data from attackers in the field of communication and computer security. To provide data protection using DNA as a carrier by utilizing its bio-molecular computational abilities and capacity, new fields for DNA-based cryptography and steganography are emerging. In this paper, according to important security parameters, we discuss different DNA-based steganography-related studies with the primary DNA-based steganography strategies. Moreover, we propose an approach that can enhance most of the previous security issues and can achieve a higher level of security.

Keywords Clinical data privacy · DNA steganography · Cryptography · Data hiding · Information security

1 Introduction

In healthcare sector, security and privacy of personal data are critical, and it is therefore a major challenge to protect the confidentiality, integrity, and availability of health data. Of all personal data, clinical data is usually considered the most sensitive and confidential data [1]. Policies relating to protection and privacy must be established and publicly coordinated. Despite many privacy and security issues, the medical sector has already adopted several innovations. There are already several

M. A. Farahat (✉) · A. Abdo
Department of Information Systems, Faculty of Computers and Artificial Intelligence, Helwan University, Helwan, Egypt

S. K. Kassim
Department of Biochemistry and Molecular Biology, Faculty of Medicine, Ain Shams University, Cairo, Egypt
e-mail: Samar_kassim@med.asu.edu.eg

D. A. Magdi et al. (eds.), *Digital Transformation Technology*, Lecture Notes in Networks and Systems 224, https://doi.org/10.1007/978-981-16-2275-5_31

privacy and security mechanisms in place, and when we use these principles in the area of healthcare information technology, we must exploit the current mechanism [2].

Nevertheless, the growing development of electronic health records (EHR) and clinical data warehouses, as well as the recent attacks on patient data, has increased the need for new technologies and ideas for data protection. The increasing amount of clinical data from multiple sources is processed in multiple locations and systems in a fragmented form. In all of these systems, security issues may cause the release of data to unauthorized persons or entities, so health information requires protection against unauthorized access and manipulation [3].

To protect secret data, cryptography and steganography are both used to guarantee data privacy. However, the main variance among them is that with cryptography, a third party can see and realize that the two parties are communicating and interactive in secret, and the ciphered data can be noted easily, while steganography is the science of invisible and unseen communication [4].

Cryptography is the way of converting the data into an incomprehensible format. The secret data can be encoded or decrypted by the party that owns the associated secret key. Cryptography techniques are classified into two main categories: substitution ciphers and transposition ciphers.

Steganography can be defined as the science of embedding sensitive information such as important text, images, or videos into other media to prevent attracting attention to the presence of the information. The name of steganography science is derived from Greek words, "stegos" means "cover" and "grafia" means "writing" to be defined as "covered writing."

Because of the speedy expansion in the use of the Internet, it has been used for sharing and transferring massive information [5]. The consequence of decreasing the chance that the data being noticed or detected during the spread is being a vital issue today. Steganography is considered one of the most robust methods to share secret information between parties by hiding the information in another medium that an attacker cannot notice or detect the sent message [6, 7].

The data hiding technique should preserve the least modification in the characteristics of the original cover media after converting the data to hide its existence [8]. The core objective behind steganography is to conceal secret data with the least variance between the original cover-medium and the modified one that can be noticed by the naked eye; accordingly, the attraction decreases, and the security increases. So, the tracker cannot expose the secret data.

Recently, the most appropriate method to secure data is to clear the trace of the existence of this data. Concealing data within the cover medium using steganographic techniques is the greatest method to achieve that. To enrich the security of steganography systems, we need robust algorithms in addition to a new cover medium. Nowadays, deoxyribonucleic acid (DNA) sequences are being used as cover mediums due to their high information density and capacity [9].

DNA steganography is enough to secure and often ideal than cryptography for two main reasons, and cryptology is not safe enough when communicating data within an insecure, public channel. It is just the science of covered writing, while steganography is hidden writing which focuses on hiding the existence of the message. Besides this, DNA is being proposed as used for many computational purposes.

The unprecedented availability of personal genomes provides a rich resource for data mining and integration toward further understanding of human biology and translation of research answers into clinical practice. Such activities call for sharing human genomic data widely and efficiently, which is vital for simplifying data-driven basic biological and translational research. Otherwise, there is a rising concern that sharing and releasing genomic sequencing poses risks to genetic privacy, that is, privacy related to the identification of individuals and sensitive data about their behaviors from anonymous genome sequences. Attacks could potentially be deployed to strategically break genetic privacy by disclosing personal identities or private traits of the individuals [10].

For achievement of extreme privacy and good protection with high capacity and low rate of change, new data embedding methods have been proposed by researchers based on DNA with the advent of biological aspects of DNA sequences. But all these researches focused on how to minimize the modification rate. Otherwise, the DNA sequence should be confidential.

The remainder of this paper is organized as follows. First, we discuss the related works and present a comparative study and author's point of view. In Sect. 6, the proposed approach and its phases are explained. Finally, the conclusion of this paper and an outlook to the future work are presented.

2 Related Work

Much researches have been developed to conceal data using DNA sequences as a cover-medium.

In comparing different steganography based on DNA, the authors used critical security parameters in [9]. They explained the main techniques for DNA-based steganography. Each approach is considered to be a building block for most of the other steganographic structures that are insertion, replacement, and complementary pair-based algorithms. They suggested proposals to assist future researchers to help design and develop DNA storage techniques to optimize the security of data storage through more effective, efficient, high capacity, and biologically preserved techniques for DNA steganography.

The increasing need for storage has generated a massive demand for new and evolving strategies for efficient data storage to be developed. As a new data carrier, DNA has now started to be used as efficient and secure storage of data. To build high-capacity protected algorithms with low cracking probability, the biomolecular computational capabilities of DNA are exploited through cryptography and stenography. The other way is to preserve the DNA by the researchers who care.

Wang et al. [11] studied how to resolve the protection deficiency of DNA steganography and developed a DNA steganography-based message hiding technique. To encrypt a message and then decompose the ciphertext into two sections, they built several procedures. To send one section, the sender uses DNA steganography. The message will be encrypted and decomposed again and again until the microdot is not assumed or contaminated whether the microdot is believed or contaminated. If it is not believed or tainted with the microdot, the corresponding section will be submitted publicly. The innovation of the proposed approach is that the decomposing message was added to solve the DNA steganography protection issue. Security analysis reveals that the security issue of DNA steganography can be solved based on the suggested system, and then, a message can be transmitted and hidden from an adversary.

In [12], authors focused on the idea of using as a primary cover a theoretical single-stranded DNA (ssDNA), which is extracted from an unremarkable cover image. On the same basis, they studied the security loopholes and efficiency problems of the current algorithm and suggested an upgraded algorithm. The efficiency of both algorithms is evaluated against many visual and statistical attacks, and both security and capability are parameterized. The comparison indicates that improved overall defense is offered by the planned improvements.

The proposed work focused on covering hidden data in various layers of media coverage. The DNA is due to the image's pixel properties. This technique thus made it more reliable than the methods that used public database reference DNAs. The multiple 2D logistic map parameters made the algorithm even more impenetrable. For the entire process, multiple keys were required from DNA building to data embedding, and a secure key exchange protocol was required for their transfer between sender and receiver.

In [13], authors addressed existing steganographic hiding methods using LSB, RSA, DNA cryptography, and DNA steganography techniques, which are used to prevent videos, audios, and photographs from being copied, falsified, and unauthorized access. Existing hiding techniques had greater key size, computational cost, speed, and input size demerit. Hyperelliptic curve cryptography (HECC) based on DNA steganography is therefore suggested, which offers a higher degree of image file protection and also ensures digital media protection.

Compared to conventional cryptographic techniques, the proposed HECC-based DNA steganography results in 30% and 42% greater processing time for the process of encryption and decryption, respectively. Finally, in terms of mean square error, peak signal to noise ratio (PSNR), embedding capacity, the proposed scheme is compared with other conventional steganographic techniques. It is concluded from the simulation results that compared to other steganographic methods, the suggested steganographic method has a lower embedding power, high MSR, and PSNR value.

In [14], authors proposed an effort to suggest the use of DNA steganography for highly protected data transmission. This was indicated by showing the results of the simulation of the image using a DNA sequence in image steganography. The DNA molecule's enormous storage capability makes it eligible for use in wireless security applications. Three phases are involved in the embedding process. Firstly, using characters in the DNA triplet conversion table, the pixel values of both the cover and the hidden message are translated to their respective DNA triplet values. Secondly, using DNA triplets, the triplet values are converted to binary values using the binary value conversion table. The two binary-valued matrices corresponding to cover and hidden messages are Ex-OR in the third and final point. The extraction method is employed at the end of the reception.

In [15], authors suggested a cryptographic DNA technique that implemented dynamic mechanisms to increase the data level of confidentiality. Although encryption, the exploitation of the dynamic sequence table containing the substitution cipher features initially ensures first-level secrecy. They used an asymmetric cryptosystem that offered confidentiality at the second level. Finally, they used dynamic DNA encoding to improve their security, retaining the properties of product cipher, providing third-level secrecy. Besides, the function of mathematical series, iterations, the DNA base sequences of the NCBI bank database, etc., is also associated with its secrecy in developing dynamic mechanisms. The attack analysis and the NIST statistical test suite ensured that the ciphertext produced was pseudorandom and significantly secure.

Hazra et al. [16] proposed a brief overview of the techniques used in DNA cryptography and the implementation of those techniques in real-time. First: DNA random one-time pad based. Using a series of randomly organized non-repeating characters, this encryption technique is applied. These serve as the ciphertext of the input. It is referred to as a one-time pad since it is not used again if an input ciphertext is used once. This helps to improve security. The single-pad length should be equal to the plain text length in this scheme. The DNA one-time pad method is used to transform short fragments of plain text messages to ciphertext. The substitution of plain text takes into account random and special codebook. The issue with this approach is that it only applies to short messages. The existing hardware is not good enough for a single pad for large messages. The increase in message size intensifies the difficulty of DNA mapping. Second: DNA chip based. A great development in the area of DNA cryptography has been the advent of DNA chip technology. It is useful for manipulating, along with storing and handling biological information, a large number of genome sequences. A microarray is commonly known as a DNA chip. The microarray is made of nucleic acid and semiconductors form its electronic circuit. Biochemical processes are used to encrypt and decrypt and can be applied to encrypt plain text messages and photos. Since the DNA chip is a biological element, other physical factors can cause its properties to change. This abrupt shift could adversely affect encrypted messages. Third: DNA steganography. The technique of hiding messages within another message is known as steganography. As DNA is used in steganography, it is known as DNA steganography. The scheme allows an image, audio, or even a video file to mask the message. In the field of DNA cryptography, it is an emerging new technique. For hiding a large amount of data, this is perfect. Biological strands are involved in the process, so the data can get contaminated because of the sudden change in the environment. Fourth: DNA fragmentation. This approach is used in the DNA sequence for library construction. It is used to divide the sequence of DNA into small bits. This is used by many encryption algorithms as the second layer of security. The encryption of the key is also enforced. The encryption systems used in the four cryptography techniques are being interchanged with each other, i.e., conventional cryptography, DNA cryptography, elliptic curve cryptography, and quantum cryptography, and new hybrid techniques are being invented as a result.

In [17], authors proposed DNA steganography as one layer and symmetric second layer DES and AES cryptography with parallel programming to improve algorithm performance. To secure data from an interceptor and suggested parallel AES and DNA steganography, DNA-based steganography and cryptography take less time to encrypt (a 20–30% reduction in time). In double-fold methods, the proposed algorithms protect knowledge. Due to the eight possible combinations for DNA steganography operation, the DNA XOR operation has more stable resistance compared to the binary XOR operation.

First, they converted a message into binary numbers in DNA steganography, while in the second step, they converted binary numbers into DNA sequences; this was regarded as an input for step 2 of DNA steganography and human DNA reference to generate cipher-based indexing as an output. These two steps are considered a middle ciphertext as one layer for translating a message into an index-based DNA sequence.

The output of algorithm 1 is used as input for both AES and 3DES algorithms. In AES, they used the previous DNA steganography cipher and the keys of AES to create the ciphertext. In 3DES, they used the same input and the keys of 3DES after applying XOR operation between the original DNA sequence and human DNA reference to create the array of ciphertext (DNACIPER []).

Khalifa [18] proposed a steganographic approach that uses DNA as a secret information carrier that can be used either for purposes of authentication or secret communication. The hiding process of the proposed system had consisted of two key stages. To encrypt the hidden message into a DNA sequence, the first stage used an 8×8 playfair ciphering algorithm. Through a random splicing process that depends on the value of a secret key, the next stage blended the secret sequence into the cover sequence. On the other side of communication, the extraction process was to break the stego-sequence into bits to separate both the message and the cover sequences. A deciphering step was then taken to disclose the original message.

Without the need for the initial cover sequence, this method can be performed blindly. The proposed method showed an excellent performance in terms of both blindness and embedding capability as compared with other methods. In terms of both blindness and ability, experimental results showed an outstanding performance for the proposed method compared to several existing techniques. A hiding capacity of 2 bits per nucleotide, which was the maximum of all approaches, was achieved by the proposed technique.

3 Comparative Study

Table 1 shows comparative study for all the mentioned studies.

Table 1 Comparison between related studies

References	Cover medium	Algorithms and techniques	Secret key	Limitations
[11]	Microdot text	Vigenere	Shared key between sender and receiver	• The microdot will be contaminated because of ciphering • The method will be traced
[12]	ssDNA image	• RC4 cipher and KE3 algorithm Least significant bit (LSB) embedding technique	Three keys • Key 1 (KE1) with six (6) parameters of 2D logistic • Key 2 (KE2)—a primer (short DNA sequence) • Key 3 (KE3)—a variable length key for RC4 encryption	• Not secured enough because of using fixed capacity (2 bits per pixel), which requires to use the same pixels for up to the third level of overlapping • Multiple keys are required for the entire process starting from DNA construction to data embedding, and their transfer between sender and receiver requires a secure key exchange protocol
[13]	DNA text	Enhanced 4 * 4 playfair cipher	Shared key between sender and receiver	Using a public key shared between the sender and receiver which can be detected easily
[14]	DNA image	LSB	No key is used	Although the PSNR ratio and embedding capacity is better than the compared techniques, it is still required a limited capacity to be hidden

(continued)

Table 1 (continued)

References	Cover medium	Algorithms and techniques	Secret key	Limitations
[15]	DNA text	Asymmetric cryptosystm named "RSA, ElGamal, and Paillier"	The receiver generates the keys • Public key for the sender • Private key for the receiver	The length of the ciphertext is many times larger than the length of the plaintext, which can be easily suspected
[17]	DNA text	AES and 3DES	Two times ciphertext generated and two keys are generated from DNA sequences	• The proposed algorithm for small files takes more time • The complexity is based on DNA reference sequence and size of data
[18]	DNA text	8 * 8 playfair cipher	Secret key shared between sender and receiver	• The secret key must be shared, and the sender and receiver need not to agree on anything in advance but the secret key • In the test sequence "SOZC01000013," the actual payload exceeds the hiding capacity

4 Discussion

For achieving the maximum protection and powerful security with high capacity and low modification rate, new data hiding methods have been proposed by researchers based on DNA with the advent of biological aspects of DNA sequences. But all these researches focused on how to minimize the modification rate concerning the hiding capacity and security issues. In [13, 18], authors used the playfair cipher technique to encrypt the message. In [13], the most provided advantage is that the proposed algorithm allowed the message and the key to be any type and any size. While in [18], their method is considered as a blind method since the sender and the receiver did not need to exchange the original cover sequence as the extraction process can be done completely without referring to it. Although their ability in [13, 18] to achieve enough level of security, the secret key must be shared, and the sender and the receiver need to only agree the secret key. In [12, 14], authors used a DNA image as a cover-medium, they used the LSB technique in the embedding process.

Although the ability to achieve acceptable values of distortion metrics, multiple keys are required in [12] for the entire process starting from DNA construction to data embedding, and their transfer between the sender and receiver requires a secure key exchange protocol. While Biswas et al. [14] enhanced the secrecy level of data since they provided a second level of security. They used dynamic DNA encoding which offers third-level secrecy to boots up the security. Some researchers tried to combine the traditional cryptography techniques with steganography but still, we may face security issues such as sharing the cipher keys. Moreover, using DNA sequences as a cover-medium requires much storage memory to be saved, and it is considered as confidential data itself. So, it should be secured too. In [15], although the three proposed cryptosystems which were named RSA, ElGamal, and Paillar passed several attack assessments. The length of the ciphertext is many times larger than the length of the plaintext, which can be easily suspected since the main factor in steganography-based techniques is to decrease the attraction. In [17], the proposed technique is robust enough since they used DNA steganography as one layer, and symmetric second-layer DES and AES cryptography with parallel programming and gene sequences are key values for DNA XOR operation which is more stable resistance compared to the binary XOR operation. The main issue in this technique is that it takes more time in small file ciphering.

5 Proposed Approach

We are aiming to provide an approach to conceal secret clinical data into genomic data objects such as DNA or RNA. This approach involves three successive stages: DNA steganography, compression, and image or video concealing [19, 20] (see Fig. 1).

The proposed approach can be considered as a multi-layered security for hiding the clinical data in a genomic data object, compressing the Stego-Genomic object then concealing it into a cover-medium.

Please try to avoid rasterized images for line-art diagrams and schemas. Whenever possible, use vector graphics instead (see Fig. 1).

6 Conclusion and Future Work

Steganography-related researches focused on how to minimize the modification rate concerning the hiding capacity and security issues. Using the genomic data as a cover-medium requires much storage memory, and the genomic data are confidential data itself.

In the upcoming research, a multi-layered security approach will be implemented to conceal secret data into DNA sequence, then concealing Stego-DNA sequence itself into another cover-medium after compressing it to decrease its size. Due to the confidentiality of genomic data, the compressed Stego-DNA object also will be

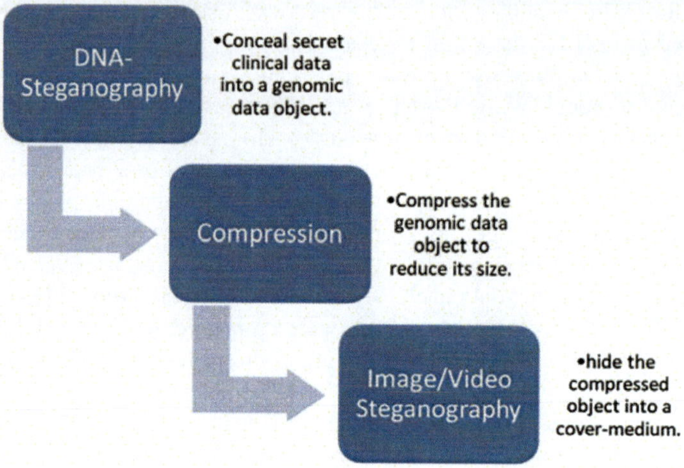

Fig. 1 Proposed multi-layered security framework to hide the clinical and genomic data

concealed into another cover-medium such as image or video utilizing an optimization algorithm such as particle swarm, coco search, firefly, and flower pollination algorithms.

References

1. Fernández-Alemán JL, Señor IC, Lozoya PÁO, Toval A (2013) Security and privacy in electronic health records: a systematic literature review. J Biomed Inform 46(3):541–562. https://doi.org/10.1016/j.jbi.2012.12.003
2. Gunter CA (2013) Building a smarter health and wellness future: privacy and security challenges. In: ICTs and the health sector: towards smarter health and wellness models, pp 141–157. https://doi.org/10.1787/9789264202863-11-en
3. Mellado D, Fernández-Medina E, Piattini M (2010) Security requirements engineering framework for software product lines. Inf Softw Technol 52(10):1094–1117. https://doi.org/10.1016/j.infsof.2010.05.007
4. Darshni P, Ghanekar U (2015) A hybrid data hiding scheme to enhance the capacity of one-third probability embedding method. In: 2015 IEEE international conference on computational intelligence and communication technology (CICT), pp 269, 272. https://doi.org/10.1109/CICT.2015.82
5. Yadav P, Mishra N, Sharma S (2013) A secure video steganography with encryption based on LSB technique. In: 2013 IEEE international conference on computational intelligence and computing research (ICCIC), pp 1–5, 26–28 Dec 2013. https://doi.org/10.1109/ICCIC.2013.6724212
6. Paul R, Acharya AK, Yadav VK, Batham S (2013) Hiding large amount of data using a new approach of video steganography. In: Confluence 2013: the next generation information technology summit (4th international conference), pp 337–343, 26–27 Sept 2013. https://doi.org/10.1049/cp.2013.2338

7. Sameh A, El Arif T, Ghaleb F, Khamis S (2015) Optimized video steganography using Cuckoo Search algorithm. In: 2015 IEEE seventh international conference on intelligent computing and information systems (ICICIS). IEEE. https://doi.org/10.1109/IntelCIS.2015.7397279
8. Wafaa H (2018) Improving image steganography using a proposed mutated levy-flight firefly algorithm. Int J Comput Appl 182:23–29. https://doi.org/10.5120/ijca2018917979
9. Hamed G, Marey M, El-Sayed SA, Tolba MF (2016) Comparative study for various DNA based steganography techniques with the essential conclusions about the future research. In: 2016 11th international conference on computer engineering and systems (ICCES), Cairo, pp 220–225. https://doi.org/10.1109/ICCES.2016.7822003
10. Shi X, Wu X (2017) An overview of human genetic privacy. Ann New York Acad Sci 1387(1):61–72. https://doi.org/10.1111/nyas.13211
11. Wang Z, Zhao X, Wang H, Cui G (2013) Information hiding based on DNA steganography. In: 2013 IEEE 4th international conference on software engineering and service science, Beijing, pp 946–949. https://doi.org/10.1109/ICSESS.2013.6615462
12. Das P, Deb S, Kar N, Bhattacharya B (2015) An improved DNA based dual cover steganography. Proc Comput Sci 46:604-611.https://doi.org/10.1016/j.procs.2015.02.103
13. Marwan S, Shawish A, Nagaty K (2016) DNA-based cryptographic methods for data hiding in DNA media. Biosystems 150:110–118. https://doi.org/10.1016/j.biosystems.2016.08.013
14. Vijayakumar P, Vijayalakshmi V, Rajashree R (2018) Increased level of security using DNA steganography. Int J Adv Intell Paradigms 10:74–82.https://doi.org/10.1504/IJAIP.2018.089490
15. Biswas MR, Alam KMR, Tamura S, Morimoto Y (2019) A technique for DNA cryptography based on dynamic mechanisms. JISA 48. https://doi.org/10.1016/j.jisa.2019.102363
16. Animesh H, Soumya G, Sampad J (2018) A review on DNA based cryptographic techniques. Int J Netw Secur 20:1093–1104. https://doi.org/10.6633/IJNS.201811_20(6).10
17. Vantigaru S, Basavaraj R (2019) Two layer encryption schemes for symmetric algorithms using DNA Sequences. Int J Math Comput Sci 14(4):953–964
18. Khalifa A (2020) A blind DNA-steganography approach using ciphering and random sequence splicing. In: 2020 10th international conference on information science and technology (ICIST), Bath, London, and Plymouth, United Kingdom, pp 86–90. https://doi.org/10.1109/ICIST49303.2020.9202036
19. Vinodhini RE, Malathi P, Kumar TG (2017) A survey on DNA and image steganography. In: 2017 4th international conference on advanced computing and communication systems (ICACCS), Coimbatore, pp 1–7. https://doi.org/10.1109/ICACCS.2017.8014670
20. Malathi P, Manoaj M, Manoj R, Vaikunth R, Vinodhini RE (2017) Highly improved DNA based steganography. Proc Comput Sci 115. https://doi.org/10.1016/j.procs.2017.09.151

Forecasting COVID-19 Pandemic Using Linear Regression Model

Heba M. Sabri, Ahmad M. Gamal El-Din, and Lamia Aladel

Abstract During the ongoing worldwide crisis, researchers, clinicians, and medical care specialists around the world continue looking for another innovation to help in handling the COVID-19 pandemic. The proof of Machine Learning (ML) and Artificial Intelligence (AI) application on the past pestilence empower scientists by giving another point to battle against the novel Coronavirus episode. This paper intends to thoroughly audit the part of AI and ML as one critical technique in anticipating SARS-CoV-2 and its related epidemic. Coronavirus is an irresistible illness, and it does serious harm to the lungs. Coronavirus causes disease in people and has executed numerous individuals in the whole world. Nonetheless, this infection is accounted for as a pandemic by the World Health Organization. (WHO) and all nations are attempting to control and lockdown all spots. This work's main principle goal is to predicting the spread of COVID-19 across Egypt and analyzing the development rates. For this aim, we access real datasets collected from Johns Hopkins University Center for Systems Science and Engineering (JHU CSSE). And European Union open dataset. We have implemented the results by using R Language.

Keywords COVID-19 · Machine learning · Linear regression · R language

1 Introduction

In December 2019, the novel COVID showed up in the Wuhan city of China and was accounted for in the World. Health Organization (W.H.O) on 31st December 2019. The infection made a worldwide danger and was named COVID-19 by W.H.O on eleventh February 2020 [1]. The COVID-19 is a group of diseases, including SARS, ARDS. W.H.O proclaimed this episode as a general wellbeing crisis and referenced the accompanying; the infection is being sent through the respiratory plot when a reliable individual comes in contact with the contaminated individual. The disease may communicate between people through different roots, which are at present indistinct. The toxic individual shows indications inside 2–14 days, contingent upon

H. M. Sabri (✉) · A. M. Gamal El-Din · L. Aladel
Department of Computers and Information Systems, Sadat Academy, Cairo, Egypt

D. A. Magdi et al. (eds.), *Digital Transformation Technology*, Lecture Notes in Networks and Systems 224, https://doi.org/10.1007/978-981-16-2275-5_32

the center east respiratory condition (MERS), and the serious intense respiratory disorder (SARS) [2]. As indicated by W.H.O the signs and indications of mellow to direct cases are a dry hack, weariness, and fever, while as in severe cases, dyspnea (windedness), fever, and sleepiness may happen. People with different illnesses like asthma, diabetes, and coronary disease are more defenseless against the infection and may become seriously sick. The individual is analyzed dependent on manifestations and his movement history. Vital signs are being noticed acutely of the customer having indications. No particular treatment has been found on tenth April 2020, and patients are being dealt with apparently. The medications like hydroxychloroquine, antipyretic against virals are utilized for the indicative treatment [3].

The COVID-19 outbreak is a unique worldwide public health challenge and trouble for governments, furthermore, people. The quantity of distinguished cases has been expanding quickly over the world, so various explores and activities confronted new late difficulties to conjecture the pinnacle of the scourge to help the legislatures settle on choices for restricting the spreading of the illness. In Egypt, the quantity of announced cases has expanded every day after the primary case was pronounced on 14 February 2020. On 15 November 2020, the number of cases is 110767 cases, and the number of deaths about 6453 cases [4].

The use of Artificial Intelligence and Machine Learning (ML) has been rapidly extending in various fields, including malware acknowledgment, adaptable malware ID, prescription, data recovery, information security [5], and image processing as an image watermarking approach to hide DNA sequence data into fingerprint templates for copyright protection-based[6]. In recent years, the Machine Learning and deep learning models are commonly used for a health information system and applications of health care researches, one of its use are COVID Prediction.

The rest of the paper is organized as follows: Sect. 2 discusses Related Work of research papers that discuss the prediction of COVID-19 Pandemic. While Sect. 3 shows the proposed solution of using Linear Regression Model. Last but not least is Sect. 4, showing the outcome results of the model on the datasets we have. Finally Sect. 5 is the conclusion.

2 Related Work

This infection is accounted for as a pandemic by the World Health Organization (WHO), and all nations are attempting to control and lockdown all spots. Several researchers presented ideas about facing such type of diseases such as:

In [7], the researchers work main fundamental target is to comprehend the five different errands, (I) Predicting the spread of COVID across locales. (II) Breaking down the development rates and the sorts of moderation across nations. (III) Predicting how the plague will end. (IV) Analyzing the transmission pace of the infection. (V) Correlating COVID and climate conditions. The benefit of managing these responsibilities to limit the disease spread by different relief, how well the alleviations are working, the number of cases have been forestalled by this alleviations, a thought regarding the number of patients that will recuperate from the contamination with old medicine. This paper proposes a novel Support Vector Regression strategy to examine five distinct errands identified with novel COVID.

Moreover, in[8] as an option in contrast to the susceptible-infected-resistant (SIR)-based models, this paper proposes a crossover AI way to deal with foresee the COVID-19, and we epitomize its potential utilizing information from Hungary. The crossover AI techniques for adaptive network-based fuzzy inference system (ANFIS) and multi-layered perceptron-imperialist competitive algorithm (MLP-ICA) are proposed to anticipate the time arrangement of contaminated people and the death rate.

The models anticipate that by late May, the outbreak and the absolute profound quality will drop significantly. The approval is performed for nine days with promising outcomes, which affirms the model precision. It is expected that the model keeps up its precision as long as no considerable interference happens. This paper gives an underlying benchmarking to show the capability of AI for future exploration.

Meanwhile, The researchers in [9] built up a deep learning approach to remove data from CT examinations. Their examination incorporated an assortment of 453 CT filters from 99 patients. They separated 195 areas of intrigue (returns on initial capital investment) of sizes extending from 395 × 223 to 636 × 533 pixels from the CT outputs of 44 Coronavirus positive pneumonia patients and 258 returns money invested from those of 50 Coronavirus negative patients. They applied an altered organization initiation model and got a precision of 82.9% for the interior approval with the explicitness of 80.5% and an affectability of 84%. The outside testing dataset displayed an absolute precision of 73.1 with 67% explicitness and an affectability of 74%.

But In [10] the researchers think about the Coronavirus information from India against a few nations just as crucial states in the US with an entire episode, and it is discovered that the main proliferation number R0 for India is in the normal scope of 1.4e3.9. Then, the ring of development of contaminations in India is near that of Washington and California. Exponential and old-style models of (SIR) rely upon current information to deliver regular short-ring and long haul expectations. From the SIR model, it is assessed that India will enter solidness before the finish of May 2020 with the last size of pestilence close to 13,000. However, if India enters the gathering transmission point, the estimate will be invalid. The impact of social separating is additionally estimated by breaking down information from different geological areas by and by with the assumption of no gathering transmission.

3 The Proposed Model

Linear Regression is one of the most commonly utilized prescient demonstrating procedures. The point of direct relapse is to locate a numerical condition for a nonstop reaction variable Y as a component of at least one X variable(s). With the goal that you can utilize this relapse model to anticipate the Y when just the X is known.

Linear Regression is used to predict Y dependent on at least one info indicator factor X. The point is to set up a linear relationship (a numerical equation) between the indicator variable(s) and the reaction variable so that we can utilize this recipe to appraise the estimation of the reaction Y when just the indicators (Xs) values are known.

This mathematical equation can be generalized as follows [11]:

$$Y = \beta_1 + \beta_2 X + \epsilon \quad (1)$$

where, $\beta 1$ is the intercept and $\beta 2$ is the slope. Collectively, they are called regression coefficients. ϵ is the error term, the part of Y the regression model is unable to explain.

In this paper, we have built the linear regression model; we have also established the relationship between the predictor and response in a mathematical formula using R Language. Moreover, after the linear model is built, we have to ensure that it is statistically significant by printing the summary statistics for linearMod. We have built the model for two datasets containing the number of confirmed cases and deaths of Egypt as a case study, the first one from Johns Hopkins University Center for Systems Science and Engineering (JHU CSSE) [12]. And the second one is from the European Union open dataset and concluded the summary for each dataset [13]. We have used the data from March 2020 till the end of October 2020 and predict the values until December 2020. Moreover, we have compared the actual values versus the predicted values during the previous period from March to October.

4 Experimental Results and Discussion

4.1 Analyzing Confirmed Cases of COVID-19 from John Hopkins Dataset

Figure 1 shows that the total number of confirmed cases accumulatively increases tremendously and keeps rising with no stationary state.

Figure 2 shows that the ds in the x-axis represents the date selected from the SQL query and must be either date or date-time data type. While the y in the y-axis represents the value, which will be forecasted, it must be a numeric data type. The forecast of the total confirmed cases accumulatively is in line with the previous figure; as shown, the total numbers of confirmed cases started increasing from May till August. Then, the increasing level begins to be stable somehow till October.

In Fig. 3, the accumulative total number of confirmed cases increases massively during the weekends, for many reasons such as the family visits and going outdoors.

As shown in Fig. 4, the model performance shows that the predicted versus the actual is consistent and too linear. The model's p-value and predictor's p-value are less than the significance level from the model summary below as shown in Table 1

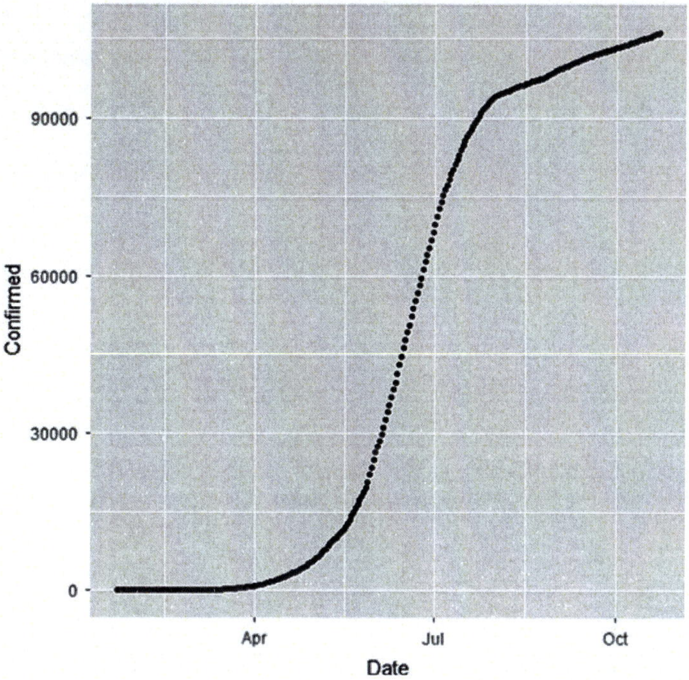

Fig. 1 The total number of confirmed cases

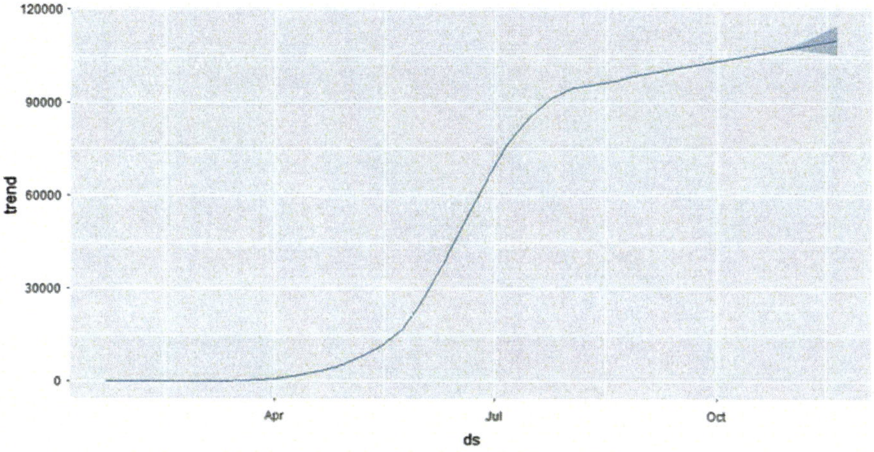

Fig. 2 The trend of total number of confirmed cases

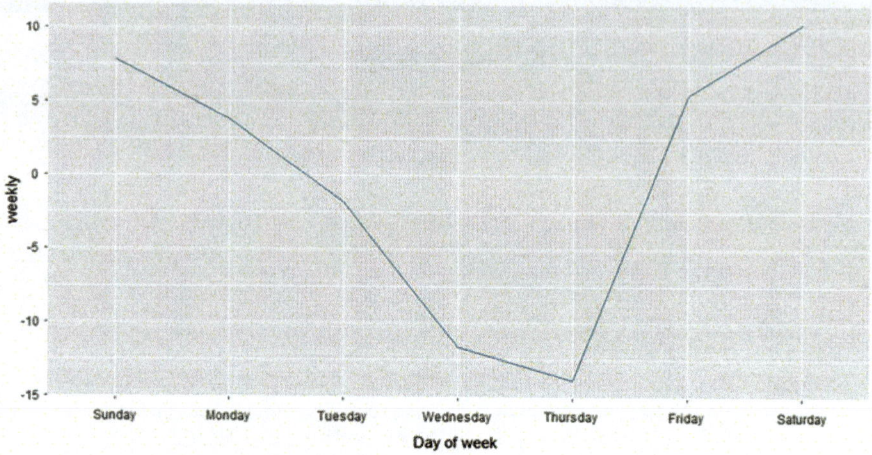

Fig. 3 The accumulative total number of confirmed cases

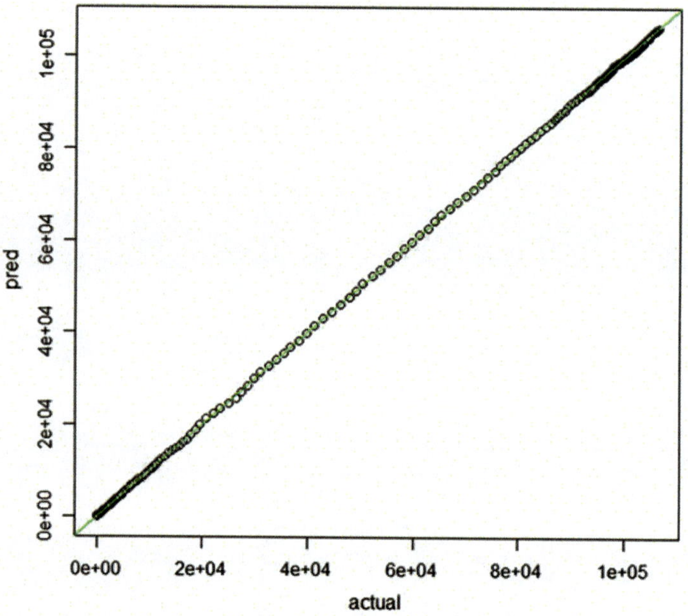

Fig. 4 The predicted data versus the actual data of confirmed cases

so the model is statistically significant. Besides, the *R*-Squared and Adj *R*-Squared are perfectly matched (each value equals 1).

Table 1 The linear regression model summary for confirmed cases according to John Hopkins dataset

Call:				
lm(formula = pred ~ actual)				
Min	1Q	Median	3Q	Max
Residuals:				
– 587.80	– 52.49	1.82	62.33	485.12
	Estimate	Std. Error	t-value	Pr(> |t|)
Coefficients				
(Intercept)	0.780983	12.641333	0.062	0.951
Actual	0.999984	0.000196	5102.255	< 2e – 16
Signif. codes	0	0.001	0.01	0.05 '.' 0.1 ' ' 1

Residual standard error: 144 on 274 degrees of freedom
Multiple R-squared: 1, Adjusted R-squared: 1
F-statistic: 2.603e + 07 on 1 and 274 DF, p-value: < 2.2e – 16

4.2 *Analyzing Confirmed Cases of COVID-19 from European Union Open Dataset*

Figure 5, the number of confirmed daily cases increases tremendously from April and keeps rising till it reaches a peak in June, then starts to decrease gradually from July to October.

Figure 6, the ds in the x-axis represents the date selected from the SQL query and must be either date or date-time data type. While the y in the y-axis represents the value, which will be forecasted, it must be a numeric data type. The forecasting of the confirmed daily cases is consistent to a certain extent with the previous figure.

Figure 7 shows that the number of confirmed daily cases started increasing tremendously from May till June, then starts to decrease gradually from July to October. However, in Fig. 8, the accumulative total number of confirmed cases increases massively during the weekends, for many reasons, such as the family visits and going outdoors and decrease during the middle of the week till Thursday.

In Fig. 9, the model performance shows that the predicted versus the actual is somehow near to each other. The model's p-value and predictor's p-value are less than the significance level from the model summary below, so the model is statistically significant. Besides, the R-Squared and Adj R-Squared are acceptable and near each other's value (Table 2).

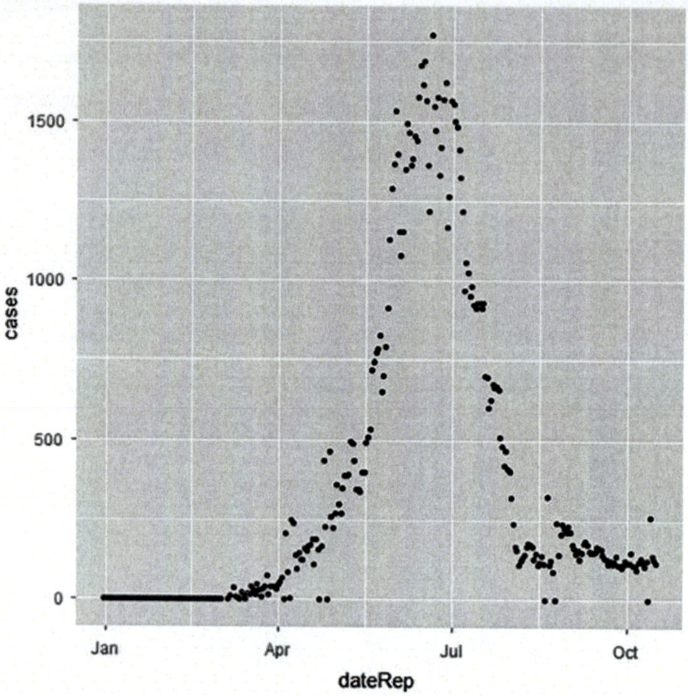

Fig. 5 The total number of confirmed cases

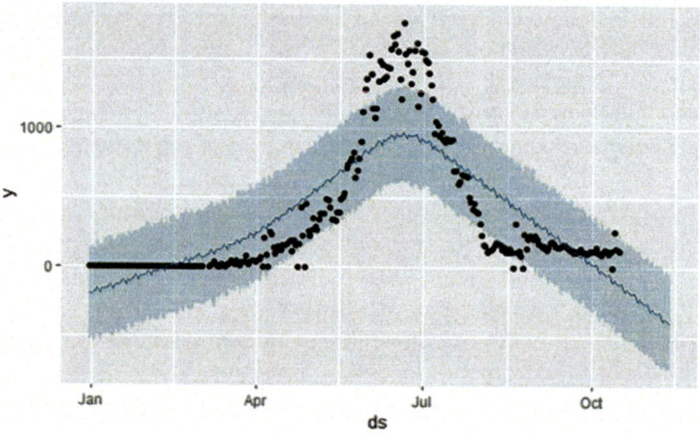

Fig. 6 The pattern of total number of confirmed cases

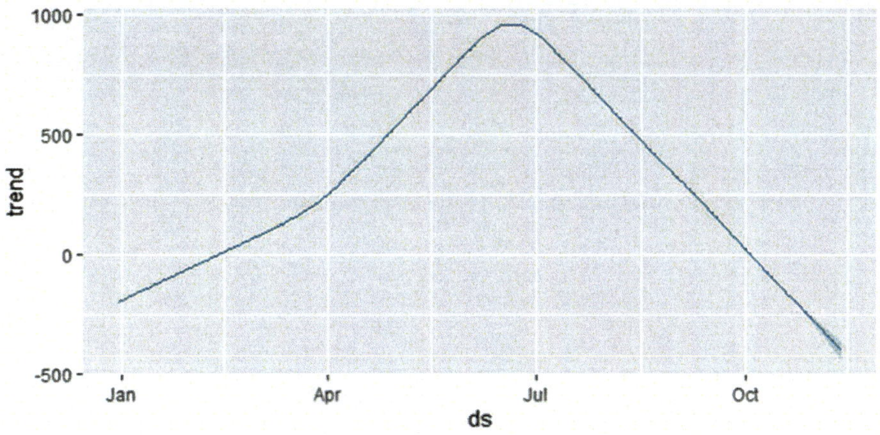

Fig. 7 The trend of total number of confirmed cases per month

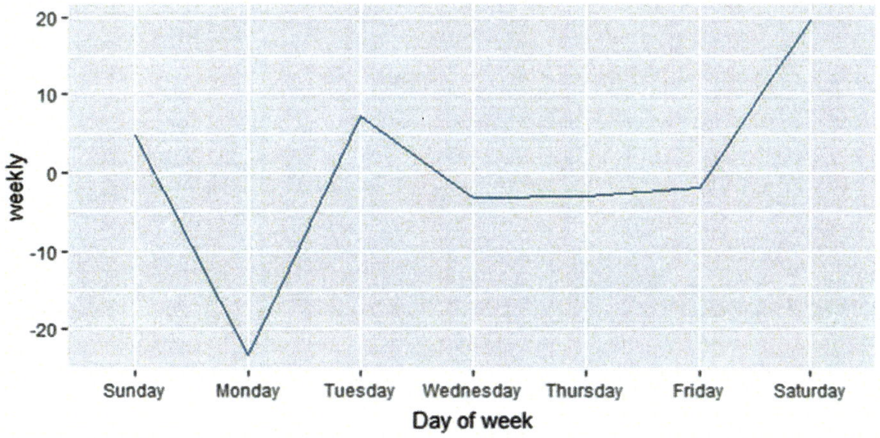

Fig. 8 The trend of total number of confirmed cases per day

4.3 Analyzing the Number of Deaths

In Fig. 10, the number of death cases increases tremendously from April and keeps rising till it reaches the peak in June, then starts to decrease gradually from July to October.

Figure 11 shows that the ds in the x-axis represents the date selected from the SQL query and must be either date or date-time data type. While the y in the y-axis represents the value, which will be forecasted, it must be a numeric data type. The forecasting of the death cases is consistent to a certain extent with the previous figure.

As shown, the trend of number of death cases started increasing tremendously from May till July, then starts to decrease gradually from August to October.

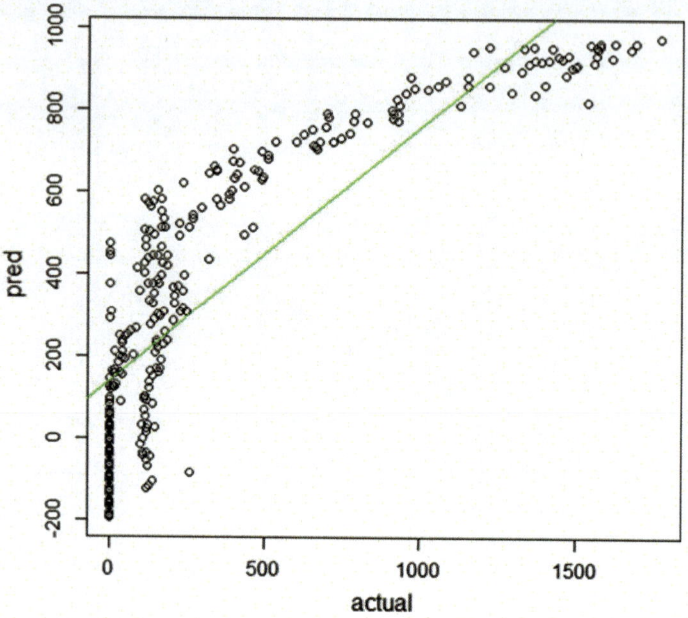

Fig. 9 The predicted data versus the actual data of confirmed cases of Egypt

Table 2 The linear regression model summary for confirmed cases according to European Union open dataset

Call				
lm(formula = pred ~ actual)				
Min	1Q	Median	3Q	Max
Residuals				
– 385.49	– 156.28	– 0.05	156.25	375.24
	Estimate	Std. Error	*t*-value	Pr(> |*t*|)
Coefficients				
(Intercept)	138.59434	13.84180	10.01	< 2e – 16
Actual	0.61923	0.02272	27.25	< 2e – 16
Signif. codes	0	0.001	0.01	0.05 '.' 0.1 ' ' 1

Residual standard error: 187.5 on 285 degrees of freedom
Multiple R-squared: 0.7227,Adjusted R-squared: 0.7217
F-statistic: 742.8 on 1 and 285 DF, p-value: < 2.2e – 16

In Fig. 12, the number of death cases increases massively during the middle of the week, decreases again till Friday, and then increases on Saturday.

In Fig. 13, the model performance shows that the predicted versus the actual is somehow near to each other. From the model summary below, the model's p-value and

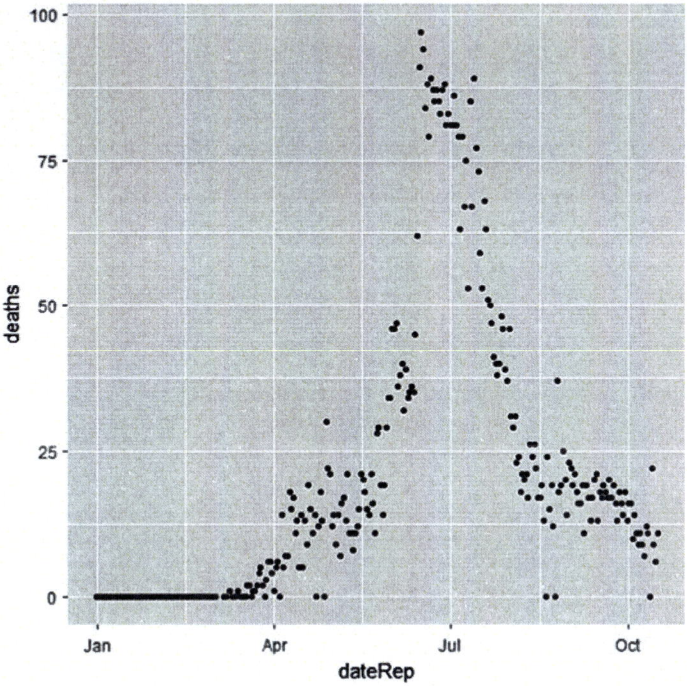

Fig. 10 The number of death cases

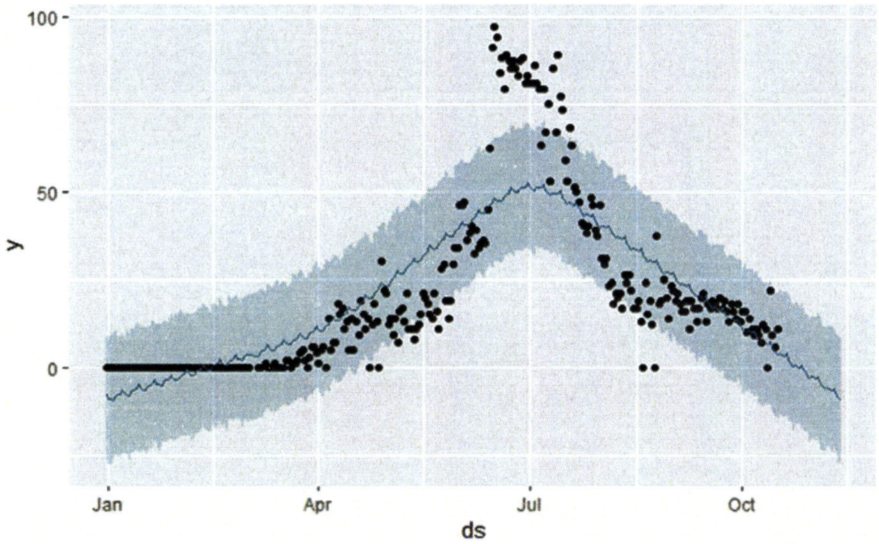

Fig. 11 The trend of number of death cases

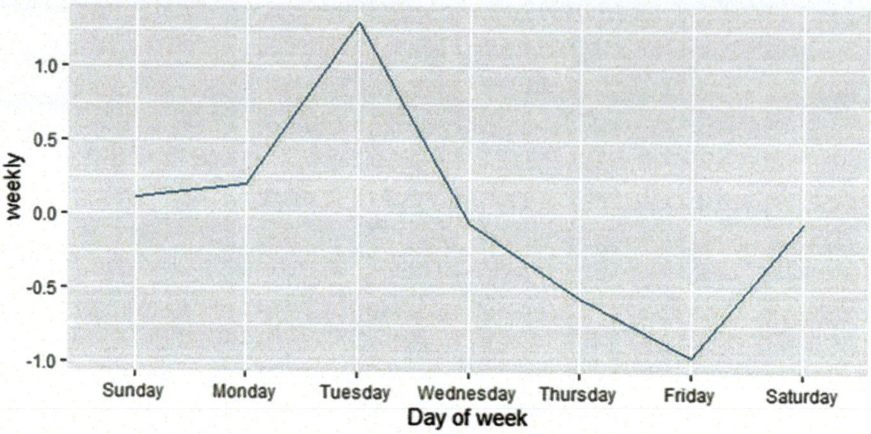

Fig. 12 The trend of number of death cases per day

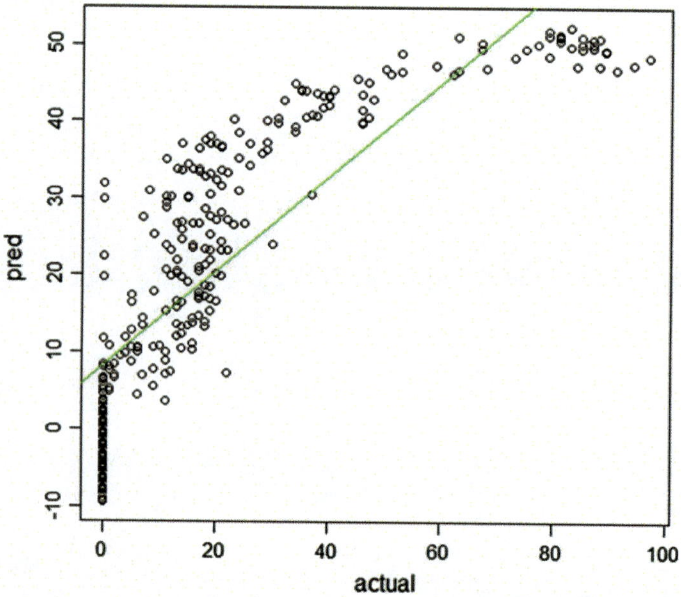

Fig. 13 The predicted data versus the actual data of death cases of Egypt

predictor's p-value are less than the significance level, so the model is a statistically significant model. Besides, the R-Squared and Adj R-Squared are acceptable and near each other's value (Table 3).

Table 3 The linear regression model summary for deaths cases

Call				
lm(formula = pred ~ actual)				
Min	1Q	Median	3Q	Max
Residuals				
– 19.157	– 7.413	– 1.198	8.192	23.628
	Estimate	Std. Error	*t*-value	Pr(> |*t*|)
Coefficients				
(Intercept)	8.23472	0.76238	10.80	< 2e – 16 ***
Actual	0.61130	0.02312	26.44	< 2e – 16***
Signif. codes	0***	0.001**	0.01*	0.05 '.' 0.1 ' ' 1

Residual standard error: 9.888 on 285 degrees of freedom
Multiple *R*-squared: 0.7104, Adjusted *R*-squared: 0.7094
F-statistic: 699.1 on 1 and 285 DF, *p*-value: < 2.2e – 16

5 Conclusion

The process of utilizing artificial intelligence and machine learning for prediction, particularly in COVID-19, demonstrated help in displaying and anticipating the end status of the infection spreading. This paper centers around applying Artificial Intelligence (AI) calculations for forecast COVID-19 using linear regression model. We applied our model to Egypt as a case study. for this aim, we use two datasets containing the number of confirmed cases and deaths of Egypt, the first one from Johns Hopkins University Center for Systems Science and Engineering (JHU CSSE), and the other one, from the European Union open dataset and concluded the summary for each dataset. We have used the data from March 2020 till the end of October 2020 and predict the values until December 2020. Moreover, we have compared the actual values versus the predicted values during the previous period from March to October.

References

1. Khanday A, Rabani S, Khan Q, Rouf N (2020) Machine learning based approaches for detecting COVID-19 using clinical text data. Int J Inf Technol 12(3):731–739
2. Alazab et al (2020) COVID-19 prediction and detection using deep learning. Int J Comput Inf Syst Ind Manage Appl
3. Narin A et al (2020) Automatic detection of coronavirus disease (COVID-19) using X-ray images and deep convolutional neural networks
4. Egyptian Ministry of Health Homepage. https://twitter.com/mohpegypt?ref_src=twsrc%5Egoogle%7Ctwcamp%5Eserp%7Ctwgr%5Eauthor. Last Accessed 5 Nov 2020
5. Ghany K, Hassan G, Hassanien A, Hefny H, Schaefer G, Ahad M (2014) A hybrid biometric approach embedding DNA data in fingerprint images. In: The 3rd IEEE international conference on informatics, electronics and vision—ICIEV14

6. Ghany K, Hassan G, Hassanien A, Hefny H, Tolba M (2013) Kekre's transform for protecting fingerprint template. In: 13th international conference on hybrid intelligent systems (HIS13). IEEE
7. Yadav M, Perumaln M, Srinivas A (2020) Analysis on novel coronavirus (COVID-19) using machine learning methods. Chaos Solitons Fract 139
8. Wang S, Kang B, Ma J, Zeng X, Xiao M, Guo J et al (2020) A deep learning algorithm using CT images to screen for Corona Virus Disease (COVID-19). medRxiv 2020
9. Pinter G, Felde I, Mosavi A, Ghamisi P, Gloaguen R (2020) COVID-19 pandemic prediction for hungary; a hybrid machine learning approach, mathematics, MDPI
10. Ranjan R (2020) Predictions for COVID-19 outbreak in India using epidemiological models. medRxiv, 3 Apr 2020
11. R-Statistics.co Homepage. http://r-statistics.co/LinearRegression.html. Last accessed 10 Nov 2020
12. John Hopkins University Center for Systems Science and Engineering (JHU CSSE). https://raw.githubusercontent.com/CSSEGISandData/COVID-19. Last accessed 30 Oct 2020
13. European Union open data set. https://data.europa.eu/euodp/en/data/. Last accessed 30 Oct 2020

Gulf Area COVID-19 Cases Prediction Using Deep Learning

Kareem Kamal A. Ghany, Hossam M. Zawbaa, and Heba M. Sabri

Abstract Coronavirus-19 (COVID-19) is the most catastrophic news of 2020. The initial onset of the virus is an unexpected extreme outlier event. Still, the human response to restrain the virus is also creating massive ripples through systems that run the world, such as health, economy, education, and tourism. This paper focuses on research and applying artificial intelligence (AI) algorithms for predicting COVID-19 propagation using the available time-series data and studies the effect of the lifestyle and the awareness of citizens on the virus in the gulf area, especially in the Kingdom of Saudi Arabia (KSA), United Arab of Emirates (UAE) and Qatar. For this aim, we access time-series real-datasets collected from Johns Hopkins University Center for Systems Science and Engineering [1]. We have implemented the proposed model based on long short-term memory (LSTM) with 100 hidden units (neurons). We confirmed that the KSA and Qatar would take the most extended period to recover from COVID-19 virus from the experimental results. The situation will be controllable in March 2021 in UAE.

Keywords COVID-19 · Time series · Long short-term memory · Deep learning

1 Introduction

Coronavirus is a large family of viruses that can cause severe illness to the human being. The first known severe epidemic is severe acute respiratory syndrome (SARS) in 2003, while the second episode of series sickness started in 2012 in Saudi Arabia with the center east respiratory syndrome (MERS). The current flare-up of disease due to COVID is accounted for in late December 2019. This new infection is exceptionally

K. K. A. Ghany (✉) · H. M. Zawbaa
Faculty of Computers and Artificial Intelligence, Beni-Suef University, Beni, Suef, Egypt

K. K. A. Ghany
College of Computing and Informatics, Saudi Electronic University,
Riyadh, Kingdom of Saudi Arabia

H. M. Sabri
Department of Computers and Information Systems, Sadat Academy, Cairo, Egypt

D. A. Magdi et al. (eds.), *Digital Transformation Technology*, Lecture Notes in Networks and Systems 224, https://doi.org/10.1007/978-981-16-2275-5_33

infectious and has rapidly spread around the world. On January 30, 2020, the World Health Organization (WHO) announced this episode a General Wellbeing Crisis of Global Concern (PHEIC) as it had spread to 18 nations. On February 11, 2020, WHO named this "COVID-19." On Walk 11, as the quantity of coronavirus cases has expanded multiple times, separated from China with more than 118,000 cases in 114 nations, what is more, more than 4000 passings, WHO pronounced this a pandemic [2].

Coronavirus disease 2019 (COVID-19) is a new contagious disease that causes respiratory sickness (like the flu or influenza) with symptoms such as coughing, sneezing, and fever. In more critical circumstances, it can cause shortness of breathing. The COVID-19 disease was first recognized in December 2019 in Wuhan, the capital of Hubei province in China. As of October 1, 2020, (according to the World Health Organization report [3], there are more than 30 million confirmed cases have been reported in more than 200 countries and more than 1 million deaths.

The World Health Organization expresses that coronavirus could be spread starting with one individual then onto the next at an extremely quick speed through contact, furthermore, respiratory spray. COVID-19 epidemic virus is mainly spread during close communication with infected people when they are coughing or sneezing. Moreover, people may also become infected by touching a contaminated surface or object with the virus on it, then touching their eyes, nose, or mouth. You can preserve yourself by personal care, such as washing the hands regularly, avoiding touching your face, and avoiding close contact (at least 1 m) with unwell people. The number of identified cases has been expanding quickly over the world, so various investigates and tasks confronted new late difficulties with forecasting the peak of the epidemic to help the government settle on choices for restricting the spreading of the disease. The challenge now is how to estimate the peak of a pandemic keeping in mind all the efforts that have been made in all directions.

Nowadays, most of the developed applications and systems for different businesses are based on artificial intelligence and machine learning. Deep learning is a part of machine learning that produces proper performance and completely surpasses classical machine learning methods, especially when data increases. Machine learning (ML) algorithms empower appraisal of the relations among info and yield of complex cycles. With regard to hazard definition, ML models produce preferred execution over conventional expectation models, attributable to their capacity to uncover nonlinear affiliations, unrivaled enhancement of multi-factorial calculations, and inherent techniques for model approval. The utilization of various ML models, including determination and grouping of multidimensional heterogeneous information, can give promising instruments to derivation in clinical practice and be proposed for pattern expectation and danger definition of patients with coronavirus [4]. The use of AI (ML) has been quickly expanding in different fields, including malware recognition, versatile malware identification, medication, and data recovery. In 2012, an advanced ML framework called profound learning was presented, which depends on a convolutional neural organization (CNN). It won the ImageNet grouping rivalry, the world's most popular PC vision rivalry. Profound learning calculations empower computational models made out of various handling layers to learn information portrayal through a few deliberation layers. They train a PC model to perform characterization assignments legitimately from pictures, messages, or sounds. As indicated by many AI researchers, their profound learning models highlight high accuracy and improve human output in specific occurrences [5].

The machine learning and deep learning models discussed above are generally utilized for time arrangement forecasting and give splendid outcomes in recent years. Several deep learning models with remarkable results across diverse application fields. Therefore, deep learning-based on time series data has become more popular and has been applied for different applications such as network intrusion detection, illegal traffic flow detection, fraud detection, and video surveillance.

The COVID-19 outbreak is an unprecedented global public health challenge and difficulty for governments, organizations, and individuals. Therefore, this research will help them react adequately, and they will have easy access to useful, clean data, and a good understanding of what can and cannot be said based on the available data.

We focused on three countries from the Gulf area, which are KSA, UAE, and Qatar, because they are sharing the same weather and almost have the same quality standard of life [6, 7].

The rest of the paper is organized as follows: Sect. 2 discusses background knowledge, including related work, the importance of the LSTM algorithm, and the quality of life in the different selected countries. While Sect. 3 shows the proposed model of using LSTM on time series data. Last but not least is Sect. 4, showing the predicted results for the selected countries. Finally, Sect. 5 is the conclusion and future work.

2 Background

2.1 Related Work

Coronavirus is a large family of viruses that cause diseases resulting from colds such as the middle east respiratory syndrome (MERS-CoV) and severe acute respiratory syndrome (SARS-CoV). Coronavirus disease (COVID-19) is a new species discovered in 2019 and has not been previously identified in humans (WHO,2020). COVID-19 causes symptoms proved to be more moderate in about 82 In Gulf Area countries, we found that the total cases are more than 900,000 cases, in which there are more than 45,000 active cases (90% in mid condition), 7227 deaths, and 847,773 recovered cases. Planning is critical to decreasing the sudden and potentially catastrophic impact of an infectious disease pandemic on society. Several researchers presented ideas about facing such type of diseases such as: In [8], the researchers showed that the SARS-CoV virus is transferred from musk cats to humans, and the MERS-CoV virus is contaminated from dromedary to humans. Several research zones have executed simulated intelligence (e.g., illness analysis in health care). One of the fundamental preferences of simulated intelligence is that it very well may be executed in a prepared model to order inconspicuous pictures. In this examination, human-made intelligence was actualized to distinguish whether a patient is positive for coronavirus utilizing their chest X-ray image. In Narin [9], the researchers proposed a deep transfer learning-based approach using chest X-ray images obtained from COVID-19 patients, and they succeeded in predicting COVID-19 patients automatically. In Xiaowei [10], the researchers presented a new method that could screen COVID-19 fully automatically by deep learning technologies. They proved that models with location-attention mechanism could be more accurately classify COVID-19 at chest radiography with an overall accuracy rate of 86.7%.

Moreover, Gozes [11] utilized a deep learning way to deal with naturally recognize coronavirus patients and look at the illness trouble evaluation on CT checks utilizing a dataset of CT examines from 157 unfamiliar patients from China and the US. Their proposed framework investigations the CT check at two certain levels: subsystems A and B. Subsystem A plays out a 3D examination, and subsystem B plays out a 2D investigation of each section of the output to recognize and find more extensive diffuse opacities, including ground-glass penetrates (which have been clinically distinguished as illustrative of coronavirus). The creators applied Resnet-50-2 to subsystem B and got a region under the bend of 99.6% to assess their framework. The affectability and explicitness were 98.2 and 92.2%, individually. In [5], the researchers utilizing real-world datasets. The framework inspects chest X-beam pictures to distinguish such patients. Their discoveries show that such an examination is essential in coronavirus determination as X-beams are helpfully accessible rapidly and at low expenses. Experimental discoveries acquired from 1000 X-beam pictures of genuine patients affirmed that their proposed framework helps identify coronavirus and accomplishes an F-measure scope of 95–99%. Also, three determining techniques—the prophet calculation (Dad), autoregressive incorporated moving normal (ARIMA) model, and long-transient memory neural organization (LSTM)—were received to anticipate the quantities of coronavirus affirmations, recuperation, and passings throughout the following seven days. The forecast outcomes show promising execution and offer a standard exactness of 94.80 Meanwhile, [12] built up a deep learning approach to remove CT examination data. Their examination incorporated an assortment of 453 CT filters from 99 patients. They separated 195 areas of intrigue (returns on initial capital investment) of sizes extending from 395 × 223 to 636 × 533 pixels from the CT outputs of 44 coronavirus positive pneumonia patients and 258 returns money invested from those of 50 coronavirus negative patients. They applied an altered organization initiation model and got a precision of 82.9% for the interior approval with the explicitness of 80.5% and an affectability of 84%. The outside testing dataset displayed an absolute precision of 73.1 with 67% explicitness and an affectability of 74%.

Moreover, [13] applied CNN to a limited dataset, which was not explicitly characterized in their examination, to assess and gage the number of reported cases in China. The creators utilized the mean supreme and root mean square blunders to contrast their model and other profound learning models, including multilayer perceptron, long momentary memory (LSTM), and gated repetitive units. The creators presumed that the got outcomes guarantee high prescient productivity.

However, [14] thinks about India's coronavirus information against a few nations just as crucial states in the US with an entire episode, and it is discovered that the main proliferation number R0 for India is in the normal scope of 1.4e3.9. Then, the ring of development of contaminations in India is near that of Washington and California. Exponential and old-style models of powerless contaminated recouped (SIR) rely upon current information to deliver regular short-ring and long-haul expectations. From the SIR model, it is assessed that India will enter solidness before the finish of May 2020, with the last size of pestilence close to 13,000. However, if India enters the gathering transmission point, the estimate will be invalid. The impact of social separating is also estimated by breaking down information from different geological areas by assuming no gathering transmission.

2.2 *Long Short-Term Memory (LSTM)*

LSTM is a novel recurrent network architecture in combination with a proper gradient-based learning algorithm. LSTM is designed to overcome these error back-flow problems. It can learn to bridge time intervals over 1000 steps, even in the noisy case, compact input sequences, without the loss of short-time delay capabilities. An efficient, gradient-based algorithm achieves this for an architecture enforcing constant error flow that is neither exploding nor disappearing through individual units' internal states [15].

In principle, an LSTMs can use its memory cells to remember long-range information and track the various attributes of text it is currently processing. For instance, it is a simple exercise to write down gadget cell weights that would allow the cell to keep track of whether it is inside a quoted string [16].

2.3 *Quality of Life*

According to the quality of life index [17], United Arab Emirates has a quality index equal to 169.17, Qatar has a quality index equal to 155.77, and Saudi Arabia has a quality index equal to 144.52. Even though the quality of life index is continuously changing, these countries' capitals are inside the top ten Arabian cities in the level of life. Although each of them is unique in its way, they all share one quality: the Arab world's best cities.

3 The Proposed Model

3.1 Data Preparation

Data preparation was managed using the COVID-19 time series data prepared and transformed into proper interpretation and coding to fit into the LSTM model for predicting and forecasting problems. The LSTM model learns a function that maps a sequence of past observations as input to an output observation as in the following steps:

1. The COVID-19 time series data for the three countries were downloaded from January 22, 2020, till December 31, 2020, from Johns Hopkins University Center for Systems Science and Engineering [1].
2. Drop/delete the date column from the time series data. The sequence of observations was as follows: [10, 7, 15, 20, 24, 36, 40, 31, 47, 53, 45]. The sequence of observations must be transformed into multiple samples from which the LSTM can learn. We divided the sequence into multiple input/output patterns called samples. In this paper, the sequence of observations was transformed using a window size equal to one week (because the spread of COVID-19 can be significant from one week to another). We have considered seven-time steps as the input and one-time step as the corresponding output for the prediction model that was being learned, as shown in Table 1.
3. We have used the long short-term memory (LSTM) as a regression model to predict and forecast the COVID-19 spread for the coming months based on the number of confirmed cases and the number of death cases.

3.2 The Proposed LSTM Model

We have implemented a simple long short-term memory (LSTM) model with an input layer, a single hidden layer, and an output layer that is used to make a prediction. The input layer has some neurons (demission) equal to seven sequence steps (for one-week COVID-19 data points). The hidden layer is an LSTM layer with 100 hidden units (neurons) and a rectified linear unit (ReLU) as an activation function.

Table 1 Inputs/output patterns for time-series data example

Inputs	Output
[10, 7, 15, 20, 24, 36, 40]	31
[7, 15, 20, 24, 36, 40, 31]	47
[15, 20, 24, 36, 40, 31, 47]	53
[20, 24, 36, 40, 31, 47, 53]	45

The output layer had a dense layer with 1 unit for predicting the output. The learning rate is set to 0.001, and it decays every 5 epochs. Moreover, we have used 1000 as the number of epochs, Adam as the optimizer, and the mean square error as the loss function. Finally, we fit the model with prepared data to make a prediction. The obtained results may vary given the stochastic nature of the LSTM model; therefore, we have run it several times. Finally, we enter the last sequence with output to forecast the next value in the series.

4 Experimental Results

The created series of prediction and forecasting models based on time-series data to check when the situation will be under control in the three countries (KSA, UAE, and Qatar) was straightforward and easy. From the LSTM model, we confirm that the KSA and Qatar would take the most extended period to recover from the virus (during the second half of 2021), as in Figs. 1 and 5. For UAE, there is a high spread of COVID-19 virus, but due to strict social distance, the situation will be controllable by March 2021, as shown in Fig. 3.

Moreover, we confirm that the UAE's total number of deaths and the situation is under control since June 2020, as shown in Fig. 4. For Qatar, the total number of deaths decreases, and the situation is controllable since August 2020, as shown in Fig. 6. For KSA, the total number of deaths decreases, and the situation will be controllable in March 2021, but it may be increased after that as shown in Fig. 2.

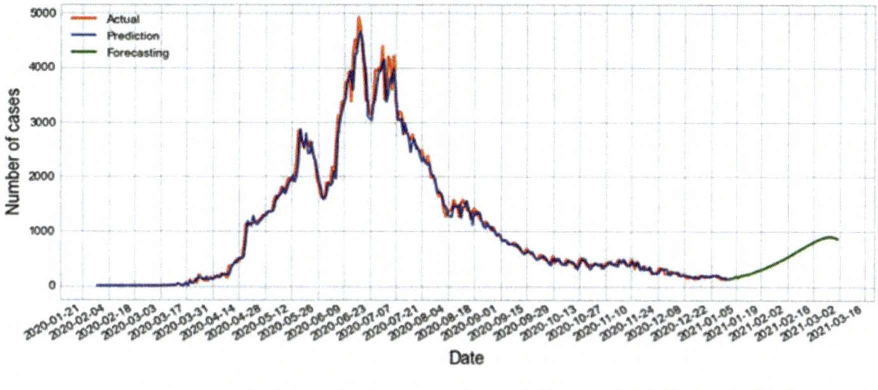

Fig. 1 COVID-19 confirmed cases in KSA

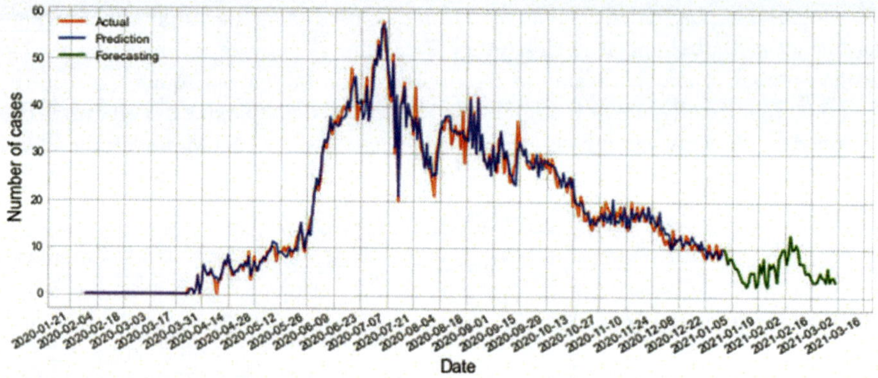

Fig. 2 COVID-19 death cases in KSA

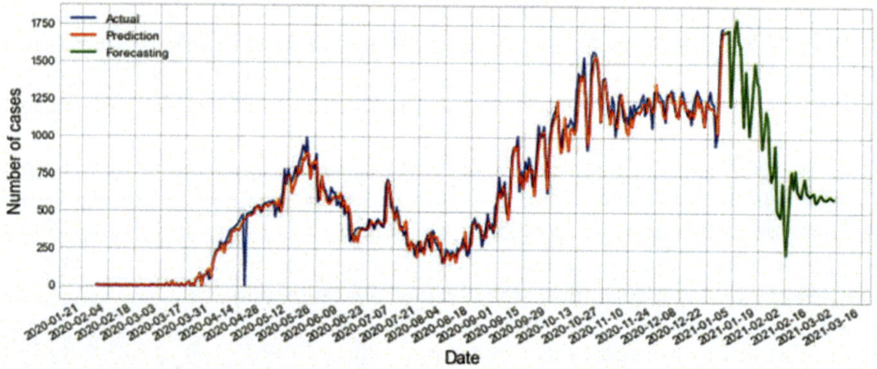

Fig. 3 COVID-19 confirmed cases in UAE

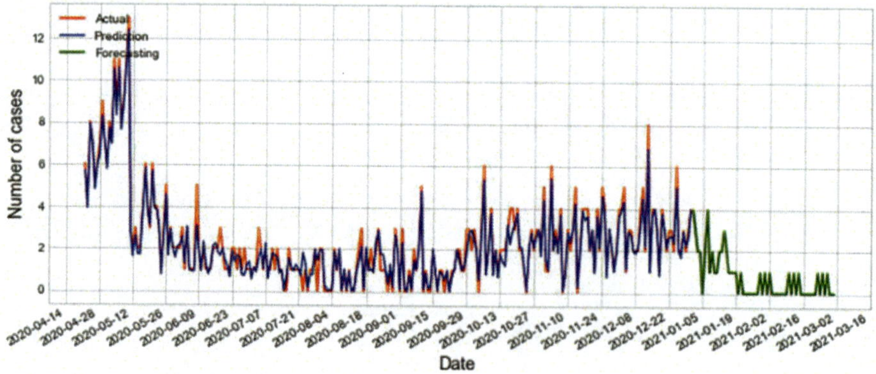

Fig. 4 COVID-19 death cases in UAE

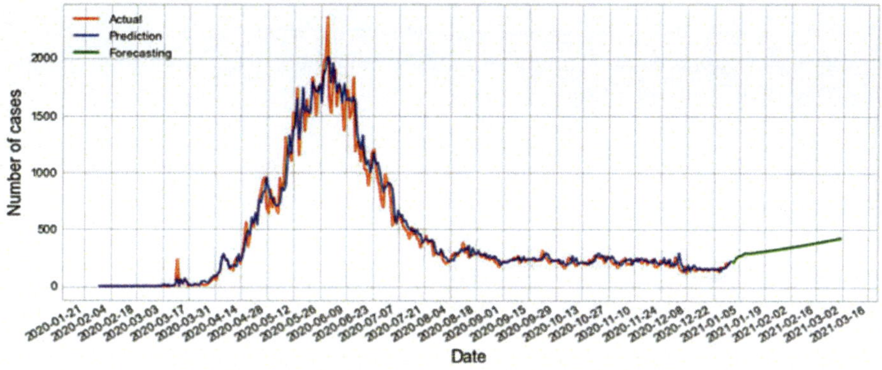

Fig. 5 COVID-19 confirmed cases in Qatar

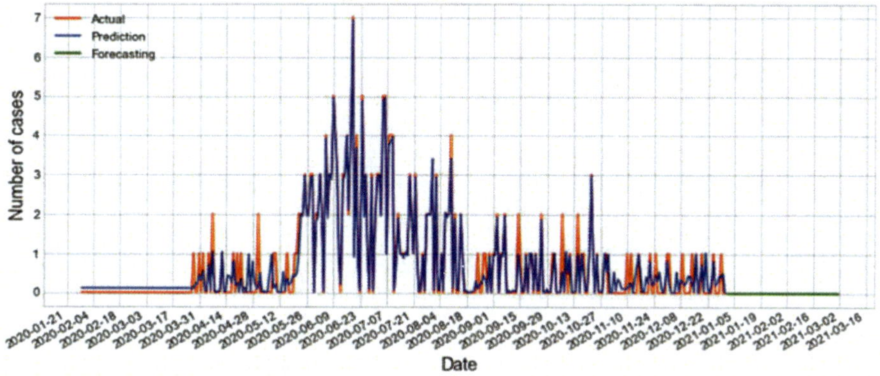

Fig. 6 COVID-19 death cases in Qatar

5 Conclusion and Future Work

The process of using machine learning for time-series prediction, especially in COVID-19, proved useful in modeling and forecasting the end status of the virus spreading. This paper focuses on applying artificial intelligence (AI) algorithms for prediction COVID-19 using the time-series data and studies the effect of the lifestyle and the awareness of citizens on the virus in the gulf area, especially in the Kingdom of Saudi Arabia (KSA), United Arab of Emirates (UAE), and Qatar. For this aim, we access time-series real-datasets collected from Johns Hopkins University Center for Systems Science and Engineering (JHU CSSE). In this paper, the obtained results from the LSTM models have confirmed that KSA and Qatar would take the most extended period to recover from the COVID-19 virus, and the situation will be controllable in March 2021 in UAE. Age and weather had a possible effect on the spread and mortality of COVID-19. Lack of social distancing has a very powerful impact on those factors. Further investigations are required. For future work, we will apply our model for all other gulf countries and compare the results according to different factors such as population, the number of tests performed, quality of life, strict rules and restrictions, and weather.

References

1. Johns Hopkins University Center for Systems Science and Engineering (JHU CSSE). https://raw.githubusercontent.com/CSSEGISandData/COVID-19/master/csse_covid_19_data/csse_covid_19_time_series/time_series_covid19_confirmed_global.csv. Accessed 10 Oct 2020
2. Punn NS, Sonbhadra SK, Agrawal S (2020) COVID-19 epidemic analysis using machine learning and deep learning algorithms
3. World Health Organization website (WHO). https://www.who.int/health-topics/coronavirus, 06.10.2020
4. Assaf D, Gutman Y, Neuman Y, Segal G, Amit S, Gefen S, Shilo N, Epstein A, Morcohen R, Biber A, Rahav G, Levy I, Tirosh A (2020) Utilization of machine-learning models to accurately predict the risk for critical COVID-19. Springer, Internal and Emergency Medicine
5. Alazab et al (2020) COVID-19 prediction and detection using deep learning. Int J Comput Inf Syst Ind Manage Appl
6. World Meters website (WM). https://www.worldometers.info/coronavirus, 06.10.2020
7. Singh H, Dhar J (2018) Mathematical population dynamics and epidemiology in temporal and spatio-temporal domains. CRC Press
8. Huang C et al (2020) Clinical features of patients infected with 2019 novel coronavirus in Wuhan, China. Lancet 395(10223):497–506
9. Narin A et al (2020) Automatic detection of coronavirus disease (COVID-19) using X-ray images and deep convolutional neural networks
10. Xiaowei X et al (2020) Deep learning system to screen coronavirus disease 2019 Pneumonia
11. Gozes O et al (2020) Rapid AI development cycle for the coronavirus (COVID-19) pandemic: initial results for automated detection and patient monitoring using deep learning CT image analysis
12. Wang S, Kang B, Ma J, Zeng X, Xiao M, Guo J et al (2020) A deep learning algorithm using CT images to screen for corona virus disease (COVID-19), medRxiv, 2020
13. Huang C, Chen Y-H, Ma Y, Kuo P-H (2020) Multiple-input deep convolutional neural network model for COVID-19 forecasting in China, medRxiv, 2020
14. Ranjan R (2020) Predictions for COVID-19 outbreak in India using epidemiological models, medRxiv, 3 Apr 2020
15. Hochreiter S, Schmidhuber J (1997) Long short-term memory. Neural Comput 9(8):1735–1780
16. Karpathy A, Johnson J, Fei-Fei L (2016) Visualizing and understanding recurrent networks. In: 4th international conference on learning representations (ICLR2016)
17. Quality of Life website (QOL). https://www.numbeo.com/quality-of-life, 06.10.2020

Diagnosis of Alzheimer's Disease: Methods and Challenges

Sarah A. Soliman, El-Sayed A. El-Dahshan, and Abdel-Badeeh M. Salem

Abstract To explore Alzheimer's disease, there were many classical machine learning techniques have been used, growing from image decomposition techniques such as principal component analysis through high-complexity algorithms. With the appearance of the deep learning methods, the extraction of high-level abstract features directly from MRI images has become promising to describe the distribution of data in low-dimensional space. Our approach includes training sparse autoencoder (SAE) approach to perform unsupervised feature leaning and prediction of AD. We spot the light on developing a SAE model to learn effective features from the AD dataset and then perform classification using the learned features.

Keywords Machine learning · Alzheimer's disease · Sparse autoencoder

1 Introduction

Dementia is a medical disorder categorized by progressive degeneration in two or more cognitive domains, including memory, language, executive and visuospatial function, personality, and behavior, which causes loss of abilities to perform instrumental and/or basic activities of daily living. By far, Alzheimer's disease (AD) is the most common cause of dementia for up to 80% of all dementia diagnoses [1, 2]. In the United States, even though the general death rate from stroke and cardiovascular disease is decreasing, the percentage of deaths related to AD is growing by 89% between 2000 and 2014 [3–6].

Medical experts are responsible for evaluating the interpretation of AD medical data, which is very difficult and restricted for a medical expert to interpret images

S. A. Soliman (✉)
Department of Computer Science, Higher Technological Institute, Cairo, Egypt

E.-S. A. El-Dahshan
Egyptian E-Learning University, Cairo, Egypt

A.-B. M. Salem
Ain Shams University, Cairo, Egypt
e-mail: absalem@cis.asu.edu.eg

D. A. Magdi et al. (eds.), *Digital Transformation Technology*, Lecture Notes in Networks and Systems 224, https://doi.org/10.1007/978-981-16-2275-5_34

due to its subjectivity and high image complexity, so the use of deep learning is seen as providing promising and reliable outputs for medical data in other areas of real-world use. With the rapid development of machine learning algorithms, the deep learning approach has been able to identify, extract, and classify high-level features which will also help in the accurate diagnosis of AD patients with less time [7]. As demonstrated by different strategies using 2D and 3D convolutional neural networks [8, 9], residual networks [10], and many other examples, the enormous revolution given by deep learning (DL) machine architectures has been essential to biomedical image processing [11, 12]. Through these architectures, sparse autoencoders (SAEs) [13, 14] have a very interesting purpose: They can model the manifold of a dataset in a self-monitored manner, which has been successfully applied to AD.

This work uses the training sparse autoencoder (SAE) approach to perform unsupervised feature leaning and prediction of AD. We spot the light on developing a SAE model to learn effective features from the AD dataset and then perform classification using the learned features. The model is optimized using the adaptive moment estimation (Adam) algorithm to achieve dynamic adjustment of different parameters, and a batch normalization technique is applied to avoid overfitting and to improve the performance, speed, and stability of the model.

This paper is organized as follows: Sect. 2 explains two types of deep learning: CNN and autoencoder. Section 3 presents the methodology of our work including the used database, image preprocessing, and the training sparse autoencoder. Section 4 discusses the challenges that face AD pathology, and finally, Sect. 5 shows the conclusion and the future work.

2 Deep Learning Methods

Nowadays, deep learning is solving many problems in every specific field, especially in medical image analysis [15, 16]. It is supervised machine learning method that uses the models of deep neural networks and neuron that is the basic computational unit in a neural network, and the concept of this has taken from human brain neurons. A typical deep neural network is consisting of three layers, each layer consists of nodes, the first layer is the input layer, the middle is hidden layer, and the last layer is called output layer which is used for learning and analyzing the data, the more hidden layer in neural network is the more accurate detection of patterns (Fig. 1).

There are many deep neural network algorithms used in the area of AD diagnosis which has been implemented by many researchers.

2.1 Convolutional Neural Network (CNN)

Convolutional neural networks (CNNs) are biologically inspired variants of multi-layer perceptrons (MLPs). They tend to recognize visual patterns, directly from raw

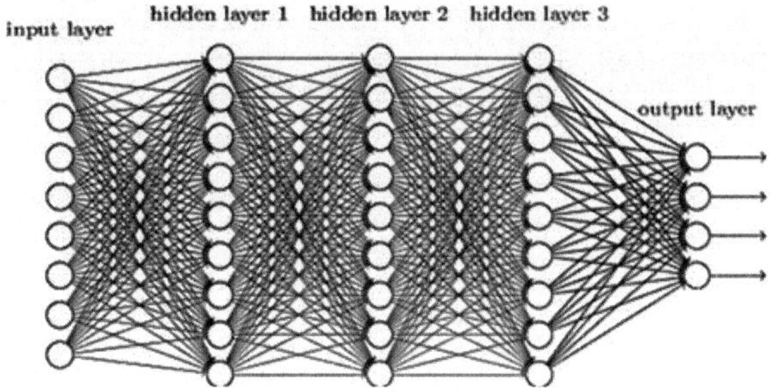

Fig. 1 Deep neural network [15]

Table 1 Convolutional neural network

Layer	Description
Convolution layer	Convolutional layers are considered the core building blocks of CNN architectures. Convolutional layers transform the input data by using a patch of locally connecting neurons from the previous layer
Pooling layer	Pooling layers provide an approach to down sampling feature maps by summarizing the presence of features on patches of the feature map
Flatten layer	Involves transforming the entire pooled feature map matrix into a single column which is then fed to the neural network for processing
Fully connected layer	Is mostly used at the end of the network for classification purpose. Unlike pooling and convolution, it is a global operation. It takes input from feature extraction stages and globally analyzes output of all the preceding layers

image pixels [17]. A typical 3D-CNN architecture generally comprises alternate layers of convolution and pooling followed by one or more fully connected layers at the end. In some cases, fully connected layer is replaced with global average pooling layer. Table 1 shows the layers of CNN [18].

2.2 *Autoencoder (AE)*

Autoencoder is one type of neural networks which is used to extract features from an input such as an image [14]. Autoencoder can grasp the features in an unsupervised way by minimizing reconstruction errors [19]. The training process in an autoencoder is based on the optimization of a cost function. The cost function computes the error between the input x and its rebuilding at the output $\hat{x}$. An autoencoder constitutes an encoder followed by decoder [13]. Similar to a general neural network, autoencoder has an input layer, a hidden layer, and an output layer as shown in Fig. 2, where the input and output layers have the same number of units.

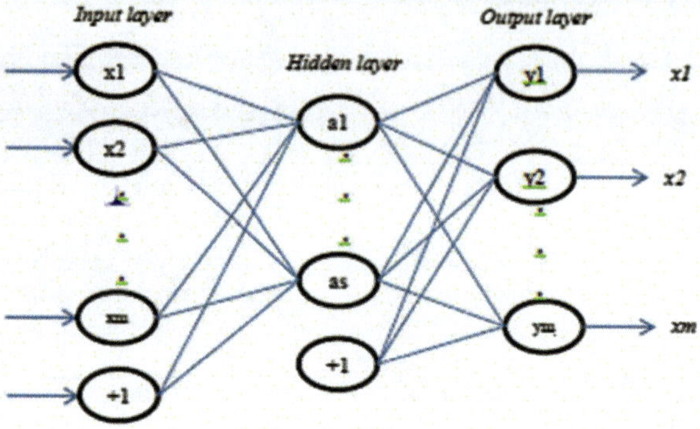

Fig. 2 Autoencoder structure [13]

Let p and n denote, respectively, the number of hidden and input units. Given an input vector $x \in \mathbb{R}n$, an autoencoder maps it to a hidden representation h through an encoder function f as

$$h = (W1x + b1) \tag{1}$$

with two parameters, i.e., weight matrix $W1 \in \mathbb{R}n \times p$ and the bias vector $b1 \in \mathbb{R}n$. The hidden representation h can be mapped back to $\hat{x}$, which is a reconstructed vector by the decoder function as

$$\hat{x} = (W^*h + b^*) \tag{2}$$

with $W* \in \mathbb{R}n \times p$ and $b* \in \mathbb{R}n$ be the weights and biases. Here, f is the sigmoid function, and g is the identity function. The cost function is mean squared error, because it is an appropriate choice for real-valued pixel intensity inputs. In details,

$$J(W, b) = \frac{1}{N}\sum_{i=1}^{N}\frac{1}{2}\left\|\hat{X}^{(i)} - x^{(i)}\right\|^2 \tag{3}$$

where N is the total number of inputs. The autoencoder is used to obtain a new representation of the input data through its hidden layer. We tried using an autoencoder with an overcompleted hidden layer, which makes autoencoder useful for feature extractors.

However, a potential problem with autoencoder is that if the error of cost function is minimized when reconstructing input, the hidden layers can just learn the identity function [20]. Therefore, the sparsity constraint was added [21], which enforces most hidden units to be close to zero when creating an autoencoder. So, the autoencoder can be get the best useful features and used them as input to our CNN model. Moreover, sparsity constraint is used because it encourages representations that may disentangle the hidden factors of MRI which control the variability of its images. After applying sparsity constraints to autoencoder, we have a new cost function as

$$J_2(W, b) = J(W, b) + \beta \sum_{j=1}^{h} \mathrm{KL}(S||\hat{S}_j) + \lambda \sum_{i,j} W_{i,j}^2 \tag{4}$$

where (W, b) is a cost function as in Eq. (3), and $\sum_{j=1}^{h} \mathrm{KL}(S||\hat{S}_j)$ are a penalty based on the concept of Kullback–Leibler divergence. $\mathrm{KL}(S||\hat{S}_j)$ quantifies the divergence between a Bernoulli distribution with mean s and one with mean $\hat{S}$, while s is a sparsity hyper-parameter which has value close to zero and $\hat{S}_j$ is the mean activation of hidden unit j, averaged over the training set. β, λ are hyper-parameters which control penalty term and weight decay term, respectively. The weight decay $\sum_{i,j} W_{i,j}^2$ is used to reduce overfitting.

3 Methodology

We implemented the deep learning framework described in this paper using Python 3 using TensorFlow and keras environment. The proposed methodology of the system is shown in Fig. 3.

3.1 Database

We used the structural brain MRI scans from the ADNI dataset [22]. The ADNI was launched in 2003 as a public–private partnership, led by Principal Investigator Michael W. Weiner, MD. The primary goal of ADNI has been to test whether serial MRI, positron emission tomography, other biological markers, and clinical and neuropsychological assessment can be combined to measure the progression of MCI and early AD. For up-to-date information, see WWW.ADNI-INFO.ORG. A total of 3888 subjects were considered in this study.

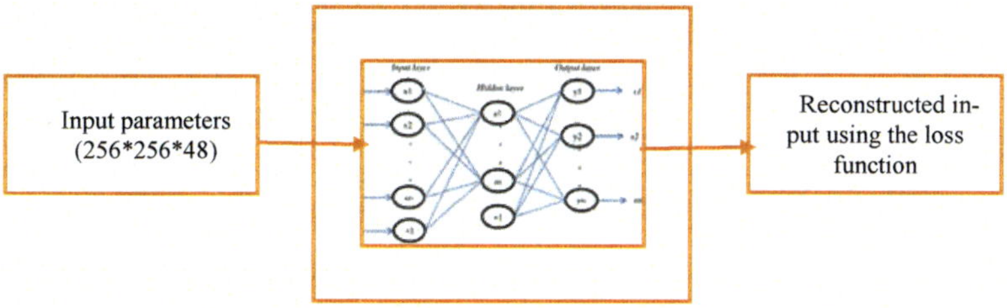

Fig. 3 Architecture of sparse autoencoder

3.2 Image Preprocessing

The images are imported from the backup folder using Collaboratory framework, the resizing and normalization processes has been applied to all of the data before putting them in any classes, and after that, the data was merged into 3D as the size was (256 × 256 × 48) length, width and number of scans for each patient to create a 3D image to feed to the model. After finishing this step, a process of collecting random patches of the data is started to get 1000 random patch of each picture of the patients we have and feed it later onto the model to increase the model reliability and keep the to control the model and prevent bias.

3.3 Training Autoencoder

The sparse autoencoder model is a Sequential model consist of three layers: the first layer is the dense1 layer of value 343, the second layer (dense 2) is with 410 with sigmoid activation function and the third layer (dense 3) with 343 with sigmoid activation function as show in Fig. 4. This autoencoder uses regularizers to hold the best weights for the current model to be used later on in another model as initial weights and to learn a sparse representation in the first layer. So, we can control the influence of these regularizers by setting various parameters: L2 Weight regularization controls the impact of an L2 regularizer for the weights of the network.

In our approach, we train an autoencoder on a set of selected 3D objects extracted from the MRI scans. We extract 1000 random patches from 3888 scans in the training

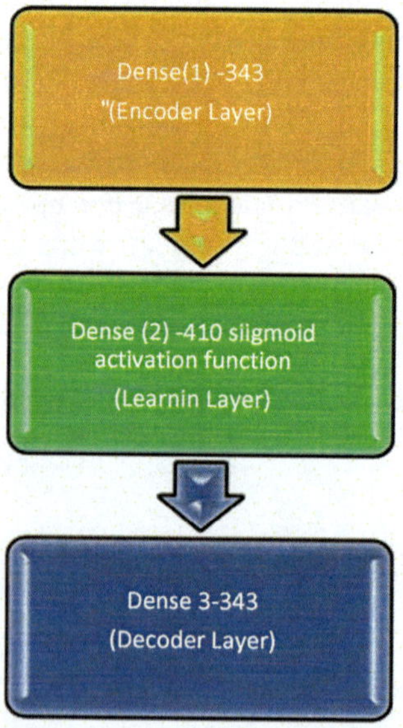

Fig. 4 Block diagram of sparse autoencoder

set. Adam optimizer was used to optimize the model output, and mean absolute error was used as loss function. The model succeeded in getting a loss of 23% which gave us an average of accuracy of 75% for this model.

4 Challenges

The diagnosis of dementia is relatively straightforward, but assessing AD at an early stage is still a clinical challenge. AD is considered a patho-clinical continuum in which, potentially 10 and 15 years before the first symptom of clinical dysfunction, neuropathological changes begin to accumulate in the brain. The last stage of the disease is in fact demetia due to AD. The ability to diagnose AD at an earlier stage will allow drug intervention that can address the pathological process specifically underlying the disease, thereby altering or even preventing the progression of the disease. As the patient may display no clinical symptoms of dementia, but subjective cognitive decline or apparently unrelated cognitive changes, personality or mood changes such as depressed condition, neuroticism or changes in social behavior, both of which may have other triggers, this poses an apparent challenge. There is therefore a need for biomarkers and new imaging methods related to pathological changes.

5 Conclusion and Future Work

Alzheimer's disease is the most common neurodegenerative disease in the world. The procedure of image biomarkers is significant in numerous automatic diagnostic systems that exist. The revolution of deep learning has provided us with innovative tools that can automatically extract image features without prior expectations about the underlying process. In this paper, we developed a sparse autoencoder (SAE) architecture which can achieve an automatic nonlinear decomposition of a very large dataset. Therefore, our SAE system can be used effectively to aid in the diagnosis of dementia, as well as to provide a new understanding of the relationships between structural damage and cognitive capacity. Our future work is to use the output of SAE model as input to new 3D-CNN model to diagnosis AD.

References

1. Crous-Bou M, Minguillón C, Gramunt N et al (2017) Alzheimer's disease prevention: from risk factors to early intervention. Alzheimers Res Ther 9(1):71
2. Association A (2017) 2017 Alzheimer's disease facts and figures. Alzheimers Dement 13(4):325–373

3. Takizawa C, Thompson PL, van Walsem A et al (2015) Epidemiological and economic burden of Alzheimer's disease: a systematic literature review of data across Europe and the United States of America. J Alzheimers Dis 43(4):1271–1284
4. Weller J, Budson A (2018) Current understanding of Alzheimer's disease diagnosis and treatment. F1000Res F1000 Faculty Rev 7:1161
5. Mossello E, Ballini E (2012) Management of patients with Alzheimer's disease: pharmacological treatment and quality of life. Ther Adv Chronic Dis 3(4):183–193
6. Pressman P, Rabinovici GD (2014) Alzheimer's disease, reference module in neuroscience and biobehavioral psychology encyclopedia of the neurological sciences, 2nd edn, pp 122–127
7. Sarraf S, DeSouza DD, Anderson J, Tofighi G (2016) DeepAD: Alzheimer's disease classification via deep convolutional neural networks using MRI and fMRI. bioRxiv, p 070441, 2017. https://doi.org/10.1101/070441
8. Pan D, Zeng A, Jia L, Huang Y, Frizzell T, Song X (2020) Early detection of Alzheimer's disease using magnetic resonance imaging: a novel approach combining convolutional neural networks and ensemble learning. Front Neurosci. https://doi.org/10.3389/fnins.2020.00259
9. Lin W, Tong T, Gao Q, Guo D, Du X, Yang Y, Guo G, Xiao M, Min D, Qu X (2018) Convolutional neural networks-based MRI image analysis for the Alzheimer's disease prediction from mild cognitive impairment. Front Neurosci.https://doi.org/10.3389/fnins.2018.00777
10. Valliani A, Soni A (2017) Deep residual nets for improved Alzheimer's diagnosis. In: The 8th ACM international conference. https://doi.org/10.1145/3107411.3108224
11. Hjelm RD, Calhoun VD, Salakhutdinov R, Allen EA, Adali T, Plis SM (2014) Restricted boltzmann machines for neuroimaging: an application in identifying intrinsic networks. Neuroimage 96:245–260
12. Vieira S, Pinaya WH, Mechelli A (2017) Using deep learning to investigate the neuroimaging correlates of psychiatric and neurological disorders: Methods and applications. Neurosci Biobehav Rev 74:58–75
13. Jha D, Kwon G-R (2017) Alzheimer"s disease detection using sparse autoencoder, scale conjugate gradient and Softmax output layer with fine tuning. Int J Mach Learn Comput 7(1)
14. Vu TD, Yang H-J, Nguyen VQ, Oh A-R, Kim M-S (2017) Multimodal learning using convolution neural network and sparse autoencoder. In: 2017 IEEE international conference on big data and smart computing (BigComp). IEEE, https://doi.org/10.1109/BIGCOMP.2017.7881683
15. https://www.kdnuggets.com/2020/02/deep-neural-networks.html. (10/6/2021)
16. Salehi AW, Baglat P, Gupta G (2020) Alzheimer's disease diagnosis using deep learning techniques. Int J Eng Adv Technol (IJEAT) 9(3). ISSN: 2249-8958
17. Jain R, Jain N, Aggarwal A, Hemanth DJ (2019) Convolutional neural network based Alzheimer's disease classification from magnetic resonance brain images. Cogn Syst Res 57:147–159
18. Soliman SA, El-Dahshan EA, Salem A-BM (2020) Predicting Alzheimer's disease with 3D convolutional neural networks. Int J Appl Fuzzy Sets Artif Intell 1:125–146. ISSN 2241-1240
19. Shin HC, Orton MR, Collins DJ, Doran SJ, Leach MO (2013) Stacked autoencoders for unsupervised feature learning and multiple organ detection in a pilot study using 4d patient data. IEEE Trans Pattern Anal Mach Intell 35:1930–1943
20. Jarrett K, Kavukcuoglu K, Ranzato M, LeCun Y (2009) What is the best multi-stage architecture for object recognition? In: 2009 IEEE 12th international conference on computer vision. IEEE, pp 2146–2153
21. Bengio Y (2012) Practical recommendations for gradient-based training of deep architectures. In: Neural networks: tricks of the trade. Springer, pp 437–478
22. http://adni.loni.usc.edu/. Retrieved Nov 2020

Sepsis Prediction Model in the Intensive Care Unit (ICU) Using Support Vector Machine (SVM)

Zeina Rayan, Marco Alfonse, and Abdel-Badeeh M. Salem

Abstract Sepsis is one of life-threatening diseases that is caused by unbalanced body response to some chemicals that the body releases into its blood stream when fighting an infection. Early prediction of sepsis can decrease mortality rates. Machine learning techniques can improve the accuracy of early sepsis prediction. This paper presents an accurate machine learning model that can predict sepsis early. This model depends on support vector machine (SVM) algorithm. This model is applied to the Medical Information Mart for Intensive Care (MIMIC-III). This model obtained an accuracy of 99%.

Keywords Sepsis prediction · Machine learning · Artificial intelligence · Intensive care unit · Medical Informatics

1 Introduction

Sepsis is one of the leading causes of mortality in hospitalized patients and is responsible for placing an enormous cost burden on the healthcare system [1]. Despite this fact, a reliable means of predicting sepsis onset remains elusive. Early and accurate sepsis onset predictions could allow more aggressive and targeted therapy while maintaining antimicrobial stewardship. Existing detection and prediction methodologies suffer from low performance and often require time-consuming laboratory test results [2].

Clinical decision support tools, using the clinical values of many tests in ICU, can help predict those at risk of future sepsis. Since there are many clinical values

Z. Rayan (✉) · M. Alfonse · A.-B. M. Salem
Computer Science Department, Faculty of Computer and Information Sciences, Ain Shams University, Cairo, Egypt
e-mail: zeinarayan@cis.asu.edu.eg

M. Alfonse
e-mail: marco_alfonse@cis.asu.edu.eg

A.-B. M. Salem
e-mail: absalem@cis.asu.edu.eg

D. A. Magdi et al. (eds.), *Digital Transformation Technology*, Lecture Notes in Networks and Systems 224, https://doi.org/10.1007/978-981-16-2275-5_35

collected from critical ill patients in the ICU, it made these data a good reference for a lot of research topics [2, 3]. This huge amount of data can be used by machine learning techniques to help in accurate prediction of sepsis [3].

This paper is constructed as follows: Sect. 2 presents some brief medical aspects of Sepsis, Sect. 3 shows a brief description of the dataset and methods used in this work, Sect. 4 includes results and discussion, and Sect. 5 contains the conclusions and future work.

2 Medical Aspects

2.1 *Intensive Care Unit*

The ICU is designed to care for patients who are seriously injured, have a critical or life-threatening illness, or have gone through a significant surgery, subsequently requiring 24-h care and monitoring [4]. ICUs have been evolved by the evolution of information and communication technology (ICT), where an ICU equipment which is found in Fig. 1, has many sensors and sensitive connections between the patient and these sensors which leads to better patient monitoring [5, 6].

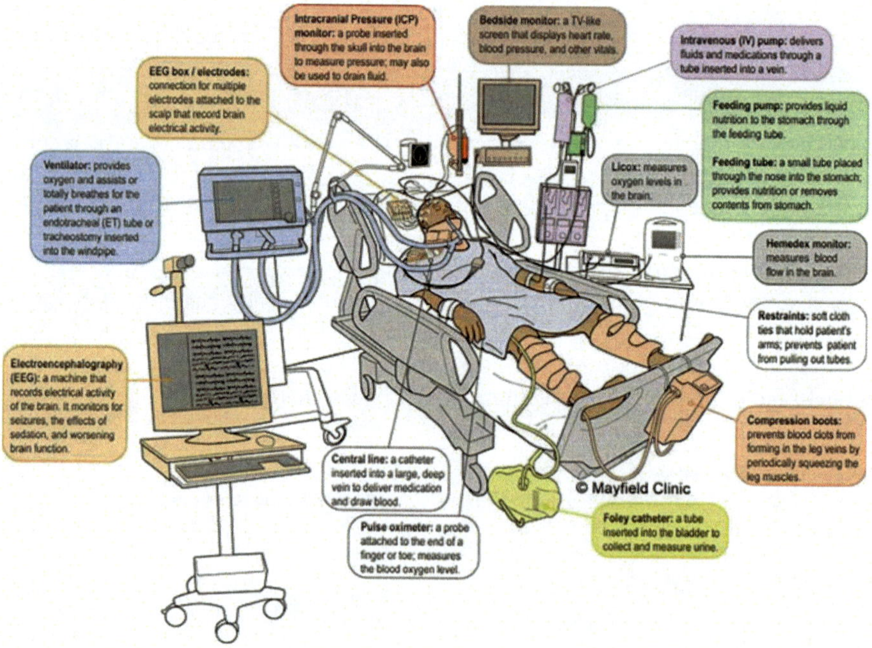

Fig. 1 Intensive care unit sensors and equipment [5]

The ICU equipment includes numerous sensors and devices responsible for various tasks; for example, patient monitoring, respiratory and cardiac support, emergency resuscitative, pain management, and other life support equipment for people who were seriously injured. The ICU equipment provides many readings such as electrical activity of heart, respiratory rate (breathing), blood pressure, body temperature, amount of oxygen and carbon dioxide in the blood, intracranial pressure (pressure of fluid in the brain) ... etc. [6]. The non-stop observing (monitoring) of ICU patients has produced an enormous amount of data, which lead to numerous opportunities and challenges [7].

2.2 *Sepsis*

Sepsis is a potentially life-threatening condition caused by the body's response to an infection [8]. The body normally releases chemicals into the bloodstream to fight an infection. Sepsis occurs when the body's response to these chemicals is out of balance, causing changes that can damage multiple organ systems. Early treatment of sepsis, usually with antibiotics and large amounts of intravenous fluids, increases the chances for survival [9].

Sepsis is caused by infection and can infect anyone. Sepsis is most common and most dangerous for older adults, pregnant women, children younger than one, people who have chronic conditions, such as diabetes, kidney or lung disease, or cancer, and people who have weakened immune systems. Sepsis stages progress from less to more severe. These stages are represented in Fig. 2.

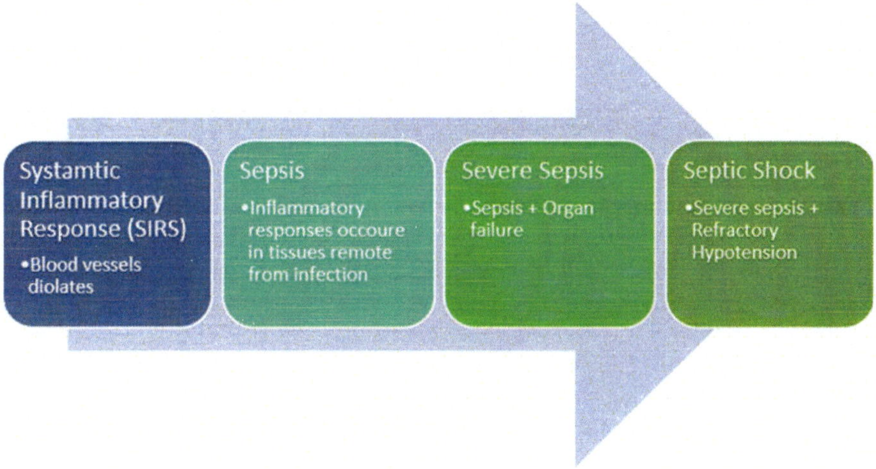

Fig. 2 Sepsis stages

As sepsis worsens, blood flow to vital organs, such as the brain, heart, and kidneys, becomes impaired. Sepsis can also cause blood clots to form in the organs and in the arms, legs, fingers, and toes, leading to severe sepsis which is different degrees of organ failure and tissue death [10]. Also, an episode of severe sepsis may place the patient at higher risk of future infections. It can also progress to septic shock. If sepsis progresses to septic shock, blood pressure drops dramatically which can lead to death [10].

3 Methods

3.1 Dataset

This work uses the Multiparameter Intelligent Monitoring in Intensive Care (MIMIC)-III version 1.4 dataset [11] which can be found on PhysioNet [12], which is a large, freely available database comprising health-related data associated with more than 40,000 patients who stayed in critical care units between 2001 and 2012. The dataset records contain vital signs, medications, laboratory measurements, observations and notes charted by care providers, fluid balance, procedure codes, diagnostic codes, imaging reports, hospital length of stay, survival data, and more [11]. The dataset is represented as relational database that can be accessed online (Through Querybuilder), or by building a special local version using PostgreSQL, in the form of twenty-six different tables [11].

3.2 Proposed Model for Sepsis Prediction

The goal of our study is to develop a machine learning model that can efficiently predict sepsis proposed model consists using minimal amount of data. The four steps are represented in Fig. 3.

The first step in this work is data extraction. A variety of data from the MIMIC-III dataset were collected to define sepsis onset. All data were extracted from the MIMIC-III dataset using custom PostgreSQL (PostgreSQL Global Development Group) queries.

The second step is selection of clinical measurements (feature selection). The learning method used in this work is flexible with regard to the patient data it uses. For the present work, we have worked on vital signs. Vital signs are utilized to measure body's fundamental (basic) functions [13]; so, we have selected systolic blood pressure, diastolic blood pressure, arterial blood pressure (mean), heart rate, respiration rate, temperature, and peripheral capillary oxygen saturation (SpO_2). All these features are considered main vital signs, usually available at the bedside, and do not rely on laboratory tests [13, 14]. A dataset sample is represented in Fig. 4.

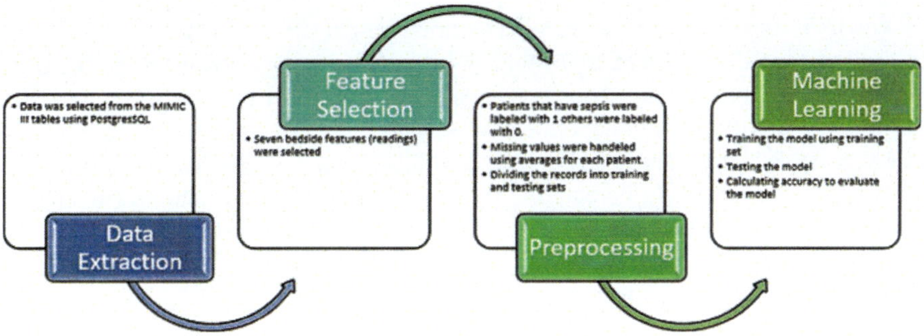

Fig. 3 Proposed model for sepsis prediction

subject_id	stc	heart_rate	arterial_bloodpressure	oxygen_saturation	bp_systolic	bp_diastolic	respiratory_rate	temperature	have_sepsis
57321	#	101	76	1	102	60	26	36.6	1
57321	#	101	76	1	105	60	26	36.6	1
57321	#	101	78	1	102	60	26	36.6	1
57321	#	101	78	1	105	60	26	36.6	1
57321	#	101	76	1	102	61	26	36.6	1
57321	#	101	76	1	105	61	26	36.6	1
57321	#	101	78	1	102	61	26	36.6	1
57321	#	101	78	1	105	61	26	36.6	1
57321	#	101	76	1	102	60	26	36.6	1
57321	#	101	76	1	105	60	26	36.6	1
57321	#	101	78	1	102	60	26	36.6	1
57321	#	101	78	1	105	60	26	36.6	1
57321	#	101	76	1	102	61	26	36.6	1
57321	#	101	76	1	105	61	26	36.6	1
57321	#	101	78	1	102	61	26	36.6	1
57321	#	101	78	1	105	61	26	36.6	1
57321	#	101	76	1	102	60	20	36.6	1
57321	#	101	76	1	105	60	20	36.6	1
57321	#	101	78	1	102	60	20	36.6	1
57321	#	101	78	1	105	60	20	36.6	1
57321	#	101	76	1	102	61	20	36.6	1
57321	#	101	76	1	105	61	20	36.6	1
57321	#	101	78	1	102	61	20	36.6	1
57321	#	101	78	1	105	61	20	36.6	1
57321	#	101	76	1	102	60	20	36.6	1
57321	#	101	76	1	105	60	20	36.6	1
57321	#	101	78	1	102	60	20	36.6	1
57321	#	101	78	1	105	60	20	36.6	1

Fig. 4 Dataset sample

The third step is pre-processing. The pre-processing operations include encoding the decision column, handling the missing values, and dividing the dataset for training and testing. For the decision column encoding, patients who have sepsis were labeled with 1, and other patients were labeled with 0. For handling the missing values, two methods were applied: The first method involves deleting the incomplete cases from the dataset, while the second method considers using averages for each patient [15]. For dividing the dataset, we divided the dataset records into training set to be used in training the machine learning model and testing set to be used to evaluate the model and calculate the accuracy.

The final step is applying a machine learning model (SVM) to the dataset in order to predict sepsis. The SVM algorithm is one of the supervised machine learning algorithms. It can be utilized for both regression and classification tasks. The aim of the SVM algorithm is to generate a hyperplane that can separate the data points. Many possible hyperplanes can be generated that could classify the data points. The objective of the SVM is to create the plane with the maximum margin (maximum distance between the records of the different classes in the dataset) [16].

4 Results and Discussion

The proposed model was evaluated using three metrics which are precision, recall, and $F1$-score. The precision refers to the ratio of correctly classified positive records to the total classified positive records. The recall refers to the ratio of correctly classified positive records to the all records in a class. The $F1$-score is also called the F-score or the F-measure. It conveys the weighted balance between the precision and the recall [17].

At first, a total of 332 records were selected excluding all the records with missing values. The support vector machine is applied with the "rbf" kernel, where 226 records were used for training the model and 67 records were used for testing. The testing dataset contains 26 patients who have sepsis and 41 that do not have sepsis. Table 1 presents the results of the SVM when applied to 332 records in terms of precision, recall, $F1$-score, and support.

The accuracy obtained is 99%. The total number of records that were correctly predicted to have sepsis is 26, the number of records that were correctly predicted to do not have sepsis is 40, and the number of records that were wrongfully predicted to do not have sepsis is 1.

Another selection was made where all records were selected whether there were missing values or not. The missing values were estimated using averages for each patient. A total of 2,762,225 records were selected. These records were divided into training set which contains 2,209,780 records and a testing set containing 552,445 record. The testing set contains 399,864 records of patients who do not have sepsis and 152,581 records of patients who have sepsis. Table 2 presents the results of the SVM when applied to 2,762,225 records in terms of precision, recall, $F1$-score, and support.

The accuracy estimated is 99%. The total number of records that were correctly predicted to have sepsis is 152569, the number of records that were wrongfully

Table 1 SVM results for the selected 332 records

	Precision	Recall	$F1$-score	Support
Do not have sepsis patients	1.00	0.98	0.99	41
Have sepsis patients	0.96	1.00	0.98	26

Table 2 SVM results for the 2,762,225 records

	Precision	Recall	F1-score	Support
Do not have sepsis patients	0.99	1.00	0.99	399,864
Have sepsis patients	0.99	1.00	0.99	152,581

predicted to do not have sepsis is 12, the number of records that were correctly predicted to do not have sepsis is 399589, and the number of records that were wrongfully predicted to do not have sepsis is 275.

From these results, we can conclude that the main vital signs can be used to predict sepsis, and that SVM with averages for missing values gives great results for sepsis prediction.

5 Conclusions and Future Work

Sepsis prediction is a challenging problem despite many years of research and development efforts because its manifestation is often unclear until later stages. In this study, a methodology is proposed for the sepsis predication. This methodology is composed of four stages: data extraction, feature selection, pre-processing, and machine learning. This methodology proves to be an effective predictor for the prediction of sepsis that uses simple and readily available patient data, by obtaining a high accuracy of 99%.

In future work, we are going to apply the proposed methodology to predict sepsis for different datasets that not only contains bedside features but also contains other features.

References

1. Angus DC, Linde-Zwirble WT, Lidicker J, Clermont G, Carcillo J, Pinsky MR (2001) Epidemiology of severe sepsis in the United States: analysis of incidence, outcome, and associated costs of care. Read Online: Crit Care Med|Soc Crit Care Med 29(7):1303–1310
2. Desautels T, Calvert J, Hoffman J, Jay M, Kerem Y, Shieh L, Shimabukuro D, Chettipally U, Feldman MD, Barton C, Wales DJ (2016) Prediction of sepsis in the intensive care unit with minimal electronic health record data: a machine learning approach. JMIR Med Inf 4(3):e28
3. Rayan Z, Alfonse M, Salem ABM (2019) Intensive care unit (ICU) data analytics using machine learning techniques. Int J Inf Theor Appl 26(1):69–82
4. Savino JS, Hanson III CW, Gardner TJ (2000) Cardiothoracic intensive care: operation and administration. Semin Thoracic Cardiovasc Surg 12(4):362–370
5. Mayfield Clinic. http://www.mayfieldclinic.com/PE-TBI.htm. Accessed on 7 Sept 2020
6. Egan M (2006) Clinical dashboards: impact on workflow, care quality, and patient safety. Crit Care Nurs Q 29(4):354–361
7. Johnson AE, Ghassemi MM, Nemati S, Niehaus KE, Clifton DA, Clifford GD (2016) Machine learning and decision support in critical care. Proc IEEE 104(2):444–466

8. Singer M, Deutschman CS, Seymour CW, Shankar-Hari M, Annane D, Bauer M, Bellomo R, Bernard GR, Chiche JD, Coopersmith CM, Hotchkiss RS (2016) The third international consensus definitions for sepsis and septic shock (Sepsis-3). JAMA 315(8):801–810
9. Gaieski DF, Mikkelsen ME, Band RA, Pines JM, Massone R, Furia FF, Shofer FS, Goyal M (2010) Impact of time to antibiotics on survival in patients with severe sepsis or septic shock in whom early goal-directed therapy was initiated in the emergency department. Crit Care Med 38(4):1045–1053
10. Schefold JC, Hasper D, Volk HD, Reinke P (2008) Sepsis: time has come to focus on the later stages. Med Hypotheses 71(2):203–208
11. Johnson AE, Pollard TJ, Shen L, Li-Wei HL, Feng M, Ghassemi M, Moody B, Szolovits P, Celi LA, Mark RG (2016) MIMIC-III, a freely accessible critical care database. Sci Data 3(1):1–9
12. Goldberger AL, Amaral LA, Glass L, Hausdorff JM, Ivanov PC, Mark RG, Mietus JE, Moody GB, Peng CK, Stanley HE (2000) PhysioBank, PhysioToolkit, and PhysioNet: components of a new research resource for complex physiologic signals. Circulation 101(23):e215–e220
13. The Cleveland Clinic. Vital Signs, http://my.clevelandclinic.org/health/diagnostics/hic_Vital_Signs [Accessed on 8/2020]
14. Holcomb JB, Salinas J, McManus JM, Miller CC, Cooke WH, Convertino VA (2005) Manual vital signs reliably predict need for life-saving interventions in trauma patients. J Trauma Acute Care Surg 59(4):821–829
15. Gupta A, Lam MS (1996) Estimating missing values using neural networks. Journal of the Operational Research Society 47(2):229–238
16. Deng N, Tian Y, Zhang C (2012) Support vector machines: optimization based theory, algorithms, and extensions. CRC Press
17. Sokolova M, Lapalme G (2009) A systematic analysis of performance measures for classification tasks. Inf Process Manage 45(4):427–437

A Hybrid Mutual Information-LASSO-Genetic Algorithm Selection Approach for Classifying Breast Cancer

Muhammed Abd-elnaby, Marco Alfonse, and Mohamed Roushdy

Abstract Breast cancer is the most common cancer among women due to many factors such as heredity and unhealthy lifestyles. Early and accurate diagnosis of this cancer improves the patient quality of life and increases the survival rate. Microarray technology provides effective way to early diagnosis cancer. However, the nature of its data complicates the classification process. A hybrid approach of mutual information (MI), least absolute shrinkage and selection operator (LASSO) and genetic algorithm (GA) is proposed to face this challenge. The proposed approach is assessed using logistic regression (LR), support vector machine (SVM), K-nearest neighbor (KNN), and random forest (RF). Compared to the state-of-the-art models, the proposed approach can effectively diagnose the breast cancer with few numbers of genes. For the benchmarked Van't veer dataset, it obtained a classification accuracy of 96% with only 23 features.

Keywords Cancer classification · Microarray data · Genetic algorithm · LASSO · Mutual information

1 Introduction

Breast cancer is the most commonly occurring cancer in women and the second most common cancer overall. Early detection and diagnosis have been proven to significantly improve patient survival rates [1], quality of life, as well as to significantly reduce the cost and complexity of cancer treatment. Microarray technology has become benchmark technique for early cancer diagnosis [2]. It monitors each gene many times under different conditions or alternately evaluates each gene in

M. Abd-elnaby (✉) · M. Alfonse
Faculty of Computer and Information Sciences, Ain Shams University, Cairo, Egypt

M. Alfonse
e-mail: marco_alfonse@cis.asu.edu.eg

M. Roushdy
Faculty of Computers and Information Technology, Future University in Egypt, New Cairo, Egypt
e-mail: mohamed.roushdy@fue.edu.eg

D. A. Magdi et al. (eds.), *Digital Transformation Technology*, Lecture Notes in Networks and Systems 224, https://doi.org/10.1007/978-981-16-2275-5_36

a single environment, but in different types of tissues, especially cancerous tissues as genes are the keystone for building protein which do cell works. In microarray experiment, mRNA value is a standard method for measuring gene expression since the structure of the protein is different from the structure of the gene, so analyzing thousands of proteins will be difficult. In addition, mRNA sequences hybridize to their complementary RNA or DNA sequences, while proteins do not possess this property [3]. Microarray data contains large number of genes with few samples, but almost of them are either redundant, noise or irrelevant [4].

However, machine learning techniques shown effective results in bioinformatics domain but when analyzing such data, the model need more samples and data to be relevant, so curse of dimensionality occurs. To build an accurate model, removing such noise and reduce the dimensions in microarray data is a mandatory step and can be achieved by the feature selection (FS) techniques. FS techniques offer biologist with genes that are related to diseases such as cancer [5]. FS techniques are categorized based on how the classifier is combined in the selection process into filter, wrapper, hybrid, and embedded approaches. Filter approach evaluates the features based on the intrinsic properties of the data. Wrapper methods use learning algorithms to select the optimal subset of features. Embedded methods learn which features best contribute to the accuracy of the model, while the model is being created. Finally, the hybrid approach can be any combination of any number of same or different methods of feature selection, to combine the advantages of both approaches and overcome or handle the drawback of each approach individually. The remainder of the paper is structured as follows: Sect. 2 presents the previous work. Section 3 outlines the used datasets and techniques. The experiment results are discussed in Sect. 4. Section 5 contains the conclusion and possible future work.

2 Related Work

Singh and Sivabalakrishnan [2] presented a hybrid selection technique that comprised mRMR with adaptive genetic algorithm (AGA). In the first phase, mRMR was effectively used to reduce the dimensions and the redundancy in the data. Subsequently produced subset was further processed via AGA to get the most relevant genes. The mRMR-AGA approach was evaluated by four classifiers on four benchmarked datasets and achieved comparable results.

Pirgazi et al. [4] proposed a hybrid selection technique for classifying cancer data. At first stage, relief method is used for weighting features then a wrapper approach; shufed frog-leaping algorithm (SFLA) and extension of incremental wrapper subset selection (IWSS) were used to select the informative features based on the weight of features. The produced method was evaluated on ten datasets using SVM classifier with tenfold cross-validation (CV). The proposed method achieved an accuracy of 93.34%.

Huynh et al. [6] proposed an ensemble of classifiers; Bag-RODS, and Boost-RODS for microarray classification. The proposed method was formed by bagging and boosting of random ensemble oblique decision stumps using linear SVM. The performance was evaluated using tenfold cross-validation. The result of Bag-RODS and Boost-RODS was compared with other classifiers including KNN, decision trees (C4.5), SVM, RF, Bagging C4.5, and Adaboost. The proposed methods achieved higher accuracy than the compared classifiers.

Purbolaksono et al. [7] proposed a filter approach to select informative genes. As a first step, the k-means algorithm was used to cluster the data and converted it into discrete. Then, mutual information was used to remove redundant genes and select the informative ones. The proposed method was applied on different datasets and evaluated by Bayesian network and Naïve Bayes (NB). Bayesian network achieved high results, but the model needs to be more accurate.

Lu et al. [8] proposed a hybrid approach (MIMGA) that combined mutual information maximization (MIM) and the AGA. Initially, MIM was used to remove redundancy and maintain only genes with strong dependency to genes of their class. AGA adjusts the probability of crossover and mutation to select optimal subset of genes. The MIMGA approach was evaluated using extreme learning machine (ELM). It had comparable results but took long time.

AlQuraishi et al. [9] proposed an ensemble classifier to accurately estimate the probability of breast cancer relapse. Initially, several FS approaches were utilized to determine the relevant gene subset for each approach. The ensemble classifier that combined deep neural network, SVM, and RF was used to evaluate each subset. However, the ensemble classifier achieved high accuracy with correlation-based feature selection with forward search (CFS-FWD) technique, but the gene subset is still not small enough.

Medjaheda et al. [10] proposed an approach to diagnosis cancer. In the first phase, support vector machines based on recursive feature elimination (SVM-RFE) was used to eliminate 40% of features. The remaining subset was processed via binary dragonfly (BDF) to retain informative genes only. The proposed method was evaluated on six microarray cancer datasets. However, the model achieved comparable result, but for breast, it was not satisfying as it achieved high accuracy but with very huge number of features.

Pashaei and Aydin [11] proposed a wrapper selection approach: binary black hole algorithm (BBHA). Six decision classifiers, RF, Bagging, C5.0, C4.5, Boosted C5.0, and CART were used to evaluate wrapper subsets. RF had the best performance and used to evaluate the proposed method on five microarray datasets. The proposed BBHA approach outperformed the performance of GA, binary particle swarm optimization, and correlation-based feature selection.

Angulo [12] proposed probabilistic attribute-value for class distinction (PAVICD) based on features value. The PAVICD was employed on fourteen datasets and effectively detect relevant features. However, the genetic bee colony (GBC) wrapper approach had better results than PAVICD, but PAVID was the faster.

Hameed et al. [13] proposed a hybrid approach to choose the informative genes. Firstly, Pearson correlation coefficient (PCC) ran ten times to select the top 100 ranked features. Then, either BPSO or GA was used for further reduction. Different classifiers were employed to test the accuracy of eleven datasets based on tenfold

CV. The result showed that SVM had higher accuracy and BPSO performing faster and have higher performance than GA with a smaller number of selected genes.

Sangaiah and Vincent [14] presented a hybrid approach to select relevant features of breast cancer dataset. A filter approach ReliefF was used initially to remove the redundant and least irrelevant feature. Then, the resultant subset was preprocessed via an entropy-based genetic algorithm (EGA) to find the optimal subset of features. The proposed approach was evaluated using KNN, NB, SVM classifiers and obtained the highest accuracy with SVM.

Hamim et al. [15] proposed a model to effectively classify breast cancer. Initially, a hybrid selection approach (FC5) of Fisher score (F) and C5.0 was used to select informative genes. The Fisher score removes redundant features and returns top 10% of features, and then, they were processed by C5.0. The model was evaluated using different classifiers; SVM, LR, artificial neural network, and C.5. The proposed model obtained high accuracy using five genes only.

Adem [16] used stacked auto-encoder (SAE) for feature reduction to remove unnecessary data from breast cancer data. The model was assessed using Softmax, SVM, and the subspace-KNN classifiers. The SAE approach compared with principal component analysis achieved better result with SVM.

Loey el al. [17] presented intelligent decision support system for diagnosing microarray cancer data. Initially, Information Gain (IG) was used to select relative genes. Then, the selected subset was reduced via gray wolf optimization (GWO). Finally, SVM was utilized for breast and colon cancer classification. However, the IG-GWO approach achieved high accuracy but with huge number of features (about 240) for breast cancer dataset.

Baliarsingh et al. [18] proposed Jaya optimized extreme learning machine (JELM) For breast cancer classification. Jaya is used to select the optimal input weights and hidden biases for ELM. The authors used Wilcoxon rank sum test to select relevant genes. The performance of JELM was compared by the performance of SVM, KNN, NB, and c4.5 and achieved higher result about 90.91%. Although the proposed model achieved high accuracy, but it selected huge subset about 505 genes, so it needs further reduction in genes subset.

Jain et al. [19] proposed a hybrid feature selection method that combined CFS and IBPSO. The IBPSO enhanced the early convergence to local optimum of BPSO. The proposed method was utilized on 11 microarray dataset and evaluated by NB with stratified 10 k CV. The model was compared with seven classifiers and outperformed them in terms of accuracy and number of selected genes in most cases.

3 The Proposed Methodology

A hybrid approach (MI-LASSO-GA) of three phases was proposed to effectively reduce the feature set while retain the accuracy high. The proposed approach is presented in Fig. 1. A filter approach, MI, was utilized to remove redundant and irrelevant features. The resultant subset was reduced by half by LASSO, and then,

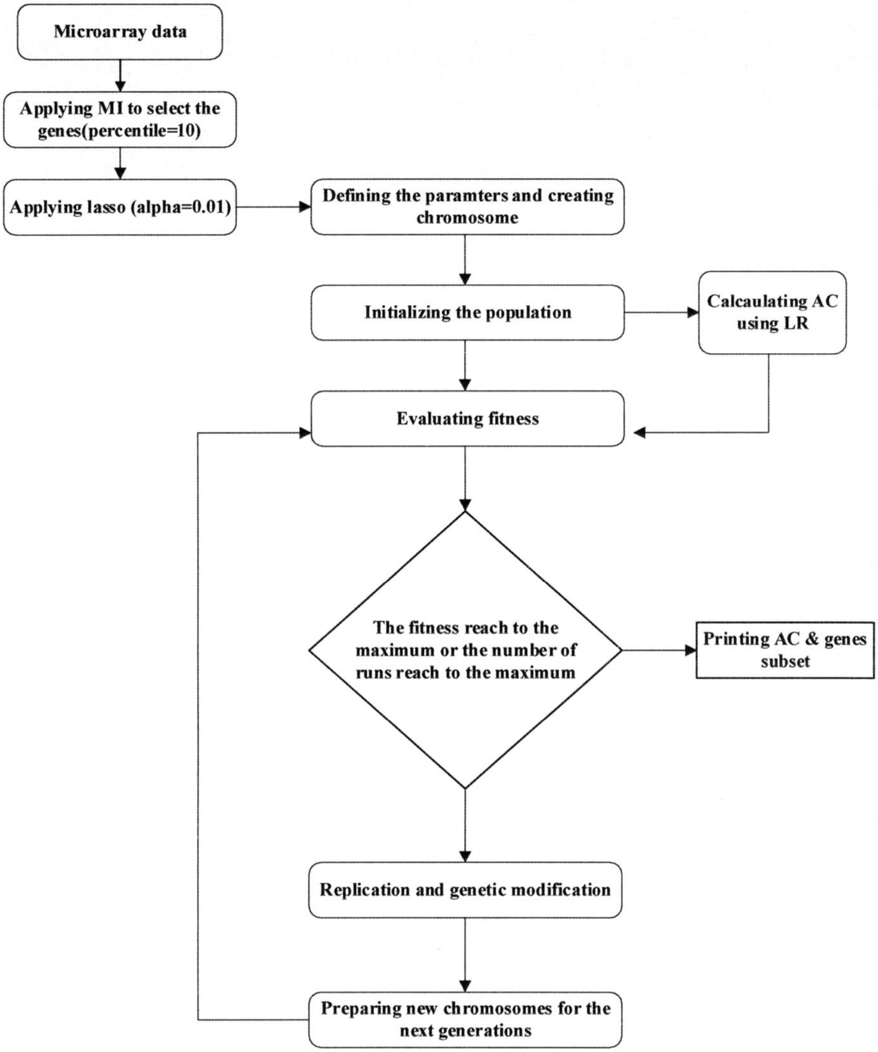

Fig. 1 Proposed model flowchart

they passed to GA to obtain the most informative genes. LR was used to evaluate each candidate subset of GA.

3.1 Normalization

Data need to be normalized to put all numeric values to common scale, so that the result does not dominated by higher value of the feature [20]. Min-max normalization makes all features in range from 0 to 1 as for each feature it will be subtracted from the min value in the dataset and then the result will be divided by the maximum value minus the minimum, as shown in Eq. 1 [21].

$$Z = \frac{X(i) - \min(X)}{\max(X) - \min(X)} \tag{1}$$

where X represents the value of the feature.

3.2 Feature Selection

The process of removing redundant data and selecting only relative features is crucial in microarray data analysis. A hybrid approach of three stage was proposed.

- Mutual information is a univariate filter technique that determines the relevant features by measuring the correlation between gene and classification category, the larger the value of MI, the more informative the genes are [22]. The genes with higher mutual information provide better information for accurate classification.
- Least absolute shrinkage and selection operator is used for regularization and feature selection. It puts a constraint on the sum of the absolute values of the model parameters, and the sum must be less than a fixed value (upper bound). To do so, the method apply a shrinking (regularization) process where it penalizes the coefficients of the regression variables shrinking some of them to zero. During features selection process, the variables that still have a nonzero coefficient after the shrinking process are selected to be part of the model. The goal of this process is to minimize the prediction error.
- Genetic algorithm is a heuristic search algorithm that iteratively produces solutions until optimal, near optimal ones are reached, or a population has converged. GA begins with a population which consist of feasible solutions. Each solution has been evaluated with fitness function, solutions with high score have greet chance to be used for better solutions. Firstly, each pair of the top solutions are selected and then are mated through crossover operation to pass their genes to the new solution. Secondly, mutation process is done in many forms for diversity maintenance. GA is more robust due to its adaptability and therefore enhances the likelihood of finding the global optimal solution [23].

3.3 Evaluation

The proposed methodology is evaluated using several classifiers: LR, SVM, KNN, and RF). The accuracy is estimated using fivefold cross-validation.

- Logistic regression is a classification algorithm used to assign observations to a discrete set of classes. It transforms its output using the logistic sigmoid function to return a probability value. This algorithm is quite simple but efficient.
- K-nearest neighbors is a supervised learning algorithm and widely used in classification. It is kind of instance-based learning method that builds no model. An object is classified to the major class among its nearest neighbors. Common distance metrics are Hamming distance, Manhattan distance, Makowski distance [24].
- Random forest is a supervised learning algorithm that randomly creates and uses ensemble approach to merge multiple tress into a forest. The random sampling method and ensemble strategies utilized in RF enable it to achieve accurate predictions as well as better generalizations [25]. RF is used in biological and microarray data as it is simple, flexible, and can handle high number of features in comparison with few samples, the heterogeneity in the genetic and noisy data [26].
- Support vector machine is commonly used for classification but also can be used for regression tasks. Firstly, data items are plotted in n-dimensional space. Then, SVM finds the hyper-plane that best differentiate between classes. SVM has been used widely in many bioinformatics' applications specifically microarray classification as it works well with high-dimensional data [27] and when number of features exceed the number of samples [28].

4 Experimental Results

The proposed MI-LASSO-GA is employed on five breast gene expression datasets, and it is evaluated by LA, RF, SVM, and KNN. Breast datasets description are summarized in Table 1.

Table 1 Breast datasets description

Dataset	No. of samples	No. of features	No. of classes
Van't veer [29]	97	24,481	2
Chin [30]	118	22,215	2
Chowdary [31]	104	22,283	2
Gravier [32]	168	2905	2
West [33]	49	7129	2

The proposed method is evaluated against MI-GA on several breast cancer datasets. Each method is run ten times, and the evaluation is based on the average of the accuracy, precision, recall, and the number of selected features using LR, SVM, RF, and KNN classifiers with fivefold CV as given in Table 2.

In the proposed method, MI removes redundant and selected only top ten percentile of features. Then, LASSO effectively reduced features subset to 37, 31, 20, 27, 19 for Van't veer [29], Chin [30], Chowdary [31], Gravier [32], and West [33] datasets, respectively, so they reduced feature subspace for GA. Through five generations, GA selects least number of the informative genes. Several classifiers are used to evaluate the GA candidate subsets, and the best result is obtained by LR. Finally, MI-LASSO-GA selects only 23, 18, 17, 13, 12 in genes contrast to 366, 337, 388, 466, 108 for MI-GA approach for the same datasets as shown in Fig. 2.

The ability of the proposed approach to select a smaller number of features enhances the performance of the classifiers. The performance of several classifier on Van't veer, Chin, Chowdary, Gravier, and West for both selection algorithms is shown in Figs. 3, 4, 5, 6, and 7, respectively. For all datasets, the MI-LASSO-GA outperforms MI-GA. LR obtains the highest accuracy ranging from 95% for Chin dataset to 100% for West dataset, while SVM obtained highest accuracy of 89% in [22] for Gravier. In comparison with state-of-the-art techniques as shown in Fig. 8, BBHA-RF, PAViCD, and FC5 selection techniques in [11, 12, 15] select a smaller number of features for Van't veer than the proposed model, but the MI-LASSO-GA achieves higher accuracy. In comparison with selection techniques in [2, 8, 13] and [23] that use meta-heuristic approaches, the proposed approach obtained higher accuracy with a smaller number of features for Van't veer due to using hybrid approach before meta-heuristic selection technique.

5 Conclusion

Classifying breast cancer at early stage is critical to avoid patient suffering. This goal can be achieved by microarray technology. However, the huge number of its' features hinder the classification process. Hence, a hybrid approach (MI-LASSO-GA) was proposed to effectively select the informative genes. MI and LASSO were used to remove noise and reduce the search space for GA to select the most informative genes. The proposed method was employed on five breast cancer datasets and evaluated by the following classifiers LR, RF, SVM, and KNN. It achieved high accuracy for all dataset ranging from 89% for Gravier dataset to 96% for the Van't veer dataset. In the future work, the proposed model will be evaluated on other cancer types.

Table 2 Result of MI-LASSO-GA VS MI-GA

Dataset	Classifiers	Measure	LR	SVM	RF	KNN
Van't veer [29]	MI-GA	Accuracy	0.813183	0.80015	0.786817	0.763459
		Precision	0.802469	0.811268	0.782413	0.782413
		Recall	0.801333	0.768	0.771556	0.771556
		No. features	366			
	MI-LASSO-GA	Accuracy	0.960401	0.860201	0.950827	0.878045
		Precision	0.989278	0.872784	0.978	0.978
		Recall	0.927556	0.833111	0.918	0.918
		No. features	23			
Chin [30]	MI-GA	Accuracy	0.906884	0.891848	0.876341	0.886739
		Precision	0.896471	0.887201	0.882867	0.882867
		Recall	0.973333	0.958667	0.941333	0.941333
		No. features	337			
	MI-LASSO-GA	Accuracy	0.955036	0.917862	0.93808	0.937971
		precision	0.941789	0.913153	0.9299	0.9299
		Recall	0.994667	0.968	0.982667	0.982667
		No. features	18			
Chowdary [31]	MI-GA	Accuracy	0.980909	0.961818	0.979091	0.963091
		Precision	0.98	0.977778	0.976727	0.976727
		Recall	0.975	0.927778	0.975	0.975
		No. features	338			
	MI-LASSO-GA	Accuracy	0.992727	0.963455	0.977182	0.981364
		Precision	0.984	0.979778	0.961172	0.961172
		Recall	1	0.932222	0.987778	0.987778
		No. features	12			
Gravier [32]	MI-GA	Accuracy	0.812395	0.775542	0.788424	0.716067
		Precision	0.845766	0.75279	0.742667	0.742667
		Recall	0.561364	0.541364	0.601667	0.601667
		No. features	46			
	MI-LASSO-GA	Accuracy	0.867304	0.805974	0.890989	0.823277
		Precision	0.955808	0.773882	0.952379	0.952379
		Recall	0.642576	0.624545	0.719394	0.719394
		No. features	13			
West [33]	MI-GA	Accuracy	0.940889	0.884667	0.914889	0.715778
		Precision	0.947905	0.918	0.9515	0.9515
		Recall	0.934	0.851	0.881	0.881
		No. features	108			

(continued)

Table 2 (continued)

Dataset	Classifiers	Measure	LR	SVM	RF	KNN
	MI-LASSO-GA	Accuracy	1	0.947556	0.991111	0.926
		Precision	1	0.962333	0.995	0.995
		Recall	1	0.931	0.985	0.985
		No. features	17			

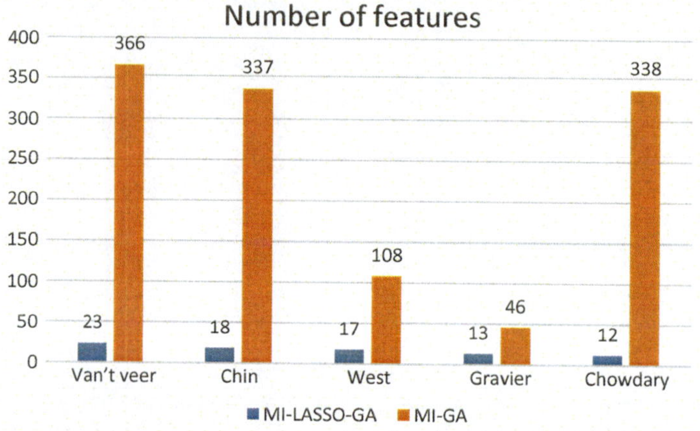

Fig. 2 Number of selected features of MI-LASSO-GA VS MI-GA

Fig. 3 Performance of MI-GA and MI-LASSO-GA on Van't veer

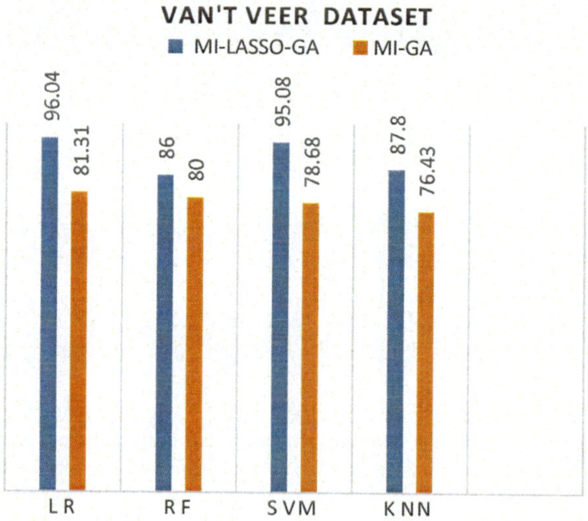

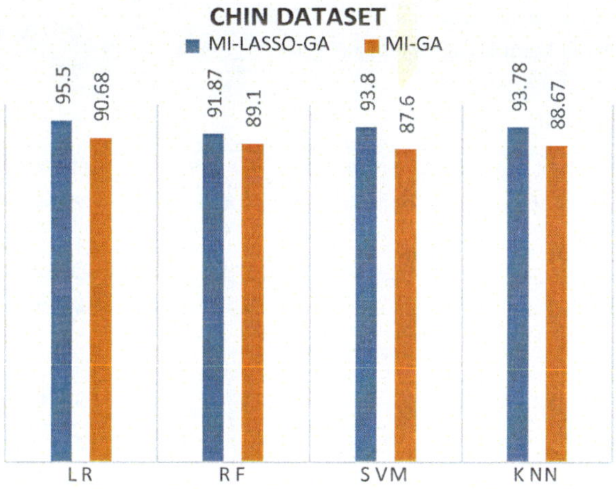

Fig. 4 Performance of MI-GA and MI-LASSO-GA on Chin

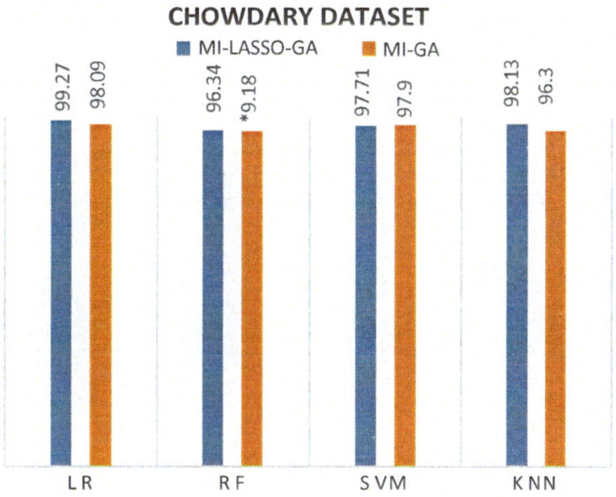

Fig. 5 Performance of MI-GA and MI-LASSO-GA on Chowdary

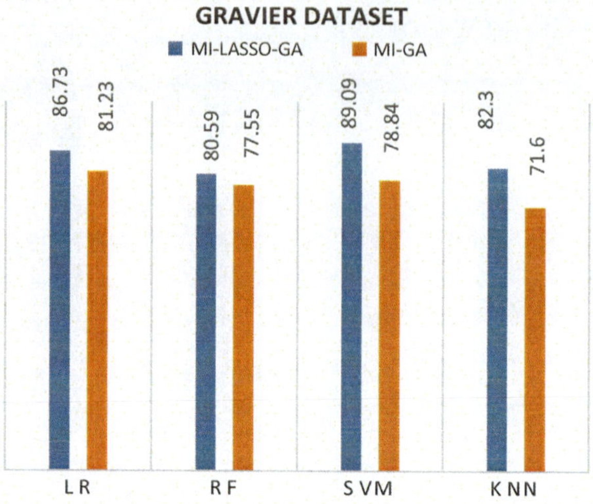

Fig. 6 Performance of MI-GA and MI-LASSO-GA on Gravier

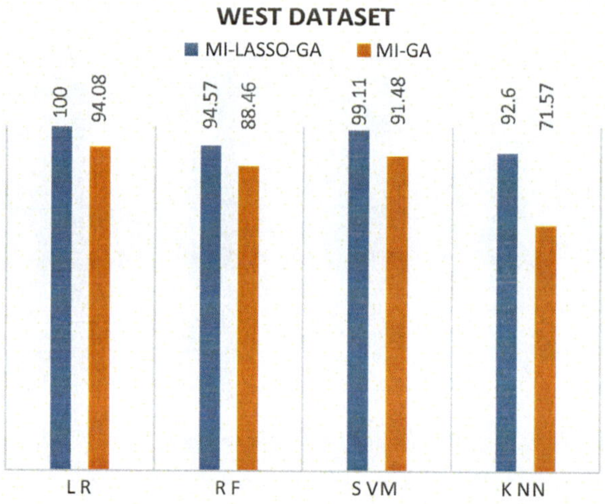

Fig. 7 Performance of MI-GA and MI-LASSO-GA on West

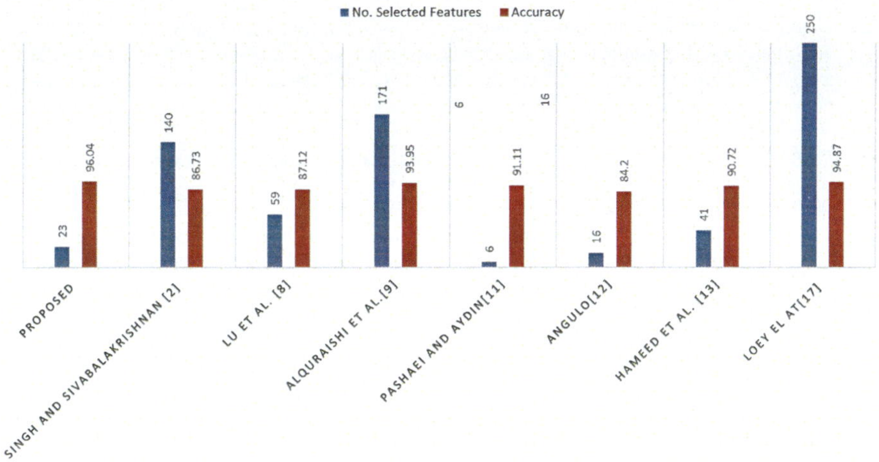

Fig. 8 Performance of the proposed approach versus other approaches

References

1. Sheikhpour R, Sarram MA, Sheikhpour R (2016) Particle swarm optimization for bandwidth determination and feature selection of kernel density estimation based classifiers in diagnosis of breast cancer. Appl Soft Comput 40:113–131
2. Singh RK, Sivabalakrishnan M (2019) Microarray gene expression data classification using a hybrid algorithm: MRMRAGA. Int J Innov Technol Explor Eng (IJITEE), Aug 2019, p 8
3. Qaraad M, Amjad S, Fathi H, Manhrawy IIM (2019) Feature selection techniques for cancer classification applied to microarray data: a survey. In: 2019 international conference on intelligent systems and advanced computing sciences (ISACS) 2019, p 1–8
4. Pirgazi J, Alimoradi M, Esmaeili Abharian T, Olyaee MH (2019) An efficient hybrid filter-wrapper metaheuristic-based gene selection method for high dimensional datasets. Sci Rep 9:18580
5. Bolón-Canedo V, Sánchez-Maroño N, Alonso-Betanzos A, Benítez JM, Herrera F (2014) A review of microarray datasets and applied feature selection methods. Inf Sci 282:111–35
6. Huynh P-H, Nguyen VH, Do T-N (2018) Random ensemble oblique decision stumps for classifying gene expression data. In: Proceedings of the ninth international symposium on information and communication technology. Association for Computing Machinery, Danang City, pp 137–144
7. Purbolaksono MD, Widiastuti KC, Mubarok MS, Ma'ruf FA (2011) Implementation of mutual information and bayes theorem for classification microarray data. J Phys Conf Ser 012011
8. Lu H, Chen J, Yan K, Jin Q, Xue Y, Gao Z (2017) A hybrid feature selection algorithm for gene expression data classification. Neurocomputing
9. Al-Quraishi T, Abawajy J, Chowdhury M, Sutharshan R, Abdalrada A (2017) Breast cancer risk assessment prediction using an ensemble classifier. In: 30th international conference on computer applications in industry and engineering (CAINE 2017), San Diego, California, USA, 2017: Int J Comput Their Appl (ISCA)
10. Medjahed SA, Saadi TA, Benyettou A, Ouali M (2017) Kernel-based learning and feature selection analysis for cancer diagnosis. Appl Soft Comput 51:39–48
11. Pashaei E, Aydin N (2017) Binary black hole algorithm for feature selection and classification on biological data. Appl Soft Comput 56:94–106

12. Angulo A (2018) Gene selection for microarray cancer data classification by a novel rule-based algorithm. Information 9:6
13. Hameed SS, Muhammad FF, Hassan R, Saeed F (2018) Gene selection and classification in microarray datasets using a hybrid approach of PCC-BPSO/GA with multi classifiers. JCS 14:868–880
14. Sangaiah I, Vincent Antony Kumar A (2019) Improving medical diagnosis performance using hybrid feature selection via relieff and entropy based genetic search (RF-EGA) approach: application to breast cancer prediction. Cluster Comput 22
15. Hamim M, El Moudden I, Moutachaouik H, Hain M (2020) Decision tree model based gene selection and classification for breast cancer risk prediction. Springer International Publishing, Cham, pp 165–177
16. Adem K (2020) Diagnosis of breast cancer with Stacked autoencoder and Subspace kNN. Physica A Stat Mech Appl 551:124591
17. Loey M, Jasim MW, EL-Bakry HM, Taha MHN, Khalifa NEM (2020) Breast and colon cancer classification from gene expression profiles using data mining techniques. Symmetry 12:408
18. Baliarsingh SK, Dora C, Vipsita S. Jaya optimized extreme learning machine for breast cancer data classification. Springer Singapore, Singapore, pp 459–67
19. Jain I, Jain VK, Jain R (2018) Correlation feature selection based improved-Binary Particle Swarm Optimization for gene selection and cancer classification. Appl Soft Comput 62:203–215
20. Moosa J, Shakur R, Kaykobad M, Rahman M (2016) Gene selection for cancer classification with the help of bees. BMC Med Genom 9
21. Palma-Mendoza R-J, Rodriguez D, De-Marcos L (2018) Distributed ReliefF-based feature selection in Spark. Knowl Inf Syst 57:1–20
22. Vergara JR, Estévez PA (2014) A review of feature selection methods based on mutual information. Neural Comput Appl 24:175–186
23. Singh P, Shukla A, Vardhan M (2017) Hybrid approach for gene selection and classification using filter and genetic algorithm. In: 2017 International conference on inventive computing and informatics (ICICI): IEEE, pp 832–837
24. Abdulrahman S, Khalifa W, Roushdy M, Salem A-BM (2020) Comparative study for 8 computational intelligence algorithms for human identification. Comput Sci Rev 36:100237
25. Boulesteix AL, Janitza S, Kruppa J, König I (2012) Overview of random forest methodology and practical guidance with emphasis on computational biology and bioinformatics. Wiley Interdiscipl Rev Data Mining Knowl Discov 2
26. Ram M, Najafi A, Shakeri MT (2017) Classification and biomarker genes selection for cancer gene expression data using random forest. Iran J Pathol 12(4):339–347
27. Han J, Kamber M, Pei J (2012) Data mining : concepts and techniques. Elsevier/Morgan Kaufmann, Amsterdam; Boston
28. Almugren N, Alshamlan H (2019) FF-SVM: new FireFly-based gene selection algorithm for microarray cancer classification. In: 2019 IEEE conference on computational intelligence in bioinformatics and computational biology (CIBCB), pp 1–6
29. Van't veer LJ, Dai H, Van De Vijver MJ, He YD, Hart AA, Mao M et al (2002) Gene expression profiling predicts clinical outcome of breast cancer. Nature 415:530-536
30. Chin K, DeVries S, Fridlyand J, Spellman PT, Roydasgupta R, Kuo W-L et al (2006) Genomic and transcriptional aberrations linked to breast cancer pathophysiologies. Cancer Cell 10:529–541
31. Chowdary D, Lathrop J, Skelton J, Curtin K, Briggs T, Zhang Y et al (2006) Prognostic gene expression signatures can be measured in tissues collected in RNAlater preservative. J Mol Diagn 8:31–39
32. Gravier E, Pierron G, Vincent-Salomon A, Gruel N, Raynal V, Savignoni A et al (2010) A prognostic DNA signature for T1T2 node-negative breast cancer patients. Genes Chromosom Cancer 49:1125–1134
33. West M, Blanchette C, Dressman H, Huang E, Ishida S, Spang R et al (2001) Predicting the clinical status of human breast cancer by using gene expression profiles. Proc Natl Acad Sci 98:11462–11467

Hybrid Model for Prediction of Treatment Response in Beta-thalassemia Patients with Hepatitis C Infection

Aisha Mohamed Hussein, Ahmed Sharaf-Eldin, Amany Abdo, and Sanaa Moharram Kamal

Abstract Patients with beta-thalassaemia major (β-TM) who get regular blood transfusions are at risk of iron overload and hepatitis C virus (HCV) infection. These double injuries together can lead to chronic liver damage. Treatment with pegylated interferon combined ribavirin (Peg-IFN/RBV) is associated with side effects that compromise the patients' quality of life. The efficacy of two anti-viral regimens (Peg-IFN/RBV) and Peg-IFN monotherapy were assessed using a machine learning model to identify patients who could achieve sustained virologic response (SVR) with HCV eradication. This paper is a follow-up study of our previous published paper that used a different method to address the same research question. A hybrid Neuro-SVM model was developed to improve the accuracy of classification that shows 98.83% in group 1 and 99.75 in group 2 and conveyed as a graphical user interface that can help the clinical support decision in the prediction of optimal treatment response. The model was compared to artificial neural network (ANN), support vector machine (SVM) and naïve Bayesian (NB). Using the hybrid model, it would be useful if we distinguish in advance those patients who may benefit from the approved direct anti-viral agents (DAAs) therapy from those who would not.

Keywords Beta-thalassemia major · Synthetic minority oversampling · Machine learning · Artificial neural networks

All authors assisted with manuscript preparation and revisions. AH and AA designed the study; SK provides the dataset with laboratory follow-up of enrolled patients. AH analyzed the data and wrote the manuscript. All authors read and approved the final version of the manuscript.

A. M. Hussein (✉) · A. Abdo
Information Systems Department, Faculty of Computers and Artificial Intelligence, Helwan University, Helwan, Egypt
e-mail: aishahussein@pg.cu.edu.eg

A. Sharaf-Eldin
Faculty of Information Technology and Computer Science, Sinai University, Cairo, Egypt

Faculty of Information Technology and Computer Science, Helwan University, Helwan, Egypt

S. M. Kamal
Tropical Medicine and Gastroenterology Department, Faculty of Medicine, Ain Shams University, Cairo, Egypt

D. A. Magdi et al. (eds.), *Digital Transformation Technology*, Lecture Notes in Networks and Systems 224, https://doi.org/10.1007/978-981-16-2275-5_37

Abbreviations

β-thalassemia	Beta-thalassemia
HCV	Hepatitis C Virus
Peg-IFN/RBV	Alpha-IFN combined with ribavirin
DAAs	Direct-Acting Anti-virals
SVM	Support vector machine
NB	Naïve Bayesian
MLP	Multilayer perceptron
SMOTE	Synthetic minority oversampling technique
PTH	Post transfusion hepatitis
SVR	Sustained virological response
AST	Aspartate aminotransferase
ALT	Alanine aminotransferase
ANN	Artificial neural network analysis; rprop+: resilient backpropagation

1 Introduction

HCV represents a major health problem in Egypt, with the highest reported worldwide prevalence [1]. Egypt ranks fifth among all countries worldwide for the burden of viral hepatitis [2] and HCV genotype 4 accounts for 8.3% of the global HCV seroprevalence but is the dominant genotype in Egypt, accounting for more than 90% of infections [1]. However, the incidence of HCV in Egypt is gradually declining after adoption of a nationwide program for prevention and treatment of HCV infection [1–3].

Beta (β)-thalassemia is a genetic condition with severe reduction or absent production of the β-globin chain resulting in hemolysis of red blood cells and severe form of anemia with iron overload that requires lifelong blood transfusion and iron chelation therapy [4]. Beta-thalassaemia major (β-TM) varies greatly (4.4% of every 10,000 live births) around the world, reaching 17–20% in some regions of eastern Saudi Arabia, southern Italy, Cyprus, and Greece [5]. In Palestine, Gaza Strip, the average incidence of thalassemia trait is 3.0–4.5% [6, 7].

In Egypt, thalassemia is the most common cause of chronic hemolytic anemia with an estimated carrier rate of 9–10% [8]. Currently, HCV transmission through blood banks decreased yet, so about 30% of patients with thalassemia may have HCV infection [9, 10]. Patients with thalassemia may acquire HCV through intrafamilial transmission or the community-related practices such as barbers or dental work [11]. Concomitant HCV infection and iron overload are major causes of advanced liver fibrosis and cirrhosis among Egyptian multi-transfused thalassemia patients [12]. Therefore, it is mandatory to monitor all patients with thalassemia for HCV and treat them as early as possible. Before 2014, the conventional therapy for HCV consisted of weekly injections of pegylated interfero-based treatment and is coupled with another anti-viral agent called ribavirin which is taken orally [13, 14]. The results of this combination treatment have not been satisfactory due to side effects of ribavirin causing hemolysis and anemia of the patients. Thus, interferon-free regimens are currently the standard of care for HCV therapy. Very few clinical trials have been conducted on the efficacy and side effects of direct anti-virals DAAs in the treatment of HCV infection in thalassaemia patients.

Machine learning (ML) is a part of Artificial Intelligence (AI) and dedicated to help make unguided decisions by taking inputs and use it to create models for identifying the activities taken by AI application. Data mining is a procedure that utilizes data analysis tools to discover knowledge and relationships within data to make valid predictions. Classification is an essential in data mining and machine learning, which intends to predict the classes of future data objects. Researchers have applied machine learning algorithms to help physicians to explore unknown patterns, to predict the progression of HCV-related liver disease and prioritize therapy to patients at risk of disease progression [7, 11, 15] especially in countries with limited recourses. Studies that focus on data mining of thalassemia are to some extent scarce, particularly in Egypt. The low number of study samples in many previous works was a major limitation factor for drawing scientific results, conclusions and outcomes. Therefore, in this follow-up work, we use the real dataset of our previous work [16] collected from β-thalassemia Egyptian patients infected with HCV and used to evaluate the efficiency of the proposed detection model.

An Oversampling technique synthetic minority oversampling (SMOTE) was applied to overcome the imbalanced class dispersion in the dataset. Classifiers function better when the response variable is equally distributed; however, issue emerged when the dataset is imbalanced [17]. The class imbalance issue occurs when there are large instances of certain classes (majority class) than others (minority class); for example, the rate and prevalence of diseases are typically lower than 50%, and positive cases are less in number than negative ones [18], so that the minority class misclassifies the standard classifiers and thereby produces high predicted accuracy than that of the majority class [19]. In medical situations, the minority class is usually the main class, which scientists need to predict with higher performance [20]. Handling imbalanced datasets occurs by two ways: At the data level, by resampling, such as oversampling and undersampling [19, 20]. At the algorithmic level, arrangements include changing the threshold level for class involvements [21].

The accuracy of three basic classifiers multilayer perceptron (MLP), support vector machine (SVM) and naïve Bayes (NB) was compared on imbalanced dataset for the prediction of treatment response. We then assessed the effect of SMOTE oversampling technique [21] on the evaluation of these classifiers. A hybrid model was built on neural network using predicted SVM output and presented by graphical user interface on R software.

2 Related Works

In study [22], the authors made a distinction between thalassemia patients, persons with thalassemia trait and normal subjects utilizing genetic programming with MLP and decision tree models. Based on characteristics of red blood cells, a reticulocyte and a platelet. The output reveals that genetic programming-based decision tree performance was almost equal to that of the MLP with one hidden layer. While using two hidden layers of MLP outperform the GP-based decision tree with accuracy of 90%. Their analysis reveals that the classification ambiguity among persons with minor thalassaemia trait and normal subjects is the main cause of classification errors. These results suggest that a combination of a MLP with a blood cell analysis may give rise to a guideline/hint for further investigation of thalassaemia classification.

In 2017, the authors [23] used five classification algorithms and vote algorithm to differentiate between three types of anemia including β-thalassemia and α-thalassemia trait and iron deficiency anemia. They focused on CBC test in WEKA software. The performance of five classification algorithms and a vote algorithm were compared. The results indicated that combining J48, IBK and Naive Bayes algorithms using voting algorithm with all the features and reduced features had the highest performance with the accuracy of 96.343 and 96.2169, respectively. Using hybrid algorithm (vote) demonstrated that hybrid algorithm increases diagnosis accuracy and decreases error rate in comparison with the single classifiers.

A comparative study done by Amendolia et al. [24], for thalassemia diagnosis using different classification techniques SVM, KNN and MLP models. SVM classification based on two-layered model. One layer was used to discriminate between normal and patient samples. While the other layer was used to discriminate between alpha and beta-thalassemia carriers. Results revealed that MLP diagnoses thalassemia carriers with 95% specificity and 92% sensitivity, while SVM achieved 83% sensitivity and 95%.

In [25], the authors compared the performance of several machine learning techniques, such as MLP, KNN, NB, Bayesian networks (BNs) and multinomial logistic regression (MLR) for screening β-thalassemia patients. Feature selection done by PCA and results showed the best algorithm was MLP with an accuracy of (86.6142%). The results of the other algorithms (KNN, NB, BNs and MLR) had the accuracy of (85.8268%, 85.0394%, 85.0394%, and 82.6772%), respectively.

The study [26], developed an algorithm for β-thalassemia major and minor classification using artificial neural network. Their practical results showed that model performed well with an accuracy of 76.9%.

Elshami and Alhalees [27] discriminate between different types of thalassemia and IDA using data mining depending on CBC. The authors applied dataset for three classifiers and results showed that MCV was the main feature with accuracy of (93.6, 93.7, and 95.7%) for decision tree (DT), naïve Bayes (NB) and artificial neural network (ANN) classifiers, respectively.

In 2019, Egejuru et al. [28] used different types of classification techniques as decision tress, MLP and Naive Bayes in WEKA software to predict risk of thalassemia in all age groups based on identified risk factors interview and questionnaire. Data collected from 51 patients including demographic variables like age, gender and social class while clinical variables included family history and spleen enlargement. The study concluded that using MLP for prediction of thalassemia could improve decision-making.

Authors in [7] conducted a study for identification of beta-thalassemia carriers in the Gaza Strip with a novel dataset collected from Palestine Avenir Foundation. The proposed model used hybrid data mining model. SMOTE oversampling technique applied to classification models, namely KNN, (NB), (DT) and (MLP) that used to differentiate between normal persons and those patients carrying β-thalassemia. The results reveal NB classifier achieved the best performance at SMOTE ratio of 400% with specificity of 99.47% and a sensitivity of 98.81%.

In [29], the authors used Pattern recognition in MATLAB for diagnosis of thalassemia minor patients from IDA and control groups. Features used RBC, (RDW) and platelet count. The first model utilized all features and the second used only three parameters, including MCV, RDW, and RBC. The results showed that the best ANN could achieve achieved slightly higher results with the three-parameter model.

In [30], authors also designed a MLP model for diagnosis of thalassemia minor and iron deficiency anemia using CBC data. The MLP backpropagation algorithm gave the best performance. The specificity, sensitivity and accuracy were 92%, 94%, and 93.9%, respectively.

Current researches are missing adequate evidence about the utilization of Peg-IFN as monotherapy or with ribavirin combination in thalassaemia patients. This study designed to reach a more precise estimation of ribavirin tolerability and the effect of its low dose administration on improving SVR in HCV infected thalassemia major patients.

3 Materials and Methods

3.1 Data Source

A dataset of 341 β-TM patients infected with HCV genotype 4 was obtained from different centers in Cairo and Upper Egypt. They were divided into two groups. The first group consisted of 206 (60.4%) patients who received Peg-IFN/Ribavirin therapy and the second group of 135 (39.5%) patients treated only with Peg-IFN monotherapy. All patients were evaluated at baseline and at different time points during treatment and follow-up till 72 weeks. The treatment endpoint was sustained virological response (SVR) defined as undetectable HCV-RNA 24 weeks following termination of therapy.

3.2 Study Variables

The pre-treatment clinical variables include 15 independent variables. The independent variables include age, gender, biochemical variables [aspartate aminotransferase (AST) and alanine aminotransferase (ALT), ferritin levels before and after treatment, baseline HCV-RNA quantity]. In addition to liver histological (Ishak score) stage (1–6) and grade (1–18), hepatocellular siderosis (iron overload) grade (1–4), transfusion requirement, splenectomy, IL28-β polymorphism (CC, CT, TT) were included. The dependent variable (SVR) which was defined into 2 categories: response and no response to treatment.

3.3 Data Preprocessing

Dataset may suffer from noise and inconsistency that lead to unreliable results. Three preprocessing steps applied which are data cleaning, normalization, and data balancing, described as follows [31]. Data cleaning includes; removing all incomplete instances. Removing irrelevant attributes, which were not important for the study and may cause confusion for the mining procedure and removing all duplicate data records. In data normalization, all numeric attributes are scaled to the range [0, 1] using Min–Max normalization. Thus, avoid the impact of attributes that have wider range of value before training Eq. (1).

$$\text{Normalized}\ (e_i) = \frac{e_i - E_{\min}}{E_{\max} - E_{\min}} \tag{1}$$

Statistical analysis used to compare the efficiency and safety of Peg-IFN/RBV combination therapy versus Peg-IFN monotherapy in treating thalassemia patients infected with HCV. The statistics were analyzed with software SPSS IBM version 20. The data were analyzed using univariate analyses (Pearson chi-square test and mean values by independent *t*-test) [32]. Table 1 shows 341 patients with major beta-thalassemia infected with HCV enrolled in the study and followed up until termination of therapy sustained virological response (SVR at 72 week). Two hundred and six patients in group 1 (Peg-IFN/RBV) of them 87 (42.2%) respond and 119 (57.7%) non-respond. Group 2 (Peg-IFN monotherapy) includes 135 patients of them 59 (43.7%) respond and 76 (56.3%) non-respond.

Variable reduction is an important preprocessing step in machine learning to decrease in the computational complexity of an algorithm. An initial filtration (15/34) pre-treatment clinical variable was done using the opinion of medical experts to clarify and eliminate unnecessary variables. Feature selection was performed further using wrapper methods. Random Forest (Fig. 1) has proven to outperform other embedded methods as a variable selector, and further confirmed by the library 'Boruta' in R package [33]. Features selected for outcomes were separated into two

Table 1 Baseline characteristics of group 1 and group 2 according to SVR at 72 weeks

Features name	Group 2: Peg-IFN monotherapy (Patients No. = 135)		Group 1: Peg-IFN/RBV (Patients No. = 206)	
	SVR 72 week (class)			
	Response $N = 59$ (43.7%)	No response $N = 76$ (56.3%)	Response $N = 87$ (42.2%)	No response $N = 119$ (57.7%)
Age (years, mean, SD) Range (18–34)	25.2 ± 4.4	26.1 ± 4.8	25.5 ± 4.7	26.1 ± 5.1
	p-value = 0.27		p-value = 0.50	
Gender (%)				
Male	40 (40.8%)	58 (59.2%)	59 (67.8%)	83 (69.7%)
Female	19 (51.3%)	18 (48.6%)	28 (32.1%)	36 (30.2%)
	ODDs = 0.65, C.I. 95% = (0.3–1.39), p-value = 0.18		ODDs = 1.09, C.I. 95% = (0.6–1.9), p-value = 0.44	
Hemoglobin (mg/dl)				
Range	(6–12)	(6–12)	(6–12)	(6–12)
M(13–16), F (12–14)	8.9 ± 1.8	9.2 ± 2.1	9.0 ± 2.1	9.1 ± 1.9
	p-value = 0.6		p-value = 0.93	
Baseline PCR (IU/ml)				
Mean ± SD	$9.4 \times 10^5 \pm 3.2 \times 10^5$	$1.1 \times 10^6 \pm 2 \times 10^5$	$3.9 \times 10^5 \pm 2.0 \times 10^5$	$1.3 \times 10^6 \pm 2.3 \times 10^6$
Range	1.4×10^5–2.3×10^6	1.4×10^5–2.3×10^6	9.8×10^4–7.8×10^5	5.9×10^4–9.3×10^6
	p-value = 0.01*		p-value = 0.33	
AST(U/L), Mean ± SD	93.4 ± 21.7	89.8 ± 24.6	93.1 ± 21.8	86.4 ± 25.6
Range	49–128	49–128	50–127	31–138
	p-value = 0.38		p-value = 0.23	
ALT(U/L), Mean ± SD	89.3 ± 21.5	84.7 ± 22.1	84.3 ± 21.6	91.7 ± 25.1
Range	50–127	50–127	50–124	39–141
	p-value = 0.6		p-value = 0.32	
Histologic_staging 1–6	1–5	1–5	1–3	1–3
Mean ± SD	2.39 ± 1.26	2.37 ± 1.08	1.9 ± 0.8	2.2 ± 0.8
	p-value = 0.21		p-value = 0.09	
Histologic_grading (Ishak score 1–18)	4–13	4–13	4–13	1–13

(continued)

Table 1 (continued)

Features name	Group 2: Peg-IFN monotherapy (Patients No. = 135)		Group 1: Peg-IFN/RBV (Patients No. = 206)	
	SVR 72 week (class)			
	Response $N = 59$ (43.7%)	No response $N = 76$ (56.3%)	Response $N = 87$ (42.2%)	No response $N = 119$ (57.7%)
Mean ± SD	8.75 ± 2.9	8.5 ± 2.7	8.8 ± 2.7	7.2 ± 3.6
	p-value = 0.48		*p*-value = 0.02	
Stage of liver siderosis				
(1)	14 (43.7%)	18 (56.2%)	19 (21.8%)	31 (26.0%)
(2)	17 (47.2%)	19 (52.7%)	17 (19.5%)	30 (25.2%)
(3)	9 (28.1%)	23 (71.8%)	21 (24.1%)	34 (28.5%)
(4)	19 (54.2%)	16 (45.7%)	30 (34.4%)	24 (20.1%)
	p-value = 0.17		*p*-value = 0.14	
Transfusion required				
Yes (1)	3 (50.0%)	3 (50.0%)	46 (52%)	69 (58.0%)
No (2)	56 (43.4%)	73 (56.5%)	41 (47.1%)	50 (42.0%)
	ODDs = 1.3, C.I. 95% = (0.25–6.7), *p*-value = 0.53		ODDs = 1.23, C.I. 95% = (0.7–2.1), *p*-value = 0.27	
Ferritin baseline				
Mean ± SD	1996.5 ± 572.4	1989.3 ± 540.6	1985.0 ± 548.4	2046.6 ± 539.5
Range (ng/mL)	1109–3006	1109–3006	1142–2955	1135–2997
	p-value = 0.53		*p*-value = 0.17	
Ferritin after treatment				
Mean ± SD	2018 ± 570.7	2001.7 ± 562.3	2097 ± 506.7	2046 ± 539.5
Range (ng/mL)	1114–3016	1114–3016	1111–2996	1125–3007
	p-value = 0.64		*p*-value = 0.16	
Splenectomy (%)				
Yes (1)	28 (43.0%)	37 (56.9%)	45 (51.7%)	51 (42.8%)
No (2)	31 (44.2%)	39 (55.7%)	42 (48.2%)	68 (57.1%)
	ODDs = 0.95, C.I. 95% = (0.48–1.88), *p*-value = 0.51		ODDs = 0.7, C.I. 95% = (0.4–1.2), *p*-value = 0.13	
IL28B polymorphism				
CC	13 (32.5%)	27 (67.5%)	25 (28.7%)	27 (22.7%)
CT/TT	46 (48.4%)	49 (51.5%)	62 (71.2%)	92 (77.3%)
	p-value = 0.14		*p*-value = 0.59	

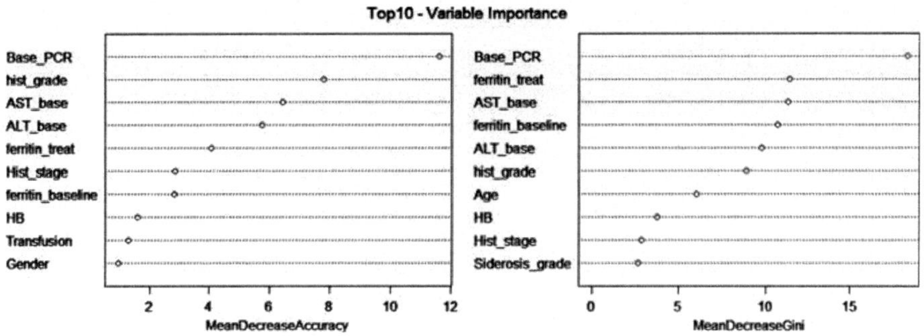

Fig. 1 Top 10-most important feature selection in group 1 by RF

refined datasets that help clinicians to predict treatment response based on a smaller amount of variables. Slightly imbalanced datasets were treated with SMOTE oversampling [17]. In the oversampling methods, the number of minority class instances is increased in the training set (100–300%). Thus, no loss of information from the original dataset. For patients in a minority class (respond to treatment), the model finds n samples (3–5) nearest in distance to the minority test (k-neighbors). The distance is determined by Euclidean distance, new samples are created (Ynew) that: the difference between samples of a minority class (Y_i) and its neighbor (Y_j) is determined. Then distance is multiplied by a number between 0 and 1 (σ) and is added to samples of minority class [17, 20].

$$Y_{\text{new}} = Y_i + \left(Y_j - Y_i\right) * \sigma \tag{2}$$

4 Results

We used the open-source software R programming version 3.4, for data analysis and machine learning. The original data was partitioned using hold-out method into 90% training set and 10% test set. The training dataset was oversampled using the SMOTE technique at 300%. Two datasets were obtained: one with- and one without oversampling. Only one test dataset was included. Then, the oversampled dataset was embedded into an ANN model where the tuning process and hyperparameters were chosen and compared using SVM and NB models (Fig. 2).

Performance metrics carried out by accuracy, sensitivity, specificity, and ROC curves. The three models MLP, SVM and NB were trained by both original and oversampled sets and then tested on the test set.

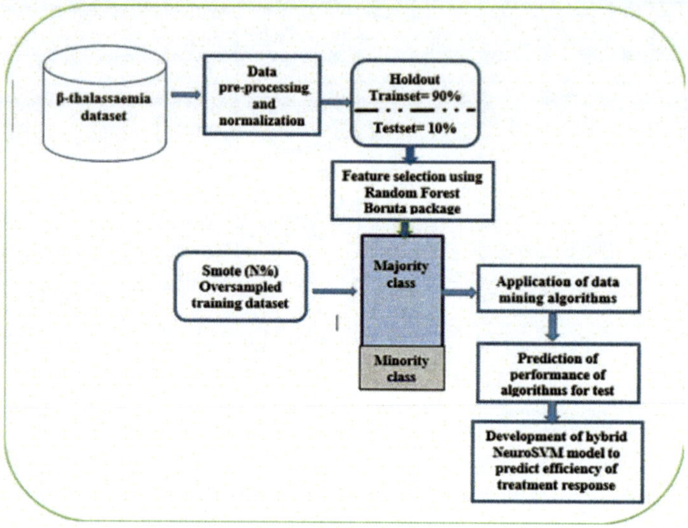

Fig. 2 Framework of the application of datasets to different classifiers

4.1 Support Vector Machine (SVM)

SVM algorithm separates the data into two categories by performing an n-dimensional hyperplane. SVM model that utilizes a sigmoid kernel function is identical to a two-layer perceptron neural network. SVM kernels are selected training method for polynomial, Radial Basis Function (RBF) networks and MLP classifiers, in which the weights of the network are found by tackling a quadratic programming issue with linear requirements, as opposed to nonconvex, as in standard neural network training [34]. The goal of SVM is to split the datasets into classes to find maximum marginal hyperplane [35], as largest margin gives better performance for test cases. In this work, we build our model using radial basis function non-linear kernel (RBF). RBF as a kernel function, which is defined by:

$$K(x, v) = \exp(-\gamma||x-v||2) \tag{3}$$

where x and v are two data vectors and $\gamma = 1/2\sigma 2$ is a training parameter. A smaller γ value makes the decision boundary smoother. The regularization factor C controls trade-off between a low training error and a large margin [18], it needs to select proper value of cost 'C' parameter and 'sigma' parameter (Fig. 3).

Different kernel types were selected in the training for SVM network. The best performance was executed for a network with a RBF kernel using CARET package in R. The performance in group 1 was 94.7% for unseen test data (Table 2).

```
#SMOTE oversampling for training set
training<-SMOTE(SVR~.,training, perc.over=300)
trctrl<- trainControl(method = "repeatedcv", number = 10, repeats = 5,
            summaryFunction = twoClassSummary,
            classProbs = TRUE,
            sampling = 'down')
#SVM Model
svm_Radial<- train(SVR~., data = training, method= "svmRadial",
          trControl= trctrl,
          metric="ROC",
          preProcess= c("center", "scale"),
          tuneLength=10)
Tuning parameter 'sigma' was held constant at a value of 0.044
Accuracy was used to select the optimal model using  the largest value.
The final values used for the model were sigma = 0.04744793 and C = 0.25.
```

Fig. 3 Implementation of SVM-RBF model for training and oversampling dataset

Table 2 Results of applying different kernels SVM

Type	Kernel type	Training data (%)	SMOTE (300%) oversampling (%)	Validation data (%)
Group 1 (Peg-IFN/RBV)	Linear	69.4	73.0	65.0
	Linear grid	71.2	84.0	80.0
	RBF	73.0	100.0	94.7
	RBF-grid	79.5	99.2	75.0
Group 2 (Peg-IFN/monotherapy)	Linear	63.0	73.0	50.6
	Linear grid	62.0	74.0	50.0
	RBF	80.0	99.6	62.5
	RBF-grid	73.0	99.5	69.0

4.2 *Artificial Neural Network Analysis*

Many studies on ANN have been published [36]. ANN refers to interconnected neurons having numeric weights that send messages to each others. Learning process for update weights and activation function that converts weighted inputs to its output. A neuralnet library of R software used standard backpropagation neural network with three layers: the input, hidden, and output layers contained (14,c(2,3),2), respectively. The transfer function is sigmoidal function and the number of the hidden neurons and hidden layers were determined via trial-and-errors process, according to the needed complexity [36]. To find local minimum of the error function (dE/dw), backpropagation modifies the weights of the network with gradient descent optimization that update weights. The backpropagation uses both forward propagation of training inputs to generate activation outputs and backward propagation to generate delta

```
#Normalize all data
maxValue<- apply(data,2,max)
minValue<- apply(data,2,min)
data<- as.data.frame(scale(data, center = minValue, scale= maxValue-minValue))
head(model.matrix(~SVR, data=df))    // factors converted to dummy
# (Intercept) SVR_0 SVR_1
#1    1    0    0
#2    1    1    0
#3    1    0    1
f<- as.formula(paste("SVR_0+SVR_1 ~", paste(n[!n %in% c("SVR_0","SVR_1")], collapse =
" + ")))
model <- neuralnet(f,data=training,hidden=c(2,3),rep=5,
          threshold = 0.01, act.fct= 'logistic',
          learningrate.limit = NULL,
          learningrate.factor =
            list(minus = 0.5, plus = 1.2),
          algorithm = "rprop+")
```

Fig. 4 Implementation of MLP model after normalization and oversampling

rule of all output and hidden neurons. Then perform weight update with chain rule differentiable to compute gradient. In this study, neuralnet R package was used that requires scaling numeric output while factor variables converted into dummy variables (0,1) using library (dummies) (Fig. 4). The dummy() function creates one new variable for every level of the factor for which we are creating dummies. The activation function between (0, 1). The training set is resilient backpropagation algorithm of which type rprop+ was selected [21, 24]. Multilayer perceptron (MLP) uses rprop algorithm is to minimize error by adding learning rate to the weights toward the opposite direction of the gradient. After the training set, the network is tested on a set of unknown vectors.

$$\Delta w_{ijr} = -\alpha r \frac{\partial y}{\partial x}. \tag{4}$$

Equation (4) used backpropagation MLP to adjust weight update using gradient descent and learning rate α.

The cost function used in our model is the logarithmic loss (cross-entropy).

$$\text{Cross Entropy Loss} = -(y \log(p) + (1 - y) \log(1 - p)) \tag{5}$$

where y is the expected output and p is the predicted probability. Training type [0, 1] means: [0 = train by epoch, 1 = train by minimum error]. Epoch: determine when the training will stop once the number of iterations exceeds epoch (mainly 5 * 106). Minimum Error: threshold [0, 0.5], default 0.01, minimum mean square error of the epoch. The performance of the network is computed (Fig. 5).

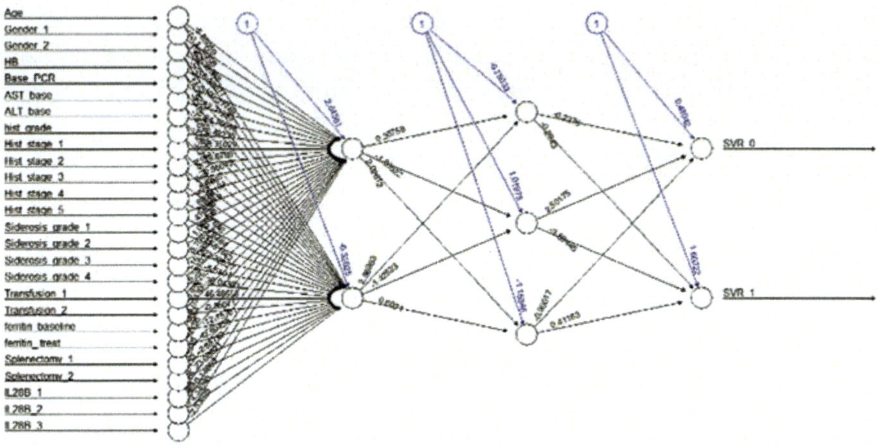

Fig. 5 Backpropagation MLP after SMOTE oversampling in group 1 trained 5 times with (rprop+) with stepmax = 10E+05, hidden layer = c (2,3) and output (respond and non-respond)

4.3 Naïve Bayes (NB)

The NB is a basic probabilistic algorithm for binary classification with independence assumptions between variables. This classifier produces likelihood estimates based on probabilistic theory. R package called 'e1071' contains NB function (Fig. 6). Laplace smoothing allows unrepresented classes to show up. NB has a strong assumption that features are conditionally independent where no feature depends on others features values for a given class label. Therefore, the following equation is applied [37]:

Fig. 6 NB model for training dataset

```
library(e1071)
library(naivebayes)
#model
e1071model<-naiveBayes(SVR~., data=training)
e1071prediction<-predict(e1071model, training)
print(confusionMatrix(e1071prediction,training$SVR, positive = "Respond",
dnn = c("Prediction", "TRUE")))
mmetric(training$SVR,e1071prediction,c("ACC","Precision","TPR","F1"))
#Predict
p<- predict(e1071model, test)
confusionMatrix(table(p,test$SVR),positive = 'Respond')
head(cbind(p,test))
#confusion matrix
p1<- predict(e1071model,train)
(tab<-table (p1,training$SVR))
1-sum(diag(tab))/sum(tab)
```

$$P(c_j|x_i) = P((c_j) * P(x_i|c_j))/P(x_i) \quad (6)$$

where; • c_j is the instance class label, • x_i is the test attribute, • $P(c_j|x_i)$ is the posterior probability of the class label given the attribute x_i, • $P(c_j)$ is the prior probability of class label (risk factor) c_j, • $P(x_i|c_j)$ is the probability of observing attribute xi given the class label c_j.

Oversampling gradually improves the sensitivity of all classifiers Tables 3 and 5. MLP and SVM in group 1 are the most improved classifiers with the oversampling of approximately 37 and 45.8% for each classifier. In addition, group 2 shows improving of MLP, SVM and NB with a ratio of 10%, 19% and 12%, respectively (Tables 4 and 6).

Estimation of ROC (AUC) curves for the best three classifiers is shown as (Fig. 7). For SVM, ROC curve was 0.9 with accuracy of 94.7% in group (1). While in group (2), MLP achieved better AUC = 0.8 and accuracy of 85.7%.

4.4 *The Conduction of Hybrid Model*

The predicted results of SVM were introduced into ANN of two input layers, three hidden layers and one output layer. The model was implemented using 'neuralnet' library in R. The resulted model shows a better performance than the individual models. Performance was estimated using the ROC curve and root mean square error (RMSE).

$$\text{RMSE} = \sqrt{\sum_{i=1}^{n} \frac{(y_i - y)^2}{n}}. \quad (7)$$

Here, y_i = Actual class, y = predicted class, n = total number of attributes.

The RMSE represents the difference between actual and predicted values that gives 1.714 in group 1 and 1.264 in group 2. Classification accuracy is shown in (Table 7) The hybrid model is shown in (Fig. 8).

The 'Shiny' package in R [38], used for plotting the hybrid (Neuro-SVM). The clinical attributes of patients are taken as input for prediction of response to treatment. The attributes of patients were used as inputs to SVM model. The predicted results were implemented into a neural network. The resulted hybrid model used for prediction of response to treatment and graphical user interface of Neuro-SVM is shown in (Fig. 9).

This application can be utilized to help decision in the prediction of optimal treatment for β-thalassemia major with liver disease. Figures 9 and 10 shows selection of group 1 dataset for training using SVM model and the model then saved. In the next step, the saved SVM model was fed into a MLP for training. The new hybrid model (Neuro-SVM) used for prediction of training set then saved.

Table 3 Performance of original and balanced training sets in group 1 (Peg-IFN/Ribavirin)

	Group 1 (Peg-IFN/Ribavirin)								
	Classifier	Sensitivity	Specificity	*F*-measure	Kappa	Precision	Accuracy	ROC	Youden index
Original training dataset *R* = 79 NR = 106	MLP	30	72	50.0	0.26	67.0	70.0	0.76	0.02
	SVM	54	76	74.8	0.3	70	73	0.66	0.3
	NB	96	29	44.2	0.22	50	57	0.47	0.25
Balanced training data with SMOTE (300%)	MLP	67.0	80.0	43.0	0.32	50.0	78.0	0.8	0.47
	SVM	99.8	98.4	100	1.0	100	100	0.99	0.97
	NB	97.8	29.7	44.9	0.36	76.1	77.8	0.5	0.26

R respond, *NR* non-respond. For SVM: Tuning parameter 'sigma' = 0.0318 and *C* = 32

Table 4 Test performance of each classifier in group 1 (Peg-IFN/Ribavirin)

	Group 1 (Peg-IFN/Ribavirin)								
	Classifiers	Sensitivity %	Specificity %	*F*-measure %	Kappa	Precision %	Accuracy %	ROC	Youden index
Balanced test dataset (10%)	MLP	90.0	85.0	89.0	0.86	87.1	85.5	0.8	0.75
	SVM	100	91.0	95.1	0.89	90.9	94.7	0.9	0.91
	NB	100	69.2	81.0	0.63	66.7	81.0	0.8	0.69

Table 5 Performance of original and balanced training dataset in group 2 (peg-IF monotherapy)

	Group 2 (peg-IF monotherapy)								
	Classifier	Sensitivity	Specificity	F-measure	Kappa	Precision	Accuracy	ROC	Youden index
Original training Dataset $R =$ 53, NR = 69	MLP	90	95.6	92.3	0.79	94.7	96.1	0.92	0.85
	SVM	81.1	78.2	80. 2	0.58	74.1	80.1	0.83	0.59
	NB	66.3	84.1	70.7	0.39	76.1	76.7	0.78	0.5
Balanced training with SMOTE (300%)	MLP	100	95.8	96.9	0.78	94.1	97.5	0.96	0.95
	SVM	100	99.4	99.4	0.9	99.2	99.6	0.99	0.99
	NB	78.7	90.6	81.2	0.54	86.2	85.6	0.84	0.68

R respond, NR non-respond. SVM: Tuning parameter 'sigma' = 0.029 and $C = 16$

Table 6 Test performance of each classifier in group 2 (Peg-IFN monotherapy)

	Group 2 (Peg-IFN—monotherapy)								
	Classifiers	Sensitivity %	Specificity %	*F*-measure %	Kappa	Precision %	Accuracy %	ROC	Youden index
Balanced test dataset (10%)	MLP	99.8	83.3	66.8	0.15	50.0	85.7	0.8	0.82
	SVM	50.0	71.4	66.6	0.22	60.0	62.5	0.64	0.21
	NB	36.8	85.7	47.9	0.20	67.1	65.3	0.61	0.22

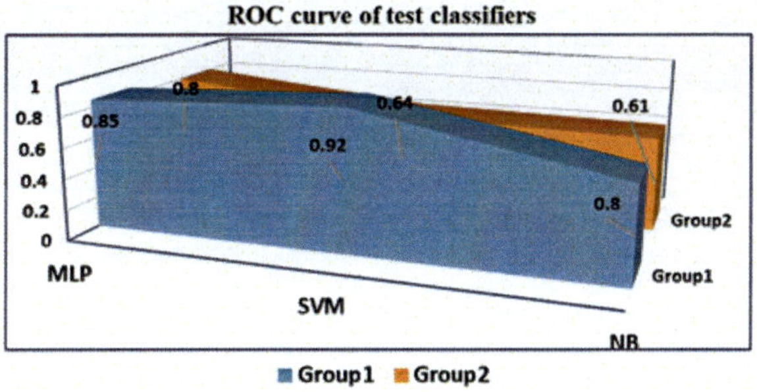

Fig. 7 Area under a curve of test set for each of three classifiers in group 1 and group 2

Table 7 Comparison between training datasets using accuracy and area under curve

Training algorithms 300% oversampled		Group (1) Peg-IFN/RBV	Group (2) Peg-IFN/monotherapy
SVM	Acc. %	100	99.6
	AUC	0.99	0.99
ANN	Acc. %	78.0	97.5
	AUC	0.8	0.96
Hybrid Neuro-SVM	Acc. %	98.8	99.87
	AUC	0.99	1.0

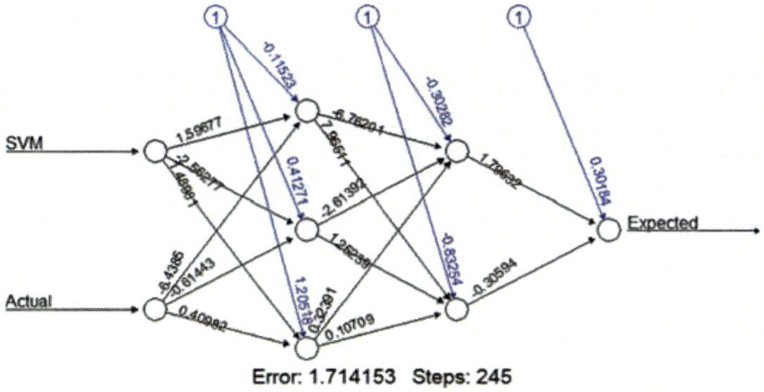

Fig. 8 The hybrid model. Acc: Accuracy, AUC: Area under curve

5 Performance Comparison Results

In classification without resampling, group (1) showed accuracy rates of 70%, 73% and 57% in MLP, SVM and NB, respectively. Sensitivity ratios of the positive class were 30%, 54% and NB of 96%. On the other hand, group (2) showed accuracy rates of 96.1%, 80.1%, 76.7% and sensitivity of 90%, 80.1% and 66.3%, respectively. The results compared with AlAgha et al. [7] who achieved high accuracy rates and best

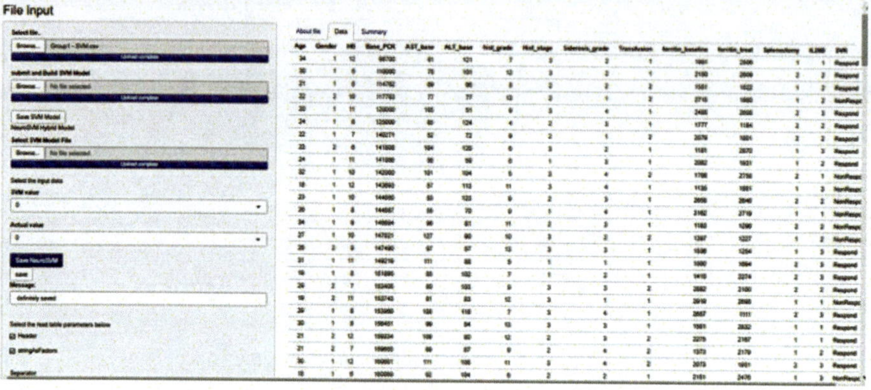

Fig. 9 GUI for prediction of Neuro-SVM training model

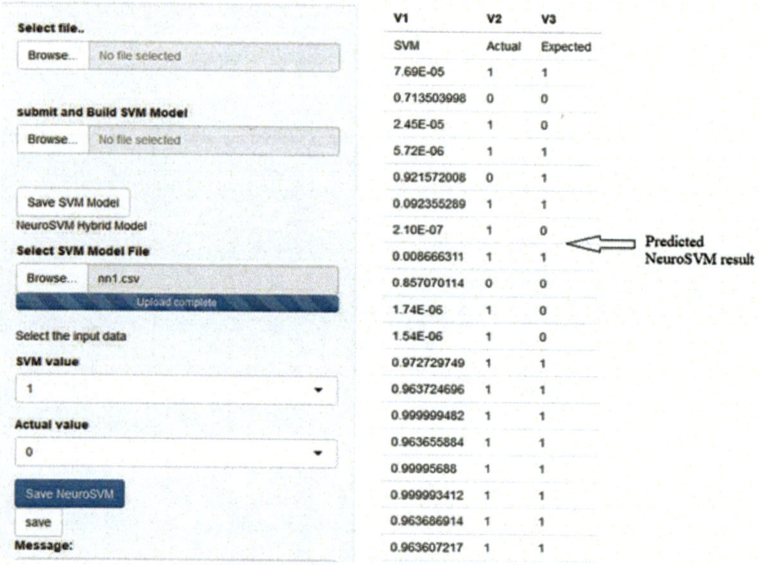

Fig. 10 GUI for prediction of test set using hybrid model

sensitivity ratios of 97.8% NB followed by 95.9% MLP, 94.6% DT and 91.1% KNN. In [28], authors reported the accuracy for DT, NB and ANN as (93.64%, 93.7%, and 95.71%), respectively. In [24], MLP was able to diagnose thalassemia carriers with 95% specificity and 92% sensitivity, while that of SVM was 83%.

In classification with 300% SMOTE oversampling: the prediction accuracy increases when a sufficient amount of minor class oversampled. Tables 3 and 5 show the evaluation results of this experiment. In group (1) MLP, SVM, NB show accuracy rates of 78%, 100%, 77.8% and sensitivity ratios of 67%, 99.8%, 97.8%.

While group (2) showed accuracy rates of 97.5%, 99.6%, 85.6% and sensitivity of 100%, 100%, 78.7%, respectively. Oversampling makes a significant effect on all classifiers in both groups. Alternatively, AlAgha et al. [7] observed that SMOTE makes significant effect on accuracy for all classifiers (NB accuracy = 100%, rate = 0.2%). Their sensitivity improved and reached best oversampling ratio for NB at 400% (98.8%) while for KNN and MLP at 500%.

The evaluation results F-score (weighted harmonic mean) of recall and precision and Kappa statistics (a measure of reliability) were used. Another scalar measure, Youden's index [39] that assumes equal weight for sensitivity and specificity for model. Evaluating the AUC using ROC for test set after 300% oversampling results shows a slightly improvement. MLP and SVM are potentially classifiers with higher ROC rates (Tables 4 and 6).

SVM model gives the best Youden's index of the three models in both the over-sampled and original training sets. Cross-validation performed in group 1 indicates that SVM is the best of the three models with 0.91. While 2, MLP outperforms other models with 0.82%.

The results of SVM predicted values were fed into neural network to develop an efficient hybrid model. The resulted model saved to file for later loading and making predictions on new data. The new model used to predict 10% of test (unseen data) as the verification dataset. The accuracy of group 1 was 98.8% and 99.87% for group.

The data of this study suggest that thalassemic patients with HCV-G4 infection, who did not achieve SVR after a course of therapy with Peg-IFN/RBV, extended 72-week therapy, may benefit from the developed DAAS (Direct-Acting Anti-virals) drugs. Although limited data exist on the use of DAAs in this population [40–43], thalassemic patients should receive these new regimens according to the current guidelines [44], even if they have apparently unfavorable predictors such as advanced hepatic fibrosis or cirrhosis. Further randomized controlled trials are required to evaluate the efficacy of such drugs with minimal side effects in thalassemic patients with HCV infection. A graphical user interface for the Neuro-SVM model was performed using R and used to help the clinicians in the prediction of optimal treatment for β-thalassemia infected with HCV.

6 Conclusion and Future Work

This study conducted with three classifiers for the prediction of β-thalassaemia major patients, using slightly imbalanced dataset. In dataset without oversampling, the training set shows, SVM gives (54%) sensitivity, (76%) specificity and accuracy of (73%) in group 1. However, MLP in group 2 outperforms other classifiers with sensitivity, specificity and accuracy of 90%, 95.6%, 96.1%, respectively. The oversampling step is performed only on the training sets using SMOTE ratio of 300%. SVM was optimal classifier in group 1 with 100% sensitivity, 91% specificity and 94.7% accuracy. Whereas in group 2, MLP proceeds the prediction task with 99.8%, 85.3% and 85.7%, respectively. A hybrid Neuro-SVM model was conducted for prediction of treatment response using ANN to assist clinicians to propose Peg-IFN therapy to patients. The approved use of DAAs has consequently changed HCV treatment due to its high efficacy (SVR > 90%) and safety, compared to IFN-based therapy. Therefore, present findings support thalassemic patients not responding to the therapy could be treated with other DAAs of newer generation or with combinations modalities.

Acknowledgments **Funding** This research received no external funding.

Conflict of Interest The authors declare that they have no conflict of interests.

References

1. El Kassas M, Alboraie M, Omran D, Salaheldin M, Wifi MN, ElBadry M, El Tahan A, Ezzat S, Moaz E, Farid AM, Omar H, Abouelkhair M, Afify S, Elsaeed K, Shazly Y, Doss W, Esmat G (2018) An account of the real-life hepatitis C management in a single specialized viral hepatitis treatment centre in Egypt: results of treating 7042 patients with 7 different direct acting antiviral regimens. Exp Rev Gastroenterol Hepatol 1-8 (2018). https://doi.org/10.1080/17474124.2018.1476137
2. Talaat M, Afifi S, Reaves E, Abu Elsood H, El-Gohary A, Refaey S, Hammad R, Abdel Fadeel M, Kandeel A (2019) Evidence of sustained reductions in the relative risk of acute hepatitis B and C virus infections, and the increasing burden of hepatitis a virus infection in Egypt: comparison of sentinel acute viral hepatitis surveillance results, 2001-17. BMC Infect Dis 19:159. https://doi.org/10.1186/s12879-019-3806-9. PMid:30764780. PMCid:PMC6376689
3. Kamal SM, Nasser IA (2008) Hepatitis C genotype 4: what we know and what we don't yet know. Hepatology 47:1371–1383. PMid:18240152. https://doi.org/10.1002/hep.22127.
4. Pennell DJ, Udelson JE, Arai AE, Bozkurt B, Cohen AR, Galanello R (2013) Cardiovascular function and treatment in ß-thalassemia major: a consensus statement from the American Heart Association. Circulation 128(3):281-308. https://doi.org/10.1161/CIR.0b013e31829b2be6. PMid:23775258
5. Rooks H, Clark B, Best S, Rushton P, Oakley M, Thein OS et al (2012) A novel 506kb deletion causing εγδβ thalassemia. Blood Cells Mol Dis 49:121–127
6. Sirdah M, Bilto YY, el Jabour S, Najjar K (1998) Screening secondary school students in the Gaza Strip for β-thalassaemia trait. Clin Lab Haematol 20:279–83
7. AlAgha AS, Faris H, Hammo BH, Al-Zoubi AA (2018) Identifying β-thalassemia carriers using a data mining approach: the case of the Gaza Strip, Palestine. Artif Intell Med 88:70–83
8. Shawky RM, Kamal TM (2012) Thalassemia intermedia: an overview. Egypt J Med Hum Genet 13: 245–255
9. Weatherall DJ (2018) The evolving spectrum of the epidemiology of thalassemia. Hematol Oncol Clin North Am 32(2):165–175
10. Williams TN, Weatherall DJ (2012) World distribution, population genetics, and health burden of the hemoglobinopathies. Cold Spring Harbor, New York, pp a011692
11. Vichinsky EP (2005) Changing patterns of thalassemia worldwide. Ann N Y Acad Sci 1054, 18–24
12. Youssry I, El Badawy A, Samy RM, Salama N, Abd Elaziz D, Rizk S (2018) Prevalence of α-thalassemia in the Egyptian population. Hemoglobin 42(4):243–246
13. El Kassas M, Elbaz T, Elsharkawy A, Omar H, Esmat G (2018) HCV in Egypt, prevention, treatment and key barriers to elimination. Expert Rev Anti Infect Ther 16(4):345–350
14. Kamal SM (2009) Hepatitis C genotype 4 therapy: increasing options and improving outcomes. Liver Int Suppl 1:39–48

15. Muncie HL Jr, Campbell J (2009) Alpha and beta thalassemia. Am Fam Physician 80:339–344
16. Hussein AM, Sharaf Eldin A, Kamal SA, Abdo A, Kassim SK, Bahnasy KA (2018) Prediction of optimum treatment in beta-thalassemia patients with hepatitis C virus infection. Int J Res Eng Sci (IJRES) 6(6), Ver. I:01–12. ISSN (Online): 2320-9364, ISSN (Print): 2320-9356 www.ijres.org
17. Chawla NV, Lazarevic A, Hall LO, Bowyer KW (2003) SMOTEBoost: improving prediction of the minority class in boosting. In: Knowledge discovery in databases: PKDD 2003. Springer, Berlin
18. Chawla N, Bowyer K, Hall L, Kegelmeyer WP (2002) SMOTE: synthetic minority over-sampling technique. J Artif Intell Res 16:321–357
19. López V, Fernández A, Moreno-Torres JG, Herrera F (2012) Analysis of preprocessing vs. cost-sensitive learning for imbalanced classification: open problems on intrinsic data characteristics. Expert Syst Appl 39(7):6585–608
20. Sun Y, Wong AK, Kamel MS (2009) Classification of imbalanced data: a review. Int J Pattern Recognit Artif Intell 23(4):687–719
21. Chawla NV, Japkowicz N, Kotcz A (2004) Editorial: special issue on learning from imbalanced data sets. SIGKDD Explor 6(1):1–6
22. Wongseree W, Chaiyaratana N, Vichittumaros K, Winichagoon P, Fucharoen S (2007) Thalassaemia classification by neural networks and genetic programming. Inf Sci 177:771–186
23. Hasani M, Hanani A (2017) Automated diagnosis of iron deficiency anemia and thalassemia by data mining techniques. IJCSNS Int J Comput Sci Netw Secur 17(4)
24. Amendolia SR, Cossu G, Ganadu M, Golosio B, Masala G, Mura GM (2003) A comparative study of k-nearest neighbour, support vector machine and multi-layer perceptron for thalassemia screening. Chemom Intell Lab Syst 69:13–20
25. Paokanta P, Harnpornchai N (2012) Risk analysis of thalassemia using knowledge representation model: diagnostic Bayesian networks. In: IEEE-EMBS international conference on biomedical and health informatics (BHI); pp. 155–158
26. Upadhyay A (2013) Beta thalassemia major and minor classification using artificial neural network. In: IJCA proceedings on international conference on recent trends in information technology and computer science 2012, ICRTITCS, pp 14–17
27. Elshami EH, Alhalees AM (2012) Automated diagnosis of thalassemia based on datamining classifiers. In: The international conference on informatics and applications (ICIA2012), pp 440–445
28. Egejuru NC, Olusanya SO, Asinobi AO, Joseph AO et al (2019) Using data mining algorithms for thalassemia risk prediction. J Biomed Sci Eng 7(2):33–44. https://doi.org/10.11648/ijbse.20190702.12
29. Barnhart-Magen G, Gotlib V, Marilus R, Einav Y: Differential diagnostics of thalassemia minor by artificial neural networks model. J Clin Lab Anal 27:481–486
30. Eshpala RH, Langarizadeh M, Mehran KH, Banafsheh T (2016) AWT-TAG, designing an expert system for differential diagnosis of ß-thalassemia minor and iron-deficiency anemia using neural network. Hormozgan Med J 20
31. Han J, Kamber M (2006) Data mining: concepts and techniques. University of Illinois at Urbana-Champaign
32. Wuensch KL (2015) Binary logistic regression with SPSS. Retrieved Mar 18 2015
33. Kursa M, Rudnicki W (2010) Feature selection with Boruta package. J Stat Softw 36(11):1–13
34. Campbell C, Cristianini N, Smola A (2000) Instance selection using support vector machines, submitted to Machine Learning
35. Witten IH, Hall MA (2011) Data mining, 3rd edn
36. Sheela GK, Deepa SN (2013) Review on methods to fix number of hidden neurons in neural networks. Math Probl Eng 1–11
37. Ozgür A (2004) Supervised and unsupervised machine learning techniques for text document categorization. Ph.D. thesis
38. Package 'shiny'. Retrieved from: http://cran.rproject.org/web/packages/shiny/shiny.pdf

39. Böhning D, Böhning W, Holling H (2008) Revisiting Youden's index as a useful measure of the misclassification error in meta-analysis of diagnostic studies. Stat Meth Med Res 17(6):543–554
40. Kamal S, Abdelhakam S, Ghoraba D, Mohsen MA, Abdel Salam A, Hassan H, Nabeigh L (2019) The course of hepatitis C infection and response to anti-viral therapy in patients with thalassemia major and hepatitis C infection: a longitudinal, prospective study. Mediterr J Hematol Infect Dis 11
41. Hezode C, Colombo M, Bourliere M, Spengler U, Ben-Ari Z, Strasser SI et al (2017) Elbasvir, grazoprevir for patients with hepatitis C virus infection and inherited blood disorders: a phase III study. Hepatology 66(3):736-745. https://doi.org/10.1002/hep.29139
42. Biliotti E, Palazzo D, Serani M, Silvestri AM, Volpicelli L, Esvan R et al (2017) Interferon free antiviral treatment of chronic hepatitis C in patients affected by beta-thalassemia major. Ann Hematol 96(6):1043–1045. https://doi.org/10.1007/s00277-017-2986-x
43. Mangia A, Sarli R, Gamberini R, Piga A, Cenderello G, Piazzolla V et al (2017) Randomised clinical trial: sofosbuvir and ledipasvir in patients with transfusion dependent thalassaemia and HCV genotype 1 or 4 infection. Aliment Pharmacol Ther 46(4):424–431. https://doi.org/10.1111/apt.14197
44. European Association for the Study of the Liver. Electronic address EEE. EASL recommendations on treatment of hepatitis C 2016. J Hepatol 66(1):153–194. https://doi.org/10.1016/j.jhep.2016.09.001

Index

D. A. Magdi et al. (eds.), *Digital Transformation Technology*, Lecture Notes in Networks and Systems 224, https://doi.org/10.1007/978-981-16-2275-5